REINFORCED CONCRETE

A Fundamental Approach

"Reflections"—High strength polymer modified concrete sculpture at Rutgers University. Work by R. H. Karol, the civil engineering class of 1982, and the author.

REINFORCED CONCRETE
A Fundamental Approach

by
Dr. Edward G. Nawy, P.E.
Distinguished Professor and Chairman

Department of Civil and Environmental Engineering
Rutgers University
The State University of New Jersey

With collaboration from Dr. Perumalsamy N. Balaguru, Associate Professor

Prentice-Hall Inc.
Englewood Cliffs, New Jersey   07632

Library of Congress Cataloging in Publication Data

Nawy, Edward G.
  Reinforced concrete.

  Includes bibliographies and index.
  1. Reinforced concrete. 2. Reinforced concrete
construction. I. Title.
TA444.N38 1984   620.1'37   84–9985
ISBN 0-13-771643-5

Editorial/production supervision: Aliza Greenblatt
Frontmatter and chapter opening design: A Good Thing, Inc.
Cover design: George Cornell
Manufacturing buyer: Anthony Caruso

PRENTICE-HALL INTERNATIONAL SERIES IN CIVIL ENGINEERING AND ENGINEERING MECHANICS
WILLIAM J. HALL, EDITOR

Printed in the United States of America

10 9 8 7 6 5 4 3 2

ISBN 0-13-771643-5 01

Prentice-Hall International, Inc., *London*
Prentice-Hall of Australia Pty. Limited, *Sydney*
Editora Prentice-Hall do Brasil, Ltda., *Rio de Janeiro*
Prentice-Hall Canada Inc., *Toronto*
Prentice-Hall of India Private Limited, *New Delhi*
Prentice-Hall of Japan, Inc., *Tokyo*
Prentice-Hall of Southeast Asia Pte. Ltd., *Singapore*
Whitehall Books Limited, *Wellington, New Zealand*

To

RACHEL E. NAWY

For her high limit state of endurance over the years
which made the writing of this book a reality.

Contents

Torsion 195   7

Serviceability of Beams & One-Way Slabs 246   8

9 Combined Compression & Bending: Columns 297

Contents

Preface

Reinforced concrete is a widely used material for constructed systems. Hence graduates of every civil engineering program must have, as a minimum requirement, a basic understanding of the fundamentals of reinforced concrete. Additionally design of the members of a total structure is achieved only by trial and adjustment: assuming a section then analyzing it. Consequently design and analysis were combined to make it simpler for the student first introduced to the subject of reinforced concrete design.

The text is an outgrowth of the author's lecture notes evolved in teaching the subject at Rutgers University over the past twenty-five years and the experience accumulated over the years in teaching and research in the areas of reinforced and prestressed concrete up to the Ph.D. level. The material is presented in such a manner that the student can be familiarized with the properties of plain concrete and its components prior to embarking on the study of structural behavior. The book is uniquely different from other textbooks at this level in that most of its contents can be covered in one semester in spite of the in-depth discussions of some of its major topics.

The concise discussion presented in Chapters 1 through 4 on the historical development of concrete, the proportioning of the constituent materials, the long-term basic behavior, and the development of safety factors should give an adequate introduction to the subject of reinforced concrete. It should also aid in developing fundamental laboratory experiments and essential knowledge of mix proportioning, strength and behavioral requirements, and the concepts of reliability of performance of structures to which every engineering student should be exposed. The discussion of quality control and quality assurance should also give the reader a good introduction to the systematic approach needed to administer the development of concrete structural systems from conception to turnkey use.

Since concrete is a nonelastic material, with the nonlinearity of its behavior starting at a very early stage of loading, only the ultimate strength approach, or what is sometimes termed the "limit state at failure approach," is given in this book. Adequate coverage is given of the serviceability checks in terms of cracking and deflection behavior as well as long-term effects. In this manner, the design should satisfy all the service-load-level requirements while ensuring that the theory used in the analysis (design) truly describes the actual behavior of the designed components.

Chapters 5, 6, 7, and 8 cover the flexural, diagonal tension, and serviceability behavior of one-dimensional members: beams and one-way slabs. Full emphasis has been placed on giving the student and the engineer a feeling for the internal strain distribution in structural reinforced concrete elements and a basic understanding of the reserve strength and the safety factors inherent in the design expressions. Chapter 9, on the analysis and design of columns and other compression members, treats the subject of strain compatibility and strain distribution in a similar manner as in Chapter 5, on flexural analysis and design of beams. It includes a detailed discussion of how to construct interaction diagrams for columns as well as proportioning columns subjected to biaxial bending and buckling. With Chapter 10, on bond and development length in reinforcement, and Chapter 12, on the design of foundations and footings, the sequence of design steps of all elements except two-way floors is complete.

It is important to mention that Chapter 6, on diagonal tension, also contains detailed coverage of the behavior of deep beams, corbels, and brackets, with sufficient design examples to supplement the theory. This topic has been included in view of the increased use of precast construction, the wider understanding of the effects of induced horizontal loads on floors, and the frequent need for including shear walls and deep beams in today's multilevel structures. Additionally, Chapter 7 treats the topic of torsion in some detail considering the space constraints of the book. The discussion ranges from the basic fundamentals of pure torsion in elastic and plastic materials to the design of reinforced concrete members subjected to combined torsion, shear, and bending. The material presented and the accompanying illustrative examples should give the background necessary for pursuing more advanced studies in this area, as listed in the selected references.

Chapter 11 presents an extensive coverage of the subject of analysis and design of slab and plate floor systems. Following a discussion of fundamental behavior, it gives detailed design examples using both the ACI procedures and yield-line theory for the flexural design of reinforced concrete floors. It also includes ultimate load solutions to most floor shapes and possible gravity loading patterns. Detailed discussion of the deflection behavior and evaluation of two-way panels as well as the cracking mechanism of such panels, with appropriate analysis examples, makes this chapter another unique feature of this concise textbook.

It is important to emphasize that in this field, the use of computers prevails today. Access to transportable personal computers and handheld computers, due to their affordable cost, has made it possible for almost every student to be equipped with such a tool. Hence Chapter 13 presents programming procedures and computer programs written for both the handheld HP41C/CV/CX and in BASIC language for the Apple IIe and IIc transportable computers for the analysis (design) of sections in flexure, shear, torsion, combined loading (including compression), and members subjected to biaxial loading, as well as deep beams and corbels. As a result, the use of handbook charts was kept to a bare minimum. The inclusion of extensive flowcharts with logical steps in each relevant topic will make it possible for the reader to develop or use, without difficulty, such programs with any handheld or desktop computer.

Selected photographs of various areas of structural behavior of concrete elements at failure are included in all the chapters. They are taken from the published research work by the author with many of his M.S. and Ph.D. students at Rutgers University over the past two decades. Additionally, photographs of landmark structures, mainly in the United States, are included throughout the book to illustrate the versatility of design in reinforced concrete.

The textbook conforms to the provisions of ACI 318-83 with an eye to stressing the basics rather than tying every step to the code, which changes once every six years. Consequently, no attempt was made to tie any design or analysis step to the particular equation numbers in the code, but rather, the student is expected to gain the habit of getting familiar with the provisions and section numbers of the ACI code on a separate basis. In this manner, the student should not only master the fundamentals presented in the textbook, but should also become well versed with the ACI code as a dynamic, ever-changing document. Conversions to SI units are included in the illustrative examples throughout the book.

The various topics have been presented in as concise a manner as possible without sacrificing the need for the instructional details of an introductory course in the subject. Hence the topic of prestressed concrete has been left for more advanced works. The major portions of this book are intended for a first course at the junior or senior level of the standard college or university curriculum in civil engineering. The contents should also serve as a valuable guideline to the practicing engineer who has to keep abreast of the state of the art in concrete, as well as the designer who is interested in a concise treatment of the fundamentals.

Rutgers University   **Edward G. Nawy**
The State University of New Jersey
New Brunswick, New Jersey

Acknowledgments

Grateful acknowledgment is due to Dr. Edward J. Bloustein, President of Rutgers University, for his continuous support and encouragement over the years, to the American Concrete Institute for contribution to the author's accomplishments and for permitting generous quotations of its ACI 318 Code and the illustrations from other ACI publications, and to his original mentor, Professor A. L. L. Baker of London University's Imperial College of Science and Technology who inspired him with the affection that he has developed for systems constructed of concrete. Grateful acknowledgment is also made to the author's many students, both undergraduate and graduate, who had much to do with generating the writing of this book, to the many who assisted in his research activities over the years, shown in the various photographs of laboratory tests throughout the book; and to his colleague at Rutgers University. Dr. P. N. Balaguru, for his valuable contributions, specifically to Chapters 9 and 13, and to the flowcharts.

Special thanks are due to the distinguished panel of authoritative reviewers: Professor William J. Hall of the University of Illinois at Urbana, Engineering Editor of Prentice-Hall Advanced Series, for his valuable input and recommendations: Professor Vitelmo V. Bertero of the University of California at Berkeley; Professor Dan E. Branson of the University of Iowa; Professor Thomas T. C. Hsu of the University of Houston; and Mr. Gerald B. Neville, Manager, Structural Codes, Portland Cement Association—with deep gratitude for agreeing to review the manuscript and for contributing extensive suggestions and advice which considerably improved the content.

Thanks to M.S. candidate Mark J. Cipolloni for his diligence and input to the manuscript, particularly for his contribution to Chapter 13, on computer programs and to its flow charts; to M.S. candidate Regina Silveira Rocha Souza for her diligent and continuous work on reviewing the computational process, details, and logic, and generating many ideas for the various chapters; to Robert M. Nawy, Rutgers engineering class of 1983, for his valuable work on the solutions, and to engineer Abe Daly and Ph.D. candidate Lily Sehayek for their critical contributions to the computer programs in BASIC. Thanks also to Mr. Charles M. Iossi, Executive Editor and Vice-President, Ms. Alice Dworkin, Executive Assistant, and Ms. Aliza Greenblatt, Production Editor, Prentice-Hall Inc., for their patience and the continuous cooperation and help received in developing the manuscript. Last, but not least, grateful acknowledgment is due to Ms. Suzanne Iazzetta, Executive Assistant, Department of Civil and Environmental Engineering at Rutgers, for her superior work on the manuscript and the invaluable help she rendered throughout.

1

Introduction

1.1 HISTORICAL DEVELOPMENT OF STRUCTURAL CONCRETE

Use of concrete and its cementatious (volcanic) constituents, such as pozzolanic ash, has been made since the days of the Greeks, the Romans, and possibly earlier ancient civilizations. However, the early part of the nineteenth century marks the start of more intensive use of the material. In 1801, F. Coignet published his statement of principles of construction, recognizing the weakness of the material in tension. J. L. Lambot in 1850 constructed for the first time a small cement boat for exhibition in the 1855 World Fair in Paris. J. Monier, a French gardener, patented in 1867 metal frames as reinforcement for concrete garden plant containers, and Koenen in 1886 published the first manuscript on the theory and design of concrete structures. In 1906, C. A. P. Turner developed the first flat slab without beams.

Thereafter, considerable progress occurred in this field such that by 1910 the German Committee for Reinforced Concrete, the Austrian Concrete Committee, the American Concrete Institute, and the British Concrete Institute were already established. Many buildings, bridges, and liquid containers of reinforced concrete were already constructed by 1920 and the era of linear and circular prestressing began.

The rapid developments in the art and science of reinforced and prestressed concrete analysis, design, and construction have resulted in very unique structural

Photo 3 Felix Candela's Xochimilco Restaurant, Mexico.

2

Photo 4   Afrikaans Languages Monument, Stellenbosch, South Africa (height of the main dynamically designed hollow columns, 186 ft).

systems, such as the Kresge Auditorium, Boston; the 1951 Festival of Britain Dome; Marina Towers and Lake Point Tower, Chicago; and many, many others.

Ultimate-strength theories were codified in 1938 in the USSR and in 1956 in England and the United States. Limit theories have also become a part of codes of several countries throughout the world. New constituent materials and composites of concrete have become prevalent, including the high-strength concretes of a strength in compression up to 20,000 psi (137.9 MPa) and 1800 psi (12.41 MPa) in tension. Steel reinforcing bars of strength in excess of 60,000 psi (413.7 MPa) and high-strength welded wire fabric in excess of 100,000 psi (689.5 MPa) ultimate strength are being used. Additionally, deformed bars of various forms have been produced. Such deformations help develop the maximum possible bond between the reinforcing bars and the surrounding concrete as a requisite for the viability of concrete as a structural medium. Prestressing steel of ultimate strengths in excess of 300,000 psi (2068 MPa) is available.

All these developments and the massive experimental and theoretical research that has been conducted, particularly in the last two decades, have resulted in rigorous theories and codes of practice. Consequently, a simplified approach has become necessary to an understanding of the fundamental structural behavior of reinforced concrete elements.

1.2 BASIC HYPOTHESIS OF REINFORCED CONCRETE

Plain concrete is formed from a hardened mixture of cement, water, fine aggregate, coarse aggregate (crushed stone or gravel), air, and often other admixtures. The plastic mix is placed and consolidated in the formwork, then cured to facilitate the acceleration of the chemical hydration reaction of the cement/water mix, resulting in hardened concrete. The finished product has high compressive strength, and low resistance to tension, such that its tensile strength is approximately one-tenth of its compressive strength. Consequently, tensile and shear reinforcement in the tensile regions of sections has to be provided to compensate for the weak-tension regions in the reinforced concrete element.

It is this deviation in the composition of a reinforced concrete section from the homogeneity of standard wood or steel sections that requires a modified approach to the basic principles of structural design, as will be explained in subsequent chapters of this book. The two components of the heterogeneous reinforced concrete section are to be so arranged and proportioned that optimal use is made of the materials involved. This is possible because concrete can easily be given any desired shape by placing and compacting the wet mixture of the constituent ingredients into suitable forms in which the plastic mass hardens. If the various ingredients are properly proportioned, the finished product becomes strong, durable, and, in combination with the reinforcing bars, adaptable for use as main members of any structural system.

1.3 ANALYSIS VERSUS DESIGN OF SECTIONS

From the foregoing discussion, it is clear that a large number of parameters have to be dealt with in proportioning a reinforced concrete element, such as geometrical width, depth, area of reinforcement, steel strain, concrete strain, steel stress, and so on. Consequently, trial and adjustment is necessary in the choice of concrete sections, with assumptions based on conditions at site, availability of the constituent materials, particular demands of the owners, architectural and headroom requirements, the applicable codes, and environmental conditions. Such an array of parameters has to be considered because of the fact that reinforced concrete is often a site-constructed composite, in contrast to the standard mill-fabricated beam and column sections in steel structures.

A trial section has to be chosen for each critical location in a structural system. The trial section has to be analyzed to determine if its nominal resisting strength is adequate to carry the applied factored load. Since more than one trial is often necessary to arrive at the required section, the first design input step generates into a series of trial-and-adjustment analyses.

The trial-and-adjustment procedures for the choice of a concrete section lead to the convergence of analysis and design. Hence every design is an analysis once a trial section is chosen. The availability of handbooks, charts, desktop and handheld per-

Photo 5 Rockefeller Empire State Plaza, Albany, New York—Ammann & Whitney design. (Courtesy of New York Office of General Services.)

Photo 6 Empire State Performing Arts Center, Albany, New York—Ammann & Whitney design. (Courtesy of New York Office of General Services.)

sonal computers and programs supports this approach as a more efficient, compact, and speedy instructional method compared with the traditional approach of treating the analysis of reinforced concrete separately from pure design.

Photo 7   Toronto City Hall, Toronto, Canada. (Courtesy Portland Cement Association.)

Photo 8   Bridge ramp system. (Courtesy of Port of New York–New Jersey Authority.)

2

Concrete-Producing Materials

Photo 9 LaGuardia Airport parking garage ramps, New York. (*See* p. 7.)

2.1 INTRODUCTION

To understand and interpret the total behavior of a composite element requires a knowledge of the characteristics of its components. Concrete is produced by the collective mechanical and chemical interaction of a large number of constituent materials. Hence a discussion of the functions of each of those components is vital prior to studying concrete as a finished product. In this manner, the designer and the materials engineer can develop skills for the choice of the proper ingredients and so proportion them as to obtain an efficient and desirable concrete satisfying the designer's strength and serviceability requirements.

This chapter presents a brief account of the concrete-producing materials: cement, fine and coarse aggregate, water, air, and admixtures. The cement manufacturing process, the composition of cement, type and gradation of fine and coarse aggregate, and the function and importance of the water, air, and admixtures are reviewed. The reader is referred to books on concrete, such as the selected references at the end of this chapter, for further information.

Photo 10 North Shore Synagogue, Glencoe, Illinois. (Courtesy Portland Cement Association.)

2.2 PORTLAND CEMENT

2.2.1 Manufacture

Portland cement is made of finely powdered crystalline minerals composed primarily of calcium and aluminum silicates. The addition of water to these minerals produces a paste which, when hardened, becomes of stonelike strength. Its specific gravity ranges between 3.12 and 3.16 and it weighs 94 lb/ft^3, which is the unit weight of a commercial sack or bag of cement.

The raw materials that make cement are:

1. Lime (CaO)—from limestone
2. Silica (SiO_2)—from clay
3. Alumina (Al_2O_3)—from clay

(with very small percentages of magnesia: MgO and sometimes some alkalis). Iron oxide is occasionally added to the mixture to aid in controlling its composition.

The process of manufacture can be summarized as follows:

1. Grinding the raw mix of CaO, SiO_2, and Al_2O_3 with the added other minor ingredients either in dry or wet form. The wet form is called a *slurry*.
2. Feeding the mixture into the upper end of a slightly inclined rotary kiln.
3. As the heated kiln operates, the material passes from its upper to its lower end at a predetermined, controlled rate.
4. The temperature of the mixture is raised to the point of incipient fusion, that is, the *clinkering temperature*. It is kept at that temperature until the ingredients combine to form at 2700°F the portland cement pellet product. The pellets, which range in size from $\frac{1}{16}$ to 2 in., are called *clinkers*.
5. The clinkers are cooled and ground to a powdery form.
6. A small percentage of gypsum is added during grinding to control or retard the setting time of the cement in the field.
7. Most of the final portland cement goes into silos for bulk shipment; some is packed in 94-lb bags for retail marketing..

Figure 2.1 illustrates schematically the manufacturing process of portland cement. The form and properties of the manufactured compound are described in the following sections.

2.2.2 Strength

The strength of cement is the result of a process of hydration. This chemical process results in recrystallization in the form of interlocking crystals producing the cement gel, which has high compressive strength when it hardens. Table 2.1 shows the relative contribution of each component of the cement toward the rate of gain in

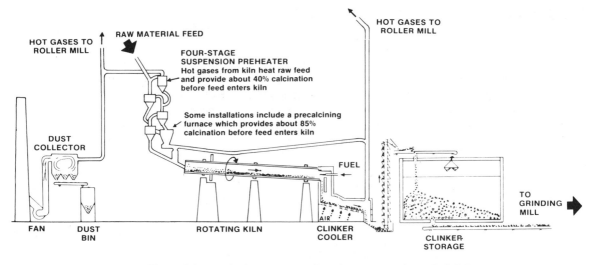

Figure 2.1 Portland cement manufacturing process. (From Ref. 2.5.)

strength. The early strength of portland cement is higher with higher percentages of C_3S. If moist curing is continuous, later strength levels will be greater, with higher percentages of C_2S. C_3A contributes to the strength developed during the first day after placing the concrete because it is the earliest to hydrate.

When portland cement combines with water during setting and hardening, lime is liberated from some of the compounds. The amount of lime liberated is approximately 20% by weight of the cement. Under unfavorable conditions, this might cause disintegration of a structure owing to leaching of the lime from the cement. Such a situation should be prevented by addition to the cement of silicious mineral such as pozzolan. The added mineral reacts with the lime in the presence of moisture to produce strong calcium silicate.

TABLE 2.1 PROPERTIES OF CEMENTS

Component	Rate of reaction	Heat liberated	Ultimate cementing value
Tricalcium silicate, C_3S	Medium	Medium	Good
Dicalcium silicate, C_2S	Slow	Small	Good
Tricalcium aluminate, C_3A	Fast	Large	Poor
Tetracalcium aluminoferrate, C_4AF	Slow	Small	Poor

TABLE 2.2   PERCENTAGE COMPOSITION OF PORTLAND CEMENTS

Type of cement	Component (%)							General characteristics
	C_3S	C_2S	C_3A	C_4AF	$CaSO_4$	CaO	MgO	
Normal: I	49	25	12	8	2.9	0.8	2.4	All-purpose cement
Modified: II	46	29	6	12	2.8	0.6	3.0	Comparative low heat liberation; used in large structures
High early strength: III	56	15	12	8	3.9	1.4	2.6	High strength in 3 days
Low heat: IV	30	46	5	13	2.9	0.3	2.7	Used in mass concrete dams
Sulfate resisting: V	43	36	4	12	2.7	0.4	1.6	Used in sewers and structures exposed to sulfates

2.2.3 Average Percentage Composition

Since there are different types of cement for various needs, it is necessary to study the percentage variation in the chemical composition of each type in order to interpret the reasons for variation in behavior. Table 2.2, studied in conjunction with Table 2.1, gives concise reasons for the difference in reaction of each type of cement when in contact with water.

2.2.4 Influence of Fineness of Cement on Strength Development

The size of the cement particles has a strong influence on the rate of reaction of cement with water. For a given weight of finely ground cement, the surface area of the particles is greater than that of the coarsely ground cement. This results in a greater rate of reaction with water and a more rapid hardening process for larger surface areas. This is one of the reasons for the high early strength type III cement giving in 3 days a strength that type I gives in 7 days, and a strength in 7 days that type I gives in 28 days.

2.2.5 Influence of Cement on the Durability of Concrete

Disintegration of concrete due to cycles of wetting, freezing, thawing, and drying and the propagation of resulting cracks is a matter of great importance. The presence of minute air voids throughout the cement paste increases the resistance of concrete to disintegration. This can be achieved by the addition of air-entraining admixtures to the concrete while mixing.

Disintegration due to chemicals in contact with the structure, such as in the case of port structures and substructures, can also be slowed down or prevented. Since the concrete in such cases is exposed to chlorides and sometimes sulfates of magnesium and sodium, it is sometimes necessary to specify sulfate-resisting cements. Usually, type II cement will be adequate for use in seawater structures.

2.2.6 Heat Generation during Initial Set

Since the different types of cement generate different degrees of heat at different rates, the type of structure governs the type of cement to be used. The bulkier and heavier in cross section the structure is, the less the generation of heat of hydration that is desired. In massive structures such as dams, piers, and caissons, type IV cement is more advantageous to use.

From the discussion above it is seen that the type of structure, the weather and other conditions under which it is built and will exist are the governing factors in the choice of the type of cement that should be used.

2.3 WATER AND AIR

2.3.1 Water

Water is required in the production of concrete in order to precipitate chemical reaction with the cement, to wet the aggregate, and to lubricate the mixture for easy workability. Normally, drinking water can be used in mixing. Water having harmful ingredients, contamination, silt, oil, sugar, or chemicals is destructive to the strength and setting properties of cement. It can disrupt the affinity between the aggregate and the cement paste and can adversely affect the workability of a mix.

Since the character of the colloidal gel or cement paste is the result only of the chemical reaction between cement and water, it is not the proportion of water relative to the whole of the mixture of dry materials that is of concern, only the proportion of water relative to the cement. Excessive water leaves an uneven honeycombed skeleton in the finished product after hydration has taken place, while too little water prevents complete chemical reaction with the cement. The product in both cases is a concrete that is weaker than and inferior to normal concrete.

2.3.2 Entrained Air

With the gradual evaporation of excess water from the mix, pores are produced in the hardened concrete. If evenly distributed, these could give improved characteristics to the product. Very even distribution of pores by artificial introduction of finely divided uniformly distributed air bubbles throughout the product is possible by adding air-entraining agents such as vinsol resin. Air entrainment increases workability, decreases density, increases durability, reduces bleeding and segregation, and reduces the required sand content in the mix. For these reasons, the percentage of entrained air should be kept at the required optimum value for the desired quality of the concrete. The optimum air content is 9% of the mortar fraction of the concrete. Air entraining in excess of 5 to 6% of the total mix proportionally weakens the concrete strength.

2.3.3 Water/Cement Ratio

To summarize the preceding discussion, strict control has to be maintained on the water/cement ratio and the percentage of air in the mix. As the water/cement ratio is the real measure of the strength of the concrete, it should be the principal criterion governing the design of most structural concretes. It is usually given as the ratio of weight of water to the weight of cement in the mix.

2.4 AGGREGATES

Aggregates are those parts of the concrete that constitute the bulk of the finished product. They comprise 60 to 80% of the volume of the concrete, and have to be so graded that the whole mass of concrete acts as a relatively solid, homogeneous, dense combination, with the smaller sizes acting as an inert filler of the voids that exist between the larger particles.

Aggregates are of two types:

1. Coarse aggregate (gravel, crushed stone, or blast-furnace slag)
2. Fine aggregate (natural or manufactured sand)

Since the aggregate constitutes the major part of the mix, the more aggregate in the mix, the cheaper is the cost of the concrete, provided that the mix is of reasonable workability for the specific job for which it is used.

2.4.1 Coarse Aggregate

Coarse aggregate is classified as such if the smallest size of the particle is greater than $\frac{1}{4}$ in. (6 mm). Properties of the coarse aggregate affect the final strength of the hardened concrete and its resistance to disintegration, weathering, and other destructive effects. The mineral coarse aggregate must be clean of organic impurities, and must bond well with the cement gel.

The common types of coarse aggregate are:

1. *Natural crushed stone:* This is produced by crushing natural stone or rock from quarries. The rock could be of igneous, sedimentary, or metamorphic type.

Although crushed rock gives higher concrete strength, it is less workable in mixing and placing than are the other types.

2. *Natural gravel:*   This is produced by the weathering action of running water on the beds and banks of streams. It gives less strength than crushed rock but is more workable.

3. *Artificial coarse aggregates:*   These are mainly slag and expanded shale, and are frequently used to produce lightweight concrete. They are the by-product of other manufacturing processes, such as blast-furnace slag or expanded shale, or pumice for lightweight concrete.

4. *Heavy-weight and nuclear-shielding aggregates:*   With the specific demands of our atomic age and the hazards of nuclear radiation due to the increasing number of atomic reactors and nuclear power stations, special concretes have had to be produced to shield against x-rays, gamma rays, and neutrons. In such concretes, economic and workability considerations are not of prime importance. The main heavy coarse aggregate types are steel punchings, barites, magnatites, and limonites.

Whereas concrete with ordinary aggregate weighs about 144 lb/ft^3, concrete made with these heavy aggregates weighs from 225 to 330 lb/ft^3. The property of heavy-weight radiation-shielding concrete depends on the density of the compact product rather than primarily on the water/cement ratio criterion. In certain cases, high density is the only consideration, whereas in others both density and strength govern.

2.4.2 Fine Aggregate

Fine aggregate is a smaller filler made of sand. It ranges in size from No. 4 to No. 100 U.S. standard sieve sizes. A good fine aggregate should always be free of organic impurities, clay, or any deleterious material or excessive filler of size smaller than No. 100 sieve. It should preferably have a well-graded combination conforming to the American Society of Testing and Materials (ASTM) sieve analysis standards. For radiation-shielding concrete, fine steel shot and crushed iron ore are used as fine aggregate.

[handwritten margin note: FINE AGG. #4 to #100 SIEVE PASSING]

2.4.3 Grading of Normal-Weight Concrete Mixes

The recommended grading of coarse and fine aggregates for normal-weight concretes is presented in Table 2.3.

2.4.4 Grading of Lightweight Concrete Mixes

The grading requirements for lightweight aggregate for structural concrete are given in Table 2.4.

TABLE 2.3 GRADING REQUIREMENTS FOR AGGREGATES IN NORMAL-WEIGHT CONCRETE (ASTM C-33)

U.S. standard sieve size	Percent passing				
	Coarse aggregate				Fine aggregate
	No. 4 to 2 in.	No 4 to $1\frac{1}{2}$ in.	No. 4 to 1 in.	No. 4 to $\frac{3}{4}$ in.	
2 in.	95–100	100	—	—	—
$1\frac{1}{2}$ in.	—	95–100	100	—	—
1 in.	25–70	—	95–100	100	—
$\frac{3}{4}$ in.	—	35–70	—	90–100	—
$\frac{1}{2}$ in.	10–30	—	25–60	—	—
$\frac{3}{8}$ in.	—	10–30	—	20–55	100
No. 4	0–5	0–5	0–10	0–10	95–100
No. 8	0	0	0–5	0–5	80–100
No. 16	0	0	0	0	50–85
No. 30	0	0	0	0	25–60
No. 50	0	0	0	0	10–30
No. 100	0	0	0	0	2–10

2.4.5 Grading of Heavy-Weight and Nuclear-Shielding Aggregates

The grading requirements to ensure heavy-weight concrete is given in Table 2.5.

2.4.6 Unit Weights of Aggregates

The unit weight of the concrete is dependent on the unit weight of the aggregate, which in turn depends on the type of aggregate: whether it is normal, lightweight, or heavyweight (for radiation shielding). Table 2.6 gives the unit weights of the various aggregates and the corresponding unit weight of the concrete.

2.5 ADMIXTURES

Admixtures are materials other than water, aggregate, or hydraulic cement which are used as ingredients of concrete and which are added to the batch immediately before or during the mixing. Their function is to modify the properties of the concrete so as "to make it more suitable for the work at hand, or for economy, or for other purposes

TABLE 2.4 GRADING REQUIREMENTS FOR AGGREGATES IN LIGHTWEIGHT STRUCTURAL CONCRETE (ASTM C-330)

Size designation	Percentages (by weight) passing sieves having square openings								
	1 in. (25.0 mm)	$\frac{3}{4}$ in. (19.0 mm)	$\frac{1}{2}$ in. (12.5 mm)	$\frac{3}{8}$ in. (9.5 mm)	No. 4 (4.75 mm)	No. 8 (2.36 mm)	No. 16 (1.18 mm)	No. 50 (300 µm)	No. 100 (150 µm)
Fine aggregate No. 4 to 0	—	—	—	100	85–100	—	40–80	10–35	5–25
Coarse aggregate									
1 in. to No. 4	95–100	—	25–60	—	0–10	—	—	—	—
$\frac{3}{4}$ in. to No. 4	100	90–100	—	10–50	0–15	—	—	—	—
$\frac{1}{2}$ in. to No. 4	—	100	90–100	40–80	0–20	0–10	—	—	—
$\frac{3}{8}$ in. to No. 8	—	—	100	80–100	5–40	0–20	0–10	—	—
Combined fine and coarse aggregate									
$\frac{1}{2}$ in. to 0	—	100	95–100	—	50–80	—	—	5–20	2–15
$\frac{3}{8}$ in. to 0	—	—	100	90–100	65–90	35–65	—	10–25	5–15

TABLE 2.5 GRADING REQUIREMENTS FOR COARSE AGGREGATE
FOR AGGREGATE CONCRETE (ASTM C-637)

	Percentage passing	
Sieve size	Grading 1: for $1\frac{1}{2}$ in. (37.5 mm) maximum-size aggregate	Grading 2: for $\frac{3}{4}$ in. (19.0 mm) maximum-size aggregate
Coarse Aggregate		
2 in. (50 mm)	100	—
$1\frac{1}{2}$ in. (37.5 mm)	95–100	100
1 in. (25.0 mm)	40–80	95–100
$\frac{3}{4}$ in. (19.0 mm)	20–45	40–80
$\frac{1}{2}$ in (12.5 mm)	0–10	0–15
$\frac{3}{8}$ in. (9.5 mm)	0–2	0–2
Fine aggregate		
No. 8 (2.36 mm)	100	—
No. 16 (1.18 mm)	95–100	100
No. 30 (600 μm)	55–80	75–95
No. 50 (300 μm)	30–55	45–65
No. 100 (150 μm)	10–30	20–40
No. 200 (75 μm)	0–10	0–10
Fineness modulus	1.30–2.10	1.00–1.60

Data in Tables 2.3 to 2.5 reprinted with permission from the American Society
for Testing and Materials, Philadelphia, PA.

TABLE 2.6 UNIT WEIGHT OF AGGREGATES

Type	Unit weight of dry-rodded aggregate $(lb/ft^3)^a$	Unit weight of concrete $(lb/ft^3)^a$
Insulating concretes (perlite, vermiculite, etc.)	15–50	20–90
Structural lightweight	40–70	90–110
Normalweight	70–110	130–160
Heavyweight	>135	180–380

^{a}1 lb/ft^3 = 16.02 kg/m^3.

such as saving energy" (Ref. 2.6). The major types of admixtures can be summarized
as follows:

1. Accelerating admixtures
2. Air-entraining admixtures
3. Water-reducing admixtures and set-controlling admixtures

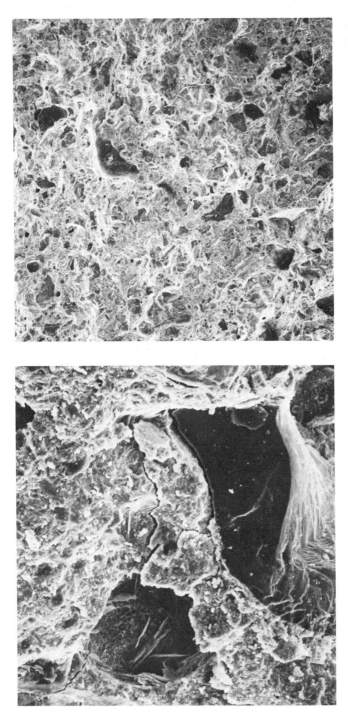

Photo 11 Scanning electron microscope photograph of polymer-cement mortar fracture surface under tension. (Tests by Nawy, Sun, and Sauer.)

Photo 12 Scanning electron microscope photograph of concrete fracture surface. (Tests by Nawy, Sun, and Sauer.)

4. Finely divided mineral admixtures
5. Admixtures for no-slump concretes
6. Polymers
7. Superplasticizers

2.5.1 Accelerating Admixtures $<2\%$

These admixtures are added to the concrete mix to reduce the time of setting and accelerate early strength development. The best known are calcium chlorides. Other accelerating chemicals include a wide range of soluble salts, such as chlorides, bromides, carbonates, silicates, and some other organic compounds, such as tri-ethanolamine.

It must be stressed that calcium chlorides should not be used where progressive corrosion of steel reinforcement can occur. The maximum dosage is 2% by weight of the portland cement.

2.5.2 Air-Entraining Admixtures $2 - 6\%$ of mix

These admixtures form minute bubbles 1 mm in diameter or smaller in the concrete or mortar during mixing, used to increase workability of the mix during placing and increasing the frost resistance of the finished product.

Most of the air-entraining admixtures are in liquid form, although a few are powders, flakes, or semisolids. The amount of the admixture required to obtain a given air content depends on the shape and the grading of the aggregate used. The finer the size of the aggregate, the larger is the percentage of admixture needed. It is also governed by several other factors, such as type and condition of the mixer, use of fly ash or other pozzolones, and the degree of agitation of the mix. It can be expected that air entrainment reduces the strength of the concrete. Maintaining cement content and workability, however, offsets the partial reduction of strength because of the resulting reduction in the water/cement ratio.

2.5.3 Water-Reducing and Set-Controlling Admixtures

These admixtures increase the strength of the concrete. They also enable reducing the cement content in proportion to the reduction in the water content.

Most admixtures of the water-reducing type are water soluble. The water they contain becomes part of the mixing water in the concrete and is added to the total weight of water in the design of the mix. It has to be emphasized that the proportion of the mortar to the coarse aggregate should always remain the same. Changes in the water content, air content, or cement content are compensated for by corresponding changes in the fine aggregate content so that the volume of the mortar remains the same.

2.5.4 Finely Divided Admixtures

These are mineral admixtures used to rectify deficiencies in concrete mix by providing missing fines from the fine aggregate; improving one or more qualities of the concrete, such as reducing permeability or expansion; and reducing the cost of concrete-making materials. Such admixtures include hydraulic lime, slag cement, fly ash, and raw or calcined natural pozzolan.

2.5.5 Admixtures for No-Slump Concrete

No-slump concrete is defined in Ref. 2.6 as a concrete with a slump of 1 in. (25 mm) or less immediately after mixing. The choice of the admixture depends on the desired properties of the finished product, such as its effect on the plasticity, setting time and strength development, freeze–thaw effects, and strength and cost.

2.5.6 Polymers $0.3 \text{ to } 0.45$

These are new types of admixtures that enable producing concretes of very high strength up to a compressive strength of 15,000 psi or higher and a tensile splitting strength of 1500 psi or higher. Such concretes are generally produced using a polymerizing material through (1) modification of the concrete property through water reduction in the field, or (2) impregnation and irradiation under elevated temperature in laboratory environment.

Polymer-modified concrete (PMC) is concrete made through the addition of resin and hardener as an "admixture." The principle is to replace part of the mixing water by the polymer so as to attain the high compressive strength and other qualities reported in detail in Ref. 2.7. The optimum polymer/concrete ratio by weight seems to lie within the range of 0.3 to 0.45 to achieve such high compressive strengths.

2.5.7 Superplasticizers

These are also new types of admixtures, which can be termed "high-range water-reducing chemical admixtures." There are three types of plasticizers:

1. Sulfonated melamine formaldehyde condensates, with a chloride content of 0.005%
2. Sulfonated naphthalene formaldehyde condensates, with negligible chloride content
3. Modified lignosulfonates, which contain no chlorides

These admixtures are made from organic sulfonates and are termed "superplasticizers" in view of their considerable ability to facilitate reducing the water content in a concrete mix while simultaneously increasing the slump up to 8 in.

(206 mm) or more. A dosage of 1 to 2% by weight of cement is advisable. Higher dosages can result in a reduction in compressive strength.

SELECTED REFERENCES

2.1 American Society for Testing and Materials, *Annual Book of ASTM Standards: Part 14, Concrete and Mineral Aggregates,* ASTM, Philadelphia, 1983, 834 pp.

2.2 Popovices, S., *Concrete-Making Materials,* McGraw-Hill, New York, 1979, 370 pp.

2.3 ACI Committee 221, "Selection and Use of Aggregate for Concrete," *Journal of the American Concrete Institute,* Proc. Vol. 58, No. 5, 1961, pp. 513–542.

2.4 American Concrete Institute, *ACI Manual of Concrete Practice 1983:* Part I: *Materials,* American Concrete Institute, Detroit, 1983, 441 p.

2.5 Portland Cement Association, *Design and Control of Concrete Mixtures,* 12th ed., Skokie, Ill., 1979, 140 pp.

2.6 ACI Committee 212, "Admixtures for Concrete," in *ACI Manual of Concrete Practice 1983,* ACI, Detroit, 1983, ACI 212.1 R-81, 29 pp.

2.7 Nawy, E. G., Ukadike, M. M., and Sauer, J. A., "High Strength Field Modified Concretes," *Journal of the Structural Division, ASCE,* Vol. 103, No. ST12, December 1977, pp. 2307–2322.

2.8 American Concrete Institute, *Super-plasticizers in Concrete,* ACI Special Publication SP-62, ACI, Detroit, 1979, 427 pp.

3

Concrete

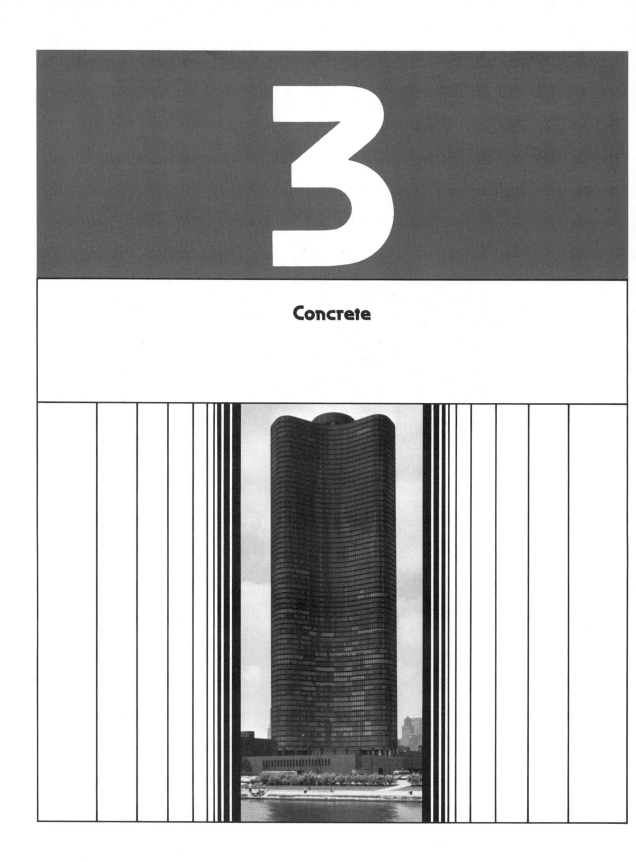

Photo 13 Lake Point Tower, Chicago. (Courtesy of Portland Cement Association.) (*See* p. 22.)

3.1 INTRODUCTION

The general knowledge gained from Chapter 2 can now be utilized to design and obtain a concrete of characteristics and functions to suit a definite purpose. As should be realized by now, the proportioning and types of ingredients establish in part the quality of the concrete and hence the quality of the total structural system. Not only must good materials be chosen, but uniformity must be maintained in the whole product.

The general characteristics of good concrete are summarized in the following sections.

3.1.1 Compactness

The space occupied by the concrete should, as much as possible, be filled with solid aggregate and cement gel free of honeycombing. Compactness may be the primary criterion for those types of concrete that intercept nuclear radiation.

3.1.2 Strength

Concrete should always have sufficient strength and internal resistance to the various types of failure.

Photo 14 Terminal building, Dallas International Airport. (Courtesy of Ammann & Whitney.)

3.1.3 Water/Cement Ratio

The water/cement ratio should be suitably controlled to give the required design strength.

3.1.4 Texture

Exposed concrete surfaces should have a dense and hard texture that can withstand adverse weather conditions.

3.1.5 Parameters Affecting Concrete Quality

To achieve the aforementioned properties, good quality control has to be exercised on the factors shown in Fig. 3.1. The following are the most important parameters:

1. Quality of cement
2. Proportion of cement in relation to water in the mix.
3. Strength and cleanliness of aggregate
4. Interaction or adhesion between cement paste and aggregate
5. Adequate mixing of the ingredients
6. Proper placing, finishing, and compaction of the fresh concrete
7. Curing at a temperature not below 50°F while the placed concrete gains strength
8. Chloride content not to exceed 0.15 percent in reinforced concrete exposed to chlorides in service and one percent for dry protected concrete

A study of these requirements shows that most of the control actions have to be taken prior to placing the fresh concrete. Since such control is governed by the proportions and the mechanical ease or difficulty in handling and placing, the development of criteria based on the theory of proportioning for each mix should be studied. Most mix design methods have become essentially only of historical and academic value.

The two universally accepted methods for mix proportioning for normalweight and lightweight concrete are the American Concrete Institute's methods of proportioning described in the recommended practice for selecting proportions for normal-weight, heavyweight, and mass concrete, and the recommended practice for selecting proportions for structural lightweight concrete (Refs. 3.1 and 3.2).

3.2 PROPORTIONING THEORY

Water/cement ratio (w/c ratio) theory states that for a given combination of materials and as long as workable consistency is obtained, the strength of concrete at a given age depends on the ratio of the weight of mixing water to the weight of cement. In other words, if the ratio of water to cement is fixed, the strength of concrete at a certain age is also essentially fixed, as long as the mixture is plastic and workable and the

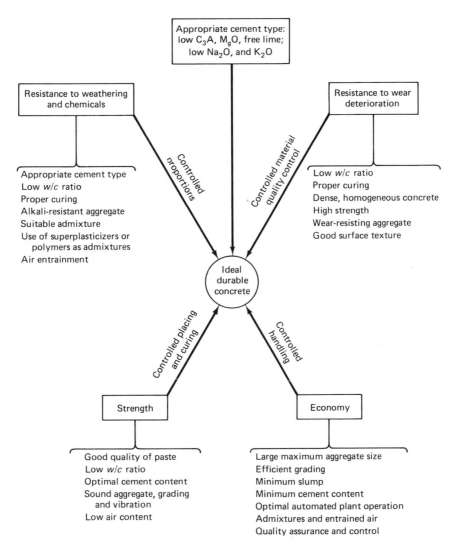

Figure 3.1 Principal properties of good concrete.

aggregate sound, durable and free of deleterious materials. Whereas strength depends on the *w/c* ratio, economy depends on the percentage of aggregate present that would still give a workable mix. The aim of the designer should always be to get concrete mixtures of optimum strength at minimum cement content and acceptable workability. The lower the *w/c* ratio, the higher the concrete strength.

Once the *w/c* ratio is established and the workability or consistency needed for the specific design is chosen, the rest should be simple manipulation with diagrams and tables based on large numbers of trial mixes. Such diagrams and tables allow an estimate of the required mix proportions for various conditions and permit predetermination on small unrepresentative batches.

3.2.1 ACI Method of Mix Design

The flowchart represented in Fig. 3.2 and the following design example best illustrate the mix design process using the ACI mix design method. One of the aims of the mix design is to produce workable concrete that is easy to place in the forms. A measure of the degree of consistency and extent of workability is the *slump*. In the slump test, the plastic concrete specimen is formed into a conical metal mold as described in ASTM Standard C-143. The mold is lifted, leaving the concrete to "slump," that is, to spread or drop in height. This drop in height is the slump measure of the degree of workability of the mix.

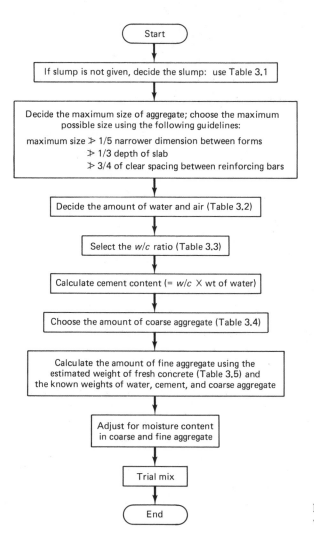

Figure 3.2 Flowchart for normal-weight concrete mix design.

Photo 15   (a) $4\frac{1}{2}$-in. slump mix; (b) $1\frac{1}{2}$-in. slump mix.

3.2.2 Example 3.1: Mix Design of Normal-Weight Concrete

Design a concrete mix using the following details:

Required strength: 4000 psi (27.6 MPa)

Type of structure: beam

Maximum size of aggregate $= \frac{3}{4}$ in. (18 mm)

Fineness modulus of sand $= 2.6$

Dry-rodded weight of coarse aggregate $= 100$ lb/ft^3

Moisture absorption 3% for coarse aggregate and 2% for fine aggregate

Solution

Required slump for beams (Table 3.1) $= 3$ in.

maximum aggregate size (given) $= \frac{3}{4}$ in.

For a slump between 3 and 4 in. and a maximum aggregate size of $\frac{3}{4}$ in.,

weight of water required per cubic yard of concrete (Table 3.2) $= 340$ lb/yd^3

TABLE 3.1   RECOMMENDED SLUMPS FOR VARIOUS TYPES OF CONSTRUCTION

	Slump (in.)[a]	
Types of construction	Maximum[b]	Minimum
Reinforced foundation walls and footings	3	1
Plain footings, caissons, and substructure walls	3	1
Beams and reinforced walls	4	1
Building columns	4	1
Pavements and slabs	3	1
Mass concrete	2	1

[a]1 in. = 25.4 mm.

[b]May be increased 1 in. for methods of consolidation other than vibration.

28

TABLE 3.2 APPROXIMATE MIXING WATER AND AIR CONTENT REQUIREMENTS FOR DIFFERENT SLUMPS AND NOMINAL MAXIMUM SIZES OF AGGREGATES

Slump (in.)	Water (lb/yd^3) of concrete for indicated nominal maximum sizes of aggregate)							
	$\frac{3}{8}$ in.[a]	$\frac{1}{2}$ in.[a]	$\frac{3}{4}$ in.[a]	1 in.[a]	1$\frac{1}{2}$ in.[a]	2 in.[a,b]	3 in.[b,c]	6 in.[b,c]
Non-Air-Entrained Concrete								
1 to 2	350	335	315	300	275	260	220	190
3 to 4	385	365	340	325	300	285	245	210
6 to 7	410	385	360	340	315	300	270	—
Approximate amount of entrapped air in non-air-entrained concrete (%)	3	2.5	2	1.5	1	0.5	0.3	0.2
Air-Entrained Concrete								
1 to 2	305	295	280	270	250	240	205	180
3 to 4	340	325	305	295	275	265	225	200
6 to 7	365	345	325	310	290	280	260	—
Recommended average total air content[d] (percent for level of exposure)								
Mild exposure	4.5	4.0	3.5	3.0	2.5	2.0	1.5[e,f]	1.0[e,f]
Moderate exposure	6.0	5.5	5.0	4.5	4.5	4.0	3.5[e,f]	3.0[e,f]
Extreme exposure[g]	7.5	7.0	6.0	6.0	5.5	5.0	4.5[e,f]	4.0[e,f]

[a]These quantities of mixing water are for use in computing cement factors for trial batches. They are maximal for reasonably well-shaped angular coarse aggregates graded within limits of accepted specifications.

[b]The slump values for concrete containing aggregate larger than $1\frac{1}{2}$ in. are based on slump tests made after removal of particles larger than $1\frac{1}{2}$ in. by wet screening.

[c]These quantities of mixing water are for use in computing cement factors for trial batches when 3-in. or 6-in. nominal maximum-size aggregate is used. They are average for reasonably well-shaped coarse aggregates, well graded from coarse to fine.

[d]Additional recommendations for air content and necessary tolerances on air content for control in the field are given in a number of ACI documents, including ACI 201, 345, 318, 301, and 302. ASTM C-94 for ready-mixed concrete also gives air content limits. The requirements in other documents may not always agree exactly, so in proportioning concrete, consideration must be given to selecting an air content that will meet the needs of the job and also meet the applicable specifications.

[e]For concrete containing large aggregates that will be wet screened over the $1\frac{1}{2}$-in. sieve prior to testing for air content, the percentage of air expected in the $1\frac{1}{2}$-in.-minus material should be tabulated in the $1\frac{1}{2}$ in. column. However, initial proportioning calculations should include the air content as a percent of the whole.

[f]When using large aggregate in low-cement-factor concrete, air entrainment need not be detrimental to strength. In most cases the mixing water requirement is reduced sufficiently to improve the water-cement ratio and thus to compensate for the strength-reducing effect of entrained-air concrete. Generally, therefore, for these large maximum sizes of aggregate, air contents recommended for extreme exposure should be considered even though there may be little or no exposure to moisture and freezing.

[g]These values are based on the criteria that 9% air is needed in the mortar phase of the concrete. If the mortar volume will be substantially different from that determined in this recommended practice, it may be desirable to calculate the needed air content by taking 9% of the actual mortar volume.

TABLE 3.3 RELATIONSHIP BETWEEN WATER/CEMENT RATIO AND COMPRESSIVE STRENGTH OF CONCRETE

Compressive strength at 28 days[a] (psi)[b]	Water/cement ratio, by weight	
	Non-air-entrained concrete	Air-entrained concrete
6000	0.41	—
5000	0.48	0.40
4000	0.57	0.48
3000	0.68	0.59
2000	0.82	0.74

[a]Values are estimated average strengths for concrete containing not more than the percentage of air shown in Table 3.2. For a constant water/cement ratio, the strength of concrete is reduced as the air content is increased.

Strength is based on 6 in. × 12 in. cylinders moist-cured 28 days at 73.4 ± 3°F (23 ± 1.7°C) in accordance with Section 9(b) of ASTM C-31, "Making and Curing Concrete Compression and Flexure Test Specimens in the Field."

Relationship assumes maximum size of aggregate about $\frac{3}{4}$ to 1 in.; for a given source, strength produced for a given water/cement ratio will increase as maximum size of aggregate decreases.

[b]1000 psi = 6.9 MPa.

TABLE 3.4 VOLUME OF COARSE AGGREGATE PER UNIT OF VOLUME OF CONCRETE

Maximum size of of aggregate (in.)[a]	Volume of dry-rodded coarse aggregate[b] per unit volume of concrete for different fineness moduli of sand			
	2.40	2.60	2.80	3.00
$\frac{3}{8}$	0.50	0.48	0.46	0.44
$\frac{1}{2}$	0.59	0.57	0.55	0.53
$\frac{3}{4}$	0.66	0.64	0.62	0.60
1	0.71	0.69	0.67	0.65
$1\frac{1}{2}$	0.75	0.73	0.71	0.69
2	0.78	0.76	0.74	0.72
3	0.82	0.80	0.78	0.76
6	0.87	0.85	0.83	0.81

[a]1 in. = 25.4 mm.

[b]Volumes are based on aggregates in dry-rodded condition as described in ASTM C-29, "Unit Weight of Aggregate." These volumes are selected from empirical relationships to produce concrete with a degree of workability suitable for usual reinforced construction. For less workable concrete, such as that required for concrete pavement construction, they may be increased about 10%. For more workable concrete, the coarse aggregate content may be decreased up to 10%—provided that the slump and water/cement ratio requirements are satisfied.

For the specified compression strength f'_c = 4000 psi, w/c ratio (Table 3.3) = 0.57

Table 3.4 is also needed if volumes instead of weights are used in the mix design calculations. Therefore,

$$\text{amount of cement required per cubic yard of concrete} = \frac{340}{0.57} = 596.5 \text{ lb/yd}^3$$

Using a sand fineness value of 2.6 and Table 3.4,

$$\text{volume of coarse aggregate} = 0.64 \text{ yd}^3$$

Using the dry-rodded weight of 100 lb/ft^3 for coarse aggregate,

$$\text{weight of coarse aggregate} = (0.64 \text{ yd}^3) \times (27 \text{ ft}^3/\text{yd}^3) \times 100 \text{ }^{lb}/_{ft^3}$$

$$= 1728 \text{ lb/yd}^3$$

estimated weight of fresh concrete for $\frac{3}{4}$ in. maximum size aggregate
(Table 3.5) = 3960 lb/yd^3

$$\text{weight of sand} = [\text{weight of fresh concrete} - \text{weights of (water} + \text{cement} + \text{coarse aggregate)}]$$

$$= 3960 - \left(340 + 596.5 + 1728\right) = 1295.5 \text{ lb}$$

net weight of sand to be taken = 1.02 × 1295.5

(moisture absorption 2%) = 1321.41 lb

net weight of gravel = 1.03 × 1728

(moisture absorption 3%) = 1779.84 lb

net weight of water = 340 $- \left(0.02 \times 1295.5\right) - \left(0.03 \times 1728\right)$

$$= 262.25 \text{ lb}$$

For 1 yd^3 of concrete:

cement = 596.5 lb. $\simeq$ 600 lb (273 kg) $1:2.2:2.97$

sand = 1321.41 lb $\simeq$ 1320 lb (600 kg)

gravel = 1779.84 lb $\simeq$ 1780 lb (810 kg)

water = 262.25 lb $\simeq$ 260 lb (120 kg)

3.3 PCA METHOD OF MIX DESIGN

The mix design method proposed by the Portland Cement Association (PCA) is essentially similar to the ACI method. Generally, results would be very close once trial batches are prepared in the laboratory. The PCA publication listed in the references gives the details of the method as well as other information on properties of the ingredients.

TABLE 3.5   FIRST ESTIMATE OF WEIGHT OF FRESH CONCRETE

Maximum size of aggregate (in.)[a]	First estimate of concrete weight[b] (lb/yd³)[c]	
	Non-air-entrained concrete	Air-entrained concrete
$\frac{3}{8}$	3840	3690
$\frac{1}{2}$	3890	3760
$\frac{3}{4}$	3960	3840
1	4010	3900
$1\frac{1}{2}$	4070	3960
2	4120	4000
3	4160	4040
6	4230	4120

[a]1 in. = 25.4 mm.

[b]Values calculated and presented below are for concrete of medium richness (550 lb of cement per cubic yard) and medium slump with aggregate specific gravity of 2.7. Water requirements are based on values for 3- to 4-in. slump in Table 5.3.2 of ASTM C-143. If desired, the estimated weight may be refined as follows if necessary information is available: for each 10-lb difference in mixing water from Table 5.3.2, values for 3- to 4-in. slump, correct the weight per cubic yard 15 lb in the opposite direction; for each 100-lb difference in cement content from 550 lb, correct the weight per cubic yard 15 lb in the same direction; for each 0.1 by which aggregate specific gravity deviates from 2.7, correct the concrete weight 100 lb in the same direction.

weight of fresh concrete per cubic yard, lb

$$= 16.85 G_a (100 - A) + C\left(1 - \frac{G_a}{G_c}\right) - W(G_a - 1)$$

where G_a = weighted average specific gravity of combined fine and coarse aggregate, bulk saturated surface dry density

G_c = specific gravity of cement (generally 3.15)

A = air content, %

W = mixing water requirement, lb/yd³

C = cement requirement, lb/yd³

[c]1 lb/yd³ = 0.6 kg/m³

3.4 MIX DESIGN FOR STRUCTURAL LIGHTWEIGHT CONCRETE

Structural lightweight concrete can best be defined as concrete having a 28-day compressive strength in excess of 2000 psi and an air-dry unit weight less than 115 lb/ft³. The coarse aggregate used is primarily expanded shale, slate, slags, and so on, and the same principles and procedures used in normalweight concrete are applicable to this type of concrete. Air entrainment is very desirable, if not mandatory. A recommended percentage of air-entraining agents of at least 6% is necessary to give the product acceptable weathering qualities.

3.5 ESTIMATING COMPRESSIVE STRENGTH OF A TRIAL MIX USING THE SPECIFIED COMPRESSIVE STRENGTH

The compressive strength for which the trial mix is designed is not the strength specified by the designer. The mix should be overdesigned to assure that the actual structure has concrete with specified minimum compressive strength. The extent of mix overdesign depends on the degree of quality control available in the mixing plant.

ACI Committee 318 specifies a systematic way of determining the compressive strength for mix designs using the specified compressive strength, f'_c. The procedure is presented in a self-explanatory flowchart form in Fig. 3.3. The cylinder compressive strength f'_c (see Section 3.7) is the test result at 28 days after casting normalweight concrete. Mix design has to be based on an adjusted higher value f'_{cr}. This adjusted cylinder compressive strength f'_{cr} for which a trial mix design is calculated depends on the extent of field data available.

1. *No cylinder test records available:*   If field-strength test records for the specified class (or within 1000 psi of the specified class) of concrete are not available, the trial mix strength f'_{cr} can be calculated by increasing the cylinder compressive strength f'_c by a reasonable value depending on the extent of spread in values expected in the supplied concrete. Such a spread can be quantified by the standard deviation values represented by the values in excess on f'_c in Table 3.6. Table 3.7 can then be used to obtain the water/cement ratio needed for the required cylinder strength value f'_c.

2. *Data available on more that 30 consecutive cylinder tests:*   If more than 30 consecutive tests results are available, Eqs. 3.1, 3.2, and 3.3a in Section 3.5.2 can be used to establish the required mix strength, f'_{cr}, from f'_c. If two groups of consecutive test results with a total of more than 30 are available, f'_{cr} can be obtained using Eqs. 3.1, 3.2, and 3.3b.

3. *Data available on fewer than 30 consecutive cylinder tests:*   If the number of consecutive test results available is fewer than 30 and more than 15, Eqs. 3.1, 3.2, and 3.3a should be used in conjunction with Table 3.8. Essentially, the designer should calculate the standard deviation s using Eq. 3.3a, multiply the s value by a magnification factor provided in Table 3.8, and use the magnified s in Eqs. 3.1 and 3.2. In this manner, the expected degree of spread of cylinder test values as measured by the standard deviation s is well accounted for.

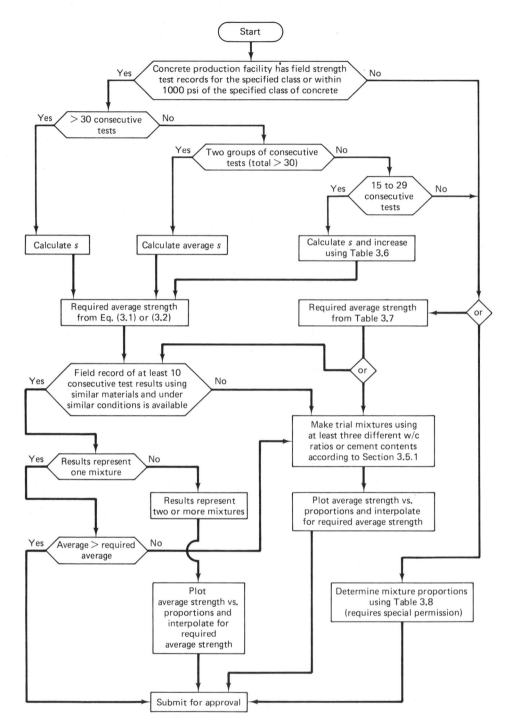

Figure 3.3 Flowchart for selection and documentation of concrete proportions.

Photo 16 Cylinder compression test.

TABLE 3.6 REQUIRED AVERAGE COMPRESSIVE STRENGTH WHEN DATA ARE NOT
AVAILABLE TO ESTABLISH A STANDARD DEVIATION

Specified compressive strength, f'_c (psi)	Required average compressive strength, f'_{cr} (psi)[a]
Less than 3000	$f'_c + 1000$
3000–5000	$f'_c + 1200$
More than 5000	$f'_c + 1400$

[a]1000 psi = 6.9 MPa.

TABLE 3.7 MAXIMUM PERMISSIBLE WATER/CEMENT RATIOS
FOR CONCRETE WHEN STRENGTH DATA FROM FIELD EXPERIENCE
OR TRIAL MIXTURES ARE NOT AVAILABLE

Specified compressive strength,[a] f'_c, (psi)[b]	Absolute water/cement ratio by weight	
	Non-air-entrained concrete	Air-entrained concrete
2500	0.67	0.54
3000	0.58	0.46
3500	0.51	0.40
4000	0.44	0.35
4500	0.38	c
5000	c	c

[a]28-day strength. With most materials, the water/cement ratios shown will provide average strengths greater than those calculated using Eqs. 3.1 and 3.2.

[b]1000 psi = 6.9 MPa.

[c]For strengths above 4500 psi for non-air-entrained concrete and 4000 psi for air-entrained concrete, mix proportions should be established using trial mixes.

TABLE 3.8 MODIFICATION FACTOR
FOR STANDARD DEVIATION WHEN FEWER
THAN 30 TESTS ARE AVAILABLE

Number of tests[a]	Modification factor for standard deviation[b]
Less than 15	Use Table 3.6
15	1.16
20	1.08
25	1.03
30 or more	1.00

[a]Interpolate for intermediate number of tests.

[b]Modified standard deviation to be used to determine required average strength f'_{cr} in Eqs. 3.1 and 3.2.

3.5.1 Recommended Proportions for Concrete Strength f'_{cr}

Once the required average strength f'_{cr} for mix design is determined, the actual mix can be established to obtain this strength using either existing field data or a basic trial mix design.

1. *Use of field data.* Field records of existing f'_{cr} values can be used if at least 10 consecutive test results are available. The test records should cover a period of time of at least 45 days. The materials and conditions of the existing field mix data should be the same as the ones to be used in the proposed work.

2. *Trial mix design.*   If the field data are not available, trial mixes should be used to establish the maximum water/cement ratio or minimum cement content for designing a mix that produces a 28-day f'_{cr} value. In this procedure, the following requirements have to be met:

   (a)   Materials used and age of testing should be the same for the trial mix and the concrete used in the structure.

   (b)   At least three water/cement ratios or three cement contents should be tried in the mix design. The trial mixes should result in the required f'_{cr}. Three cylinders should be tested for each w/c ratio and each cement content tried.

   (c)   The slump and air content should be within ±0.75 in. and 0.5% of the permissible limits.

   (d)   A plot is constructed of the compressive strength at the designated age versus the cement content or water/cement ratio, from which one can then choose the w/c ratio or the cement content that can give the average f'_{cr} value required.

3.5.2 Trial Mix Design for Average Strength When Prior Field Strength Data Are Available

If field test data are available for more than 30 consecutive tests, the trial mix should be designed for compressive strength f'_{cr} calculated from

$$f'_{cr} = f'_c + 1.34s \tag{3.1}$$

or

$$f'_{cr} = f'_c + 2.33s - 500 \tag{3.2}$$

The larger value of f'_{cr} from Eqs. 3.1 and 3.2 should be used in designing the mix, with the expectation of attaining the minimum f'_c specified design compressive strength. The standard deviation s is defined by the expression

$$s = \left[\frac{\Sigma(f_{ci} - \bar{f}_c)^2}{n - 1} \right]^{1/2} \tag{3.3a}$$

where f_{ci} = individual strength
      $\bar{f}_c$ = average of the n specimens

If two test records are used to determine the average strength, the standard deviation becomes

$$s = \left[\frac{(n_1 - 1)s_1^2 + (n_2 - 1)s_2^2}{n_1 + n_2 - 2} \right]^{1/2} \tag{3.3b}$$

where s_1, s_2 = standard deviations calculated from two tests records, 1 and 2, respectively
      n_1, n_2 = number of tests in each test record, respectively

If the number of test results available is fewer than 30, and more than 15 the value of s used in Eqs. 3.1 and 3.2 should be multiplied by the appropriate modification factor value given in Table 3.8.

3.5.3 Example 3.2: Calculation of Design Strength for Trial Mix

Calculate the average compressive strengths f'_{cr} for the design of a concrete mix if the specified compressive strength f'_c is 5000 psi (34.5 MPa) such that (a) the standard deviation obtained using more than 30 consecutive tests is 500 psi (3.45 MPa), (b) the standard deviation obtained using 15 consecutive tests is 450 psi (3.11 MPa), and (c) records of prior cylinder test results are not available.

Solution

(a) Using Eq. 3.1: $f'_{cr} = f'_c + 1.34s$ s- STD. DEV

No, of Tests ≥ 30

$$f'_{cr} = 5000 + 1.34 \times 500$$
$$= 5670 \text{ psi}$$

Using Eq. 3.2:   CHOOSE HIGHER OF TWO

$$f'_{cr} = f'_c + 2.33s - 500$$
$$f'_{cr} = 5000 + 2.33 \times 500 - 500$$
$$= 5665 \text{ psi}$$

Hence the required trial mix strength $f'_{cr} = 5670$ psi (39.12 MPa).

(b) $s = 450$ psi in 15 tests. From Table 3.8, the modification factor for s is 1.16. Hence the value of standard deviation to be used in Eqs. 3.1 and 3.2 is 1.16 × 450 = 522 psi (3.6 MPa).

Using Eq. 3.1:

No. of Tests ≥ 15

$$f'_{cr} = 5000 + 1.34 \times 522$$
$$= 5700 \text{ psi}$$

Using Eq. 3.2:

$$f'_{cr} = 5000 + 2.33 \times 522 - 500$$
$$= 5716 \text{ psi}$$

Hence the required trial mix strength $f'_{cr} = 5716$ psi (39.44 MPa).

(c) Records of prior test results are not available. Using Table 3.6:

No. of Tests = ?

$$f'_{cr} = f'_c + 1200 \text{ for 5000-psi concrete}$$

Hence the trial mix strength = 5000 + 1200 = 6200 psi (42.78 MPa).

It can be observed that if the mixing plant keeps good records of its cylinder test results over a long period, the required trial mix strength f'_{cr} can be reduced as a result of such quality control, hence reducing costs for the owner.

3.6 MIX DESIGNS FOR NUCLEAR-SHIELDING CONCRETE

Whereas from the foregoing discussion it is seen that the design criterion was the water/cement ratio, in concrete used for shielding against x-rays, gamma rays, and neutrons, the criterion is compactness or density of mix, regardless of workability. To achieve maximum density, tests have been conducted on various mixes using crushed magnatite ore or fine steel shot instead of sand, and steel punchings, magnatites, barites, or limonites instead of stone, as discussed previously. Results of these tests for both compactness and strength have shown that the *w/c* ratio has to be limited to 3.5 to 4.0 gal of water per bag of cement.

3.7 QUALITY TESTS ON CONCRETE

3.7.1 Workability or Consistency

Possible tests for workability or consistency include:

1. Slump test by means of the standard ASTM code. The slump in inches recorded in the mix indicates its workability.
2. Remolding tests using Power's flow table.
3. Kelley's ball apparatus.

The first method is the accepted ASTM standard.

3.7.2 Air Content

Measurement of the air content in fresh concrete is always necessary, especially when air-entraining agents are used.

3.7.3 Compressive Strength of Hardened Concrete

This is done by loading cylinders 6 in. in diameter and 12 in. high in compression perpendicular to the axis of the cylinder.

3.7.4 Flexural Strength of Plain Concrete Beams

This test is performed by three-point loading of plain concrete beams of size 6 in. $\times$ 6 in. $\times$ 18 in. which have spans three times their depth.

3.7.5 Tensile Splitting Tests

These tests are performed by loading the standard 6 in. $\times$ 12 in. cylinder by a line load perpendicular to its longitudinal axis, with the cylinder placed horizontally on the

Photo 17   Tensile splitting test.

testing machine platten. The tensile splitting strength can be defined as

$$f'_t = \frac{2P}{\pi DL}$$

P- LOAD
D- DIA
L - LENGTH .

(3.4)

where P = total value of the line load registered by the testing machine
D = diameter of the concrete cylinder
L = cylinder height

The results of all these tests give the designer a measure of the expected strength of the designed concrete in the built structure.

3.8 PLACING AND CURING OF CONCRETE

3.8.1 Placing

The techniques necessary for placing concrete depend on the type of member to be cast: that is, whether it is a column, a beam, a wall, a slab, a foundation, a mass concrete dam, or an extension of previously placed and hardened concrete. For beams, columns, and walls, the forms should be well oiled after cleaning them, and the reinforcement should be cleared of rust and other harmful materials. In foundations, the earth should be compacted and thoroughly moistened to about 6 in. in depth to avoid absorption of the moisture present in the wet concrete. Concrete should always be placed in horizontal layers which are compacted by means of high-frequency power-driven vibrators of either the immersion or external type, as the case requires, unless it is placed by pumping. It must be kept in mind, however, that overvibration can be harmful since it could cause segregation of the aggregate and bleeding of the concrete.

3.8.2 Curing

As seen in Chapter 2, hydration of the cement takes place in the presence of moisture at temperatures above 50°F. It is necessary to maintain such a condition in order that the chemical hydration reaction can take place. If drying is too rapid, surface cracking takes place. This would result in reduction of concrete strength due to cracking as well as the failure to attain full chemical hydration.

To facilitate good curing conditions, any of the following methods can be used:

1. Continuously sprinkling with water.
2. Ponding with water.
3. Covering the concrete with wet burlap, plastic film, or waterproof curing paper.
4. Using liquid membrane-forming curing compounds to retain the original moisture in the wet concrete.
5. Steam curing in cases where the concrete member is manufactured under factory conditions, such as in cases of precast beams and pipes, and prestressed girders and poles. Steam-curing temperatures are about 150°F. Curing time is usually 1 day, compared to the 5 to 7 days necessary when using the other methods.

3.9 PROPERTIES OF HARDENED CONCRETE

The mechanical properties of hardened concrete can be classified as (1) short-term or instantaneous properties, and (2) long-term properties. The short-term properties can be enumerated as (1) strength in compression, tension, and shear, and (2) stiffness measured by modulus of elasticity. The long-term properties can be classified in terms of creep and shrinkage. The following sections present some details of the afore-mentioned properties.

3.9.1 Compressive Strength

Depending on the type of mix, the properties of aggregate, and the time and quality of the curing, compressive strengths of concrete can be obtained up to 14,000 psi or more. Commercial production of concrete with ordinary aggregate is usually in the range 300 to 10,000 psi, with the most common concrete strengths in the range 3000 to 6000 psi.

The compressive strength, f_c', is based on standard 6 in. by 12 in. cylinders cured under standard laboratory conditions and tested at a specified rate of loading at 28 days of age. The standard specifications used in the United States are usually taken from ASTM C-39. It should be mentioned that the strength of concrete in the actual structure may not be the same as that of the cylinder because of the difference in compaction and curing conditions.

The ACI code specifies for a strength test the average of two cylinders from the same sample tested at the same age, which is usually 28 days. As for the frequency

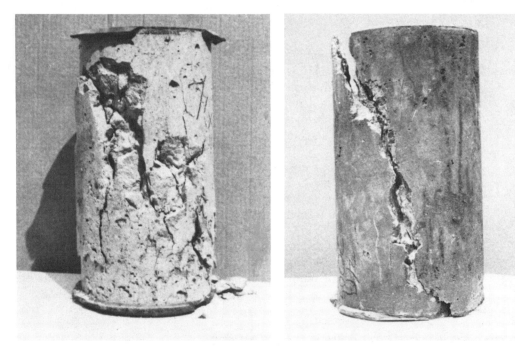

Photo 18   Concrete cylinders tested to failure in compression. Specimen A, low-epoxy-cement content; specimen B, high-epoxy-cement content. (Tests by Nawy, Sun, and Sauer.)

of testing, the code specifies that the strength level of an individual class of concrete can be considered as satisfactory if (1) the average of all sets of three consecutive strength tests equal or exceed required f'_c, and (2) no individual strength test (average of two cylinders) falls below the required f'_c by more than 500 psi. The average concrete strength for which a concrete mix must be designed should exceed f'_c by an amount that depends on the uniformity of plant production, as explained in Section 3.5.

It must be emphasized that the design f'_c should not be the average cylinder strength. The design value should be chosen as the conceivable minimum cylinder strength.

3.9.2 Tensile Strength

The tensile strength of concrete is relatively low. A good approximation for the tensile strength f_{ct} is $0.10f'_c < f_{ct} < 0.20f'_c$. It is more difficult to measure tensile strength than compressive strength because of the gripping problems with testing machines. A number of methods are available for tension testing, the most commonly used method being the cylinder splitting test or Brazilian test.

For members subjected to bending, the value of the modulus of rupture f_r rather

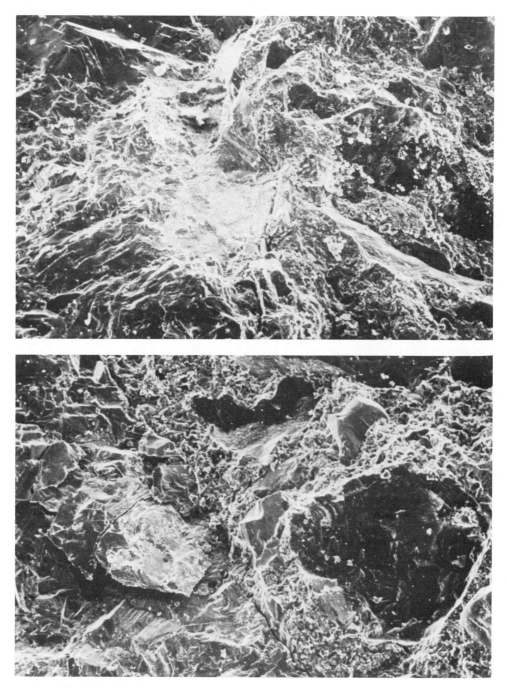

Photo 19 Electron microscope photographs of concrete from specimens A and B in the preceding photograph. (Tests by Nawy et al.)

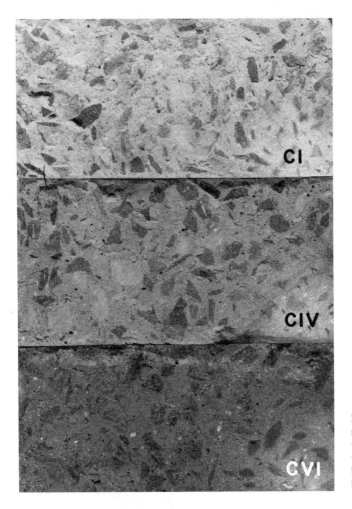

Photo 20 Fracture surfaces in tensile splitting tests of concretes with different w/c contents. Specimens CI and CIV with higher w/c content, hence more bond failures than specimen CVI. (Tests by Nawy et al.)

than tensile splitting strength f_t' is used in design. The modulus of rupture is measured by testing to failure plain concrete beams 6 in. square in cross section, having a span of 18 in. and loaded at its third points (ASTM C-78). The modulus of rupture has a higher value than the tensile splitting strength. The ACI specifies a value of $7.5\sqrt{f_c'}$ for the modulus of rupture of normalweight concrete.

In most cases, lightweight concrete has a lower tensile strength than does normalweight concrete. The following are the code stipulations for lightweight concrete.

1.   If the splitting tensile strength f_{ct} is specified,

$$f_r = 1.09 f_{ct} \le 7.5\sqrt{f_c'}$$

2. If f_{ct} is not specified, use a factor of 0.75 for all lightweight concrete and 0.85 for sand-lightweight concrete. Linear interpolation may be used for mixtures of natural sand and lightweight fine aggregate.

3.9.3 Shear Strength

Shear strength is more difficult to determine experimentally than the tests discussed previously because of the difficulty in isolating shear from other stresses. This is one of the reasons for the large variation in shear-strength values reported in the literature, varying from 20% of the compressive strength in normal loading to a considerably higher percentage of up to 85% of the compressive strength in cases where direct shear exists in combination with compression. Control of a structural design by shear strength is significant only in rare cases, since shear stresses must ordinarily be limited to continually lower values in order to protect the concrete from failure in diagonal tension.

3.9.4 Stress–Strain Curve

Knowledge of the stress–strain relationship of concrete is essential for developing all the analysis and design terms and procedures in concrete structures. Figure 3.4(a) shows a typical stress–strain curve obtained from tests using cylindrical concrete specimens loaded in uniaxial compression over several minutes. The first portion of the curve, to about 40% of the ultimate strength f'_c, can essentially be considered linear for all practical purposes. After approximately 70% of the failure stress, the material loses a large portion of its stiffness, thereby increasing the curvilinearity of the diagram. At ultimate load, cracks parallel to the direction of loading become distinctly visible, and most concrete cylinders (except those with very low strengths) fail suddenly shortly thereafter. Figure 3.4(b) shows the stress–strain curves of concrete of various strengths reported by the Portland Cement Association. It can be observed that

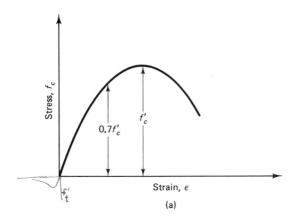

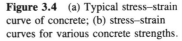

Figure 3.4 (a) Typical stress–strain curve of concrete; (b) stress–strain curves for various concrete strengths.

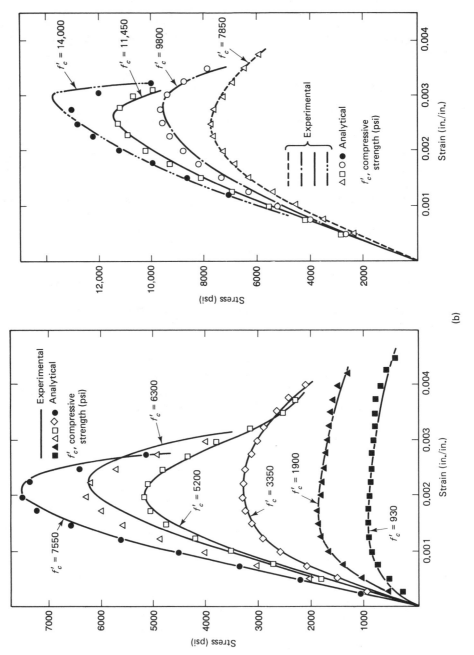

Figure 3.4 *(cont.)*

(b)

(1) the lower the strength of concrete, the higher the failure strain; (2) the length of the initial relatively linear portion increases with the increase in the compressive strength of concrete; and (3) there is an apparent reduction in ductibility with increased strength.

3.9.5 Modulus of Elasticity

Since the stress–strain curve shown in Fig. 3.5 is curvilinear at a very early stage of its loading history, Young's modulus of elasticity can be applied only to the tangent of the curve at the origin. The initial slope of the tangent to the curve is defined as the initial tangent modulus and it is also possible to construct a tangent modulus at any point of the curve. The slope of the straight line that connects the origin to a given stress (about $0.4f_c'$) determines the secant modulus of elasticity of concrete. This value, termed in design calculation the *modulus of elasticity,* satisfies the practical assumption that strains occurring during loading can be considered basically elastic (completely recoverable on unloading), and that any subsequent strain due to the load is regarded as creep.

The ACI building code gives the following expressions for calculating the secant modulus of elasticity of concrete (E_c).

$$E_c = 33w_c^{1.5}\sqrt{f_c'} \qquad \text{for } 90 < w_c < 155 \text{ lb/ft}^3$$

where w_c is the density of concrete in pounds per cubic foot (1 lb/ft^3 = 16.02 kg/m^3) and f_c' is the compressive cylinder strength in psi. For normal-weight concrete,

$$E_c = 57,000\sqrt{f_c'} \text{ psi} \qquad \text{or} \qquad E_c = 4730\sqrt{f_c'} \text{ N/mm}^2$$

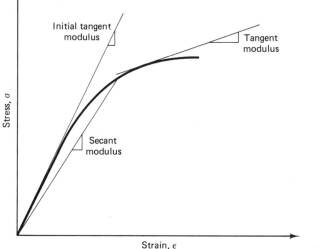

Figure 3.5 Tangent and secant moduli of concrete.

It has to be pointed out that these expressions are valid only in general terms, since the value of the modulus of elasticity is also affected by factors other than loads, such as moisture in the concrete specimen, the water/cement ratio, age of the concrete, and temperature. Therefore, for special structures such as arches, tunnels, and tanks, the modulus of elasticity needs to be determined from test results.

Limited work exists on the determination of the modulus of elasticity in tension because the low tensile strength of concrete is normally disregarded in calculations. It is, however, valid to assume within those limitations that the value of the modulus in tension is equal to that in compression.

3.9.6 Shrinkage

Basically, there are two types of shrinkage: plastic shrinkage and drying shrinkage. *Plastic shrinkage* occurs during the first few hours after placing fresh concrete in the forms. Exposed surfaces such as floor slabs are more easily affected by exposure to dry air because of their large contact surface. In such cases, moisture evaporates faster from the concrete surface than it is replaced by the bleed water from the lower layers of the concrete elements. *Drying shrinkage,* on the other hand, occurs after the concrete has already attained its final set and a good portion of the chemical hydration process in the cement gel has been accomplished.

Drying shrinkage is the decrease in the volume of a concrete element when it loses moisture by evaporation. The opposite phenomenon, that is, volume increase through water absorption, is termed *swelling*. In other words, shrinkage and swelling represent water movement out of or into the gel structure of a concrete specimen due to the difference in humidity or saturation levels between the specimen and the surroundings irrespective of the external load.

Shrinkage is not a completely reversible process. If a concrete unit is saturated with water after having fully shrunk, it will not expand to its original volume. Figure 3.6 relates the increase in shrinkage strain ϵ_{sh} with time. The rate decreases with time since older concretes are more resistant to stress and consequently undergo less shrinkage, such that the shrinkage strain becomes almost asymptotic with time.

Several factors affect the magnitude of drying shrinkage:

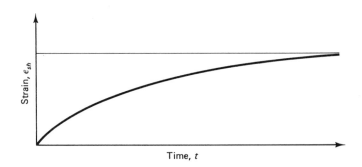

Figure 3.6   Shrinkage–time curve.

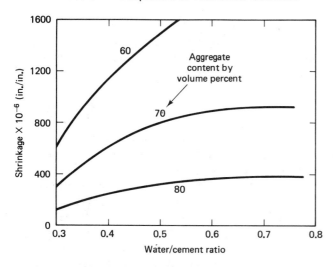

Figure 3.7   *w/c* ratio and aggregate content effect on shrinkage.

1. *Aggregate.*   The aggregate acts to restrain the shrinkage of the cement paste; hence concretes with high aggregate content are less vulnerable to shrinkage. In addition, the degree of restraint of a given concrete is determined by the properties of aggregates; those with high modulus of elasticity or with rough surfaces are more resistant to the shrinkage process.

2. *Water/cement ratio.*   The higher the water/cement ratio, the higher the shrinkage effects. Figure 3.7 is a typical plot relating aggregate content to water/cement ratio.

3. *Size of the concrete element.*   Both the rate and total magnitude of shrinkage decrease with an increase in the volume of the concrete element. However, the duration of shrinkage is longer for larger members since more time is needed for drying to reach the internal regions. It is possible that 1 year may be needed for the drying process to begin at a depth of 10 in. from the exposed surface, and 10 years to begin at 24 in. below the external surface.

4. *Medium ambient conditions.*   The relative humidity of the medium affects greatly the magnitude of shrinkage; the rate of shrinkage is lower at high states of relative humidity. The environment temperature is another factor, in that shrinkage becomes stabilized at low temperatures.

5. *Amount of reinforcement.*   Reinforced concrete shrinks less than plain concrete; the relative difference is a function of the reinforcement percentage.

6. *Admixtures.*   This effect varies depending on the type of admixture. An accelerator such as calcium chloride, used to accelerate the hardening and setting of the concrete, increases the shrinkage. Pozzolans can also increase the drying shrinkage, whereas air-entraining agents have little effect.

7. *Type of cement.*   Rapid-hardening cement shrinks somewhat more than other types, while shrinkage-compensating cements minimize or eliminate shrinkage cracking if used with restraining reinforcement.

8. *Carbonation.*   Carbonation shrinkage is caused by the reaction between carbon dioxide (CO_2) present in the atmosphere and that present in the cement paste. The

amount of the combined shrinkage varies according to the sequence of occurrence of carbonation and drying processes. If both phenomena take place simultaneously, less shrinkage develops. The process of carbonation, however, is dramatically reduced at relative humidities below 50%.

3.9.7 Creep

Creep or lateral material flow is the increase in strain with time due to a sustained load. Initial deformation due to load is the *elastic strain*, while the additional strain due to the same sustained load is the *creep strain*. This practical assumption is quite acceptable since the initial recorded deformation includes few time-dependent effects.

Figure 3.8 illustrates the increase in creep strain with time, and as in the case of shrinkage, it can be seen that creep decreases with time. Creep cannot be observed directly and can be determined only by deducting elastic strain and shrinkage strain from the total deformation. Although shrinkage and creep are not independent phenomena, it can be assumed that superposition of strains is valid; hence

$$\text{total strain } (\epsilon_t) = \text{elastic strain } (\epsilon_e) + \text{creep } (\epsilon_c) + \text{shrinkage } (\epsilon_{sh})$$

An example of the relative numerical values of strain due to the foregoing three factors is presented for a normal concrete specimen subjected to 900 psi in compression:

$$
\begin{aligned}
\text{Immediate elastic strain, } \epsilon_e &= 250 \times 10^{-6} \text{ in.-in.} \\
\text{Shrinkage strain after 1 year, } \epsilon_{sh} &= 500 \times 10^{-6} \text{ in.-in.} \\
\text{Creep strain after 1 year } \epsilon_c &= \underline{750 \times 10^{-6} \text{ in.-in.}} \\
\epsilon_t &= 1500 \times 10^{-6} \text{ in./in.}
\end{aligned}
$$

These relative values illustrate that stress–strain relationships for short-term loading lose their significance and long-term loadings become dominant in their effect on the behavior of a structure.

Figure 3.9 qualitatively shows in a three-dimensional model the three types of strain discussed resulting from sustained compressive stress and shrinkage. Since creep is time dependent, this model has to be such that its orthogonal axes are deformation, stress, and time.

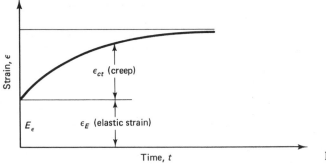

Figure 3.8   Strain-time curve.

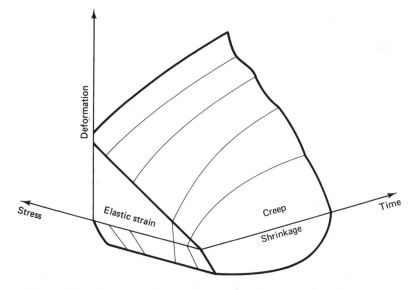

Figure 3.9   Three-dimensional model of time-dependent structural behavior.

Numerous tests have indicated that creep deformation is proportional to the applied stress, but the proportionality is valid only for low stress levels. The upper limit of the relationship cannot be determined accurately but can vary between 0.2 and 0.5 of the ultimate strength f_c'. This range in the limit of the proportionality is expected due to the large extent of microcracks at about 40% of the ultimate load.

Figure 3.10(a) shows a section of the three-dimensional model in Fig. 3.9 parallel to the plane containing the stress and deformation axes at time t_1. It indicates that both elastic and creep strains are linearly proportional to the applied stress. In a

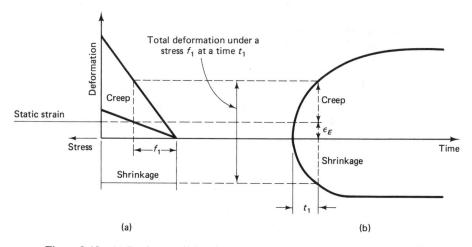

Figure 3.10   (a) Section parallel to the stress-deformation plane; (b) section parallel to the deformation-time plane.

similar manner, Fig. 3.10(b) illustrates a section parallel to the plane containing the time and strain axes at a stress f_1; hence it shows the familiar creep–time and shrinkage–time relationships.

As in the case of shrinkage, creep is not completely reversible. If a specimen is unloaded after a period under a sustained load, an immediate elastic recovery is obtained which is less than the strain precipitated on loading. The instantaneous recovery is followed by a gradual decrease in strain, called *creep recovery*. The extent of the recovery depends on the age of the concrete when loaded with older concretes presenting higher creep recoveries, while residual strains or deformations become frozen in the structural element (see Fig. 3.11).

Creep is closely related to shrinkage, and as a general rule, a concrete that is resistant to shrinkage also presents a low creep tendency, as both phenomena are related to the hydrated cement paste. Hence creep is influenced by the composition of the concrete, the environmental conditions, and the size of the specimen, but principally creep depends on loading as a function of time.

The composition of a concrete specimen can be essentially defined by the water/cement ratio, aggregate and cement types, and agggregate and cement contents. Therefore, like shrinkage, an increase in the water/cement ratio and in the cement content increases creep. Also, as in shrinkage, the aggregate induces a restraining effect such that an increase in aggregate content reduces creep.

3.9.8 Creep Effects

As in shrinkage, creep increases the deflection of beams and slabs, and causes loss of prestress. In addition, the initial eccentricity of a reinforced concrete column increases with time due to creep, resulting in the transfer of the compressive load from the concrete to the steel in the section.

Once the steel yields, additional load has to be carried by the concrete. Consequently, the resisting capacity of the column is reduced and the curvature of the column increases further, resulting in overstress in the concrete, leading to failure.

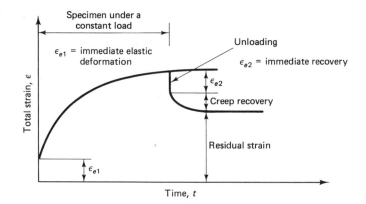

Figure 3.11 Creep recovery versus time.

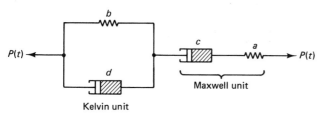

Figure 3.12   Burgers model.

3.9.9 Rheological Models

Rheological models are mechanical devices that portray the general deformation behavior and flow of materials under stress. A model is basically composed of elastic springs and ideal dashpots denoting stress, elastic strain, delayed elastic strain, irrecoverable strain, and time. The springs represent the proportionality between stress and strain, and the dashpots represent the proportionality of stress to the rate of strain. A spring and a dashpot in parallel form a Kelvin unit; and in series they form a Maxwell unit.

Two rheological models will be discussed: the Burgers model and the Ross model. The Burgers model in Fig. 3.12 is shown since it can approximately simulate the stress–strain–time behavior of concrete at the limit of proportionality with some limitations. This model simulates the instantaneous recoverable strain, (a); the delayed recoverable elastic strain in the spring, (b); and the irrecoverable time-dependent strain in dashpots, (c) and (d). The weakness in this model is that it continues to deform at a uniform rate as long as the load is sustained by the Maxwell dashpot—a behavior not similar to concrete, where creep reaches a limiting value with time, as shown in Fig. 3.8.

A modification in the form of the Ross rheological model in Fig. 3.13 can eliminate this deficiency. *A* in this model represents the Hookian direct proportionality of stress-to-strain element, *D* represents the Newtonian element, and *B* and *C* are the elastic springs that can transmit the applied load $P(t)$ to the enclosing cylinder walls by direct friction. Since each coil has a defined frictional resistance, only those coils whose resistance equals the applied load $P(t)$ are displaced; the others remain unstressed, symbolizing the irrecoverable deformation in concrete. As the load continues to increase, it overcomes the spring resistance of unit *B*, pulling out the spring from the dashpot and signifying failure in a concrete element. More rigorous models have been used, such as Roll's model to assist in predicting the creep strains. Mathematical

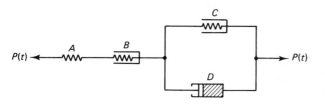

Figure 3.13   Ross model.

expressions for such predictions can be very rigorous. One convenient expression due to Ross defines creep C under load after a time interval t as follows:

$$C = \frac{t}{a + bt} \tag{3.5}$$

where a and b are constants determinable from tests.

Work by Branson (Refs. 3.10 and 3.11) has simplified creep evaluation. The additional strain ϵ_{cu} due to creep can be defined as

$$\epsilon_{cu} = \rho_u f_{ci} \tag{3.6a}$$

where ρ_u = unit creep coefficient, generally called *specific creep*
f_{ci} = stress intensity in the structural member corresponding to unit strain ϵ_{ci}

If C_u is the ultimate creep coefficient,

$$C_u = \rho_u E_c \tag{3.6b}$$

An average value of $C_u \simeq 2.35$.

Branson's model, verified by extensive tests, relates the creep coefficient C_t at any time to the ultimate creep coefficient as follows:

$$C_t = \frac{t^{0.6}}{10 + t^{0.6}} C_u \tag{3.7}$$

or alternatively,

$$\rho_t = \frac{t^{0.6}}{10 + t^{0.6}} \tag{3.8}$$

where t is the time in days.

The selected references at the end of the chapter give detailed information on the Creep coefficients and constants to be used to evaluate creep effect. The brief discussion in this section is intended to provide exposure to the procedures considered in any fundamental study of creep and shrinkage behavior.

SELECTED REFERENCES

3.1   ACI Committee 211, "Standard Practice for Selecting Proportions for Normal, Heavyweight, and Mass Concrete," ACI 211.1-81, American Concrete Institute, Detroit, 32 pp.

3.2   ACI Committee 211, "Standard Practice for Selecting Proportions for Structural Lightweight Concrete," ACI 211.1-81, American Concrete Institute, Detroit, 18 pp.

3.3   Portland Cement Association, *Design and Control of Concrete Mixtures,* 12th ed., Skokie, Ill. 1979, 140 pp.

3.4   ACI Committee 318, "Building Code Requirements for Reinforced Concrete 318-83," American Concrete Institute, Detroit, 1983, 111 pp. and the "Commentary on Building Code Requirements for Reinforced Concrete," 1983, 155 pp.

3.5   American Society for Testing and Materials, *Significance of Tests and Properties of Concrete and Concrete Making Materials,* Special Technical Publication 169B, ASTM, Philadelphia, 1978, 882 pp.

3.6   Ross, A. D., "The Elasticity, Creep and Shrinkage of Concrete," in *Proceedings of the Conference on Non-metallic Brittle Materials,* Interscience Publishers, London, 1958, pp. 157–174.

3.7   Neville, A. M., *Properties of Concrete,* 3rd ed., Pitman Books, London, 1981, 779 pp.

3.8   Freudenthal, A. M., and Roll, F., "Creep and Creep Recovery of Concrete under High Compressive Stress," *Journal of the American Concrete Institute,* Proc. Vol. 54, June 1958, pp. 1111–1142.

3.9   Ross, A. D., "Creep Concrete Data," *Proceedings, Institution of Structural Engineers,* London, Vol. 15, 1937, pp. 314–326.

3.10   Branson, D. E., *Deformation of Concrete Structures,* McGraw-Hill, New York, 1977, 546 pp.

3.11   Branson, D. E., "Compression Steel Effects on Long Term Deflections," *Journal of the American Concrete Institute,* Proc. Vol. 68, August 1971, pp. 555–559.

3.12   Mindess, S., and Young, J. F., *Concrete,* Prentice-Hall, Inc., Englewood Cliffs, N.J., 1981, 671 pp.

3.12   Nawy, E. G., and Balaguru, P. N., "High Strength Concrete," Chapter 5 in *Handbook of Structural Concrete,* Pitman Books, London/McGraw-Hill, New York, 1983, 1968 pp.

PROBLEMS FOR SOLUTION

3.1   Design a concrete mix using the following data:

Required strength f'_c = 5000 psi (34.5 MPa)

Type of structure: beam

Maximum size of aggregate = $\frac{3}{4}$ in. (18 mm)

Fineness modulus of sand = 2.6

Dry-rodded weight of coarse aggregate = 100 lb/ft^3

Moisture absorption: 2% for coarse aggregate and 2% for fine aggregate

3.2   Using the data of Ex. 3.1, design a 6% air-entrained mix.

3.3   Repeat Exs. 3.1 and 3.2 for a mix design strength f'_c = 3000 psi (20.7 MPa).

3.4 Estimate the strength of the trial mix, f'_{cr}, for the following cases.

(a) f'_c = 3500 psi (24.15 MPa); s (using 40 consecutive tests) = 300 psi (2.07 MPa).

(b) f'_c = 3000 psi (20.7 MPa); s (using 20 consecutive tests) = 250 psi (1.73 MPa).

(c) f'_c = 3000 psi (20.7 MPa); test results are not available.

(d) f'_c = 4000 psi (27.6 MPa); s (using 15 tests) = 375 psi (2.59 MPa).

4

Reinforced Concrete

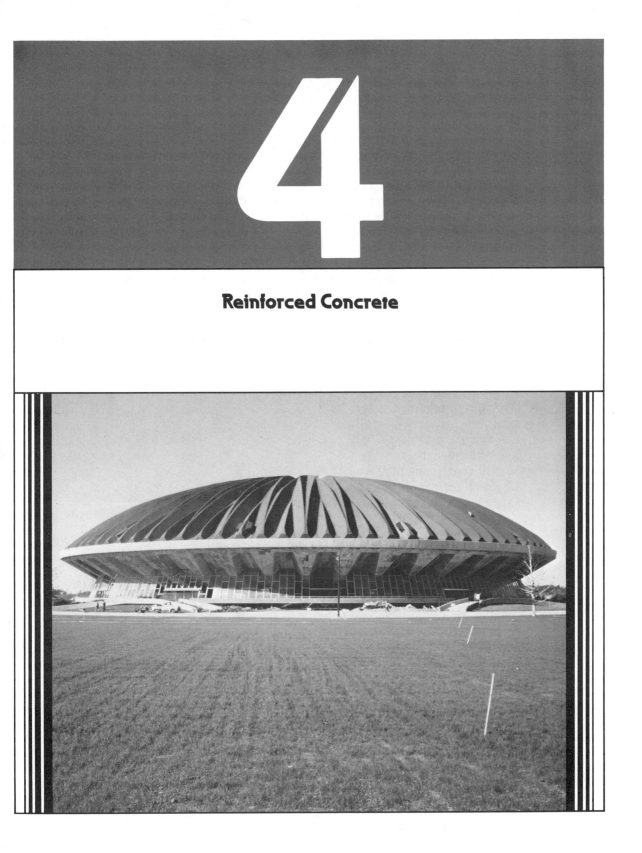

4.1 INTRODUCTION

Concrete is strong in compression but weak in tension. Therefore, reinforcement is needed to resist the tensile stresses resulting from the induced loads. Additional reinforcement is occasionally used to reinforce the compression zone of concrete beam sections. Such steel is necessary for heavy loads in order to reduce long-term deflections. Whereas Chapters 2 and 3 dealt with plain concrete and its constituent materials, this chapter discusses composite reinforced concrete, which can withstand high tensile as well as compressive forces. A discussion of the types of reinforcing material, the variety of structural systems, and their components is presented.

Additionally, concrete structures have to perform adequately under service-load conditions in addition to having the necessary reserve strength to resist ultimate load. The subjects of reliability, safety, and load factors are also presented.

4.2 TYPES AND PROPERTIES OF STEEL REINFORCEMENT

Steel reinforcement for concrete consists of bars, wires, and welded wire fabric, all of which are manufactured in accordance with ASTM standards. The most important properties of reinforcing steel are:

1. Young's modulus, E_s
2. Yield strength, f_y
3. Ultimate strength, f_u
4. Steel grade designation
5. Size or diameter of the bar or wire

To increase the bond between concrete and steel, projections called *deformations* are rolled on the bar surface as shown in Fig. 4.1 in accordance with ASTM specifications. The deformations shown must satisfy ASTM Specification A616-76 to be accepted as deformed bars. The deformed wire has indentations pressed into the wire or bar to serve as deformations. Except for wire used in spiral reinforcement in columns, only deformed bars, deformed wires, or wire fabric made from smooth or deformed wire may be used in reinforced concrete under approved practice.

Figure 4.2 shows typical stress–strain curves for grade 40, 60, and 80 steels. They have corresponding yield strengths of 40,000, 60,000, and 80,000 psi (276, 345, and 517 N/mm^2, respectively) and generally have well-defined yield points. For steels that lack a well-defined yield point, the yield-strength value is taken as the strength corresponding to a unit strain of 0.005 for grades 40 and 60 steels and 0.0035 for grade 80 steel. The ultimate tensile strengths corresponding to the 40, 60, and 80 grade steels are 70,000, 90,000, and 100,000 psi (483, 621, and 690 N/mm^2), and some steel types are given in Table 4.1. The percent elongation at fracture, which

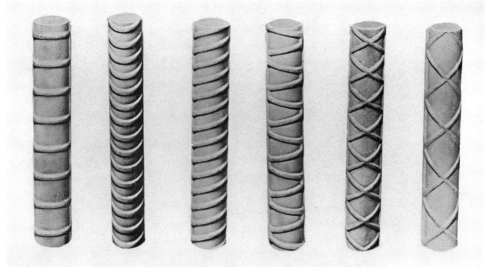

Figure 4.1   Various forms of ASTM-approved deformed bars.

varies with the grade, bar diameter, and manufacturing source, ranges from 4.5 to 12% over an 8-in. (203.2-mm) gage length.

For most steels, the behavior is assumed to be elastoplastic and Young's modulus is taken as 29×10^6 psi (200×10^6 MPa). Table 4.1 presents the reinforcement-grade strengths, and Table 4.2 presents geometrical properties of the various sizes of bars.

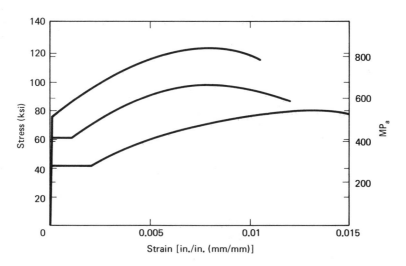

Figure 4.2   Typical stress–strain diagrams for various steels.

TABLE 4.1   REINFORCEMENT GRADES AND STRENGTHS

1982 standard type	Minimum yield point or yield strength, f_y (psi)	Ultimate strength, f_u (psi)
Billet steel (A615)		
Grade 40	40,000	70,000
Grade 60	60,000	90,000
Axle steel (A617)		
Grade 40	40,000	70,000
Grade 60	60,000	90,000
Low-alloy steel (A706): Grade 60	60,000	80,000
Deformed wire		
Reinforced	75,000	85,000
Fabric	70,000	80,000
Smooth wire		
Reinforced	70,000	80,000
Fabric	65,000, 56,000	75,000, 70,000

TABLE 4.2   WEIGHT, AREA, AND PERIMETER OF INDIVIDUAL BARS

Bar Designation Number	Weight per foot (lb.)	1982 standard nominal dimensions		
		Diameter, d_b[in. (mm)]	Cross-sectional area, A_b (in.^{2})	Perimeter (in.)
3	0.376	0.375 (9)	0.11	1.178
4	0.668	0.500 (13)	0.20	1.571
5	1.043	0.625 (16)	0.31	1.963
6	1.502	0.750 (19)	0.44	2.356
7	2.044	0.875 (22)	0.60	2.749
8	2.670	1.000 (25)	0.79	3.142
9	3.400	1.128 (28)	1.00	3.544
10	4.303	1.270 (31)	1.27	3.990
11	5.313	1.410 (33)	1.56	4.430
14	7.65	1.693 (43)	2.25	5.32
18	13.60	2.257 (56)	4.00	7.09

Welded wire fabric is increasingly used for slabs because of the ease of placing the fabric sheets, control of reinforcement spacing, and better bond. The fabric reinforcement is made of smooth or deformed wires which run in perpendicular directions and are welded together at intersections. Table 4.3 presents geometrical properties of some standard wire reinforcement.

TABLE 4.3 STANDARD WIRE REINFORCEMENT

W&D size		U.S. customary			Area (in.²/ft of width for various spacings) Center-to-center spacing (in.)						
Smooth	Deformed	Nominal diameter (in.)	Nominal area (in.²)	Nominal weight (lb/ft)	2	3	4	6	8	10	12
W31	D31	0.628	0.310	1.054	1.86	1.24	0.93	0.62	0.465	0.372	0.31
W30	D30	0.618	0.300	1.020	1.80	1.20	0.90	0.60	0.45	0.366	0.30
W28	D28	0.597	0.280	0.952	1.68	1.12	0.84	0.56	0.42	0.336	0.28
W26	D26	0.575	0.260	0.934	1.56	1.04	0.78	0.52	0.39	0.312	0.26
W24	D24	0.553	0.240	0.816	1.44	0.96	0.72	0.48	0.36	0.288	0.24
W22	D22	0.529	0.220	0.748	1.32	0.88	0.66	0.44	0.33	0.264	0.22
W20	D20	0.504	0.200	0.680	1.20	0.80	0.60	0.40	0.30	0.24	0.20
W18	D18	0.478	0.180	0.612	1.08	0.72	0.54	0.36	0.27	0.216	0.18
W16	D16	0.451	0.160	0.544	0.96	0.64	0.48	0.32	0.24	0.192	0.16
W14	D14	0.422	0.140	0.476	0.84	0.56	0.42	0.28	0.21	0.168	0.14
W12	D12	0.390	0.120	0.408	0.72	0.48	0.36	0.24	0.18	0.144	0.12
W11	D11	0.374	0.110	0.374	0.66	0.44	0.33	0.22	0.165	0.132	0.11
W10.5		0.366	0.105	0.357	0.63	0.42	0.315	0.21	0.157	0.126	0.105
W10	D10	0.356	0.100	0.340	0.60	0.40	0.30	0.20	0.15	0.12	0.10
W9.5		0.348	0.095	0.323	0.57	0.38	0.285	0.19	0.142	0.114	0.095
W9	D9	0.338	0.090	0.306	0.54	0.36	0.27	0.18	0.135	0.108	0.09
W8.5		0.329	0.085	0.289	0.51	0.34	0.255	0.17	0.127	0.102	0.085
W8	D8	0.319	0.080	0.272	0.48	0.32	0.24	0.16	0.12	0.096	0.08
W7.5		0.309	0.075	0.255	0.45	0.30	0.225	0.15	0.112	0.09	0.075
W7	D7	0.298	0.070	0.238	0.42	0.28	0.21	0.14	0.105	0.084	0.07
W6.5		0.288	0.065	0.221	0.39	0.26	0.195	0.13	0.097	0.078	0.065
W6	D6	0.276	0.060	0.204	0.36	0.24	0.18	0.12	0.09	0.072	0.06
W5.5		0.264	0.055	0.187	0.33	0.22	0.165	0.11	0.082	0.066	0.055
W5	D5	0.252	0.050	0.170	0.30	0.20	0.15	0.10	0.075	0.06	0.05
W4.5		0.240	0.045	0.153	0.27	0.18	0.135	0.09	0.067	0.054	0.045
W4	D4	0.225	0.040	0.136	0.24	0.16	0.12	0.08	0.06	0.048	0.04
W3.5		0.211	0.035	0.119	0.21	0.14	0.105	0.07	0.052	0.042	0.035
W3		0.195	0.030	0.102	0.18	0.12	0.09	0.06	0.045	0.036	0.03
W2.9		0.192	0.029	0.098	0.174	0.116	0.087	0.058	0.043	0.035	0.029
W2.5		0.178	0.025	0.085	0.15	0.10	0.075	0.05	0.037	0.03	0.025
W2		0.159	0.020	0.068	0.12	0.08	0.06	0.04	0.03	0.024	0.02
W1.4		0.135	0.014	0.049	0.084	0.056	0.042	0.028	0.021	0.017	0.014

4.3 BAR SPACING AND CONCRETE COVER FOR STEEL REINFORCEMENT

It is necessary to guard against honeycombing and ensure that the wet concrete mix passes through the reinforcing steel without separation. Since the graded aggregate size in structural concrete often contains $\frac{3}{4}$ in. (19 mm diameter) coarse aggregate, minimum allowable bar spacing and minimum required cover are needed. Additionally, to protect the reinforcement from corrosion and loss of strength in case of fire, codes specify a minimum required concrete cover. Some of the major requirements of the ACI 318 code are:

1. Clear distance between parallel bars in a layer must not be less than the bar diameter d_b or 1 in. (25.4 mm).
2. Clear distance between longitudinal bars in columns must not be less than $1.5d_b$ or 1.5 in. (38.1 mm).
3. Minimum clear cover in cast-in-place concrete beams and columns should not be less than 1.5 in. (38.1 mm) when there is no exposure to weather or contact with the ground; this same cover requirement also applies to stirrups, ties, and spirals.

In the case of slabs, plates, shells, and folded plates, where concrete is not exposed to a severe environment and where the reinforcement size does not exceed a No. 11 bar diameter (85.8 mm), the clear cover should not be less than $\frac{3}{4}$ in. (19 mm). Detailed requirements as to thickness of cover for various conditions can be found in various codes of practice, such as the Underwriters' National Building Code and the ACI code.

4.4 CONCRETE STRUCTURAL SYSTEMS

Every structure is proportioned as to both architecture and engineering to serve a particular function. Form and function go hand in hand and the best structural system is the one that fulfills most of the needs of the user while being serviceable, attractive, and, hopefully, economically cost-efficient. Although most structures are designed for a life span of 50 years, the durability performance record indicates that properly proportioned concrete structures have generally had longer useful lives.

Numerous concrete landmarks can be cited where major credit is due to the art and science of structural design applied with ingenuity, logic, and imagination. Such concrete structural systems as the TWA Terminal, New York; the Newark Terminal, New Jersey; Symphony Hall, Melbourne, Australia; Chicago's Marina Towers and Water Tower Place; the Dallas Super Dome; and many others are a testimony to the marriage of form and function with superior engineering judgment. Photographs of several such landmarks appear throughout the book.

Such concrete systems are composed of a variety of concrete structural elements that, when synthesized, produce a total system. The components can be broadly classified into (1) floor slabs, (2) beams, (3) columns, (4) walls, and (5) foundations.

4.4.1 Floor Slabs

Floor slabs are the main horizontal elements that transmit the moving live loads as well as the stationary dead loads to the vertical framing supports of a structure. They can be slabs on beams, as in Fig. 4.3, or waffle slabs, slabs without beams (flat plates) resting directly on columns, or composite slabs on joists. They can be proportioned such that they act in one direction (one-way slabs) or so that they act in two perpendicular directions (two-way slabs and flat plates). A detailed discussion of the analysis and design of such floor systems is given in subsequent chapters.

4.4.2 Beams

Beams are the structural elements that transmit the tributory loads from floor slabs to vertical supporting columns. They are normally cast monolithically with the slabs and are structurally reinforced on one face, the lower tension side, or both the top and bottom faces. As they are cast monolithically with the slab, they form a T-beam section for interior beams or an L beam at the building exterior, as seen in Fig. 4.3. The plan dimensions of a slab panel determine whether the floor slab behaves essentially as a one-way or a two-way slab.

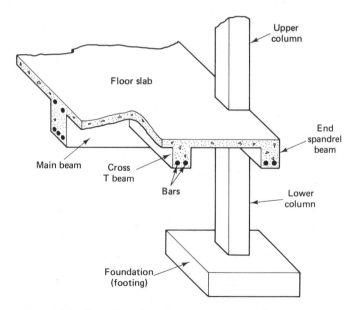

Figure 4.3 Typical reinforced concrete structural framing system.

4.4.3 Columns

The vertical elements support the structural floor system. They are compression members subjected in most cases to both bending and axial load, and are of major importance in the safety considerations of any structure. If a structural system is also composed of horizontal compression members, such members would be considered as beam–columns.

4.4.4 Walls

Walls are the vertical enclosures for building frames. They are not usually or necessarily made of concrete but of any material that aesthetically fulfills the form and functional needs of the structural system. Additionally, structural concrete walls are often necessary as foundation walls, stairwell walls, and shear walls that resist horizontal wind loads and earthquake-induced loads.

4.4.5 Foundations

Foundations are the structural concrete elements that transmit the weight of the superstructure to the supporting soil. They could be in many forms, the simplest being the isolated footing shown in Fig. 4.3. It can be viewed as an inverted slab transmitting a distributed load from the soil to the column. Other forms of foundations are piles driven to rock; combined footings supporting more than one column; mat foundations; and rafts, which are basically inverted slab and beam construction.

   The results of the analysis and design process of a structure have to be presented in concise and standardized form which the constructor can use for building the entire system. Hence knowledge and easy reading of working drawings is important. A typical layout drawing of a multilevel parking garage structure is shown in Fig. 4.4. The ACI Manual of Detailing gives an adequate coverage of typical working drawings for various structural systems and of the layout and detailing of reinforcement.

4.5 RELIABILITY AND STRUCTURAL SAFETY OF CONCRETE COMPONENTS

Three developments in recent decades have had a major influence on present and future design procedures. They are the vast increase in experimental and analytical evaluation of concrete elements, the probabilistic approach to the interpretation of behavior, and the digital computational tools available for rapid analysis of safety and reliability of systems. Until recently, most safety factors in design have had an empirical background based on local experience over an extended period of time. As additional experience is accumulated and more knowledge is gained from failures as well as familiarity with the properties of concrete, factors of safety are adjusted and in most cases lowered by the codifying bodies.

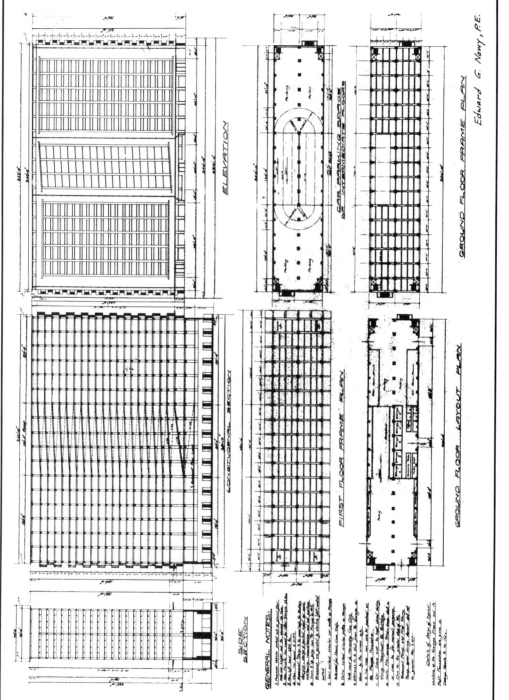

Figure 4.4 Typical working drawing for a reinforced concrete parking structure. (Design by E. G. Nawy.)

A. L. L. Baker in 1956 proposed a simplified method of safety factor determination, as shown in Table 4.4, based on probabilistic evaluation. This method expects the design engineer to make critical choices regarding the magnitudes of safety margins in a design. The method takes into consideration that different weights should be assigned to the various factors affecting a design. The weighted failure effects W_t for the various factors of workmanship, loading conditions, results of failure, and resistance capacity are tabulated in Table 4.4.

The safety factor against failure is

$$\text{S.F.} = 1.0 + \frac{\Sigma W_t}{10} \tag{4.1}$$

where the maximum total weighted value of all parameters affecting performance equals 1.0. In other words, for the worst combination of conditions affecting structural performance, the safety factor S.F. = 2.0.

This method assumes adequate information on prior performance data similar to a design in progress. Such data in many instances are not readily available for determining safe weighted values W_t in Eq. 4.1. Additionally, if the weighted factors are numerous, a probabilistic determination of them is more difficult to codify. Hence an undue value-judgment burden is probably placed on the design engineer if the full economic benefit of the approach is to be achieved.

Another method with a smaller number of probabilistic parameters deals primarily with loads and resistances. Its approach for both steel and concrete structures is

TABLE 4.4 BAKER'S WEIGHTED SAFETY FACTOR

Weighted failure effect		Maximum W_t
1. Results of failure: 1.0 to 4.0		
Serious, either human or economic		4.0
Less serious, only the exposure of		
nondamageable material	1.0	
2. Workmanship: 0.5 to 2.0		
Cast in place		2.0
Precast "factory manufactured"	0.5	
3. Load conditions: 1.0 to 2.0		2.0
(high for simple spans and overload		
possibilities; low for load combina-		
tions such as live loads and wind)		
4. Importance of member in structure		0.5
(beams may use lower value than columns)		
5. Warning of failure		1.0
6. Depreciation of strength		0.5
	Total = ΣW_t =	10.0

$$\text{S.F.} = 1.0 + \frac{\Sigma W_t}{10}$$

generally similar: both the load and resistance factor design methods (LRFD) and first-order second-moment method (FOSM) propose general reliability procedures for evaluating probability-based factored load design criteria, as in Refs. 4.7, 4.8, and 4.10. They are intended for use in proportioning structural members on the basis of load types such that the resisting strength levels are *greater* than the factored load or moment distributions. As these approaches are basically load oriented, they reduce the number of individual variables that have to be considered, such as those listed in Table 4.4.

Assume that ϕ_i represents the resistance factors of a concrete element and that γ_i represents the load factors for the various types of load. If R_n is the nominal resistance of the concrete element and W_i represents the load effect for various types of superimposed load,

$$\phi_i R_n \geq \gamma_i W_i \qquad (4.2)$$

where i represents the various types of load, such as dead, live, wind, earthquake, or time-dependent effects.

Figures 4.5(a) and (b) show a plot of the separate frequency distributions of the actual load W and the resistance R with mean values $\overline{R}$ and $\overline{W}$. Figure 4.5(c) gives the two distributions superimposed and intersecting at point C.

It can be recognized that safety and reliable integrity of the structure can be expected to exist if the load effect W falls at a point to the left of intersection C, on the W curve and to the right of intersection C, on the resistance curve R. Failure, on the other hand, would be expected to occur if the load effect or the resistance fall within the shaded area in Fig. 4.5(c). If β is a safety index, then

$$\beta = \frac{\overline{R} - \overline{W}}{\sqrt{\sigma_R^2 + \sigma_W^2}} \qquad (4.3)$$

where σ_R and σ_W are the standard deviations of the resistance and the load, respectively.

A plot of the safety index β for a hypothetical structural system is shown in Fig. 4.6 against the probability of failure of the system. One can observe that such a probability is reduced as the difference between the mean resistance $\overline{R}$ and load effect $\overline{W}$ is increased, or the variability of resistance and load effect as measured by their standard deviations σ_R and σ_W is decreased, thereby reducing the shaded area under intersection C in Fig. 4.5(c).

The extent of increasing the $(\overline{R} - \overline{W})$ difference or decreasing the degree of scatter of σ_R or σ_W is naturally dictated by economic considerations. It is economically unreasonable to design a structure for zero failure, particularly since types of risk other than load are an accepted matter, such as the risks of severe earthquake, hurricane, volcanic eruption, or fire. Safety factors and corresponding load factors would thus have to disregard those types or levels of load, stress, and overstress whose probability of occurrence is very low. In spite of this, it is still possible to achieve reliable safety conditions by choosing such a safety index value β through a proper choice of R_n and W_i values using the appropriate resistance factors ϕ_i; and load factors γ_i in Eq. 4.2. A

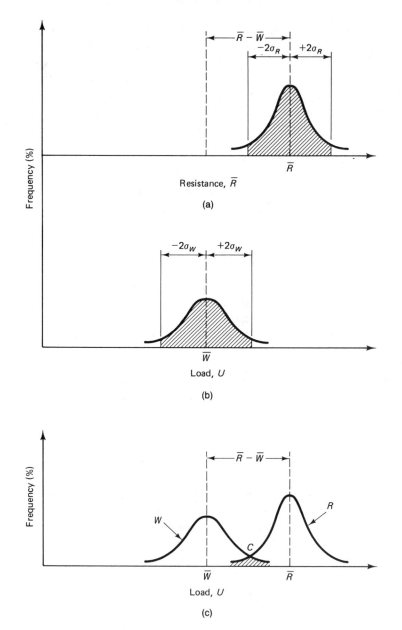

Figure 4.5 Frequency distribution of loads versus resistance.

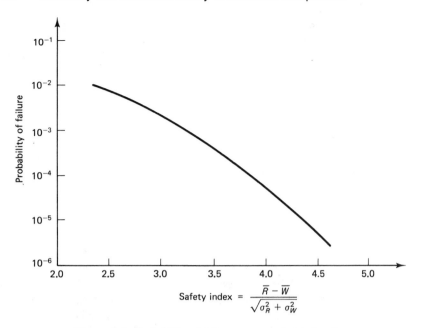

Figure 4.6   Probability of failure versus safety index β.

safety index β having the value 1.75 to 3.2 for concrete structures is suggested in Ref. 4.8, where the lower value accounts for load contributions from wind and earthquake.

If the factored external load is expressed as U_i, then $\Sigma \, \gamma_i \, W = U_i$ for the different loading combinations. The following are U_i values recommended in Ref. 4.8 to choose a maximum U for use in Eq. 4.2, that is, $\phi_i R_n \geq \gamma_i W_i \geq U_{i(\max)}$:

$$
\gamma_i U_w = \text{max.} \begin{cases}
1.4D_n \\
1.2D_n + 1.6L_n \\
1.2D_n + 1.6S_n + (0.5L_n \text{ or } 0.8W_n) \\
1.2D_n + 1.3W_n + 0.5L_n \\
1.2D_n + 1.5E_n + (0.5L_n \text{ or } 0.2S_n) \\
0.9D_n - (1.3W_n \text{ or } 1.5E_n)
\end{cases} \tag{4.4}
$$

where the subscript n stands for the nominal value of the variable working load:

D_n = dead load L_n = live load S_n = snow load

W_n = wind load E_n = earthquake load

ϕ_i and γ_i are considered to have optimal values; a resistance factor ϕ value of 0.7 to 0.85 is recommended. As discussed in Section 4.6, in this probabilistic approach the present ACI code would require that

$$
\phi_i R_n = \text{max.} \begin{cases}
1.4D_n + 1.7L_n \\
0.9D_n - 1.3W_n
\end{cases} \tag{4.5}
$$

As more substantive records of performance are compiled with time, the details of the foregoing approach to reliability, safety, and reserve strength evaluation of structural components can be more universally accepted and extended beyond treatment of the component elements to the treatment of the total structural system.

4.6 ACI LOAD FACTORS AND SAFETY MARGINS

The general concept of safety and reliability of performance presented in the preceding sections is inherent in a more simplified but less accurate fashion in the ACI code. The γ load factors and the ϕ strength reduction factors give an overall safety factor based on load types such that

$$\text{S.F.} = \frac{\gamma_1 D + \gamma_2 L}{D + L} \times \frac{1}{\phi} \tag{4.6}$$

[handwritten annotation: 1.4 and 1.7 pointing to γ_1 and γ_2; ϕ – strength Reduction factor]

where ϕ is the strength reduction factor and γ_1 and γ_2 are the respective load factors for the dead load D and the live load L. Basically, a single common factor is used for dead load and another for live load. Variation in resistance capacity is considered in the ϕ reduction factor. Hence the method is a simplified empirical approach to safety and reliability of structural performance that is not economically efficient for every case and not fully adequate in other instances, such as combinations of dead and wind loads.

[handwritten annotations in left margin:
$\phi = 0.9$ for bending
$\phi = 0.85$ for shear & Torsion
$\phi = 0.70$ Tied Column
$\phi = 0.75$ Spiral Column*]*

The ACI factors are termed *load factors,* as they restrict the estimation of reserve strength to the loads only as compared to the other parameters listed in Table 4.4. The estimated service or working loads are magnified by the coefficients, such as a coefficient of 1.4 for dead loads and 1.7 for live load. The types of normally occurring loads can be identified as (1) dead load, D; (2) live load, L; (3) wind load, W; (4) loads due to lateral pressure such as from soil in a retaining wall, H; (5) lateral fluid pressure loads, F; (6) loads due to earthquake, E; and (7) loads due to time-dependent effects, such as creep or shrinkage.

The basic combination of vertical loads is dead load plus live load. The *dead load,* which constitutes the weight of the structure and other relatively permanent features, can be estimated more accurately than the live load. The *live load* is estimated using the weight of nonpermanent loads, such as people and furniture. The transient nature of live loads makes them difficult to estimate more accurately. Therefore, a higher load factor is normally used for live loads than for dead loads. If the combination of loads consists only of live and dead loads, the ultimate load can be taken as

$$U = 1.4D + 1.7L \tag{4.7a}$$

Structures are seldom subjected to dead and live loads alone; wind load is often present. For structures in which wind load should be considered, the recommended combination is

$$U = 0.75(1.4D + 1.7L + 1.7W) \tag{4.7b}$$

Maximum dead, live, and wind loads rarely, if ever, occur simultaneously. Hence the total factored load has to be reduced using a reduction factor of 0.75. Since wind load is applied laterally, it is possible that the absence of vertical live load while wind load is present can produce maximum stress. The following load combination should also be used to arrive at the maximum value of the factored load U:

$$U = 0.9D + 1.3W \qquad (4.7c)$$

Structures that have to resist lateral pressure due to earth fill or fluid pressure should be designed for the worst of the following combination of factored loads:

$$U = 1.4D + 1.7L + 1.7H \qquad (4.8a)$$

$$U = 0.9D + 1.7H \qquad (4.8b)$$

$$U = 1.4D + 1.7L \qquad (4.8c)$$

$$U = 1.4D + 1.7L + 1.4F \qquad (4.8d)$$

$$U = 0.9D + 1.4F \qquad (4.8e)$$

$$U = 1.4D + 1.7L \qquad (4.8f)$$

The following combinations must be considered for earthquake loading:

$$U = 0.75(1.4D + 1.7L + 1.87E) \qquad (4.9a)$$

$$U = 0.9D + 1.43E \qquad (4.9b)$$

or

$$U \geq 1.4D + 1.7L \qquad (4.9c)$$

whichever is largest. The philosophy used for combining the various load components for earthquake loading is essentially the same as that used for wind loading.

4.7 DESIGN STRENGTH VERSUS NOMINAL STRENGTH: STRENGTH REDUCTION FACTOR ϕ

The strength of a particular structural unit calculated using the current established procedures is termed *nominal strength*. For example, in the case of a beam, the resisting moment capacity of the section calculated using the equations of equilibrium and the properties of concrete and steel is called the *nominal resisting moment capacity* M_n of the section. This nominal strength is reduced using a strength reduction factor, ϕ, to account for inaccuracies in construction, such as in the dimensions or position of reinforcement or variations in properties. The reduced strength of the member is defined as the design strength of the member.

$\phi M_n \geq M_u$

For a beam, the design moment strength ϕM_n should be at least equal to or slightly greater than the external factored moment M_u for the worst condition of factored load U. The factor ϕ varies for the different types of behavior and for the different types of structural elements. For beams in flexure, for instance, the reduction factor is 0.9.

Photo 22 Flexural behavior of a reinforced concrete beam. (Test by Nawy et al.)

For tied columns that carry dominant compressive loads, the ϕ factor equals 0.7. The smaller strength reduction factor used for columns is due to the structural importance of the columns in supporting the total structure compared to other members and to guard against progressive collapse and brittle failure with no advance warning of collapse. Beams, on the other hand, are designed to undergo excessive deflections before failure. Hence the inherent capability of the beam for advanced warning of failure permits the use of a higher strength reduction factor or resistance factor.

Table 4.5 summarizes the resistance factors ϕ for various structural elements as given in the ACI code. A comparison of these values to those given in Ref. 4.8 indicates that the ϕ values in this table, as well as the load factors of Eq. 4.8, are in some cases more conservative than they should be. In cases of earthquakes, wind, and shear forces, the probability of load magnitude and reliability of performance is subject to higher randomness and hence a higher coefficient of variation than the other types of loading.

TABLE 4.5 RESISTANCE OR STRENGTH
REDUCTION FACTOR ϕ

Structural element	Factor ϕ
Beam or slab: Bending or flexure	0.9
Columns with ties	0.7
Columns with spirals	0.75
Columns carrying very small axial loads (refer to Chapter 9 for more details)	0.7–0.9, or 0.75–0.9
Beam: Shear and torsion	0.85

4.8 QUALITY CONTROL AND QUALITY ASSURANCE

Quality control assures the reliability of performance of the designed system in accordance with assumed and expected reserve strengths in the design. To exercise "quality control" and achieve "quality assurance" encompasses monitoring the roles and performance of all participants: the client or owner, the designer, the concrete producer, the laboratory tester, the constructor, and the user.

Most of the different phases of the total process of construction are affected by complex standards and regulations of the various codifying agencies. Also, in contrast to mechanized production such as in the case of machines, building construction does not follow the moving-belt or chain-production process, where the products move but the workers are relatively stationary; the contrary is true. Consequently, complications are more profound in constructed systems such as those made with concrete. This is due partially to the fact that concrete is a nonhomogeneous material with properties dependent on many variables, requiring extra effort in quality control due to the greater effect of the human factor on the quality of the finished product.

The reliability of performance of personnel involved in the various stages of creating a concrete structural system from conception through design, construction, and use depends on knowledge, training, and communication at all levels. A smooth flow of correct information among all participants and a shared systematic understanding of the developing problems lead to increased motivation toward improved solutions, hence improved quality control and a resulting high level of quality assurance. In summary, a quality assurance system needs to be provided based on the exercise of quality control at the various phases and interacting parameters of a total system, as shown in Fig. 4.7.

4.8.1 The User

Construction of a designed system is governed basically by five primary tasks: planning, design, materials selection, construction, and use (including maintenance).

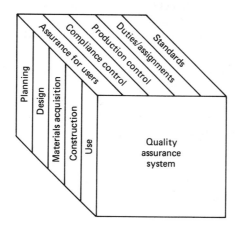

Figure 4.7 Components of a quality assurance system. (From Ref. 4.12.)

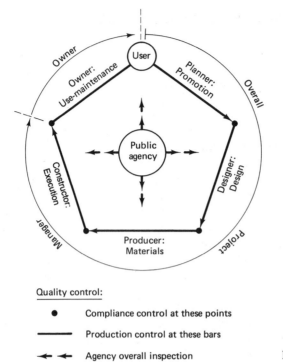

Quality control:

●   Compliance control at these points

──────   Production control at these bars

◄─ ◄─   Agency overall inspection

Figure 4.8 Quality control schematic.

Figure 4.8 presents schematically the sequence of these enumerated tasks and the respective divisions of responsibility. As seen from this diagram, the process starts with the user, since the principal aim of a project is to satisfy the user's needs, and it culminates with the user as the primary beneficiary of the final product.

Quality assurance is necessary to satisfy the user's needs and rights. It ensures that the activities influencing the final quality of a concrete structure are

1. Based on clearly defined fundamental requirements that satisfy the operational, environmental, and boundary conditions set for the project at the outset
2. Properly presented in accurate, well-dimensioned engineering working drawings based on optimal design procedures
3. Correctly and efficiently carried out by competent personnel in accordance with predetermined plans and working drawings well supervised during the design stage
4. Systematically executed in accordance with detailed specifications that conform to the applicable codes and local regulations

To achieve these aims, the expertise denoted by the other components of the polygon in Fig. 4.8 are called upon—starting with the planner and designer and culminating with the constructor.

4.8.2 Planning

In order to plan the successful execution of a proposed constructed system, all main and subactivities have to be clearly defined. This is accomplished through dividing the total project into a network plan of separately defined activities, relating each activity with time, analyzing the input control and the resulting output control, and expressing these conditions in the form of a checklist. In such a manner, the successful decision-making process concerning performance requirements becomes easier to accomplish Such a process usually entails decisions on *what* function needs to be accomplished in a construction project, *where* and *when* that function will be executed, *how* the system will be constructed, and *who* the user will be. Correct determination of these factors leads to a decision as to the level of quality control needed and the degree of quality assurance that is expected to result.

4.8.3 Design

Quality control in design aims at verifying that the designed system has the safety, serviceability, and durability required for the use to which the system is intended as required by the applicable codes, and that such a design is correctly presented in the working drawings and the accompanying specifications. The degree of quality control depends on the type of system to be constructed: the more important the system, the more control that is required.

As a minimum, a design must always be checked by an engineer other than the originating design engineer. Usually, one of three types of verifications is used, depending on the practice of the designing agency: (1) total direct checking, in which all computations are verified; (2) total parallel checking, in which calculations are *independently* made and the two sets of calculations compared; and (3) partial checking, in which selected parts are checked in both direct and parallel checking.

Quality control of the design calculations can generally be achieved through assuring that

1. A clear understanding exists of the structural concept that applies to the particular system.
2. There is knowledge and compliance with the relevant fundamental requirements of the design and the environmental, operational, and boundary conditions.
3. Where possible, applicable calculation models utilizing available computer programs are used for checks.
4. No discrepancies exist between the different phases or parts of the total design computations.
5. All expected load cases and load combinations as described in Section 4.6 are considered.

6. The appropriate safety factors are adopted and the required reliability levels verified.
7. Verifiable computer programs are used in the design, and the experienced designer is well acquainted with the programming steps and background of the programs, particularly when total computer-aided designs are used.

Since engineering working drawings are the primary link between the design process and the construction process, they should be a major object of design quality assurance. Consequently, the student has to be well acquainted with reading and interpreting working drawings and must be able to produce clear sketches that accurately express the design details if the constructed system is to reflect the actual design. Figure 4.4, as well as Figs. 10.11 to 10.20, are intended to give general guidance on the systematic detailing necessary for composing sets of logical engineering working drawings.

Quality control of working drawings normally encompasses a verification of whether the following parameters are included in the project set of drawings:

1. General definition of the structure
2. Consistency among the working drawings
3. Compliance with the site boundary conditions, including soil test boring requirements
4. Listing of the type, grade, quality, and structural strength of the various construction materials involved, such as cement, concrete mix proportion and strength, reinforcing steel, formwork, and so on
5. No ambiguity and risk of misunderstanding of the details in the drawings
6. Compliance with the design calculation results and correct dimensioning
7. Adequate cross-sectional and construction details, as well as explicit dimensional tolerances
8. Sequence of formwork placing and removal

4.8.4 Materials Selection

It has to be emphasized at the outset that the quality of materials such as reinforced concrete is not determined only by compressive or tensile strength tests. As seen from previous sections of this book, many other factors affect the quality of the finished product such as water/cement ratio, cement content, creep and shrinkage characteristics, freeze and thaw properties, and other durability aspects and conditions.

Two types of quality are involved: (1) *required quality,* which is the specified contractural requirement for the material, and (2) *usage quality,* which is the ability of the material to satisfy the needs of the user (Ref. 4.12).

Required quality of the material, such as ready-mix concrete, is assured by production control. Such a process involves:

1. General organization of the production staffing and operation.
2. Production sequence and supply line of the constituent materials, such as stone, fine aggregate, cement, and additives.
3. Internal control, involving frequency of verifications and tests, analysis of test results, the recording and observation methods used, and the procedures applied in dealing with discrepancies and deviations.
4. Use of statistical control charts to classify the specified requirements of quality levels into measurable main variables and nonmeasurable variables, selection of the main variables to be controlled by the control charts, and the preparation of a mean chart and a range chart for each variable selected. The measurable variables are to be controlled using $\bar{X}$ and $\bar{R}$ charts, while the nonmeasurable variables are controlled by $\bar{p}$ and $\bar{c}$ charts, denoting the averages of means and variations. An action limit and a warning limit with an upper and a lower boundary need to be specified.
5. Classification of defects as a measure of the nondefinable variables.

Usage quality is determined by the compliance control set in the specifications. These are prescribed by the client for the mix proportioning and design and the constituent materials of the concrete as well as the quality of the reinforcement, whether normal or prestressing reinforcement. A projection of the expected quality control record and hence quality assurance of the concrete, for instance, can normally be made if the concrete producer has maintained good statistical quality control of strength test results over a lengthy period of operation. Hence compliance control levels can vary depending on the reliability and confidence in the effectiveness of quality management of the material producer to conform with the specifications of the delivered lots.

4.8.5 Construction

Construction is the execution stage of a project, which can be used to satisfy all design and specification requirements within prescribed time limits at minimum cost. To achieve the desired quality assurance, the construction phase has to be preceded by an elaborate and correct preparation stage, which can be part of the design phase. The preparation or planning phase is very critical since it gives an overall clear view of the various activities involved and the possible problems that could arise at the various phases of execution. The use of computers in the planning phase is essential today for large projects in order that relevant input as to product quality, output, time scheduling, and costs can be charted and monitored.

The human factor is of major significance at the construction phase. In most instances, the major site activities involve labor-force use and path scheduling of its utilization. An improved information flow system, clear delineation of the chain of command, and reward for superior performance increase motivation and lead to overall improvement in the entire quality assurance system and an optimization of the efficiency/cost ratio of a project. The steps to be carried out to achieve quality assurance, as proposed in Ref. 4.12, are summarized briefly below and are represented graphically in Fig. 4.9.

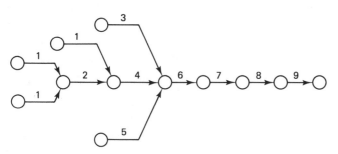

Figure 4.9 Operations sequence network for concrete structures.

1. Organizational preparation covering planning, time scheduling, contract details, and definition and assignment of duties
2. On-site preparation, involving access roads, site trailers and offices, energy provisions, amenities, and so on
3. Formwork acquisition, type, and preparation for installation
4. Reinforcement procurement, fabrication, and planning
5. Concrete mix proportioning, laboratory mix designing, and coordination with design engineers
6. Concrete delivery, placing in the forms, and field slump testing
7. Curing and surface treatment of the hardened concrete
8. Quality control tests of the concrete at 7- and 28-day intervals
9. Removal of formwork, sequential removal of shoring supports, then reshoring

The probability of errors in execution for quality control can normally be expected. The extent and importance of such errors depend on a variety of factors described in previous sections. In order to apply corrective measures, a logical sequence of steps has to be followed for the detection and analysis of the undesired occurrence. A quantitative analysis of the error impact can often be made provided that the probabilities of occurrence of all the basic events are known to the investigators.

The flowchart in Fig. 4.10 depicts the cause–effect sequence that can be followed in identifying an undesired event in a quality assurance program. γ in the chart is the safety margin factor available in the original design.

In summary, the brief discussion in Section 4.8 should provide the reader with an introduction to a continuously evolving topic that has a profound impact on the strength and durability of constructed systems: that is, quality control and quality assurance. It should complement Section 4.5 on reliability and structural safety and Sections 4.6 and 4.7 on load factors and design strengths.

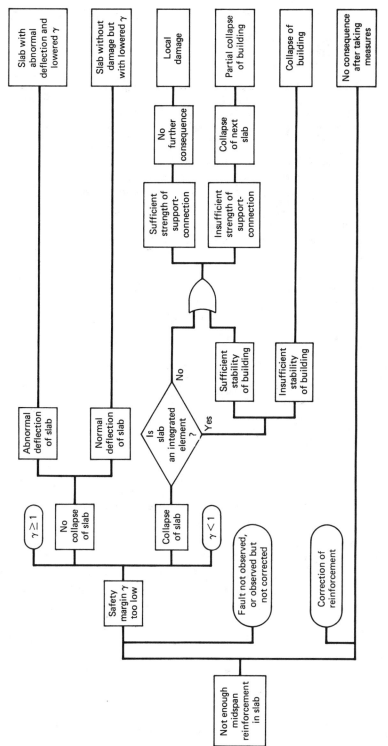

Figure 4.10 Cause–effect flowchart. (From Ref. 4.12.)

SELECTED REFERENCES

4.1   American Society for Testing and Materials, "Standard Specification for Deformed and Plain Billet-Steel Bars for Concrete Reinforcement, A6 15-79," ASTM, Philadelphia, 1980, pp. 588–599.

4.2   American Society for Testing and Materials, "Standard Specification for Rail-Steel Deformed and Plain Bars for Concrete Reinforcement, A6 16-79," ASTM, Philadelphia, 1980, pp. 600–605.

4.3   American Society for Testing and Materials, "Standard Specification for Axle Steel Deformed and Plain Bars for Concrete Reinforcement, A6 17-79," ASTM, Philadelphia, 1980, pp. 606–611.

4.4   American Society for Testing and Materials, "Standard Specification for Cold-Drawn Steel Wire for Concrete Reinforcement, A8 2-79," ASTM, Philadelphia, 1980, pp. 154–157.

4.5   American Society for Testing and Materials, "Standard Specification for Low-Alloy Steel Deformed Bars for Concrete Reinforcement, A706-79," ASTM, Philadelphia, 1980, pp. 755–760.

4.6   Baker, A. L. L., *The Ultimate Load Theory Applied to the Design of Reinforced and Prestressed Concrete Frames*, Concrete Publications, London, 1956, 91 pp.

4.7   Galambos, T. V., "Proposed Criteria for Load and Resistance Factor Design of Steel Building Structures," *Steel Research for Construction Bulletin No. 27*, American Iron and Steel Institute, January 1978.

4.8   Ellingwood, B., McGregor, J. G., Galambos, T. V., and Cornell, C. A., "Probability Based Load Criteria: Load Factors and Load Combinations," *Journal of the Structural Division, ASCE*, Vol. 108, No. ST5, May 1982, pp. 978–997.

4.9   National Bureau of Standards, *Development of a Probability Based Load Criterion for American National Standard A58—Building Code Requirements for Minimum Design Loads in Buildings and Other Structures*," NBS Special Publication 577, Washington, D.C., June 1980, 221 pp.

4.10 American National Standards Institute, "Minimum Design Loads for Buildings and Other Structures ANSI A58," ANSI, New York, 1982, p. 100.

4.11 ACI Committee 318, "Building Code Requirements for Reinforced Concrete, 318-83," 111 pp. and the "Commentary on Building Code Requirements for Reinforced Concrete," American Concrete Institute, Detroit, 1983, 155 pp.

4.12 Comité Euro-International du Béton "Quality Control and Quality Assurance for Concrete Structures," CEB Commission I (Dr. A. G. Meseguer, Chairman), *Bulletin d'Information No. 157*, Paris, March 1983, 98 pp.

4.13 Riggs, J. L., *Production Systems: Planning, Analysis, and Control*, Wiley, New York, 1981, 649 pp.

5

Flexure in Beams

5.1 INTRODUCTION

Loads acting on a structure, be they live gravity loads or other types, such as horizontal wind loads or those due to shrinkage and temperature, result in bending and deformation of the constituent structural elements. The bending of the beam element is the result of the deformational strain caused by the flexural stresses due to the external load.

As the load is increased, the beam sustains additional strain and deflection, leading to development of flexural cracks along the span of the beam. Continuous increases in the level of the load lead to failure of the structural element when the external load reaches the capacity of the element. Such a load level is termed the *limit state of failure in flexure*. Consequently, the designer has to design the cross section of the element or beam such that it would not develop excessive cracking at service load levels and have adequate safety and reserve strength to withstand the applied loads or stresses without failure.

Flexural stresses are a result of the external bending moments. They control in most cases the selection of the geometrical dimensions of a reinforced concrete section. The design process through the selection and analysis of a section is usually started by satisfying the flexural (bending) requirements, except for special components such as footings. Thereafter, other factors, such as shear capacity, deflection, cracking, and bond development of the reinforcement, are analyzed and satisfied.

While the input data for the analysis of sections differ from the data needed for design, every design is essentially an analysis. One assumes the geometrical properties of a section in a design and proceeds to analyze such a section to determine if it can safely carry the required external loads. Hence a good understanding of the fundamental principles in the analysis procedure significantly simplifies the task of designing sections. The basic mechanics of materials principles of equilibrium of internal couples have to be adhered to at all stages of loading.

If a beam is made up of homogeneous, isotropic, and linearly elastic material, the maximum bending stress can be obtained using the well-known beam flexure formula, $f = Mc/I$. At ultimate load, the reinforced concrete beam is neither homogeneous nor elastic, thereby making that expression inapplicable for evaluating the stresses. However, the basic principles of the theory of bending can still be used to analyze reinforced concrete beam cross sections. Figure 5.1 shows a typical simply supported reinforced concrete beam. If the beam is so proportioned that all its constituent materials attain their capacity prior to failure, both the concrete and the steel fail simultaneously at midspan when the ultimate strength of the beam is reached. The corresponding strain and stress diagrams are shown in Fig. 5.2.

The following assumptions are made in defining the behavior of the section:

Photo 24   Empire State Performing Arts Center (Albany, New York) during construction.

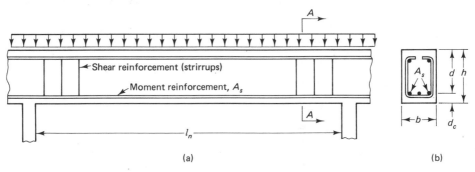
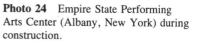

Figure 5.1   Typical reinforced concrete beam: (a) elevation; (b) section *A-A*.

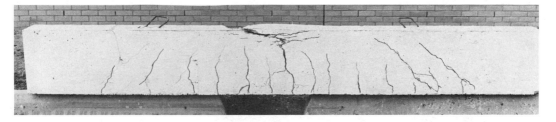

Photo 25 Simply supported beam in flexural failure. (Tests by Nawy.)

1. Strain distribution is assumed to be linear. This assumption is based on Bernoulli's hypothesis that plane sections before bending remain plane and perpendicular to the neutral axis after bending.
2. Strain in the steel and the surrounding concrete is the same prior to cracking of the concrete or yielding of the steel.
3. Concrete is weak in tension. It cracks at an early stage of loading at about 10% of its limit compressive strength. Consequently, concrete in the tension zone of the section is neglected in the flexural analysis and design computations, and the tension reinforcement is assumed to take the total tensile force.

To satisfy the equilibrium of the horizontal forces, the compressive force C in the concrete and the tensile force T in the steel should balance each other, that is,

$$C = T \tag{5.1}$$

The terms in Fig. 5.2 are defined as follows:

b = width of the beam at the compression side

d = depth of the beam measured from the extreme compression fiber to the centroid of steel area

h = total depth of the beam

A_s = area of the tension steel

ϵ_c = strain in extreme compression fiber

ϵ_s = strain at the level of tension steel

f_c' = compressive strength of the concrete

f_s = stress in the tension steel

f_y = yield strength of the tension reinforcement

c = depth of the neutral axis measured from extreme compression fibers

5.2 THE EQUIVALENT RECTANGULAR BLOCK

The actual distribution of the compressive stress in a section has the form of a rising parabola, as shown in Fig. 5.2c. It is time consuming to evaluate the volume of the compressive stress block if it has a parabolic shape. An equivalent rectangular stress block due to Whitney can be used with ease and without loss of accuracy to calculate

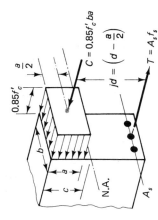

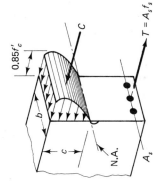

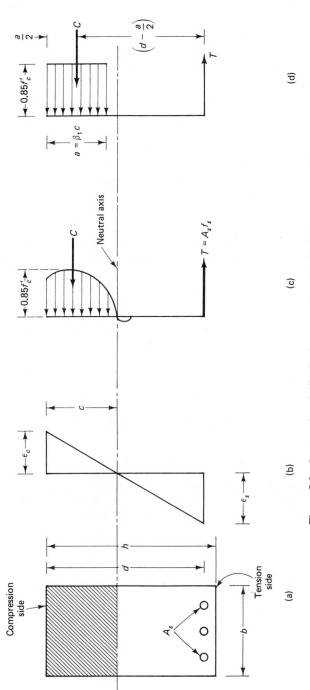

Figure 5.2 Stress and strain distribution across beam depth: (a) beam cross section; (b) strains; (c) actual stress block; (d) assumed equivalent stress block.

the compressive force and hence the flexural moment strength of the section. This equivalent stress block has a depth a and an average compressive strength $0.85f'_c$. As seen from Fig. 5.2d, the value of $a = \beta_1 c$ is determined using a coefficient β_1 such that the area of the equivalent rectangular block is approximately the same as that of the parabolic compressive block, resulting in a compressive force C of essentially the same value in both cases.

The $0.85f'_c$ value for the average stress of the equivalent compressive block is based on the core test results of concrete in the structure at a minimum age of 28 days. Based on exhaustive experimental tests, a maximum allowable strain of 0.003 in./in. was adopted by the ACI as a safe limiting value. Even though several forms of stress blocks including trapezoidal have been proposed to date, the simplified equivalent rectangular block is accepted as the standard in the analysis and design of reinforced concrete. The behavior of the steel is assumed to be elastoplastic, as shown in Fig. 5.3a.

Using all of the assumptions above, the stress distribution diagram shown in Fig. 5.2c can be redrawn as shown in Fig. 5.2d. One can easily deduce that the compression force C can be written as $0.85f'_c\,ba$, that is, the *volume* of the compressive block at or near the ultimate when the tension steel has yielded, $\epsilon_s > \epsilon_y$. The tensile force T can be written as $A_s f_y$. Thus equilibrium Eq. 5.1 can be rewritten as

$$0.85f'_c ba = A_s f_y \qquad (5.2)$$

or

$$a = \frac{A_s f_y}{0.85f'_c b} \qquad (5.3)$$

The moment of resistance of the section, that is, the nominal strength M_n, can be expressed as

$$M_n = (A_s f_y)jd \qquad \text{or} \qquad M_n = (0.85f'_c ba)jd \qquad (5.4a)$$

where jd is the lever arm, denoting the distance between the compression and tensile forces of the internal resisting couple. Using the simplified equivalent rectangular stress block from Fig. 5.2d, the lever arm is

$$jd = d - \frac{a}{2}$$

Hence the nominal moment of resistance becomes

$$M_n = A_s f_y \left(d - \frac{a}{2} \right) \qquad (5.4b)$$

Since $C = T$, the moment equation can also be written as

$$M_n = 0.85f'_c ba \left(d - \frac{a}{2} \right) \qquad (5.4c)$$

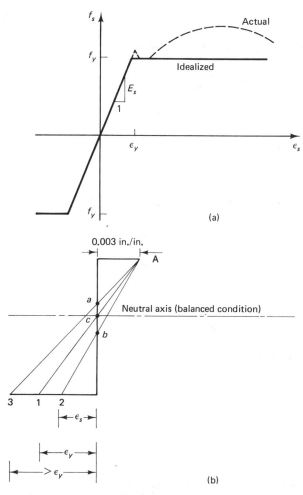

(a)

0.003 in./in.

A

a

c

Neutral axis (balanced condition)

b

3 1 2

ϵ_s

ϵ_y

$> \epsilon_y$

(b)

Figure 5.3 Strain distribution across depth: (a) idealized stress-strain diagram of the reinforcement; (b) strain distribution for various modes of flexural failure.

Photo 26 Closeup of flexural cracks in the preceding photograph.

87

$$\rho = \frac{A_s}{bd}$$

If the reinforcement ratio $\rho = A_s/bd$, Eq. 5.3 can be rewritten as

$$a = \frac{\rho d f_y}{0.85 f_c'}$$

If $r = b/d$, Eq. 5.4c becomes

$$M_n = \rho r d^2 f_y \left(d - \frac{\rho d f_y}{1.7 f_c'} \right) \tag{5.5a}$$

or

$$M_n = [\omega r f_c' (1 - 0.59\omega)] d^3 \tag{5.5b}$$

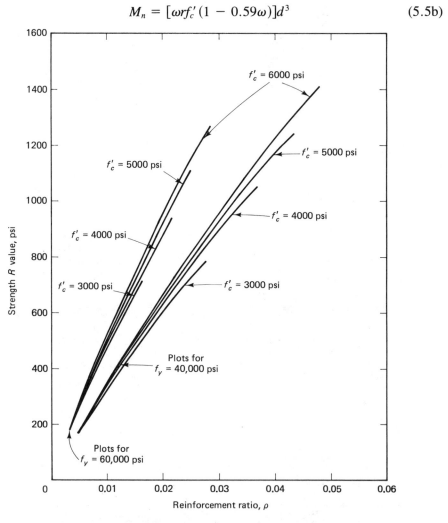

Figure 5.4   Strength–R curves for singly reinforced beams.

where $\omega = \rho f_y / f_c'$. Equation 5.5b is sometimes expressed as

$$M_n = R b d^2 \qquad (5.6a)$$

where

$$R = \omega f_c' (1 - 0.59\omega) \qquad (5.6b)$$

Equations 5.5 and 5.6 are useful for the development of charts. A plot of the R value for singly reinforced beams is shown in Fig. 5.4.

If f_c', f_y, b, d, and A_s are given for a rectangular section and the beam is so proportioned and reinforced that failure occurs by simultaneous yielding of the tension steel and crushing of the concrete at the compression side, the resisting moment strength can be obtained using Eq. 5.4 or 5.5, but using the balanced steel area A_{sb} and the balanced rectangular block depth a_b instead of A_s and a. However, beams have to be designed to fail in tension by initial yielding of the reinforcement, for reasons explained in subsequent sections.

Depending on the type of failure, that is, yielding of the steel or crushing of the concrete, three types of beams can be identified.

1. *Balanced section:*   The steel starts yielding when the concrete just reaches its ultimate strain capacity and commences to crush. At the start of failure, the permissible extreme fiber compressive strain is 0.003 in./in., while the tensile strain in the steel equals the yield strain $\epsilon_y = f_y / E_s$. The "balanced" strain distribution follows line $Ac1$ in Fig. 5.3b across the depth of the beam.
2. *Over-reinforced section:*   Failure occurs by initial crushing of the concrete. At the initiation of failure, the steel strain ϵ_s will be lower than the yield strain ϵ_y, as in line $Ab2$, Fig. 5.3b; hence the steel stress f_s will be lower than its yield

Photo 27   Beam at failure subjected to combined compression and bending. (Tests by Nawy et al.)

Photo 28 Cracking level at rupture.

strength f_y. Such a condition is accomplished by using more reinforcement at the tension side than that required for the balanced condition.

3. *Under-reinforced section:* Failure occurs by initial yielding of the steel, as in line $Aa3$, Fig. 5.3b. The steel continues to stretch as the steel strain increases beyond ϵ_y. This condition is accomplished when the area of the tension reinforcement used in the beam is less than that required for the balanced strain condition.

It is to be noted from positions c, b, and a of the neutral axis that the axis rises toward the compressive fibers in the under-reinforced beam as the limit state of failure is reached. This behavior is easily identified in tests as the flexural cracks propagate toward the compression fibers until the concrete crushes. It should also be recognized that the vertical distances between points c, b, and a of the neutral axis from the extreme compression fibers for the three types of failure depend largely on the percentage ratio $\rho = A_s/bd$, but do not differ significantly since low values of strain are involved.

Concrete failure is sudden since it is a brittle material. Therefore, almost all codes of practice recommend designing under-reinforced beams to provide sufficient warning, such as excessive deflection before failure. In the case of statically indeterminate structures, ductile failure is essential for proper moment redistribution. Hence for beams, the ACI code limits the maximum amount of steel to 75% of that required for a balanced section. For practical purposes, however, the reinforcement ratio A_s/bd should not normally exceed 50%, to avoid congestion of the reinforcement and facilitate proper placing of the concrete. If the actual reinforcement ratio and the balanced reinforcement ratio are denoted as ρ and $\bar{\rho}_b$, respectively, then

$$\rho \leq 0.75\bar{\rho}_b \tag{5.7a}$$

The code also stipulates the minimum steel requirement as

$$\rho > \frac{200}{f_y} \tag{5.7b}$$

where f_y is expressed in psi, to account for temperature stresses and to ensure ductile failure in tension.

5.3 BALANCED REINFORCEMENT RATIO $\bar{\rho}_b$

To analyze a given beam, one has to determine first the maximum allowable reinforcement ratio $0.75\bar{\rho}_b$. For rectangular sections reinforced only at the tension side, $\bar{\rho}_b$ is a function of only concrete strength and properties of steel, that is, modulus of elasticity E_s and yield strength f_y, irrespective of the section geometry. Using the strain distribution diagram in Fig. 5.2 for the balanced strain condition and from similar triangles, the relationship between the depth c (c_b for the balanced condition) of the neutral axis and the effective depth d can be written as

$$\frac{c_b}{d} = \frac{0.003}{0.003 + f_y/E_s}$$

If E_s is taken as 29×10^6 psi,

$$\frac{c_b}{d} = \frac{87,000}{87,000 + f_y} \tag{5.8a}$$

The relationship between the depth a of the equivalent rectangular stress block and the depth c of neutral axis is

$$a = \beta_1 c \tag{5.8b}$$

The value of the stress block depth factor β_1 is as follows:

$$\beta_1 = \begin{cases} 0.85 & \text{for } 0 < f_c' \leq 4000 \text{ psi} \\ 0.85 - 0.05\left(\dfrac{f_c' - 4000}{1000}\right) & \text{for } 4000 \text{ psi} < f_c' \leq 8000 \text{ psi} \\ 0.65 & \text{for } f_c' > 8000 \text{ psi} \end{cases}$$

Hence, for the balanced strain condition, the depth of the rectangular stress block is

$$a_b = \beta_1 c_b$$

For equilibrium of the horizontal forces,

$$A_{sb}f_y = 0.85f_c' ba_b$$

or

$$\bar{\rho}_b = \frac{A_{sb}}{bd} = \frac{0.85f_c'}{f_y}\frac{a_b}{d}$$

From Eq. 5.8, the balanced steel ratio becomes

$$\bar{\rho}_b = \beta_1 \frac{0.85f_c'}{f_y}\frac{87,000}{87,000 + f_y} \tag{5.9}$$

where f_c' and f_y are expressed in psi. Thus if f_c' and f_y are known, $\bar{\rho}_b$ and thus $0.75\,\bar{\rho}_b$ can be readily obtained regardless of the geometry of the concrete section.

Representative values of the maximum permissible reinforcement ratio ρ for singly reinforced beams are given in Table 5.1 both in pounds and in SI units. These values are 75% of the balanced reinforcement ratio $\bar{\rho}_b$ and should aid the student in eliminating tedious computations of these frequently used values.

TABLE 5.1 MAXIMUM PERMISSIBLE REINFORCEMENT RATIO ($0.75\,\bar{\rho}_b \times 10^4$) FOR BEAMS WITH TENSION REINFORCEMENT ONLY (SINGLY REINFORCED BEAMS)[a]

f_y (psi)	$f_c' = 3000$ $\beta_1 = 0.85$	$f_c' = 4000$ $\beta_1 = 0.85$	$f_c' = 5000$ $\beta_1 = 0.80$	$f_c' = 6000$ $\beta_1 = 0.75$
40,000	278	371	437	491
50,000	206	275	324	364
60,000	160	214	252	283

[a]The values for the reinforcement ratio ρ in both the standard and SI units are almost identical since the ρ is a dimensionless quantity. ACI 318 M–83, for example, rounds up the conversion values such that for $f_y = 60,000$ psi $\simeq 400$ MPa and for $f_c' = 3000$ psi $\simeq 21$ MPa, $\rho = 160$, as above.

5.4 ANALYSIS OF SINGLY REINFORCED RECTANGULAR BEAMS FOR FLEXURE

The sequence of calculations presented in the flowchart of Fig. 5.5 can be used for the analysis of a given beam for both longhand and machine computations. The flowchart was developed using the method of analysis presented in Section 5.2. The following examples illustrate typical analysis calculations following the flowchart logic in Fig. 5.5.

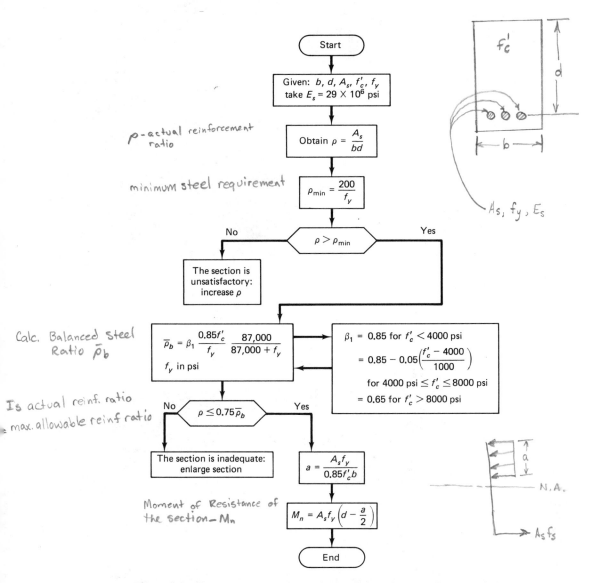

The flowchart contains the following elements:

Start

Given: b, d, A_s, f'_c, f_y
take $E_s = 29 \times 10^6$ psi

ρ-actual reinforcement ratio

Obtain $\rho = \dfrac{A_s}{bd}$

minimum steel requirement

$\rho_{min} = \dfrac{200}{f_y}$

$\rho > \rho_{min}$ No / Yes

The section is unsatisfactory: increase ρ

Calc. Balanced steel Ratio $\overline{\rho}_b$

$\overline{\rho}_b = \beta_1 \dfrac{0.85 f'_c}{f_y} \dfrac{87{,}000}{87{,}000 + f_y}$

f_y in psi

$\beta_1 = 0.85$ for $f'_c < 4000$ psi

$= 0.85 - 0.05 \left(\dfrac{f'_c - 4000}{1000} \right)$

for 4000 psi $\leq f'_c \leq 8000$ psi

$= 0.65$ for $f'_c > 8000$ psi

Is actual reinf. ratio ≤ max. allowable reinf ratio

$\rho \leq 0.75\overline{\rho}_b$ No / Yes

The section is inadequate: enlarge section

$a = \dfrac{A_s f_y}{0.85 f'_c b}$

Moment of Resistance of the section — M_n

$M_n = A_s f_y \left(d - \dfrac{a}{2} \right)$

End

(Diagram annotations: f'_c, d, b, A_s, f_y, E_s, a, N.A., $A_s f_s$)

Figure 5.5 Flowchart for analysis of singly reinforced rectangular beams in bending.

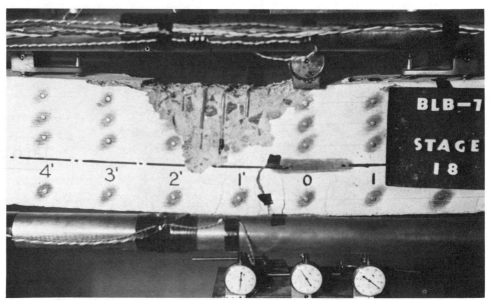

Photo 29 Beam subjected to combined axial load and bending. The neutral axis is at 70% of depth. (Tests by Nawy et al.)

Photo 30 Flexural cracking and deflection of beam subjected to flexure only prior to failure. (Tests by Nawy et al.)

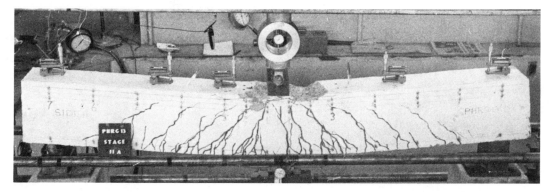

Photo 31   Crushing of concrete at compression side of beam subjected to flexure.

5.4.1 Example 5.1: Flexural Analysis of a Singly Reinforced Beam (Tension Reinforcement Only)

A singly reinforced concrete beam ($f_c' = 4000$ psi or 27.58 MPa) has the cross section shown in Fig. 5.6. Determine if the beam is over-reinforced or under-reinforced and if it satisfies the ACI code requirements for maximum and minimum reinforcement ratios for (a) $f_y = 60,000$ psi (413.4 MPa), and (b) $f_y = 40,000$ psi (275.6 MPa).

Solution

(a) From Eq. 5.9 for the balanced condition,

$$\bar{\rho}_b = \beta_1 \frac{0.85 f_c'}{f_y} \frac{87,000}{87,000 + f_y}$$

$$f_c' = 4000 \text{ psi}$$

$$f_y = 60,000 \text{ psi}$$

$$\beta_1 = 0.85 \qquad \text{for } f_c' = 4000 \text{ psi}$$

Therefore,

$$\bar{\rho}_b = 0.85 \left(\frac{0.85 \times 4000}{60,000}\right) \frac{87,000}{87,000 + 60,000} = 0.029$$

$$A_{sb} = \bar{\rho}_b bd = 0.029 \times 10 \times 18$$

$$= 5.22 \text{ in.}^2 \ (3367 \text{ mm}^2) < A_s = 6 \text{ in.}^2 \ (3870 \text{ mm}^2)$$

$$\rho = \frac{6.0}{10 \times 18} = 0.033$$

Thus this section is over-reinforced because $A_s > A_{sb}$ or $\rho > \bar{\rho}_b$ and does not satisfiy ACI code requirements for ductility and maximum allowable reinforcement.

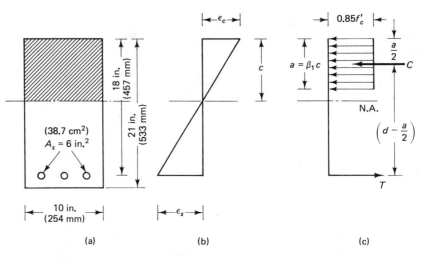

Figure 5.6 Stress and strain distribution in a typical singly reinforced rectangular section: (a) cross section; (b) strains; (c) stresses.

(b) In a new trial with f_y = 40,000 psi, one finds that

$$\bar{\rho}_b = 0.85 \left(\frac{0.85 \times 4000}{40,000}\right) \frac{87,000}{87,000 + 40,000} = 0.0495$$

$$A_{sb} = 8.91 \text{ in.}^2 \ (5746.95 \text{ mm}^2) > 6 \text{ in.}^2 \ (3367 \text{ mm}^2)$$

Therefore, the section is under-reinforced

$$\text{minimum allowable reinforcement ratio } \rho_{\min} = \frac{200}{f_y} = \frac{200}{40,000} = 0.005 << \text{actual } \rho$$

$$\text{maximum allowable steel area} = 0.75 \times 8.91 = 6.68 \text{ in.}^2 > 6 \text{ in.}^2$$

Therefore, the cross section satisfies ACI code requirements for maximum and minimum reinforcement. Note that the actual steel area in case (b) is only slightly less than the maximum allowable 75% of the balanced steel area. Hence congestion of steel is likely and the design can be improved by increasing the beam section size and reducing A_s.

5.4.2 Example 5.2: Nominal Resisting Moment in a Singly Reinforced Beam

For the beam cross section shown in Fig. 5.7, calculate the nominal moment strength if f_y is 60,000 psi (413.4 MPa) and f'_c is (a) 3000 psi (20.68 MPa), (b) 5000 psi (34.47 MPa), (c) 9000 psi (62.10 MPa).

Solution

$$b = 10 \text{ in. } (254.0 \text{ mm})$$

$$d = 18 \text{ in. } (457.2 \text{ mm})$$

$$A_s = 4 \text{ in.}^2 \ (2580 \text{ mm}^2)$$

$$f_y = 60,000 \text{ psi}$$

For beam:

$$\rho_{min} = \frac{200}{f_y} = \frac{200}{60,000} = 0.003 \quad (psi)$$

$$\rho = \frac{A_s}{bd} = \frac{4}{10 \times 18} = 0.0222 > 0.003 \quad \text{O.K.}$$

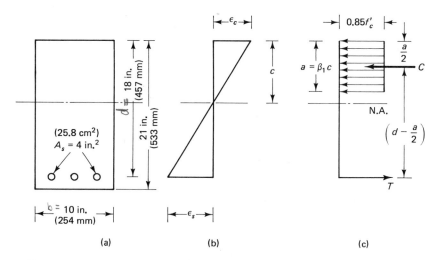

Figure 5.7   Beam cross-section strain and stress diagrams, Ex. 5.2: (a) cross section; (b) strains; (c) stresses.

Note that f_y should be in psi units in the ρ_{min} expression.
   (a) $f'_c = 3000$ psi (20.68 MPa).

$$\beta_1 = 0.85$$

Using Eq. 5.9,

$$\bar{\rho}_b = \beta_1 \frac{0.85 f'_c}{f_y} \frac{87,000}{87,000 + f_y} \quad (psi)$$

$$= 0.85 \left(\frac{0.85 \times 3,000}{60,000} \right) \frac{87,000}{87,000 + 60,000} = 0.021$$

$$0.75 \bar{\rho}_b = 0.016$$

$$\rho > 0.75 \bar{\rho}_b$$

Hence the beam is considered over-reinforced and does not satisfy the ACI requirements for ductility and maximum allowable reinforcement ratio.

(b) $f'_c = 5000$ psi (34.47 MPa).

$$\beta_1 = 0.85 - 0.05 \left(\frac{5000 - 4000}{1000} \right)$$

$$= 0.8$$

$$\bar{\rho}_b = 0.8 \left(\frac{0.85 \times 5000}{60,000} \right) \frac{87,000}{87,000 + 60,000}$$

$$= 0.034$$

steel Ratio $0.75\bar{\rho}_b = 0.025 > \rho = 0.0222$ O.K.

$$A_s = 4 \text{ in.}^2$$

$a = \dfrac{A_s f_y}{0.85 f'_c b}$

$$a = \frac{4 \times 60,000}{0.85 \times 5000 \times 10}$$

$$= 5.65 \text{ in.}$$

$M_n = A_s f_y \left(d - \dfrac{a}{2} \right)$

$$M_n = 4 \times 60,000 \left(18 - \frac{5.65}{2} \right) = 3,642,000 \text{ in.-lb}$$

$$= 303,500 \text{ ft-lb (411.52 kN-m)}$$

Or, using Eq. 5.5 gives

$\omega = \dfrac{\rho f_y}{f'_c}$

$$\omega = \frac{0.0222 \times 60,000}{5000} = 0.267$$

$$M_n = \left[0.267 \times \frac{10}{18} \times 5000 \, (1 - 0.59 \times 0.267) \right] 18^3$$

$$= 3,644,019 \text{ in.-lb (411.77 kN-m)}$$

(c) $f'_c = 9000$ psi (62.10 MPa).

$$\beta_1 = 0.65 \text{ for } f'_c \geq 8000 \text{ psi}$$

$$\bar{\rho}_b = 0.65 \left(\frac{0.85 \times 9000}{60,000} \right) \frac{87,000}{87,000 + 60,000}$$

$$= 0.049$$

$$0.75\bar{\rho}_b = 0.037 > \rho = 0.022 \qquad \text{O.K.}$$

$$a = \frac{4.0 \times 60,000}{0.85 \times 9000 \times 10} = 3.14 \text{ in. (79.8 mm)}$$

$$M_n = 4.0 \times 60,000 \left(18 - \frac{3.14}{2} \right) = 3,943,200 \text{ in.-lb}$$

$$= 328,600 \text{ ft-lb (445.6 kN-m)}$$

5.5 TRIAL-AND-ADJUSTMENT PROCEDURES FOR THE DESIGN OF SINGLY REINFORCED BEAMS

In Ex. 5.2, the geometrical properties of the beam, that is, b, d, and A_s, were given. In a design example, an assumption of width b (or the ratio b to d) and the level of reinforcement ratio ρ have to be made. The ratio b/d varies between 0.25 and 0.6 in usual practice. Although the ACI code permits a reinforcement ratio ρ up to $0.75\bar{\rho}_b$, a ratio of $0.5\bar{\rho}_b$ is advisable to prevent congestion of steel, secure a good bond between the reinforcement and the adjacent concrete, and provide good deflection control.

Studies on cost optimum design indicate that cost-effective sections can be obtained using a minimum practical b/d ratio and a maximum practical reinforcement ratio ρ within the above-stated limitations. Hence one could use the following steps to design the beam cross section following the flowchart logic of Fig. 5.8.

1. Calculate the external factored moment. To obtain the beam self-weight, an assumption has to be made for the value of d. The minimum depth for deflection specified in the ACI code can be used as a guide. Assume a b/d ratio r between 0.25 and 0.6 and calculate $b = rd$. A first trial assumption $b \simeq d/2$ is recommended.
2. Choose a reinforcement ratio of approximately $0.5\bar{\rho}_b$.
3. (a) Select a value of moment factor R based on an assumed ρ value $\simeq 0.5\bar{\rho}_b$. Assuming that $b \simeq d/2$, calculate d for $M_n = Rbd^2$ and proceed to analyze the section.
   (b) Alternatively, choose d on the basis of minimum deflection requirement. Choose a width b as in 3(a). Assume a moment arm $jd \simeq 0.85d$ to $0.90d$. Calculate A_s as a first trial, then analyze the section using $b = d/2$.

[handwritten margin note:] Trial $b = \frac{1}{2}d$ $d = 2b$

The process of arriving at the final section is highly convergent even by longhand computations in that it should not require more than three trial cycles. The use of handheld or desktop personal computers (see Chapter 13) enormously simplifies the design–analysis process and permits the student or engineer to proportion sections at a fraction of the time needed when using handbooks, charts, or longhand computations, easy as these other means can be.

For designers who prefer charts, Eq. 5.6 ($M_n = Rbd^2$) can be used for the first trial in design. The value of R can be obtained from charts (see Fig. 5.4) for various values of ρ, f_c', and f_y available in handbooks.

5.5.1 Example 5.3: Design of a Singly Reinforced Simply Supported Beam for Flexure

A reinforced concrete simply supported beam has a span of 30 ft (9.14 m) and is subjected to a service uniform load $w_u = 1500$ lb/ft (21.9 kN/m), as shown in Fig. 5.9. Design a beam section to resist the factored external bending load. Given:

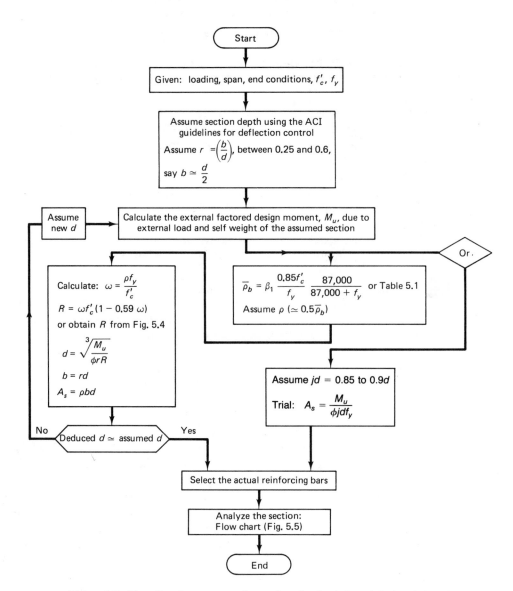

Figure 5.8 Flowchart for sequence of operations for the design of singly reinforced rectangular sections.

$$f'_c = 4000 \text{ psi } (27.58 \text{ MPa})$$
$$f_y = 60,000 \text{ psi } (413.4 \text{ MPa})$$

Solution

Assume a minimum thickness from the ACI code deflection table:

$$\frac{l_n}{16} = \frac{30 \times 12}{16} = 22.5 \text{ in.}$$

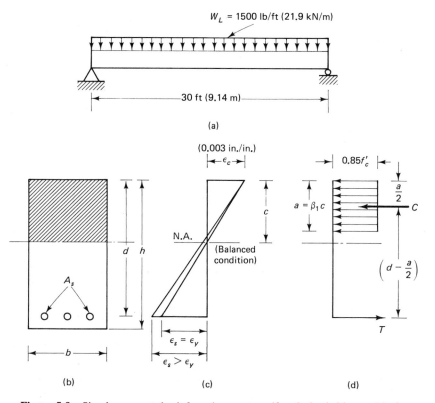

Figure 5.9 Simply supported reinforced concrete uniformly loaded beam: (a) elevation; (b) cross section; (c) strains; (d) stresses.

For the purpose of estimating the preliminary self-weight, assume total thickness $h = 24.0$ in., effective depth $d = 20$ in., and width of the beam $b = 10$ in. ($r = b/d = 0.5$).

$$\text{beam self-weight} = \frac{24 \times 10}{144} \times 150 = 250 \text{ lb/ft}$$

$$\text{factored load } W_u = 1.4D + 1.7L = 1.4 \times 250 + 1.7 \times 1500 = 2900 \text{ lb/ft}$$

$$\text{required moment } M_u = \frac{w_u l_n^2}{8} = \frac{2900 \times 30^2}{8} \times 12 = 3,915,000 \text{ in.-lb}$$

$$\text{required nominal resisting moment } M_n = \frac{M_u}{\phi} = \frac{3,915,000}{0.9} = 4,350,000$$

Get $\bar{\rho}_b$ from Table 5.1 which gives $0.75\,\bar{\rho}_b$ or calculate:

$$\bar{\rho}_b = \beta_1 \frac{0.85f_c'}{f_y} \frac{87,000}{87,000 + f_y} = 0.85 \left(\frac{0.85 \times 4000}{60,000}\right) \frac{87,000}{87,000 + 60,000} = 0.0285$$

Assume a reinforcement ratio $\rho = 0.5\overline{\rho}_b = 0.0143$.

$$\omega = \frac{\rho f_y}{f'_c} = \frac{0.0143 \times 60,000}{4000} = 0.215$$

Using Eq. 5.6b yields

$$R = \omega f'_c(1 - 0.59\omega) = 0.215 \times 4000(1 - 0.59 \times 0.215)$$

$$\approx 750$$

The value of R can also be obtained from the chart in Fig. 5.4 using the chosen ρ and the given values for f'_c and f_y.

Using Eq. 5.6a, one has $M_n = Rbd^2$ and assuming that $b = 0.5d$,

$$d = \sqrt[3]{\frac{M_n}{0.5R}} = \sqrt[3]{\frac{4,350,000}{0.5 \times 750}} = 22.64 \text{ in.}$$

$$b = 0.5 \times 22.64 = 11.32 \text{ in.}$$

Based on practical considerations, try a section with $b = 12$ in., $d = 23$ in., and $h = 26$ in.

$$\text{revised self-weight} = \frac{12 \times 26}{144} \times 150 = 325 \text{ lb/ft}$$

$$\text{factored load } W_u = 1.4 \times 325 + 1.7 \times 1500 = 3005 \text{ lb/ft}$$

$$\text{factored moment } M_u = \frac{3005(30)^2}{8} \times 12 = 4,056,750 \text{ in.-lb}$$

$$\text{required resisting moment, } M_n = \frac{M_u}{\phi} = \frac{4,056,750}{0.9} = 4,507,500 \text{ in.-lb}$$

$$A_s = \rho b d = 0.0143 \times 12 \times 23 = 3.95 \text{ in.}^2$$

Try three No. 10 bars (32.3 mm diameter) with $A_s = 3.81$ in.^{2}

$$\rho = \frac{3.81}{12 \times 23} = 0.0138 < 0.75\overline{\rho}_b > \rho_{\min} \qquad \text{O.K.}$$

Check the nominal moment strength of the assumed section:

$$a = \frac{A_s f_y}{0.85 f'_c b} = \frac{3.81 \times 60,000}{0.85 \times 4000 \times 12} = 5.60 \text{ in.}$$

$$M_n = 3.81 \times 60,000\left(23 - \frac{5.60}{2}\right) = 4,617,720 \text{ in.-lb } (521.8 \text{ kN-m})$$

$$> \text{required } M_n = 4,507,500 \text{ in.-lb}$$

Adopt the section. Note that the designed section resists a slightly larger moment than the required moment:

$$\text{percent overdesign} = \frac{4,617,720 - 4,507,500}{4,507,500} = 2.45\%$$

which is a reasonable level expected in proportioning concrete elements. It is always necessary also to check that the web width can accommodate the number of bars in each layer based on concrete cover and minimum spacing requirements. In this example, the minimum web width to accommodate three No. 10 bars = 10.5 in. $< b = 12.0$ in., which is O.K.

Alternative Solution by Trial and Adjustment

Assume that moment arm $jd \simeq 0.85d$:

$$\text{minimum thickness } h = \frac{l_n}{16} = \frac{30 \times 12}{16} = 22.5 \text{ in.}$$

Try $h = 26$ in. (660.4 mm), $d = 23.0$ in (584.2 mm), and $b \simeq \frac{1}{2} d \simeq 12$ in. (304.8 mm).

$$\text{self-weight} = \frac{12 \times 26}{144} \times 150 = 325.0 \text{ lb/ft}$$

$$\text{factored load } U = 1.4 \times 325.0 + 1.7 \times 1500 = 3005 \text{ lb/ft}$$

$$\text{factored moment } M_u = \frac{3005(30.0)^2}{8} \times 12$$

$$= 4,056,750 \text{ in.-lb (458.1 kN-m)}$$

$$\text{required nominal resisting moment } M_n = \frac{M_u}{\phi} = \frac{4,056,750}{0.9}$$

$$= 4,507,500 \text{ in.-lb (509.3 kN-m)}$$

$$\text{moment arm } jd \simeq 0.85 \simeq 0.85 \times 23.0 = 19.55 \text{ in.}$$

$$M_n = A_s f_y \left(d - \frac{a}{2} \right) = A_s f_y jd \text{ or } 4,507,500 = A_s \times 60,000 \times 19.55.$$

Hence

$$A_s = \frac{4,507,500}{60,000 \times 19.55} = 3.84 \text{ in.}^2$$

Try three No. 10 bars (32.3 mm diameter = 3.81 in.^{2} (2457.5 mm^2). Continue the design following the flowchart in Fig. 5.8.

5.5.2 Arrangement of Reinforcement

Figure 5.10 shows the cross section of the beam at midspan. In arranging the reinforcing bars, one should satisfy the minimum cover requirements explained in Chapter 3. The required clear cover for beams is 1.5 in. (38 mm).

The stirrups shown in Fig. 5.10 should be designed to satisfy the shear requirements of the beam explained in Chapter 6. Two bars called *hangers* are placed on the

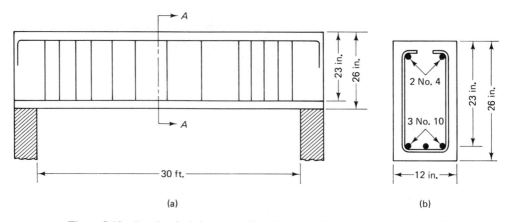

Figure 5.10   Details of reinforcement, Ex. 5.3: (a) sectional elevation (not to scale); (b) midspan section A–A.

compression side to support the stirrups. Reinforcement detailing provisions and bar development length requirements are discussed in Chapter 10.

5.6 ONE-WAY SLABS

One-way slabs are concrete structural floor panels for which the ratio of the long span to the short span equals or exceeds a value of 2.0. When this ratio is less than 2.0, the floor panel becomes a two-way slab or plate, as discussed in Chapter 11. A one-way slab is designed as a singly reinforced 12-in. (304.8-mm)-wide beam strip using the same design and analysis procedure discussed earlier for singly reinforced beams. Figure 5.11 shows a one-way slab floor system.

Loading for slabs is normally specified in pounds per square foot (psf). One has to distribute the reinforcement over the 12-in. strip and specify the center-to-center spacing of the reinforcing bars. In slab design, a thickness is normally assumed and the reinforcement is calculated using a trial lever arm $(d - a/2)$ or $0.9d$.

Supported slabs, namely, slabs not on grade, do not normally require shear reinforcement for typical loads. Transverse reinforcement has to be provided perpendicular to the direction of bending in order to resist shrinkage and temperature stresses. Shrinkage and temperature reinforcement should not be less than 0.002 times the gross area for grade 40 or 50 bars and 0.0018 for grade 60 steel and welded fire fabric.

5.6.1 Example 5.4: Design of a One-Way Slab for Flexure

A one-way single-span reinforced concrete slab has a simple span of 10 ft (3.05 m) and carries a live load of 120 psf (5.75 kPa) and a dead load of 20 psf (0.96 kPa) in addition to its self-weight. Design the slab and the size and spacing of the reinforcement at midspan assuming a simple support moment. Given:

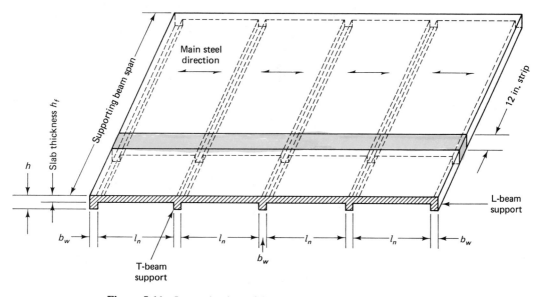

Figure 5.11 Isometric view of four-span continuous one-way-slab floor system.

$f'_c = 4000$ psi (27.5 MPa), normalweight concrete

$f_y = 60,000$ psi (413.4 MPa)

Minimum thickness for deflection $= l/20$

Solution

$$\text{minimum depth for deflection, } h = \frac{l}{20} = \frac{10 \times 12}{20} = 6 \text{ in. (152.4 mm)}$$

Assume for flexure an effective depth $d = 5$ in. (127 mm).

$$\text{self-weight of a 12-in. strip} = \frac{6 \times 12}{144} \times 150 = 75 \text{ lb/ft (3.59 kN/m)}$$

Therefore,

$$\text{factored external load } U = 1.7 \times 120 + 1.4(20 + 75) = 337 \text{ lb/ft}$$

$$\text{factored external moment } M_u = \frac{337 \times 10^2}{8} \times 12 \text{ in.-lb}$$

$$= 50,550 \text{ in.-lb (5712.15 kN-m)}$$

Assume that the arm $(d - a/2) = 0.9d = 0.9 \times 5 = 4.50$ in.

$$M_u = \phi A_s f_y \left(d - \frac{a}{2} \right)$$

Therefore,

$$50,550 = 0.9 \times A_s \times 60,000(4.50)$$

or $A_s = 0.21$ in.^{2} per 12 in. of slab.

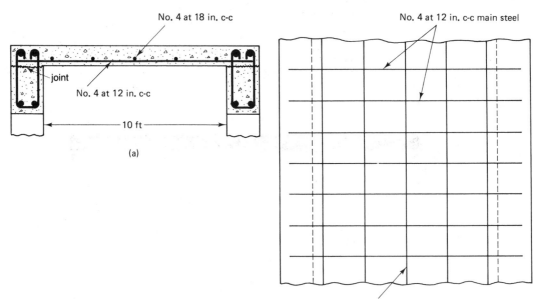

Figure 5.12 Reinforcement details of the one-way slab in Ex. 5.4: (a) sectional elevation; (b) reinforcement plan.

Trial-and-adjustment check for assumed moment arm:

$$a = \frac{A_s f_y}{0.85 f'_c b} = \frac{0.21 \times 60{,}000}{0.85 \times 4000 \times 12} = 0.31 \text{ in.} \quad (8.6 \text{ mm})$$

$$50{,}550 = 0.9 \times A_s \times 60{,}000\left(5 - \frac{0.31}{2}\right)$$

$$A_s = 0.193 \text{ in.}^2 \text{ per 12-in. slab strip}$$

Use No. 4 bars at 12 in. center-to-center spacing (13-mm-diameter bars at 304.8 mm center to center) with an area of 0.20 in.^{2} or No. 3 bars at $6\frac{1}{2}$ in. center to center spacing.

$$\rho = \frac{0.20}{5.0 \times 12} = 0.0033 \quad \rho_{min} = \frac{200}{60{,}000} = 0.0033 = \rho \qquad \text{O.K.}$$

$$\rho_{max} = 0.75\,\bar{\rho}_b = 0.75\left(\frac{0.85 \times 4{,}000}{60{,}000} \times 0.85 \times \frac{87{,}000}{87{,}000 + 60{,}000}\right)$$

$$= 0.0214 > \rho \qquad \text{O.K.}$$

Shrinkage and temperature reinforcement:

$$\rho = 0.0018$$

area of steel $= 0.0018 \times 6 \times 12 = 0.13 \text{ in.}^2 = $ No. 4 bars at 18″ c-c

Provide No. 4 bars at 18 in. center to center (maximum allowable spacing = 5h = 5 × 6 = 30 in.).

Hence this design can be adopted with slab thickness h = 6 in. (152.4 mm) and effective depth d = $6.0 - (0.75 + 0.25) = 5$ in. (127.0 mm) to satisfy the $\frac{3}{4}$-in. minimum concrete cover requirement. Use for main reinforcement No. 4 bars at 12 in. center to center and for temperature reinforcement No. 4 bars at 18 in. center to center, as shown in Fig. 5.12.

5.7 DOUBLY REINFORCED SECTIONS

Doubly reinforced sections contain reinforcement both at the tension and at the compression face, usually at the support section only. They become necessary when either architectural limitations restrict the beam web depth at midspan, or the midspan section is not adequate to carry the support negative moment even when the tensile steel at the support is sufficiently increased. In such cases, most of the bottom bars at midspan are extended and well anchored at the supports to act as compression reinforcement. The bar development length has to be well established and the compressive and tensile steel at the support section well tied with closed stirrups to prevent buckling of the compressive bars.

In analysis or design of beams with compression reinforcement A'_s, the analysis is so divided that the section is theoretically split into two parts, as shown in Fig. 5.13. The two parts of the solution comprise (1) the singly reinforced part involving the equivalent rectangular block, as discussed in Section 5.2, with the area of tension reinforcement being $(A_s - A'_s)$; and (2) the two areas of equivalent steel A'_s at both the tension and compression sides to form the couple T_2 and C_2 as the second part of the solution.

It can be seen from Fig. 5.13 that the total nominal resisting moment $M_n = M_{n1} + M_{n2}$, that is, the summation of the moments for parts 1 and 2 of the solution.

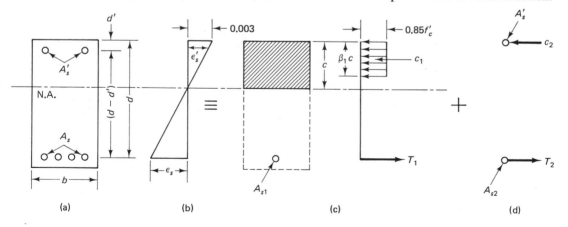

Figure 5.13   Doubly reinforced beam design: (a) cross section; (b) strains; (c) part 1 of the solution—singly reinforced part; (d) part 2 of the solution—contribution of compression reinforcement.

Part 1

The tension force $T_1 = A_{s1}f_y = C_1$. But $A_{s1} = A_s - A_s'$ since equilibrium requires that A_{s2} at the tension side be balanced by an equivalent A_s' at the compression side. Hence the nominal resisting moment

$$M_{n1} = A_{s1}f_y\left(d - \frac{a}{2}\right) \quad \text{or} \quad M_{n1} = (A_s - A_s')f_y\left(d - \frac{a}{2}\right) \tag{5.10a}$$

where

$$a = \frac{A_{s1}f_y}{0.85f_c'b} = \frac{(A_s - A_s')f_y}{0.85f_c'b}$$

Part 2

$$A_s' = A_{s2} = (A_s - A_{s1})$$

$$T_2 = C_2 = A_{s2}f_y$$

Taking the moment about the tension steel, we have

$$M_{n2} = A_{s2}f_y(d - d') \tag{5.10b}$$

Adding the moments for parts 1 and 2 yields

$$M_n = M_{n1} + M_{n2} = (A_s - A_s')f_y\left(d - \frac{a}{2}\right) + A_s'f_y(d - d') \tag{5.11a}$$

The design moment strength ϕM_n must be equal to or greater than the external factored moment M_u such that

$$M_u = \phi\left[(A_s - A_s')f_y\left(d - \frac{a}{2}\right) + A_s'f_y(d - d')\right] \tag{5.11b}$$

This equation is valid *only* if A_s' yields. Otherwise, the beam has to be treated as a singly reinforced beam neglecting the compression steel, or one has to find the actual stress f_s' in the compression reinforcement A_s' and use the actual force in the moment equilibrium equation.

Strain-Compatibility Check

It is always necessary to verify that the strains across the depth of the section follow the linear distribution indicated in Fig. 5.13. In other words, a check is necessary to ensure that strains are compatible across the depth at the strength design levels. Such a verification is called a *strain-compatibility check*.

For A_s' to yield, the strain ϵ_s' in the compression steel should be greater than or equal to the yield strain of reinforcing steel, which is f_y/E_s. The strain ϵ_s' can be calculated from similar triangles. Referring to Fig. 5.13b, one has

$$\epsilon_s' = \frac{0.003(c - d')}{c}$$

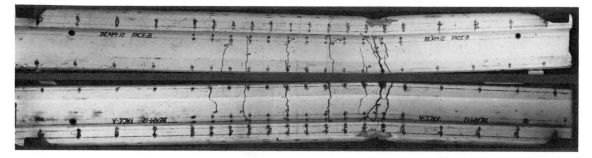

Photo 33   Flexural cracking in lightly reinforced beam. (Tests by Nawy et al.)

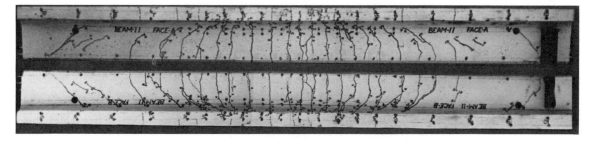

Photo 32   Flexural cracking in heavily reinforced beam. (Tests by Nawy, Pot-yondy, et al.)

or

$$\epsilon'_s = 0.003\left(1 - \frac{d'}{c}\right)$$

Since

$$c = \frac{a}{\beta_1} = \frac{(A_s - A'_s)f_y}{\beta_1 \times 0.85f'_c\,b} = \frac{(\rho - \rho')f_y d}{\beta_1 \times 0.85f'_c}$$

$$\epsilon'_s = 0.003\left[1 - \frac{0.85\beta_1 f'_c d'}{(\rho - \rho')df_y}\right] \tag{5.12}$$

As mentioned earlier, for compression steel to yield, the following condition must be satisfied:

$$\epsilon'_s \geq \frac{f_y}{E_s} \qquad \text{or} \qquad \epsilon'_s \geq \frac{f_y}{29 \times 10^6}$$

The compression steel yields if

$$0.003\left[1 - \frac{0.85\beta_1 f'_c}{(\rho - \rho')f_y}\frac{d'}{d}\right] \geq \frac{f_y}{29 \times 10^6} \tag{5.13}$$

or
$$-\frac{0.85\beta_1 f_c' d'}{(\rho - \rho')f_y d} \geq \frac{f_y - 87,000}{87,000}$$

or
$$\rho - \rho' \geq \frac{0.85\beta_1 f_c' d'}{f_y d} \frac{87,000}{87,000 - f_y} \tag{5.14}$$

If ϵ_s' is less than ϵ_y the stress in the compression steel, f_s', can be calculated as

$$f_s' = E_s \epsilon_s' = 29 \times 10^6 \epsilon_s' \tag{5.15}$$

Using Eqs. 5.12 and 5.15 yields

$$f_s' = 29 \times 10^6 \times 0.003 \left[1 - \frac{0.85\beta_1 f_c' d'}{(\rho - \rho')f_y d} \right] \tag{5.16}$$

This value of f_s' can be used as a first approximation in the strain compatibility check in cases where the compression reinforcement did *not* yield. The reinforcement ratio for the balanced section can be written as

$$\rho_b = \bar{\rho}_b + \rho' \frac{f_s'}{f_y} \tag{5.17a}$$

where $\bar{\rho}_b$ corresponds to the balanced steel ratio for a singly reinforced beam that has a tension steel area A_{s1}.

The singly reinforced part of the solution in a doubly reinforced section normally utilizes the maximum allowable reinforcement ratio, $0.75\rho_b$. Consequently, the maximum allowable reinforcement ratio for a doubly reinforced beam can be expressed as

$$\rho \leq 0.75\bar{\rho}_b + \rho' \frac{f_s'}{f_y} \tag{5.17b}$$

In this discussion, adjustment for the concrete area replaced by the compression reinforcement is disregarded as being insignificant for practical design purposes. It is to be noted that in cases where the compression reinforcement A_s' did not yield, the depth of the rectangular compressive block should be calculated using the actual stress in the compression steel from the calculated strain value ϵ_s' at the compression reinforcement level so that

$$a = \frac{A_s f_y - A_s' f_s'}{0.85 f_c' b} \tag{5.18}$$

Eq. 5.16 can be used for the f_s' value in the first trial in order to obtain an "a" value and hence the first trial neutral axis depth value c. Once c is known, ϵ_s' can be evaluated from similar triangles in Fig. 5.13b, thereby obtaining the first approximation of f_s' to be used in recalculating a more refined value. More than one or two additional trials for calculating f_s' are not justified since undue refinement has negligeable practical effect on the true value of the nominal moment strength M_n.

The nominal moment strength in Eq. 5.11 becomes in this case

$$M_n = (A_s f_y - A'_s f'_s)\left(d - \frac{a}{2}\right) + A'_s f'_s (d - d') \qquad (5.19)$$

The flowchart, Fig. 5.14, can be used for the sequence of calculations in the analysis of doubly reinforced beams. Examples 5.5 and 5.6 illustrate the analysis and design of doubly reinforced sections.

5.7.1 Example 5.5: Analysis of a Doubly Reinforced Beam for Flexure

Calculate the nominal moment strength M_n of the doubly reinforced section shown in Fig. 5.15. Given:

> f'_c = 5000 psi (34.46 MPa), normal-weight concrete
> f_y = 60,000 psi (413.4 MPa)
> d' = 2.5 in. (63.5 mm)

Solution

$$A_s = 5.08 \text{ in.}^2 \qquad \rho = \frac{A_s}{bd} = \frac{5.08}{14 \times 21} = 0.0173$$

$$A'_s = 1.2 \text{ in.}^2 \qquad \rho' = \frac{A'_s}{bd} = \frac{1.2}{14 \times 21} = 0.0041$$

$$A_s - A'_s = A_{s1} = 5.08 - 1.2 = 3.88 \text{ in.}^2$$

$$\rho - \rho' = 0.0173 - 0.0041 = 0.0132$$

To check whether the compression steel has yielded, use Eq. 5.14:

$$\rho - \rho' \geq \frac{0.85 \beta_1 f'_c d'}{f_y d} \frac{87,000}{87,000 - f_y}$$

$$\geq \frac{0.85 \times 0.80 \times 5000 \times 2.5}{60,000 \times 21} \frac{87,000}{87,000 - 60,000}$$

$$\geq 0.0217$$

The actual $(\rho - \rho') = 0.0132 < 0.0217$. Therefore, the compression steel did not yield and f'_s is less than f_y. For the first trial in cases where the compression steel did not yield

$$f'_s = 87,000\left[1 - \frac{0.85 \beta_1 f'_c}{(\rho - \rho')f_y} \frac{d'}{d}\right]$$

$$= 87,000\left(1 - \frac{0.85 \times 0.80 \times 5000}{0.0132 \times 60,000} \times \frac{2.5}{21}\right) = 42,538 \text{ psi}$$

$$a = \frac{A_s f_y - A'_s f'_s}{0.85 f'_c b} = \frac{5.08 \times 60,000 - 1.2 \times 42,538}{0.85 \times 5000 \times 14} = 4.26 \text{ in. (108.33 mm)}$$

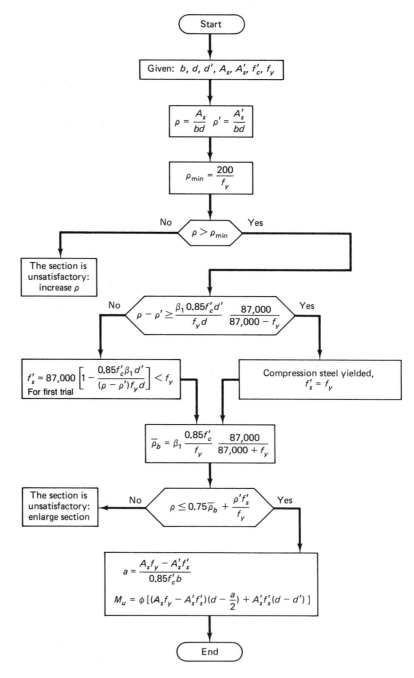

Figure 5.14 Flowchart for the analysis of doubly reinforced rectangular beam.

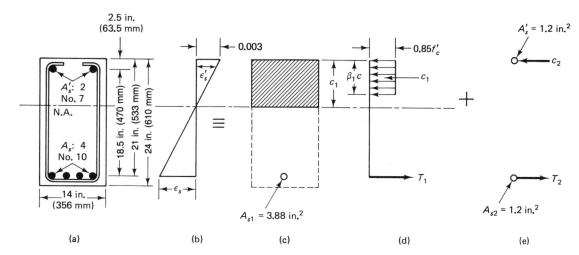

Figure 5.15 Doubly reinforced cross-section geometry and stress and strain distribution: (a) cross section; (b) strains; (c) part 1 section; (d) part 1 forces; (e) part 2 forces.

$$\text{Neutral axis depth } c = \frac{4.26}{0.80} = 5.325 \text{ in.}$$

From similar triangles in Fig. 5.15b, the strain ϵ'_s at the compression steel level $= 0.0019$ in./in. giving $f'_s = 0.0019 \times 29 \times 10^6 = 46,155$ psi. An additional trial cycle for a more refined value of $a = 4.21$ in., hence $c = 5.26$ in. gives $f'_s = 45,650$ psi (314.76 kN).

$$\bar{\rho}_b = \beta_1 \frac{0.85 f'_c}{f_y} \frac{87,000}{87,000 + f_y}$$

$$= 0.8 \frac{0.85 \times 5000}{60,000} \frac{87,000}{87,000 + 60,000} = 0.0335$$

Hence $0.75\bar{\rho}_b = 0.0252$

From Eq. 5.12, the maximum allowable reinforcement ratio

$$\rho \le 0.75\bar{\rho}_b + \rho' \frac{f'_s}{f_y}$$

$$0.75\bar{\rho}_b + \rho' \frac{f'_s}{f_y} = 0.0252 + 0.0041 \times \frac{45,650}{60,000}$$

$$= 0.0283 > \rho = 0.0173 \qquad \text{O.K.}$$

$$a = \frac{5.08 \times 60,000 - 1.2 \times 45,650}{0.85 \times 5000 \times 14} = 4.20 \text{ in. (106.73 mm)}$$

$$M_n = (A_s f_y - A'_s f'_s)\left(d - \frac{a}{2}\right) + A'_s f'_s (d - d')$$

$$M_n = (5.08 \times 60,000 - 1.2 \times 45,650)\left(21.0 - \frac{4.20}{2}\right)$$

$$+ \; 1.2 \times 45,650(21.0 - 2.5) = 5,738,808 \text{ in.-lb (648.49 kN-m)}$$

Note that if the first trial value of $f'_s = 42,538$ psi from Eq. 5.16 is used, $M_n = 5,732,689$ in.-lb, which differs by less than one percent from the final M_n value. Such a low percent difference can justify using the value of f'_s obtained from Eq. 5.16 for all practical purposes without resort to additional trials.

$$M_u = \phi M_n = 0.9 \times 5,738,808 = 5,164,927 \text{ in.-lb (583.64 kN-m)}$$

5.7.2 Trial-and-Adjustment Procedure for the Design of Doubly Reinforced Sections for Flexure

1. *Midspan section:* The trial-and-adjustment procedure described in Section 5.5 is followed in order to design the section at midspan if it is a rectangular section; otherwise, follow the same procedure as that for the design of T beams and L beams (Section 5.10).

2. *Support section:* The width b and the effective depth d is already known from part 1 together with the value of the external negative factored moment M_u.

   (a) Find the strength M_{n1} of a singly reinforced section using the already established b and d dimensions of the section at midspan and a reinforcement ratio $\rho \leqslant 0.75 \bar{\rho}_b$.

   (b) From step (a), find $M_{n2} = M_n - M_{n1}$ and determine the resulting $A_{s2} = A'_s$. The total steel area at the tension side would be $A_s = A_{s1} + A'_s$.

   (c) *Alternatively,* determine how many bars are extended from the midspan to the support to give the A'_s to be used in calculating M_{n2}.

   (d) From step (c), find the value of $M_{n1} = M_n - M_{n2}$. Calculate A_{s1} for a singly reinforced beam as the first part of the solution. Then determine total $A_s = A_{s1} + A'_s$. Verify that A_{s1} does not exceed $0.75 \bar{\rho}_b$ if it is revised in the solution.

   (e) Check for the compatibility of strain in both alternatives to verify whether the compression steel yielded or not and use the corresponding stress in the steel for calculating the forces and moments.

   (f) Check for satisfactory minimum reinforcement requirements.

   (g) Select the appropriate bar sizes.

If it is necessary to design a doubly reinforced rectangular precast continuous beam, alternative method 3 (a) or 3 (b) of Section 5.5 for singly reinforced beams can be followed. An assumption is made of an R value higher than the R value that is used for singly reinforced beams for selection of the first trial section. Since it is not advisable to use an A'_s value larger than $\frac{1}{3} A_s$ or $\frac{1}{2} A_s$, assume that $R' \simeq 1.3$ to $1.5\ R$.

5.7.3 Example 5.6: Design of a Doubly Reinforced Beam for Flexure

A doubly reinforced concrete beam section has a maximum effective depth $d = 25$ in. (635 mm) and is subjected to a total factored moment $M_u = 9.4 \times 10^6$ in.-lb (1062 kN-m), including its self-weight. Design the section and select the appropriate reinforcement at the tension and the compression faces to carry the required load. Given:

$f'_c = 4000$ psi (27.58 MPa)
$f_y = 60,000$ psi (413.4 MPa)
Minimum effective cover $d' = 2.5$ in. (63.5 mm)

Solution

Assume that $b = 14$ in. $\simeq 0.55d$.

$$\overline{p}_b = \beta_1 \frac{0.85 f'_c}{f_y} \frac{87,000}{87,000 + 60,000} = 0.85\left(\frac{0.85 \times 4000}{60,000}\right)\frac{87,000}{87,000 + 60,000}$$

$$= 0.0285$$

Or obtain $0.75\overline{p}_b$ from Table 5.1. Assume a tension reinforcement ratio 0.016 ($\simeq 0.5\overline{p}_b$) for the simply reinforced part of the solution

tension reinforcement area $A_{s1} = (A_s - A'_s) = 0.016 \times 14 \times 25 = 5.6$ in.^{2}

The resisting strength of a singly reinforced section of dimensions 14 in. $\times$ 25 in. and a tensile steel area $A_{s1} = 5.6$ in.^{2} is

$$M_{n1} = 5.6 \times 60,000\left(25.0 - \frac{5.6 \times 60,000}{2 \times 0.85 \times 4000 \times 14}\right)$$

$$= 7,214,118 \text{ in.-lb (815.2 kN-m)}$$

ϕM_{n1} is less than the total factored $M_u = 9.4 \times 10^6$ in.-lb; that is, the section is too small to support the required factored moment M_u. Therefore, the section should be designed as doubly reinforced. The resisting moment corresponding to the singly reinforced part

$$M_{n1} = 7,214,118 \text{ in.-lb}$$

The moment to be resisted by the doubly reinforced part is

$$\frac{9,400,000}{0.9} - 7,214,118 = 3,230,326 \text{ in.-lb (365.0 kN-m)}$$

To verify if the compression steel A'_s has yielded, check

$$\rho - \rho' \geq \frac{0.85 f'_c \beta_1 d'}{f_y d} \frac{87,000}{87,000 - f_y}$$

$$\geq \frac{0.85 \times 4000 \times 0.85 \times 2.5}{60,000 \times 25} \frac{87,000}{87,000 - 60,000} \qquad (5.14)$$

$$\geq 0.0155$$

The actual $\rho - \rho' = 0.016 > 0.0155$; hence compression steel yielded

$$f'_s = f_y$$

Since $M_{n2} = A'_s f_y (d - d')$, $3,230,326 = A'_s \times 60,000(25.0 - 2.5)$, to give $A'_s = 2.39$ in.^{2}, corresponding to A_{s2}.

$$A_s = A_{s1} + A'_s = 5.6 + 2.39 = 7.99 \text{ in.}^2$$

Use eight No. 9 (28.6-mm-diameter) bars in *two layers* at the tension side ($A_s = 8.0$ in.^{2}) and four No. 7 (22.2-mm-diameter) bars in one layer at the compression side ($A'_s = 2.4$ in.^{2}), as shown in Fig. 5.16. Check if compression steel yielded in the final design:

$$\rho = \frac{8.0}{14 \times 25} = 0.02286 \qquad \rho' = \frac{2.4}{14 \times 25} = 0.00686$$

$$\rho - \rho' = 0.0160 < 0.75 \bar{\rho}_b \qquad \text{O.K.}$$

$$\rho - \rho' > 0.0155 \qquad \text{hence } f'_s = f_y \qquad \text{O.K.}$$

$$\text{minimum reinforcement ratio} = \frac{200}{f_y} = \frac{200}{60,000} = 0.0033 < \rho \qquad \text{O.K.}$$

$$A_s - A'_s = 8.00 - 2.4 = 5.6 \text{ in.}^2$$

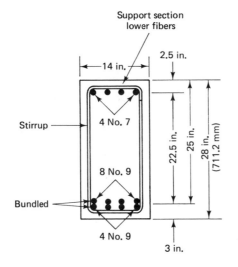

Figure 5.16 Reinforcing details of the doubly reinforced beam in Ex. 5.6.

$$\text{design moment } M_u = 0.9\left[5.6 \times 60,000\left(25.0 - \frac{5.6 \times 60,000}{2 \times 0.85 \times 4,000 \times 14}\right)\right.$$

$$\left.+ 2.4 \times 60,000(25.0 - 2.5)\right]$$

$$= 9,408,706 \text{ in.-lb} > 9,400,000 \text{ in.-lb (1063 kN-m} > 1062.0 \text{ kN-m.)}$$

Adopt the design.

Alternative Solution

Assume an R' value $\approx 1.5R \approx 1350$ (larger than $R = 900$ for singly reinforced beams of the same material properties).

$$M_n = \frac{9.4 \times 10^6}{0.9} = 10.44 \times 10^6 \text{ lb-in.}$$

$$M_n = Rbd^2 \quad \text{or} \quad 10.44 \times 10^6 = 1350bd^2$$

$$bd^2 = \frac{10.44 \times 10^6}{1350} = 7737 \text{ in.}^3$$

Assume that $b \approx \frac{1}{2}d$; hence $d^3 = 15,474$ in.^{3} and $d = 24.92$ in. Assume a trial section with $b = 14$ in., $d = 25$ in., and proceed to analyze the section in the usual manner, first choosing $\rho - \rho' \leq \bar{\rho}_b$, as given in the preceding alternative solution.

5.8 NONRECTANGULAR SECTIONS

T beams and L beams are the most commonly used flanged sections. Because slabs are cast monolithically with the beams as shown in Fig. 5.17, additional stiffness or strength is added to the rectangular beam section from participation of the slab. Based on extensive tests and longstanding engineering practice, a segment of the slab can be considered to act as a monolithic part of the beam across the beam flange. It should

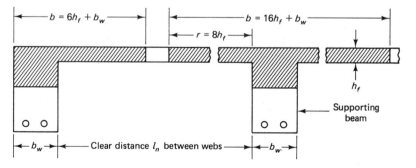

Figure 5.17   T and L beams as part of a slab beam floor system (cross section at beam midspan).

Photo 34   Structural behavior of simply supported prestressed flanged beam. (Tests by Nawy et al.)

be noted that in the case of composite sections, if the beam and slab are continuously shored during construction (supported continuously), the slab and beam can be assumed to act together in supporting all loads, including their self-weight. However, if the beam is not shored, the beam must carry its weight plus the weight of the slab while it hardens. After the slab has hardened, the two together will support the additional loads.

The flange width accepted for inclusion with the beam in forming the flanged section has to satisfy the following requirements:

T beams:

Effective overhang $\not> 8h_f$
Overhang width on each side $\not> \frac{1}{2}$ the clear distance to the face of the next web ($\not> \frac{1}{2} l_n$)
Flange width $b \not> \frac{1}{4}$ of supporting beam span $= \frac{1}{4}L$

Spandrel or edge beams (beams with a slab on one side only):

The effective overhang $\not> 6h_f$ nor $\not> \frac{1}{2}$ the clear distance to the next web ($\not> \frac{1}{2} l_n$) nor $\frac{1}{12}$ the span length of the beam.
Beams with overhang on one side are called *L beams*.

5.9 ANALYSIS OF T AND L BEAMS

Flanged beams are considered primarily for use as sections at midspans, as shown in Fig. 5.17. This is because the flange is in compression at midspan and can contribute to the moment strength of the midspan section. At the support, the flange is in tension; consequently, it is disregarded in the flexural strength computations of the support section. In other words, the support section would be an inverted doubly reinforced section having the compressive steel A_s' at the bottom fibers and tensile steel A_s at the top fibers. Figure 5.18 shows an elevation of a continuous beam with sections taken at midspan and at the supports to illustrate this discussion.

The basic principles used for the design of rectangular beams are also valid for the flanged beams. The major difference between the rectangular and flanged sections is in the calculation of compressive force C_c. Depending on the depth of the neutral axis, c, the following cases can be identified.

Case 1: Depth of Neutral Axis c Less Than Flange Thickness h_f (Fig. 5.19)

This case can be treated similar to the standard rectangular section provided that the depth a of the equivalent rectangular block is less than the flange thickness. The flange width b of the compression side *should be used* as the beam width in the analysis.

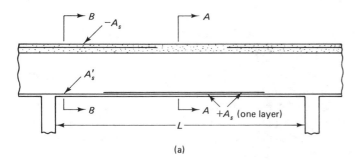

(a)

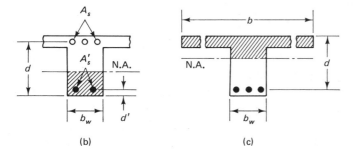

(b)   (c)

Figure 5.18 Elevation and sections of a monolithic continuous beam: (a) beam elevation; (b) support section B–B (inverted doubly reinforced beam); (c) midspan section A–A (real T-beam).

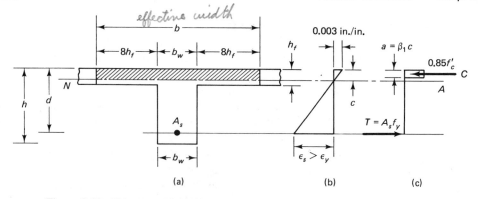

Figure 5.19 T-beam section with neutral axis within the flange ($c < h_f$): (a) cross section; (b) strains; (c) stresses.

Referring to Fig. 5.19 for force equilibrium where C is equal to T gives us

$$0.85f'_c ba = A_s f_y \qquad \text{or} \qquad a = \frac{A_s f_y}{0.85f'_c b}$$

The nominal moment strength would thus be $M_n = A_s f_y (d - a/2)$. This expression is the same as that of Eq. 5.4 for the rectangular section. Since the force contribution of concrete in the tension zone is neglected, it does not matter whether part of the flange is in the tension zone.

Case 2: Depth of Neutral Axis c Larger Than Flange Thickness h_f (Fig. 5.20)

In this case, $(c > h_f)$, the depth of the equivalent rectangular stress block a could be smaller or larger than the flange thickness h_f. If c is greater than h_f and a is less than h_f, the beam could still be considered as a rectangular beam for design purposes. Hence the design procedure explained for case 1 is applicable to this case.

If both c and a are greater than h_f, the section has to be considered as a T section. This type of T beam $(a > h_f)$ can be treated in a manner similar to that for a doubly reinforced rectangular cross section (Fig. 5.20). The contribution of the flange overhang compressive force is considered analogous to the contribution of imaginary compressive reinforcement. In Fig. 5.20, the compressive force C_n is equal to the average concrete strength f'_c multiplied by the cross-sectional area of the flange overhangs.

Thus $C_n = 2r'h_f \times 0.85f'_c = 0.85f'_c(b - b_w)h_f$, where r' is the overhang length on each side of the web. The compressive force C_n is *equated* to a tensile force T_n for equilibrium such that $T_n = (A_{sf} \times f_y)$, where A_{sf} is an imaginary compressive steel area whose force capacity is equivalent to the force capacity of the compression flange overhang. Consequently, an equivalent area A_{sf} of compression reinforcement to develop the overhang flanges would have a value

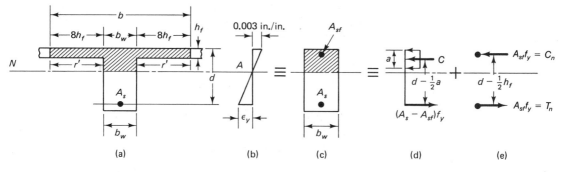

Figure 5.20 Stress and strain distribution in flanged sections design (T-beam transfer): (a) cross section; (b) strains; (c) transformed section; (d) part 1 forces; (e) part 2 forces.

$$A_{sf} = \frac{0.85f'_c(b - b_w)h_f}{f_y} \qquad (5.20)$$

For a beam to be considered as a *real* T beam, the tension force $A_s f_y$ generated by the steel should be greater than the compression force capacity of the total flange area $0.85 f'_c b h_f$. Hence

$$a = \frac{A_s f_y}{0.85f'_c b} > h_f \qquad (5.21a)$$

or

$$h_f < (1.18\overline{\omega}d = a) \qquad (5.21b)$$

where $\overline{\omega} = (A_s/bd)\,(f_y/f'_c)$.

The concrete stress block is in reality parabolic and extends to the neutral-axis depth c. Consequently, from a theoretical viewpoint, if one were using a parabolic stress block, Eq. 5.21b for a T beam can also be written as

$$h_f < \frac{1.18\overline{\omega}d}{\beta_1} \qquad (5.21c)$$

The percentage for the balanced condition in a T beam can be written as

$$\rho_b = \frac{b_w}{b}(\overline{\rho}_b + \rho_f) \qquad (5.22)$$

where $\overline{\rho}_b = \dfrac{0.85\beta_1 f'_c}{f_y}\dfrac{87,000}{87,000 + f_y}$

$\rho_f =$ reinforcement ratio for tension steel area necessary to develop the compressive strength of the overhanging flanges

$$\rho_f = 0.85f'_c(b - b_w)\frac{h_f}{f_y b_w d}$$

As in the case of singly and doubly reinforced beams, the maximum allowable percentage of the steel ρ at the tension side should not exceed 75% of the balanced steel percentage ρ_b to ensure ductile failure. Hence, in the case of a T beam,

$$\rho = \frac{A_s}{bd} \leq 0.75\rho_b \tag{5.23}$$

A strain-incompatibility check is not needed since the imaginary steel area A_{sf} is assumed to yield in all cases. To satisfy the requirement of minimum reinforcement so that the beam does not behave as nonreinforced,

$$\rho_w = \frac{A_s}{b_w d} \geq \frac{200}{f_y} \tag{5.24}$$

It is to be noted that b_w is used in Eq. 5.24 instead of width b, which is used in the case of singly or doubly reinforced beams.

As in the case of design and analysis of doubly reinforced sections, the reinforcement at the tension side is considered to be composed of two areas: A_{s1} to balance the rectangular block compressive force on area $b_w a$, and A_{s2} to balance the imaginary steel area A_{sf}. Consequently, the total nominal moment strength for parts 1 and 2 of the solution is

$$M_n = M_{n1} + M_{n2} \tag{5.25a}$$

$$M_{n1} = A_{s1}f_y\left(d - \frac{a}{2}\right) = (A_s - A_{sf})f_y\left(d - \frac{a}{2}\right) \tag{5.25b}$$

$$M_{n2} = A_{s2}f_y\left(d - \frac{h_f}{2}\right) = A_{sf}f_y\left(d - \frac{h_f}{2}\right) \tag{5.25c}$$

The design moment strength ϕM_n, which has to be at least equal to the external factored moment M_u, becomes

$$M_u = \phi M_n = \phi\left[(A_s - A_{sf})f_y\left(d - \frac{a}{2}\right) + A_{sf}f_y\left(d - \frac{h_f}{2}\right)\right] \tag{5.26}$$

The flowchart, Fig. 5.21, presents the sequence of calculations for the analysis of the T beam. The following analysis example illustrates the nominal moment strength calculations for a typical precast T beam.

5.9.1 Example 5.7: Analysis of a T Beam for Moment Capacity

Calculate the nominal moment strength and the design ultimate moment of the precast T beam shown in Fig. 5.22 if the beam span is 30 ft (9.14 m). Given:

$f'_c = 4000$ psi (27.58 MPa), normalweight concrete
$f_y = 60,000$ psi (413.4 MPa)

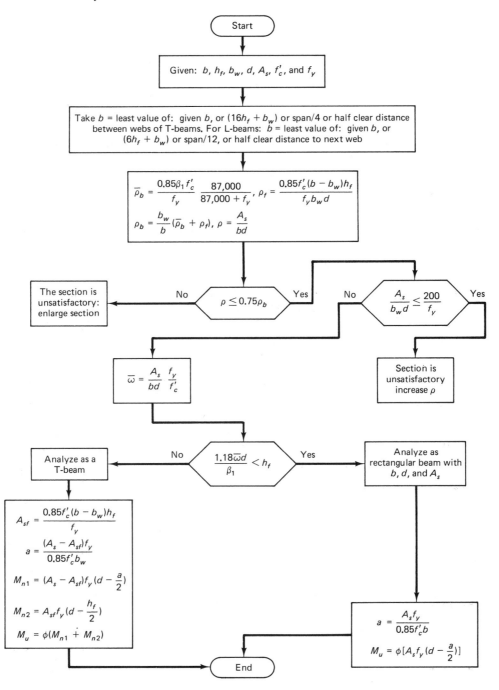

Figure 5.21   Flowchart for the analysis of T and L beams.

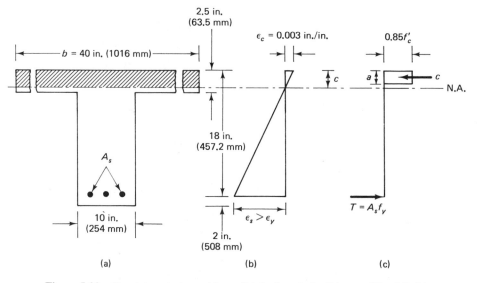

Figure 5.22 Geometry, strain, and force distributions in the T-beam of Ex. 5.7: (a) cross section; (b) strains; (c) stresses.

Reinforcement area at the tension side:
(a) $A_s = 4.0$ in.^{2} (2580 mm^2)
(b) $A_s = 6.0$ in.^{2} (3870 mm^2)

Solution

Flange width check is not necessary for a precast beam since the precast section can act independently depending on the construction system.

Check for ρ_{max}:

$$\rho_{max} \leq 0.75\rho_b$$

$$\bar{\rho}_b = \frac{0.85\beta_1 f_c'}{f_y} \frac{87,000}{87,000 + f_y}$$

$$= \frac{0.85 \times 0.85 \times 4000}{60,000} \frac{87,000}{87,000 + 60,000} = 0.029$$

$$\rho_f = \frac{0.85 f_c'(b - b_w)h_f}{f_y b_w d}$$

$$= \frac{0.85 \times 4000(40 - 10) \times 2.5}{60,000 \times 10 \times 18} = 0.024$$

$$\rho_b = \frac{b_w}{b}(\bar{\rho}_b + \rho_f) = \frac{10}{40}(0.029 + 0.024) = 0.013$$

$$\rho_{max} = 0.75\rho_b = 0.010$$

$$\rho_{min} = \frac{200}{60,000} = 0.0033$$

(a) $A_s = 4$ in.^{2}

$$\rho_w = \frac{A_s}{b_w d} = \frac{4.0}{10 \times 18} = 0.0222 > \rho_{min} = 0.0033 \qquad \text{O.K.}$$

$$\rho = \frac{A_s}{bd} = \frac{4.0}{40 \times 18} = 0.006 < \rho_{max} \qquad \text{O.K.}$$

Check whether the section will act as a T beam.

$$\overline{\omega} = \frac{A_s\, f_y}{bd\, f_c'} = \frac{4.0}{40 \times 18}\left(\frac{60}{4}\right) = 0.083$$

$$c = \frac{1.18\overline{\omega}d}{\beta_1} = \frac{1.18 \times 0.083 \times 18}{0.85} = 2.10 \text{ in.} < (h_f = 2.5 \text{ in.})$$

Therefore, the beam can be analyzed as a rectangular beam using b, d, and A_s.

$$M_n = A_s f_y\left(d - \frac{a}{2}\right)$$

$$a = \frac{4.0 \times 60{,}000}{0.85 \times 4000 \times 40} = 1.765 \text{ in.}$$

$$M_n = 4.0 \times 60{,}000\left(18 - \frac{1.765}{2}\right) = 4{,}108{,}200 \text{ in.-lb}$$

$$M_u = \phi M_n = 0.9 \times 4{,}108{,}200 \text{ in.} = 3{,}697{,}380 \text{ in.-lb (417.8 kN-m)}$$

(b) $A_s = 6.0$ in.^{2}.

$$\rho_w = \frac{A_s}{b_w d} = \frac{6.0}{10 \times 18} = 0.033 > \rho_{min} = 0.0033 \qquad \text{O.K.}$$

$$\rho = \frac{A_s}{bd} = \frac{6.0}{40 \times 18} = 0.0083 < \rho_{max} = 0.010 \qquad \text{O.K.}$$

$$\overline{\omega} = \frac{6.0}{40 \times 18}\left(\frac{60}{4}\right) = 0.0125$$

$$\frac{1.18\overline{\omega}d}{\beta_1} = \frac{1.18 \times 0.0125 \times 18}{0.85} = 3.124 > (h_f = 2.5)$$

Therefore, the neutral axis is *below* the flange. The beam has to be treated as a T beam or equivalent doubly reinforced rectangular beam with imaginary compression steel area A_{sf}.

$$A_{sf} = \frac{0.85 f_c'(b - b_w)h_f}{f_y}$$

$$= \frac{0.85 \times 4000(40 - 10) \times 2.5}{60{,}000} = 4.25 \text{ in.}^2$$

$$a = \frac{(A_s - A_{sf})f_y}{0.85f_c'b_w}$$

$$a = \frac{(6.00 - 4.25) \times 60,000}{0.85 \times 4000 \times 10}$$

$$= 3.09 \text{ in.}$$

$$M_{n1} = (A_s - A_{sf})f_y\left(d - \frac{a}{2}\right)$$

$$= (6.00 - 4.25) \times 60,000\left(18 - \frac{3.09}{2}\right)$$

$$= 1,727,780 \text{ in.-lb}$$

$$M_{n2} = A_{sf}f_y\left(d - \frac{h_f}{2}\right) = 4.25 \times 60,000\left(18 - \frac{2.5}{2}\right) = 4,271,250 \text{ in.-lb}$$

$$M_n = M_{n1} + M_{n2} = (1,727,780 + 4,271,150) = 6,000,000 \text{ in.-lb}$$

$$= 6000.00 \text{ in.-kips}$$

$$M_u = \phi M_n = 0.9 \times 6,000,000 = 5,400,000 \text{ in.-lb } (610.2 \text{ kN-m})$$

5.10 TRIAL-AND-ADJUSTMENT PROCEDURE FOR THE DESIGN OF FLANGED SECTIONS

The slab thickness h_f of the flange overhang is known at the outset since the slab is designed first. Also available is the external factored moment M_u at midspan. The trial-and-adjustment steps for proportioning the web of the beam section can be summarized as follows.

1. Choose a singly reinforced beam section that can resist the external factored moment M_u and the moment due to self-weight. Remember that a T section or an L section would have a smaller size or depth than a singly reinforced section.
2. Check whether the span/depth ratio is reasonable, namely, between 12 and 18. If not, adjust the preliminary section.
3. Calculate the flange width on the basis of the criteria in Section 5.8.
4. Determine if the neutral axis is within or outside the flange, where the neutral-axis depth $c = 1.18\bar{\omega}d/\beta_1$ for rectangular singly reinforced sections.
    (a) If $c < h_f$, the beam has to be treated as a singly reinforced beam with a width b equivalent to the flange width determined in step 2.
    (b) If $c > h_f$ and the equivalent block depth a is also $> h_f$, design as a T beam or an L beam, as the case may be.
5. Find the equivalent compressive steel area A_{sf} for the flange overhang and analyze the assumed section as in Ex. 5.7(b). Calculate the nominal resisting capacities M_{n1} and M_{n2}.

6. Repeat steps 4 and 5 until the calculated $\phi M_n = \phi(M_{n1} + M_{n2})$ is close in value to the factored moment M_u and verify that the assumed self-weight of the web is correct.

7. *Alternatively*, the first trial section can be chosen using a moment factor $R'' > R$ in step 3(a) of Section 5.5 for singly reinforced beams such that $R'' \simeq 1.35 - 1.50R$. Select a trial section depth from $M_n = R''bd^2$ and proceed to analyze the section.

5.10.1 *Example 5.8:* Design of an End-Span L Beam

A roof-garden floor is composed of a monolithic one-way slab system on beams as in Fig. 5.23. The effective beam span is 35 ft (10.67 m) and all beams are spaced at 7 ft 6 in. (2.29 m) center to center. The floor supports a 6 ft 4 in. (1.52 m) depth of soil in addition

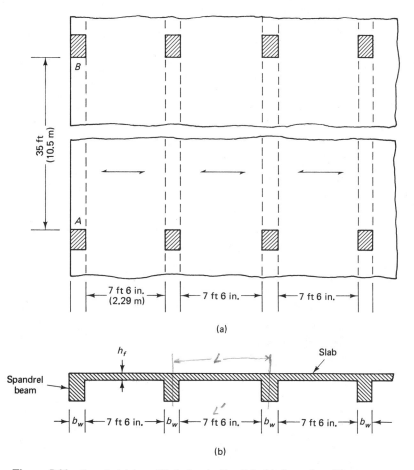

Figure 5.23 Spandrel beam *AB* design in Ex. 5.8: (a) floor plan; (b) transverse section.

to its self-weight. Assume also that the slab edges support a 12-in.-wide, 7-ft wall weighing 840 lb per linear foot. Design the midspan section of the edge spandrel L-beam *AB* assuming that the moist soil weighs 125 lb/ft³ (2.56 tons/m³). Given:

$f'_c = 3000$ psi (20.68 MPa), normal-weight concrete
$f_y = 60,000$ psi (413.7 MPa)

Solution

Slab design

The weight of soil $= 6.33 \times 125 = 791$, say 800 psf (38.32 kPa). Assume a slab thickness $h = 4$ in. (101.6 mm) $= 4/12 \times 150 = 50$ psf. $d = h - (\frac{3}{4}$ in. cover $+ \frac{1}{2}$ diameter of No. 4 bars) $= 4.0 - 1.0 = 3.0$ in.

factored load intensity $w_u = 1.4(800 + 50) = 1190$ lb/ft² (57.0 kPa)

From the ACI code, the negative moment for the first interior support of a continuous slab is

$$-M_u = \frac{w_u l_n^2}{12} = \frac{1190(7.5)^2}{12} \times 12 = 66,938 \text{ in.-lb}$$

The required slab negative moment strength

$$-M_n = \frac{66,938}{0.9} = 74,376 \text{ in.-lb}$$

$$M_n = A_s f_y \left(d - \frac{a}{2} \right)$$

Assume that $(d - a/2) \simeq 0.9d = 0.9 \times 3.0 = 2.7$ in.

$$A_s = \frac{74,376}{60,000 \times 2.7} = 0.459 \text{ in.}^2 \text{ on a 12-in. strip}$$

Try $b = 12$ in., $d = 3$ in., and No. 5 bars at 7.5 in. center to center. $A_s = 0.496$ in².

$$a = \frac{A_s f_y}{0.85 f'_c b} = \frac{0.496 \times 60,000}{0.85 \times 3,000 \times 12} = 0.97 \text{ in.}$$

Nominal resisting moment $M_n = 0.496 \times 60,000 \left(3.0 - \frac{0.97}{2} \right) = 74,846$ in.-lb

$> $ required $M_n = 74,376$ in.-lb   O.K.

$$\rho = \frac{0.496}{12 \times 3} = 0.0138$$

$$\bar{\rho}_b = \frac{0.85\beta_1 f'_c}{f_y} \left(\frac{87,000}{87,000 + f_y} \right) = \frac{0.85 \times 0.85 \times 3000}{60,000} \left(\frac{87,000}{87,000 + f_y} \right) = 0.0214$$

maximum allowable $\rho = 0.75\bar{\rho}_b = 0.016 > 0.0138$ O.K.

Similarly, for the positive moment, $+M_u = w_u l_n^2/16$, requiring No. 5 bars at 10 in. c-c. Use No 5 bars at $7\frac{1}{2}$ in. center to center main negative reinforcement (15.9 mm diameter at 190.5-mm spacing) and No. 5 bars at 10 in. center to center for main positive reinforcement.

$$\text{temperature steel} = 0.0018bh = 0.0018 \times 12 \times 4.0 = 0.0864 \text{ in.}^2$$

$$\text{maximum allowable spacing} = 3h = 3 \times 4 = 12 \text{ in. (304.8 mm)}$$

Use No. 3 bars at 12 in. $= 0.11$ in.^{2} (25.4 mm diameter) for temperature.

Beam web design

In order to choose a trial web section, assume that $d = l_n/18$ for deflection or

$$d = \frac{35.0 \times 12}{18} = 23.33 \text{ in.}$$

Assume that $h = 26$ in. (660.4 mm), $d = 22.5$ in. (571.6 mm), and $b_w = 14$ in. (355.6 mm).

$$\text{load area on L beam } AB = \frac{7.5}{2} + \frac{14}{12} = 4.92 \text{ ft}$$

$$\text{superimposed working } w_w = (4.92 - 1.0) \times 800 = 3136 \text{ lb/ft}$$

$$\text{slab weight} = \frac{4.0}{12} \times 150 \times 4.92 = 246 \text{ lb/ft}$$

$$\text{weight of beam web} = \frac{14(26 - 4)}{144} \times 150 = 321 \text{ lb/ft}$$

$$\text{7-ft wall weight} = 840 \text{ lb/ft}$$

$$\text{total service load} = 3136 + 246 + 321 + 840 = 4543 \text{ lb/ft}$$

$$\text{factored load } w_u = 1.4 \times 4543 = 6360 \text{ lb/ft}$$

$$\text{factored external moment } M_u = \frac{w_u l_n^2}{11} = \frac{6360(35.0)^2}{11} \times 12 = 8{,}499{,}273 \text{ in.-lb}$$

$$M_n = \frac{M_u}{\phi} = \frac{8{,}499{,}273}{0.9} = 9{,}443{,}637 \text{ in.-lb (1067.06 kN-m)}$$

To determine whether the beam is an actual L beam or not, it is necessary to find if the neutral axis falls outside the flange. Consequently, the area of the tension steel A_s at midspan has to be assumed. If a rectangular section is initially assumed with an appropriate moment arm $jd \simeq 0.85d = 0.85 \times 22.5 = 19.3$ in.,

$$M_n = A_s f_y jd \quad \text{or} \quad 9{,}443{,}637 = A_s \times 60{,}000 \times 19.3$$

$$A_s = \frac{9{,}443{,}637}{60{,}000 \times 19.3} = 8.16 \text{ in.}^2$$

Assume eight No. 9 bars in two layers $= 8.0$ in.^{2} (5160 mm^2).

$$\rho = \frac{A_s}{bd} \qquad \text{where } b = b_w + 6h_f = 14 + 6 \times 4.0 = 38 \text{ in. (965.2 mm)}$$

$$\rho = \frac{8.0}{38 \times 22.5} = 0.00936$$

$$\overline{\omega} = \rho \frac{f_y}{f_c'} = 0.00936 \times \frac{60,000}{3000} = 0.187$$

$$\text{depth of neutral axis } c = \frac{1.18\overline{\omega}d}{\beta_1} = \frac{1.18 \times 0.187 \times 22.5}{0.85}$$

or

$$c = 5.84 \text{ in. } > 4.0 \text{ in.}$$

$$a = \beta_1 c = 0.85 \times 5.84 = 4.96 > 4.0 \text{ in.}$$

Hence the section is an L beam since the neutral axis is below the flange, as shown in Fig. 5.24.

For $f_c' = 3000$ psi and $f_y = 60,000$ psi, $\overline{\rho}_b = 0.0214$ and

$$A_{sf} = \frac{h_f(b - b_w)0.85f_c'}{f_y} = \frac{4.0(38 - 14) \times 0.85 \times 3000}{60,000}$$

$$= 4.08 \text{ in.}^2 \text{ (2631.6 mm}^2\text{)}$$

$$\rho_f = \frac{A_{sf}}{b_w d} = \frac{4.08}{14 \times 22} = 0.01295$$

$$\rho_b = (\overline{\rho}_b + \rho_f)\frac{b_w}{b} = (0.0214 + 0.01295)\frac{14}{38} = 0.01266$$

$$0.75\rho_b = 0.00949 > \text{actual } \rho = 0.00936$$

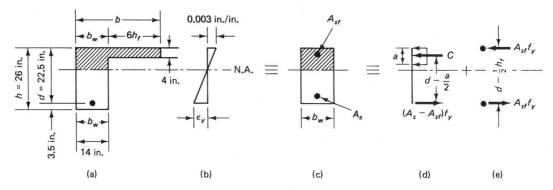

Figure 5.24 Forces and stresses in L-beams: (a) cross section; (b) strain diagram; (c) transformed section; (d) part 1 forces; (e) part 2 forces.

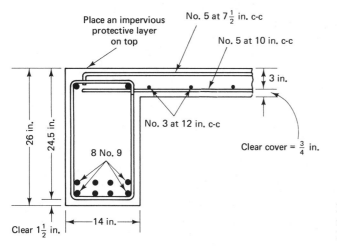

Place an impervious protective layer on top

No. 5 at $7\frac{1}{2}$ in. c-c

No. 5 at 10 in. c-c

3 in.

No. 3 at 12 in. c-c

Clear cover = $\frac{3}{4}$ in.

8 No. 9

26 in.

24.5 in.

14 in.

Clear $1\frac{1}{2}$ in.

Figure 5.25 Midspan section flexural reinforcement details for beam *AB* of Ex. 5.8.

Hence the section is under-reinforced and satisfies the ACI code requirements.

$$a = \frac{(A_s - A_{sf})f_y}{0.85f'_c b_w} = \frac{(8.0 - 4.08)60,000}{0.85 \times 3000 \times 14} = 6.59 \text{ in. } (167.4 \text{ mm})$$

$$M_n = (A_s - A_{sf})f_y\left(d - \frac{a}{2}\right) + A_{sf}f_y\left(d - \frac{1}{2}h_f\right)$$

$$= (8.0 - 4.08)60,000\left(22.5 - \frac{6.59}{2}\right) + 4.08 \times 60,000\left(22.5 - \frac{4.0}{2}\right)$$

$$= 9,535,416 \text{ in.-lb}$$

design moment $M_u = 0.9 \times 9,535,416 \text{ in.-lb} = 8,581,874 \text{ in.-lb}$

actual factored $M_u = 8,499,273 \text{ in.-lb} < 8,581,874 \text{ in.-lb}$

Adopt the design. Flexural reinforcement details for the L-beam *AB* are shown in Fig. 5.25.

5.10.2 Example 5.9: Design of an Interior Continuous Floor Beam for Flexure

Design a rectangular interior beam having a clear span of 25 ft (7.62 m) and carrying a working live load of 8000 lb per linear foot (35.58 kN) in addition to its self-weight, as shown in Fig. 5.26. Assume the beam to have a 4-in. (101.6-mm) slab cast monolithically with it. Given:

$f'_c = 4000$ psi (27.58 MPa), normal-weight concrete
$f_y = 60,000$ psi (413.4 MPa)

Assume no wind or earthquake.

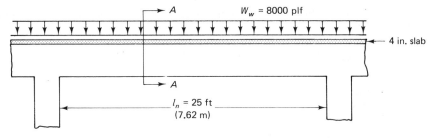

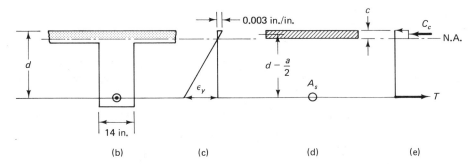

Figure 5.26 Continuous beam midspan section in Ex. 5.9: (a) beam elevation; (b) section A–A; (c) strain distribution; (d) N.A. inside flange; (e) force and stress.

Solution

Assume that the web self-weight = 400 lb/ft

$$\text{factored load } U = 1.4 \times 400 + 1.7 \times 8000$$

$$= 14{,}160 \text{ lb/ft}$$

The positive factored moment M_u for interior midspan lower fibers (ACI) is

$$+M_u = \frac{w_u l_n^{\,2}}{16} = \frac{14{,}160 \times (25.0)^2}{16} \times 12 = 6{,}637{,}500 \text{ in.-lb. (750.04 kN-m)}$$

The negative factored moment M_u at support (tension at top fibers) is

$$-M_u = \frac{14{,}160(25.0)^2}{11} \times 12 = 9{,}654{,}546 \text{ in.-lb (1091.0 kN-m)}$$

Section at midspan (T beam)

Assume that $b_w = 14$ in. (0.3556 m).

$$b \not> 16 \times 4 + 14 = 78 \text{ in.}$$

$$\not> \frac{25 \times 12}{4} = 75 \text{ in.}$$

$$\not> \text{ half clear distance to next web not known}$$

Therefore, $b = 75$ in. (1.905 m) controls.

If the depth a of the stress block is assumed equivalent to the flange thickness $h_f = 4$ in., the compressive force C_n (volume of the compressive block) is

$$C_n = 0.85f_c'bh_f = 0.85 \times 4000 \times 75 \times 4.0 = 1,020,000 \text{ lb}$$

$$\text{required positive} + M_n = \frac{6,637,500}{\phi = 0.90} = 7,375,000 \text{ in.-lb}$$

In order to obtain on first trial a reasonable area A_s of steel at the tension side and also determine if the beam section is flanged, find for a first trial an A_s for a rectangular section that can resist $M_n = 7,375,000$ in.-lb.

For deflection purposes assume that

$$d = \frac{l}{12} = \frac{25.0 \times 12}{12} = 25 \text{ in. (635.0 mm)}$$

For a self-weight of 400 lb/ft, try $b_w = 14$ in. (355.6 mm) and $h = 28.0$ in. (711.2 mm).

$$\text{self-weight} = \frac{14 \times 28}{144} \times 150 = 408 \text{ lb/ft}$$

$$\text{revised factored } U = 1.4 \times 408 + 1.7 \times 8000 = 14,171 \text{ lb/ft}$$

$$\text{revised} + M_n = 7,375,500 \times \frac{14,171}{14,160} = 7,381,230 \text{ in.-lb}$$

$$\text{revised} - M_n = \frac{9,654,546}{0.9} \times \frac{14,171}{14,160} = 10,736,607 \text{ in.-lb}$$

Assume that moment arm $jd \simeq 0.9d = 0.9 \times 25 = 22.5$ in.

$$A_s = \frac{M_n}{f_y \times 0.9d} = \frac{7,381,270}{60,000 \times 22.5} = 5.47 \text{ in.}^2$$

For a T beam, a smaller A_s is needed. Try four No. 10 bars $= 5.08$ in.^{2}

$$T_n = 5.08 \times 60,000 = 304,800 \text{ lb} \ll 1,020,000 \text{ lb}$$

The neutral axis has to be well within the flange in order to balance $T_n = C_n$. Hence treat the T beam as a singly reinforced section having a compression flange width $b = 75$ in. and

$$\rho = \frac{5.08}{b \times d} = \frac{5.08}{75 \times 25} = 0.0027$$

Another check for the neutral-axis position can be accomplished using the following expression, where

$$\bar{\omega} = \rho\frac{f_y}{f_c'} = 0.0027 \times \frac{60,000}{4000} = 0.0406$$

$$c = \frac{1.18\bar{\omega}d}{\beta_1} = 1.18 \times 0.0406 \times \frac{25}{0.85} = 1.41 \text{ in.} \ll h_f = 4.0 \text{ in.}$$

The nominal resisting moment $M_n = A_sf_y(d - a/2)$ or

$$M_n = 5.08 \times 60,000(25 - \tfrac{1}{2} \times 0.85 \times 1.41) = 7,437,349 \text{ in.-lb (840.4 kN-m)}$$

The actual M_n is larger than the required $M_n = 7,381,230$ in.-lb.

$$\frac{A_s}{b_w d} = \frac{5.08}{14 \times 25} = 0.0145 > \rho_{min} = \frac{200}{f_y} = 0.0033 \quad \text{O.K.}$$

Adopt a midspan section with $b_w = 14$ in. (355.6 mm), $h = 28$ in. (685.8 mm), $d = 25.0$ in. (635.0 mm), and $A_s =$ four No. 10 bars (diameter 32.25 mm).

Section at support (doubly reinforced rectangular section)

This section is subjected to moments similar to the moments acting on the section in Ex. 5.6 and has the same cross-sectional dimensions. The required nominal moment of resistance $M_n = 10,736,607$ in.-lb (1212.1 kN-m). Assume that two No. 10 bars extend from the midspan to the support $= 2.54$ in.^{2}. $M_{n2} = A'_s f_y (d - d')$, assuming that A'_s has yielded since the area is so close to 2.4 in.^{2} in Ex. 5.6, to be subsequently verified, or

$$M_{n2} = 2.54 \times 60,000(25 - 2.5) = 3,429,000 \text{ in.-lb}$$

$$M_{n1} = 10,736,607 - 3,429,000 = 7,307,607 \text{ in.-lb}$$

Assume that moment arm $jd \simeq 0.85d = 25 \times 0.85 = 21.5$ in.

$$\text{trial } A_{s1} = \frac{7,307,607}{60,000 \times 21.5} = 5.73 \text{ in.}^2$$

$$\text{total } A_s = A_{s1} + A_{s2} = 5.73 + 2.54 = 8.27 \text{ in.}^2$$

Try seven No. 10 bars (50.0 mm) in *two* layers, $A_s = 8.89$ in.^{2}.

$$A_{s1} = 8.89 - 2.54 = 6.35 \text{ in.}^2$$

$$\rho - \rho' = \frac{6.35}{14 \times 25} = 0.0181 > 0.0155$$

from Ex. 5.6, hence the assumption that A'_s yielded is valid.

$$a = \frac{(A_s - A'_s)f_y}{0.85f'_c b} = \frac{6.35 \times 60,000}{0.85 \times 4000 \times 14} = 8.00 \text{ in.}$$

$$M_{n1} = (A_s - A'_s)f_y \left(d - \frac{a}{2}\right) = 6.35 \times 60,000 \left(25.0 - \frac{8.00}{2}\right)$$

$$= 8,000,200 \text{ in.-lb}$$

$$\text{available } M_n = M_{n1} + M_{n2} = 8,000,200 + 3,429,000$$

$$= 11,429,200 \text{ in.-lb} \quad (1291.5 \text{ kN-m})$$

The available nominal moment strength M_n is larger than the required $M_n = 10,736,607$ in.-lb (1215.4 kN-m). Hence the design is adequate. Therefore, use seven No. 10 bars on top at the support in *two* layers and two No. 10 bars at the bottom fibers of the section at the support. Provide closed stirrups to tie the tension and the compression steel at the support. It is to be noted that bar sizes larger than No. 11 should be avoided where possible in superstructure beams, as they are difficult to cut and are less efficient for crack control.

For the design to be complete, a check of diagonal tension capacity and serviceability and bar development checks have to be made, as discussed in Chapters 6, 8, and 10. Details of the reinforcement over the span are shown in Fig. 5.27.

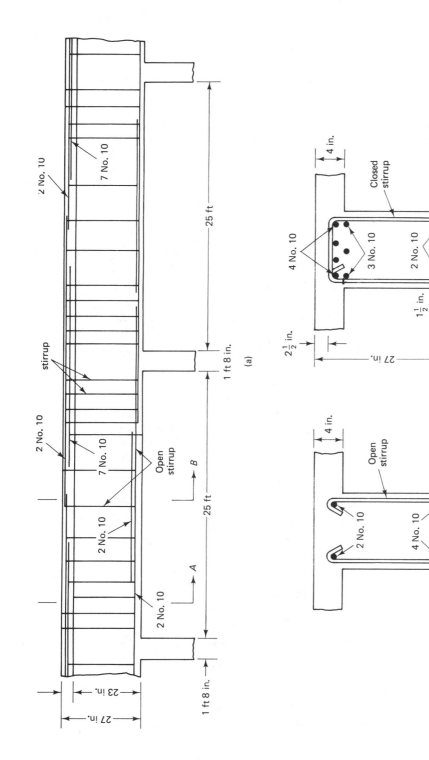

Figure 5.27 Reinforcement arrangement for the continuous beam in Ex. 5.9: (a) sectional elevation (not to scale); (b) midspan section *B–B* (c) support section *A–A*.

135

SELECTED REFERENCES

5.1 Whitney, C. S., "Plastic Theory of Reinforced Concrete Design," *Transactions of the ASCE,* Vol. 107, 1942, pp. 251–326.

5.2 Hognestad, E. N., Hanson, N. W., and McHenry, D., "Concrete Stress Distribution in Ultimate Strength Design," *Journal of the American Concrete Institute,* Proc. Vol. 52, December 1955, pp. 455–479.

5.3 Whitney, C. S., and Cohn, E., "Guide for Ultimate Strength Design of Reinforced Concrete," *Journal of the American Concrete Institute,* Proc. Vol. 53, November 1956, pp. 455–475.

5.4 Mattock, A. H., Kriz, L. B., and Hognestad, E. N., "Rectangular Stress Distribution in Ultimate Strength Design", *Journal of the American Concrete Institute,* Proc. Vol. 58, February 1961, pp. 825–928.

5.5 ACI-ASCE Joint Committee: "Report of ASCE-ACI Joint Committee on Ultimate Strength Design", *ASCE, Proceedings,* Vol. 81, October, 1955, 68 pp. See also *Journal of the American Concrete Institute,* Proc. Vol. 52, January 1956, pp. 502–524.

5.6 Balaguru, P., "Cost Optimum Design of Singly Reinforced Sections," *Journal of Civil Engineering Design,* Vol. 2, No. 2, 1981, pp. 149–169.

5.7 Concrete Reinforcing Steel Institute, *CRSI Handbook,* CRSI, Chicago, 1975, 928 pp.

5.8 ACI Committee 340, *Design Handbook,* Vol. 1, Special Publication No. 17, American Concrete Institute, Detroit, 1978, 508 pp.

5.9 Nawy, E. G., "Strength, Serviceability and Ductility," Chapter 12 in *Handbook of Structural Concrete,* Pitman Books, London/McGraw-Hill, New York, 1983, 1968 pp.

PROBLEMS FOR SOLUTION

5.1 For the beam cross section shown in Fig. 5.28, determine whether the failure of the beam will be initiated by crushing of concrete or yielding of steel. Given:

$$f'_c = 4000 \text{ psi (27.58 MPa) for case (a), } A_s = 10 \text{ in.}^2$$
$$f'_c = 7000 \text{ psi (48.26 MPa) for case (b), } A_s = 5 \text{ in.}^2$$
$$f_y = 60,000 \text{ psi (413.7 MPa)}$$

Also determine whether the section satisfies ACI code requirements.

5.2 Calculate the nominal moment strength of the beam sections shown in Fig. 5.29. Given:

$$f'_c = 3000 \text{ psi (20.68 MPa) for case (a)}$$
$$f'_c = 6000 \text{ psi (41.36 MPa) for case (b)}$$

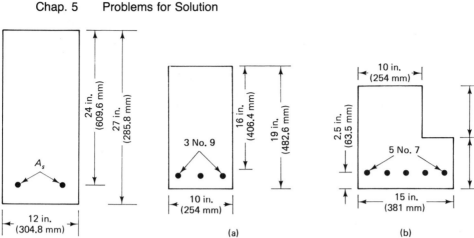

Figure 5.28   **Figure 5.29**

$f'_c = 9000$ psi (62.10 MPa) for case (c)
$f_y = 60,000$ psi (413.7 MPa)

5.3   Calculate the safe load that the beam shown in Fig. 5.30 can carry. Given:

$f'_c = 4000$ psi (27.58 MPa), normalweight concrete
$f_y = 60,000$ psi (413.7 MPa)

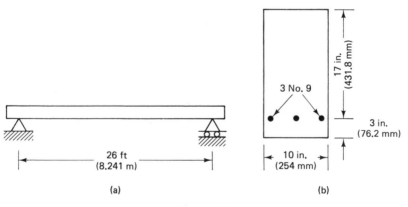

Figure 5.30

5.4   Design a one-way slab to carry a live load of 100 psf and an external dead load of 50 psf. The slab is simply supported over a span of 12 ft. Given:

$f'_c = 4000$ psi (27.58 MPa), normalweight concrete
$f_y = 60,000$ psi (413.7 MPa)

5.5   Design the simply supported beams shown in Fig. 5.31. Given:

$f'_c = 5000$ psi (34.47 MPa), normalweight concrete
$f_y = 60,000$ psi (413.7 MPa)
$\rho = 0.5\rho_b$

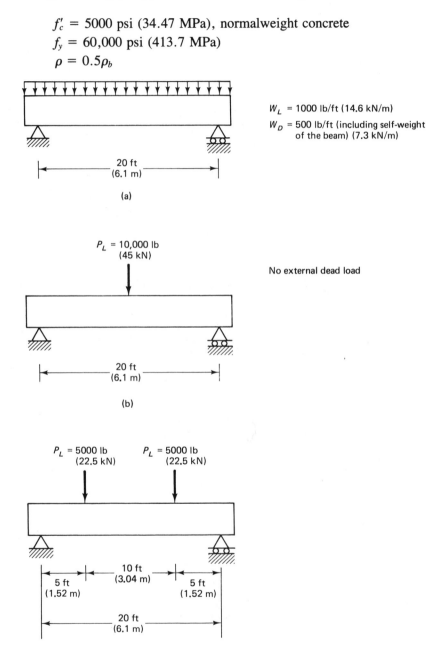

$W_L = 1000$ lb/ft (14.6 kN/m)

$W_D = 500$ lb/ft (including self-weight of the beam) (7.3 kN/m)

20 ft
(6.1 m)

(a)

$P_L = 10,000$ lb
(45 kN)

No external dead load

20 ft
(6.1 m)

(b)

$P_L = 5000$ lb
(22.5 kN)

$P_L = 5000$ lb
(22.5 kN)

10 ft
(3.04 m)

5 ft
(1.52 m)

5 ft
(1.52 m)

20 ft
(6.1 m)

(c)

Figure 5.31

5.6   Check whether the sections shown in Fig. 5.32 satisfy ACI (318-83) code requirements for maximum and minimum reinforcement. Given:

f'_c = 5000 psi (34.48 MPa), normalweight concrete
f_y = 60,000 psi (413.7 MPa)

The compression fibers in all the figures are the top fibers of the sections.

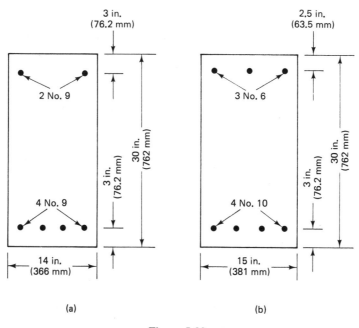

(a)   (b)

Figure 5.32

5.7   Calculate the stresses in the compression steel, f'_s, for the cross sections shown in Fig. 5.33. Given:

f'_c = 6000 psi (41.37 MPa), normalweight concrete
f_y = 60,000 psi (413.7 MPa)

5.8   Calculate the ultimate moment capacity of the beam sections of Problem 5.2. Assume two No. 6 bars for compression reinforcement.

5.9   Solve Problem 5.3 if two No. 6 bars are added as compression reinforcement.

5.10   Solve Problem 5.5 using an additional constraint on the thickness h of the beam, which should not exceed 20 in. Double the load for Fig. 5.31(a) and (b) and triple the load for part (c).

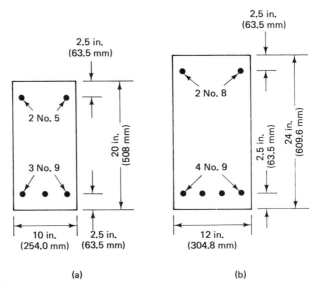

Figure 5.33

5.11 At failure, determine whether the precast sections shown in Fig. 5.34 will act similarly to rectangular sections or as flanged sections. Given:

$$f'_c = 4000 \text{ psi } (27.58 \text{ MPa}), \text{ normalweight concrete}$$
$$f_y = 60,000 \text{ psi } (413.7 \text{ MPa})$$

5.12 Check whether the sections of Problem 5.11 satisfy ACI (318-83) code requirements.

5.13 Calculate the nominal moment strength of the sections shown for Problem 5.11.

5.14 Repeat Problem 5.5 using a T section instead of a rectangular section. Use a flange thickness of 3 in. (76.2 mm) and a flange width of 30 in. (762 mm).

5.15 Using the details of Problem 5.4, design a reinforced concrete T beam for the slab floor system shown in Fig. 5.35. The floor area is 30 ft × 60 ft (9.14 m × 18.29 m) with an effective T-beam span of 30 ft (9.14 m).

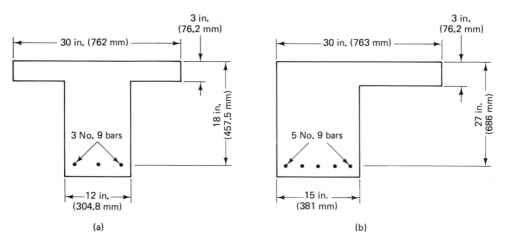

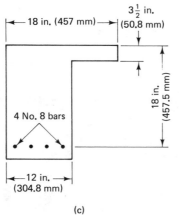

Figure 5.34

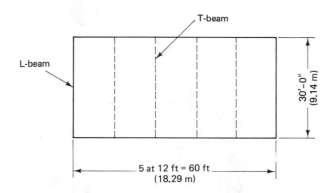

Figure 5.35 Plan of one-way-slab floor system.

6

Shear & Diagonal Tension in Beams

6.1 INTRODUCTION

This chapter presents procedures for the analysis and design of reinforced concrete sections to resist the shear forces resulting from externally applied loads. Since the strength of concrete in tension is considerably lower than its strength in compression, design for shear becomes of major importance in concrete structures.

The behavior of reinforced concrete beams at failure in shear is distinctly different from their behavior in flexure. They fail abruptly without sufficient advanced warning and the diagonal cracks that develop are considerably wider than the flexural cracks. The accompanying photographs show typical beam shear failure in diagonal tension as discussed in the subsequent sections. Because of the brittle nature of such failures, the designer has to design sections that are adequately strong to resist the external factored shear loads without reaching their shear strength capacity. Shear is also a significant parameter in the behavior of brackets, corbels, and deep beams. Consequently, the design of these elements is also discussed in detail.

6.2 BEHAVIOR OF HOMOGENEOUS BEAMS

Consider the two infinitesimal elements A_1 and A_2 of a rectangular beam in Fig. 6.1a made of homogeneous, isotropic, and linearly elastic material. Figure 6.1b shows the bending stress and shear stress distributions across the depth of the section. The tensile

Photo 36   Typical diagonal tension failure at rupture load level. (Test by Nawy et al.)

Photo 37   Simply supported beam prior to developing diagonal tension crack (load stage 11). (Test by Nawy et al.)

Photo 38   Principal diagonal tension crack at failure of beam in the preceding photograph (load stage 12).

normal stress f_t and the shear stress v are the values in element A_1 across plane a_1–a_1 at a distance y from the neutral axis. From the principles of classical mechanics, the normal stress f and the shear stress v for element A_1 can be written as

$$f = \frac{My}{I} \tag{6.1}$$

and

$$v = \frac{VA\overline{y}}{Ib} \tag{6.2}$$

where M and V = bending moment and shear force at section a_1–a_1
    A = cross-sectional area of the section at the plane passing through the centroid of element A_1
    y = distance from the element to the neutral axis
    $\overline{y}$ = distance from the centroid of A to the neutral axis
    I = moment of inertia of the cross section
    b = width of the beam

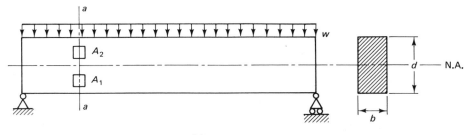

(a)

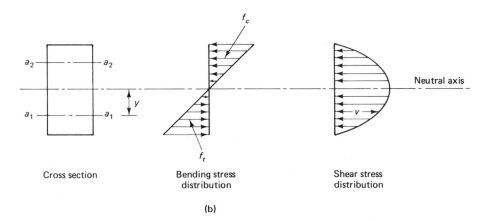

| Cross section | Bending stress distribution | Shear stress distribution |

(b)

Figure 6.1   Stress distribution for a typical homogeneous rectangular beam.

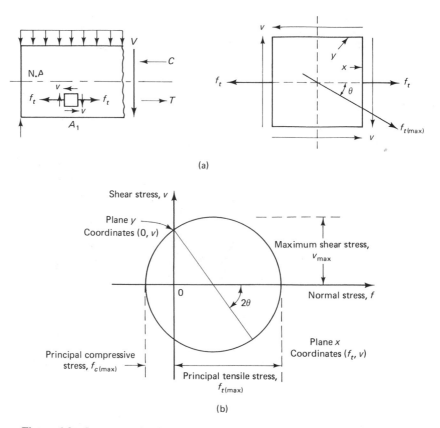

Figure 6.2   Stress state in elements A_1 and A_2: (a) stress state in element A_1; (b) Mohr's circle representation, element A_1; (c) stress state in element A_2; (d) Mohr's circle representation, element A_2.

Figure 6.2 shows the internal stresses acting on the infinitesimal elements A_1 and A_2. Using Mohr's circle in Fig. 6.2b, the principal stresses for element A_1 in the tensile zone below the neutral axis become

$$f_{t(max)} = \frac{f_t}{2} + \sqrt{\left(\frac{f_t}{2}\right)^2 + v^2} \qquad \text{principal tension} \qquad (6.3a)$$

$$f_{c(max)} = \frac{f_t}{2} - \sqrt{\left(\frac{f_t}{2}\right)^2 + v^2} \qquad \text{principal compression} \qquad (6.3b)$$

and

$$\tan 2\theta_{max} = \frac{v}{f_t/2} \qquad (6.3c)$$

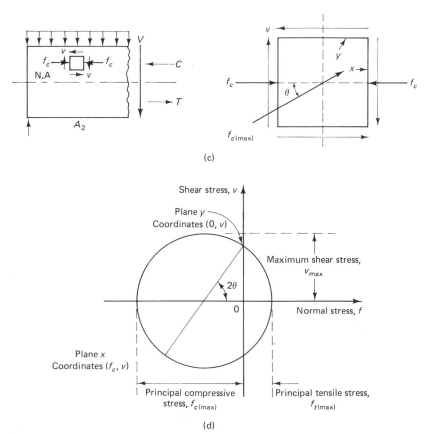

(c)

Figure 6.2   *(cont.)*

6.3 BEHAVIOR OF REINFORCED CONCRETE BEAMS AS NONHOMOGENEOUS SECTIONS

The behavior of reinforced concrete beams differs in that the tensile strength of concrete is about one-tenth of its strength in compression. The compression stress f_c in element A_2 of Fig. 6.2b above the neutral axis prevents cracking, as the maximum principal stress in the element is in compression. For element A_1 below the neutral axis, the maximum principal stress is in tension; hence cracking ensues. As one moves toward the support, the bending moment and hence f_t decreases, accompanied by corresponding increase in the shear stress. The principal stress $f_{t(max)}$ in tension acts at an approximately 45° plane to the normal at sections close to the support, as seen in

Fig. 6.3. Because of the low tensile strength of concrete, diagonal cracking develops along planes perpendicular to the planes of principal tensile stress—hence the term *diagonal tension cracks*. To prevent such cracks from opening, special "diagonal tension" reinforcement has to be provided.

If f_t close to the support in Fig. 6.3 is assumed equal to zero, the element becomes nearly in a state of pure shear and the principal tensile stress, using Eq. 6.3b, would be equal to the shear stress v on a 45° plane. It is this diagonal tension stress that causes the inclined cracks.

Definitive understanding of the correct shear mechanism in reinforced concrete is still incomplete. However, the approach of ACI-ASCE Joint Committee 426 gives a systematic empirical correlation of the basic concepts developed from extensive test results.

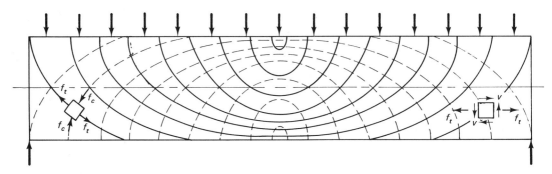

Figure 6.3 Trajectories of principal stresses in a homogeneous isotropic beam. Solid lines: tensile trajectories; dashed lines: compressive trajectories.

6.4 REINFORCED CONCRETE BEAMS WITHOUT DIAGONAL TENSION REINFORCEMENT

In regions of large bending moments, cracks develop almost perpendicular to the axis of the beam. These cracks are called *flexural cracks*. In regions of high shear due to the diagonal tension, the inclined cracks develop as an extension of the flexural crack and are termed *flexure shear cracks*. Figure 6.4 portrays the types of cracks expected in a reinforced concrete beam with or without adequate diagonal tension reinforcement.

6.4.1 Modes of Failure of Beams without Diagonal Tension Reinforcement

The slenderness of the beam, that is, its shear span/depth ratio, determines the failure mode of the beam. Figure 6.5 demonstrates schematically the failure patterns. The shear span a for concentrated load is the distance between the point of application of

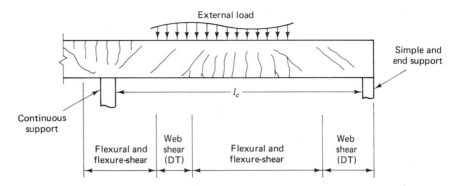

Figure 6.4 Crack categories.

the load and the face of support. For distributed loads, the shear span l_c is the clear beam span. Fundamentally, three modes of failure or their combinations occur: (1) flexural failure, (2) diagonal tension failure, and (3) shear compression failure. The more slender the beam, the stronger the tendency toward flexural behavior, as seen from the following discussion.

6.4.2 Flexural Failure (F)

In this region, cracks are mainly vertical in the middle third of the beam span and perpendicular to the lines of principal stress. These cracks result from a very small shear stress v and a dominant flexural stress f of a value close to an almost horizontal principal stress $f_{t(\max)}$. In such a failure mode, few very fine vertical cracks start to develop in the midspan area at about 50% of the failure load in flexure. As the external load increases, additional cracks develop in the central region of the span and the initial cracks widen and extend deeper toward the neutral axis and beyond, with a marked increase in the deflection of the beam. If the beam is under-reinforced, failure occurs in a ductile manner by initial yielding of the main longitudinal flexural reinforcement. This type of behavior gives ample warning of the imminence of collapse of the beam. The shear span/depth ratio for this behavior exceeds a value of 5.5 in the case of concentrated loading and in excess of 16 for distributed loading.

6.4.3 Diagonal Tension Failure (DT)

This failure precipitates if the strength of the beam in diagonal tension is lower than its strength in flexure. The shear span/depth ratio is of *intermediate* magnitude, with the ratio a/d varying between 2.5 and 5.5 for the case of concentrated loading. Such beams can be considered of intermediate slenderness. Cracking starts with the development of a few fine vertical flexural cracks at midspan, followed by the destruction of the bond between the reinforcing steel and the surrounding concrete at the support. Thereafter, without ample warning of impending failure, two or three diagonal cracks

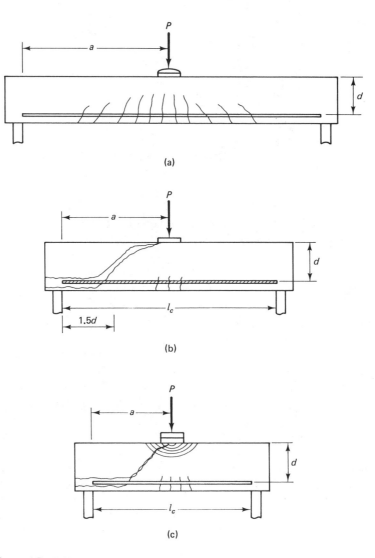

Figure 6.5   Failure patterns as a function of beam slenderness: (a) flexural failure; (b) diagonal tension failure; (c) shear compression failure.

develop at about $1\frac{1}{2}d$ to $2d$ distance from the face of the support. As they stabilize, one of the diagonal cracks widens into a principal diagonal tension crack and extends to the top compression fibers of the beam, as seen in Fig. 6.5b or 6.7. Notice that the flexural cracks do not propagate to the neutral axis in this essentially brittle failure mode, which has relatively small deflection at failure.

6.4.4 Shear Compression Failure (SC)

These beams have a small shear span/depth ratio, a/d, of magnitude 1 to 2.5 for the case of concentrated loading and less than 5.0 for distributed loading. As in the diagonal tension case, few fine flexural cracks start to develop at midspan and stop propagating as destruction of the bond occurs between the longitudinal bars and the surrounding concrete at the support region. Thereafter, an inclined crack steeper than in the diagonal tension case suddenly develops and proceeds to propagate toward the neutral axis. The rate of its progress is reduced with crushing of the concrete in the top compression fibers and a redistribution of stresses within the top region. Sudden failure takes place as the principal inclined crack dynamically joins the crushed concrete zone, as illustrated in Fig. 6.5c. This type of failure can be considered relatively less brittle than the diagonal tension failure due to the stress redistribution. Yet it is, in fact, a brittle type of failure with limited warning, and such a design should be avoided completely.

A reinforced concrete beam or element is not homogeneous and the strength of the concrete throughout the span is subject to a normally distributed variation. Hence one cannot expect that a stabilized failure diagonal crack occurs at both ends of the beam. Also, because of these properties, overlapping combinations of flexure–diagonal tension failure and diagonal tension–shear compression failure can occur at overlapping shear span/depth ratios. If the appropriate amount of shear reinforcement is provided, brittle failure of horizontal members can be eliminated with little additional cost to the structure. Table 6.1 summarizes the effect of the slenderness values on the mode of failure.

TABLE 6.1   BEAM SLENDERNESS EFFECT ON MODE OF FAILURE

Beam category	Failure mode	Shear span/depth ratio as a measure of slenderness[a]	
		Concentrated load, a/d	Distributed load, l_c/d
Slender	Flexure (F)	Exceeds 5.5	Exceeds 16
Intermediate	Diagonal tension (DT)	2.5–5.5	11–16[b]
Deep	Shear compression (SC)	1–2.5	1–5[b]

[a] a = shear span for concentrated loads

l_c = shear span for distributed loads

d = effective depth of beam

[b] For a uniformly distributed load, a transition develops from deep beam to intermediate beam effect.

6.5 DIAGONAL TENSION ANALYSIS OF SLENDER AND INTERMEDIATE BEAMS

The occurrence of the first inclined crack determines the shear strength of a beam without web reinforcement. As crack development is a function of the tensile strength of the concrete in the beam web, a knowledge of the principal stress in the critical sections is necessary, as discussed in Sections 6.2 and 6.3. The controlling principal stress in concrete is the result of the shearing stress v_u due to the external factored shear V_u and the horizontal flexural stress f_t due to the external factored bending moment M_u. The ACI code provides an empirical model based on results of extensive tests to failure of a large number of beams without web reinforcement. The model is a regression solution to the basic equation for two-dimensional principal stress at a point.

$$f_{t(\max)} = f'_t = \frac{f_t}{2} + \sqrt{\left(\frac{f_t}{2}\right)^2 + v^2} \qquad (6.3a)$$

where $f_{t(\max)}$ is the principal stress in tension and can be assumed to be equal to a constant multiplied by the tensile splitting strength f'_t of plain concrete. Since f'_t has been proven to be a function of $\sqrt{f'_c}$, Eq. 6.3a becomes

$$\sqrt{f'_c} = K_1\left[\frac{f_t}{2} + \sqrt{\left(\frac{f_t}{2}\right)^2 + v^2}\right] \qquad (6.4)$$

where K_1 is a constant.

The flexural stress f_t in the concrete is a function of the steel stress in the longitudinal reinforcement or the moment of resistance of the section, or

$$f_t \propto \frac{E_c}{E_s} f_s \propto \frac{E_c M_n}{E_s A_s d}$$

But the reinforcement ratio $\rho_w = A_s/bd$ at the tension side and E_c/E_s has a constant value. Hence

$$f_t = F_1 \frac{M_n}{\rho_w bd^2} \qquad (6.5)$$

where F_1 is a constant to be determined by test and M_n is the nominal moment strength of a given section. The shear stress v at the specific cross section bd due to the vertical external factored shear force V_u is

$$v = F_2 \frac{V_n}{bd} \qquad (6.6)$$

where V_n is the nominal shear resistance at the section under consideration and F_2 is the other constant, to be determined from the beam tests. Coefficients F_1 and F_2 both depend on several variables, including the geometry of the beam, type of loading, amount and arrangement of reinforcement, and the interaction between the steel reinforcement and the concrete.

Substituting f_t of Eq. 6.5 and v of Eq. 6.6, rearranging terms, and evaluating the constants K_1, F_1, and F_2 of the experimental model yields the following regression expression:

$$\frac{V_n}{bd\sqrt{f_c'}} = 1.9 + 2500\rho_w \frac{V_n d}{M_n \sqrt{f_c'}} \le 3.5 \qquad (6.7)$$

A plot of Eq. 6.7 is shown in Fig. 6.6. It is to be noted that $M_n/V_n d = a/d$; consequently, Eq. 6.7 accounts indirectly for the shear span/depth ratio, hence it accounts for the slenderness of the member. If the nominal shear resistance of the *plain* concrete web is termed V_c, V_n in the left-hand side of Eq. 6.7 has to be expressed as V_c. Transforming Eq. 6.7 into a force format for evaluation of the nominal shear resistance of the web of a beam of normal concrete and having no diagonal tension steel gives

$$V_c = 1.9 b_w d \sqrt{f_c'} + 2500\rho_w \frac{V_n d}{M_n} b_w d \le 3.5 b_w d \sqrt{f_c'} \qquad (6.8)$$

It is to be emphasized that the ratio $V_u d/M_u$ or $V_n d/M_n$ cannot exceed 1.0, where $V_n = V_u/\phi$ and $M_n = M_u/\phi$ as the values of shear and moment at the section for which V_c is being evaluated.

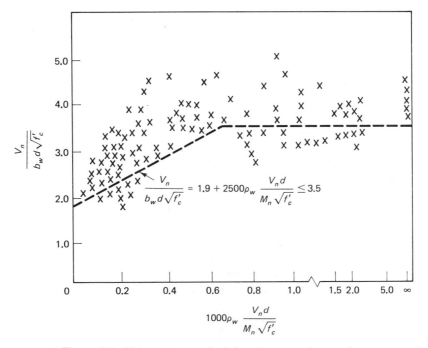

Figure 6.6 Shear resistance of reinforced concrete beam webs.

The first critical values of V_n and M_n are taken at a distance d from the face of the support since the stabilized (principal) diagonal tension cracks develop in that zone, as seen from Fig. 6.5b. As one moves toward the midspan of the beam, the values of M_n and V_n will change. The appropriate moments M_n and shears V_n have to be calculated for the particular section that is being analyzed for web steel reinforcement.

For simplicity of calculations, a more conservative ACI expression can be applied, particularly if the same beam section is not repetitively used in the structure:

$$V_c = \lambda \times 2.0\sqrt{f_c'}\, b_w d \tag{6.9}$$

where λ is a factor dependent on the type of concrete, with values of 1.0 for normal-weight concrete, 0.85 for sand-lightweight concrete, and 0.75 for all lightweight concrete. When axial compression also exists, V_c in Eq. 6.9 becomes

$$V_c = 2\lambda\left(1 + \frac{N_u}{2000A_g}\right)\sqrt{f_c'}\, b_w d \tag{6.10a}$$

When significant axial tension exists,

$$V_c = 2\lambda\left(1 + \frac{N_u}{500A_g}\right)\sqrt{f_c'}\, b_w d \tag{6.10b}$$

N_u/A_g is expressed in psi, where N_u is the axial load on the member and A_g is the gross area of the section; N_u is negative in tension.

6.6 WEB STEEL PLANAR TRUSS ANALOGY

As discussed previously, web reinforcement has to be provided to prevent failure due to diagonal tension. Theoretically, if the necessary steel bars in the form of the tensile stress trajectories shown in Fig. 6.3 are placed in the beam, no shear failure can occur. However, practical considerations eliminate such a solution, and other forms of reinforcing are improvised to neutralize the principal tensile stresses at the critical shear failure planes. The mode of failure in shear reduces the beam to a simulated arched section in compression at the top and tied at the bottom by the longitudinal beam tension bars, as seen in Fig. 6.7a. If one isolates the main concrete compression element shown in Fig. 6.7b, it can be considered as the compression member of a triangular truss, as shown in Fig. 6.7c with the polygon of forces C_c, T_b, and T_s representing the forces acting on the truss members—hence the expression *truss analogy*. Force C_c is the compression in the simulated concrete strut, force T_b is the tensile force increment of the main longitudinal tension bar, and T_s is the force in the bent bar. Figure 6.8a shows the analogy truss for the case of using vertical stirrups instead of inclined bars, with the forces polygon having a vertical tensile force T_s instead of the inclined one in Fig. 6.7c.

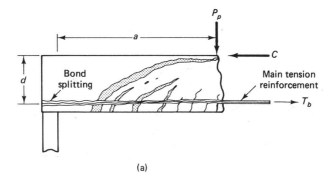

(a)

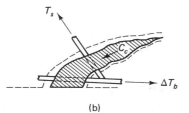

(b)

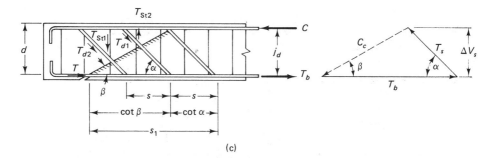

(c)

Figure 6.7   Diagonal tension failure mechanism: (a) failure pattern; (b) concrete simulated strut; (c) planar truss analogy.

As can be seen from the previous discussion, the shear reinforcement basically performs four main functions:

1. Carries a portion of the external factored shear force V_u
2. Restricts the growth of the diagonal cracks
3. Holds the longitudinal main reinforcing bars in place so that they can provide the dowel capacity needed to carry the flexural load
4. Provides some confinement to the concrete in the compression zone if the stirrups are in the form of closed ties

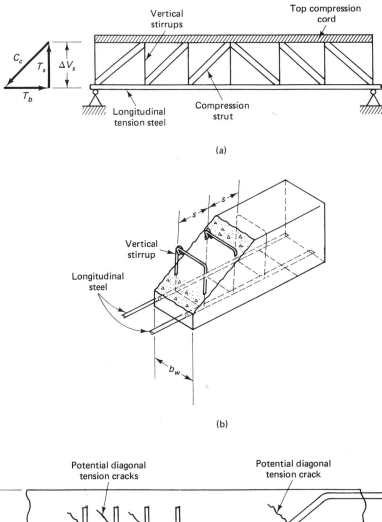

(a)

(b)

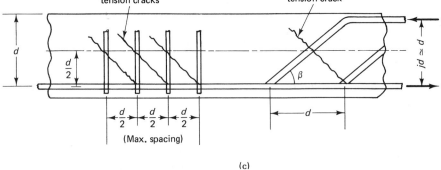

(c)

Figure 6.8 Web steel arrangement: (a) truss analogy for vertical stirrups; (b) three-dimensional view of vertical stirrups; (c) spacing of web steel.

6.6.1 Web Steel Resistance

If V_c, the nominal shear resistance of the plain web concrete, is less than the nominal total vertical shearing force $V_u/\phi = V_n$, web reinforcement has to be provided to carry the difference in the two values; hence

$$V_s = V_n - V_c \tag{6.11}$$

The nominal resisting shear V_c can be calculated from Eq. 6.8 or 6.9 and V_s can be determined from equilibrium analysis of the bar forces in the analogous triangular truss cell. From Fig 6.7c,

$$V_s = T_s \sin \alpha = C_c \sin \beta \tag{6.12a}$$

where T_s is the force resultant of all web stirrups across the diagonal crack plane and n is the number of spacings s. If $s_1 = ns$ in the bottom tension chord of the analogous truss cell, then

$$s_1 = jd(\cot \alpha + \cot \beta) \tag{6.12b}$$

Assuming that moment arm $jd \simeq d$, the stirrup force per unit length from Eqs. 6.12a and 6.12b, where $s_1 = ns$, becomes

$$\frac{T_s}{s_1} = \frac{T_s}{ns} = \frac{V_s}{\sin \alpha} \frac{1}{d(\cot \beta + \cot \alpha)} \tag{6.12c}$$

If there are n inclined stirrups within the s_1 length of the analogous truss chord, and if A_v is the area of one inclined stirrup,

$$T_s = nA_v f_y \tag{6.13a}$$

Hence

$$nA_v = \frac{V_s ns}{d \sin \alpha(\cot \beta + \cot \alpha)f_y} \tag{6.13b}$$

But it can be assumed that in the case of diagonal tension failure the compression diagonal makes an angle $\beta = 45°$ with the horizontal; Eq. 6.13b becomes

$$V_s = \frac{A_v f_y d}{s}[\sin \alpha(1 + \cot \alpha)]$$

to get

$$V_s = \frac{A_v f_y d}{s}(\sin \alpha + \cos \alpha) \tag{6.14a}$$

or

$$s = \frac{A_v f_y d}{V_n - V_c}(\sin \alpha + \cos \alpha) \tag{6.14b}$$

If the inclined web steel consists of a single bar or a single group of bars all bent at the same distance from the face of the support,

$$V_s = A_v f_y \sin \alpha \leq 3.0\sqrt{f_c'}b_w d$$

If vertical stirrups are used, angle α becomes 90°, giving

$$V_s = \frac{A_v f_y d}{s} \tag{6.15a}$$

or

$$s = \frac{A_v f_y d}{(V_u/\phi) - V_c} = \frac{A_v \phi f_y d}{V_u - \phi V_c} \tag{6.15b}$$

6.6.2 Limitations on Size and Spacing of Stirrups

Equations 6.14 and 6.15 give inverse relationships between the spacing of the stirrups and the shear force or shear stress they resist, with the spacing s decreasing with the increase in $(V_n - V_c)$. In order for every *potential* diagonal crack to be resisted by a vertical stirrup as seen in Fig. 6.7c, maximum spacing limitations are to be applied as follows for vertical stirrups:

1. $V_n - V_c > 4\sqrt{f_c'}b_w d$: $s_{max} = d/4 \leq 24$ in.
2. $V_n - V_c \leq 4\sqrt{f_c'}b_w d$: $s_{max} = d/2 \leq 24$ in.
3. $V_n - V_c > 8\sqrt{f_c'}b_w d$: enlarge section

The minimum web steel area A_v has to be provided if the factored shear force V_u exceeds one-half the shear strength ϕV_c of the plain concrete web. This precaution is necessary to prevent brittle failure, thus enabling both the stirrups and the beam compression zone to continue carrying the increasing shear after the formulation of the first inclined crack.

$$\text{minimum } A_v = \frac{50 b_w s}{f_y} \tag{6.16}$$

where A_v is the area of all the vertical stirrup legs in the cross section.

6.7 WEB REINFORCEMENT DESIGN PROCEDURE FOR SHEAR

The following is a summary of the recommended sequence of design steps.

1. Determine the critical section and calculate the factored shear force V_u. When the reaction, in the direction of applied shear, introduces compression into the end regions of a member, the critical section can be assumed at a distance of d from the support, provided that no concentrated load acts between the support face and distance d thereafter.
2. Check whether

$$V_u \le \phi(V_c + 8\sqrt{f_c'}b_w d)$$

    where b_w is the web width or diameter of the circular section. If this condition is not satisfied, the cross section has to be enlarged.
3. Use minimum shear reinforcement A_v if V_u is larger than one-half ϕV_c, with the following exceptions:
    (a) Concrete joist construction
    (b) Slabs and footings
    (c) Small shallow beams of depth not exceeding 10 in. (254 mm) or $2\frac{1}{2}$ times the flange thickness:

$$\text{minimum } A_v = \frac{50 b_w s}{f_y}$$

    Good construction practice dictates that some stirrups always be used to facilitate proper handling of the reinforcement cage.
4. If $V_u > \phi V_c$, shear reinforcement must be provided such that $V_u \le \phi(V_c + V_s)$, where

$$V_s = \begin{cases} \dfrac{A_v f_y d}{s} & \text{for vertical stirrups} \\[2ex] \dfrac{A_v f_y d}{s}(\sin \alpha + \cos \alpha) & \text{for inclined stirrups} \end{cases}$$

5. Maximum spacing s must be $s = d/2 \le 24$ in., except that in cases where $V_s > 4\sqrt{f_c'}b_w d$, the spacing then becomes $s \le d/4 \le 24$ in.

Figure 6.9 presents a flowchart for the performance of the sequence of calculations necessary for the design of vertical stirrups. Simple corresponding modifications of this chart can be made so that the chart can be used in the design of inclined web steel.

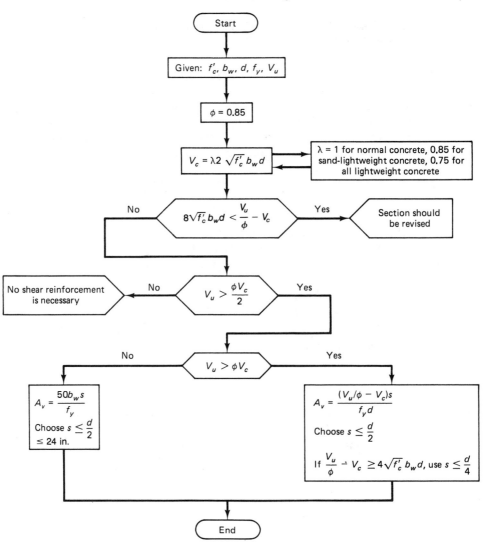

Figure 6.9 Flowchart for web reinforcement design procedure.

6.8 EXAMPLES ON THE DESIGN OF WEB STEEL FOR SHEAR

6.8.1 Example 6.1: Design of Web Stirrups

A rectangular isolated beam has an effective span of 25 ft (7.62 m) and carries a working live load of 8000 lb per linear foot (10.85 kN/m) and no external dead load except its self-weight. Design the necessary shear reinforcement. Use the simplified term of Eq. 6.9 for calculating the capacity V_c of the plain concrete web. Given:

$f_c' = 4000$ psi (27.6 MPa), normal weight concrete
$f_y = 60,000$ psi (414 MPa)
$b_w = 14$ in. (356 mm)
$d = 28$ in. (712 mm)
$h = 30$ in. (762 mm)
Longitudinal tension steel: six No. 9 bars (diameter 28.6 mm)
No axial force acts on the beam

Solution

Factored shear force (Step 1)

$$\text{beam self-weight} = \frac{14 \times 30}{144} \times 150 = 437.5 \text{ lb/ft}$$

$$\text{total factored load} = 1.7 \times 8000 + 1.4 \times 437.5 = 14,212.5 \text{ lb/ft}$$

The factored shear force at the face of the support is

$$V_u = \frac{25}{2} \times 14,212.5 = 177,656 \text{ lb}$$

The first critical section is at a distance $d = 28$ in. from the face of the support of this beam (half-span = 150 in.).

$$V_u \text{ at } d = \frac{150 - 28}{150} \times 177,656 = 144,494 \text{ lb}$$

Shear capacity (Step 2)

The shear capacity of the plain concrete in the web from the simplified equation for normal-weight concrete ($\lambda = 1.0$) is

$$V_c = 2.0\lambda \sqrt{f_c'} b_w d = 2 \times 1.0\sqrt{4000} \times 14 \times 28 = 49,585 \text{ lb}$$

Check for adequacy of section for shear:

$$(8 + 2.0)\sqrt{f_c'} b_w d = 10\sqrt{f_c'} b_w d = 247,923 \text{ lb} \qquad 129620$$

$$V_n = \frac{V_u}{\phi} = \frac{144,494}{0.85} = 169,993 \text{ lb} \qquad \text{cross section O.K.}$$

$$V_n > V_c, \qquad \text{hence stirrups are necessary}$$

Shear reinforcement (Steps 3 to 5)

Try No. 4 two-legged stirrups (area per leg = 0.20 in.²).

$$A_v = 2 \times 0.2 = 0.40 \text{ in.}^2$$

From Eq. 6.15b,

$$s = \frac{A_v f_y d}{(V_u/\phi) - V_c} = \frac{0.4 \times 60,000 \times 28}{169,993 - 49,584}$$

$$= 5.58 \text{ in. } (141.7 \text{ mm})$$

Since $V_n - V_c > 4\sqrt{f_c'} b_w d$, the maximum allowable spacing $s = d/4 = 28/4 = 7$ in. At the critical section, $d = 28$ in. from the face of the support, the maximum allowable spacing would in this case be 5.58 in.

The shear force for distributed load decreases linearly from the support to midspan of the beam. Hence the web reinforcement can be reduced accordingly after determining the zone where minimum reinforcement is necessary and the zone where no web reinforcement is needed. The same size and spacing of stirrups needed at the critical section d from face of support should be continued to the support. Figure 6.10 illustrates the various values being calculated:

Critical Phase x_d (consider the midspan as the origin): $V_n = 169,993$ lb and from before, $s = 5.58$ in. x_d from the midspan point $= 150 - 28 = 122$ in.

Plane x_1 at $s = d/4$ maximum spacing:

$$V_{s1} = 4\sqrt{f_c'} b_w d = 4\sqrt{4000} \times 14 \times 28 = 99,169 \text{ lb}$$

$$V_{n1} = 99,169 + 49,585 = 148,754 \text{ lb}$$

$$x_1 \text{ from midspan point} = (150 - 28) \times \frac{148,754}{169,993} = 106.76 \text{ in.}$$

Plane x_2 at $s = d/2$ maximum spacing:

$$s = \frac{A_v f_y d}{V_n - V_c} \quad \text{or} \quad \frac{28}{2} = \frac{0.4 \times 60,000 \times 28}{V_s}$$

or

$$V_{s2} = 48,000 \text{ lb}$$

$$V_{n2} = 48,000 + 49,585 = 97,585 \text{ lb}$$

$$x_2 \text{ from midspan point} = 122 \times \frac{97,585}{169,993} = 70.03 \text{ in.}$$

From Fig. 6.10a, the distance 36.73 in. is the transition zone from $s = 7$ in. to $s = 14$ in.; hence a stirrup spacing of 8 in. center to center is shown in Fig. 6.10b.

Plane x_3 at shear force V_c:

$$V_c = 2\sqrt{f_c'} b_w d = 49,585 \text{ lb}$$

$$x_3 \text{ from midspan point} = 122 \times \frac{49,585}{169,993} = 35.59 \text{ in.}$$

Extend stirrups to a distance $d = 28$ in. beyond the point where they are theoretically not required.

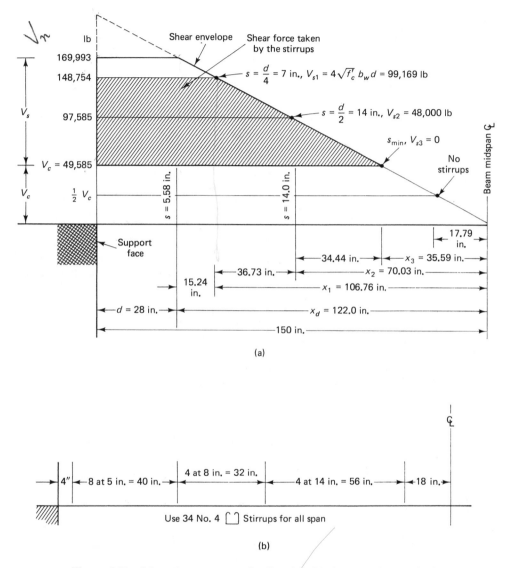

Figure 6.10 Stirrups arrangements for Ex. 6.1: (a) shear envelope and stirrup design segments; (b) vertical stirrups spacing.

Minimum web steel: Test when $V_u > \frac{1}{2}\phi V_c$ or $V_n > \frac{1}{2}V_c$

$$V_u = 144,494$$

$$\tfrac{1}{2}V_c = \tfrac{1}{2} \times 49,585 = 24,793 \text{ lb}$$

$$A_v = \frac{50 b_w s}{f_y} = \frac{50 \times 14 \times 14}{60,000} = 0.16 \text{ in.}^2$$

$$< \text{ actual } A_v = 0.40 \text{ in.}^2 \quad \text{O.K.}$$

or   maximum allowed $s = \dfrac{A_v f_y}{50 b_w} = \dfrac{0.40 \times 60,000}{50 \times 14} = 34.3 \text{ in.}$

$$\text{vs. maximum used } s = \frac{d}{2} = 14 \text{ in.} \quad \text{O.K.}$$

$$x_y = 122.0 \times \frac{24,793}{169,993} = 17.79 \text{ in. from midspan}$$

Proportion the spacing of the vertical stirrups accordingly.

The shaded area in Fig. 6.10a is the shear force area for which stirrups must be provided. The spacing of the stirrups in Fig. 6.10b is based on the practical consideration of the desirability to use whole spacing dimensions and to vary the spacing as little as possible.

6.8.2 Example 6.2: Alternative Solution to Ex. 6.1

Find the force V_c and the change in stirrup spacing for the beam in Ex. 6.1 if the more refined Eq. 6.8 is used where the separate contribution of the main longitudinal steel at the tension side is more accurately reflected.

Solution

The shear capacity of the plain concrete in the web is

$$V_c = 1.9\sqrt{f_c'}\, b_w d + 2500 \rho_w \frac{V_u d}{M_u} b_w d \leq 3.5\sqrt{f_c'}\, b_w d$$

where ρ_w is the longitudinal steel in the web at the tension side only.

$$\rho_w = \frac{6.0}{14 \times 28} = 0.0153$$

V_u at d from support $= 144,494 \text{ lb}$ (Ex. 6.1)

$V_u d = 144,499 \times 28 = 4,045,972 \text{ in.-lb}$

M_u at d from support $= V_u d - \dfrac{w_u d^2}{2} = \left(14,212.5 \times \dfrac{25}{2}\right) \times 28 - \dfrac{14,212.5(28)^2}{12 \times 2}$

$$= 4,510,100 \text{ in.-lb}$$

$$\frac{V_u d}{M_u} \quad \text{or} \quad \frac{V_n d}{M_n} = \frac{4,045,972}{4,510,100} = 0.9 < 1.0 \quad \text{use } 0.9$$

$$V_c = 1.9\sqrt{4000} \times 14 \times 28 + 2500 \times 0.0153 \times 0.9 \times 14 \times 28$$

$$= 47,105.4 \text{ lb} + 13,494.6 = 60,600 \text{ lb}$$

Use a two-legged No. 4 size vertical stirrup, as in Ex. 6.1.

$$s = \frac{A_v f_y d}{V_u/\phi - V_c} = \frac{0.4 \times 60,000 \times 28}{169,933 - 60,600} = 6.14 \text{ in. (156.0 mm)}$$

$$\underline{s = \frac{d}{4} = 7 \text{ in.}}$$

By trial and adjustment and applying the expression in Eq. 6.8 in the trials,

$$V_{c1} = 57,738 \text{ lb}, \quad V_{n1} = 152,941 \text{ lb}$$

$$x_1 = (150 - 28)\frac{152,941}{169,993} = 109.86 \text{ in.}$$

$$\underline{s = \frac{d}{2} = 14 \text{ in.}}$$

$$V_{c2} = 50,718 \text{ lb} \qquad V_{n2} = 98,824 \text{ lb}$$

$$x_2 = 122. \times \frac{98,824}{169,993} = 70.92 \text{ in.}$$

At the point in the shear envelope where $V_s = 0$, the value of $V_u d/M_u$ in Eq. 6.8 is close to zero for uniformly distributed loads. Hence assume that

$$V_c \simeq 2.0\sqrt{f_c'}\, b_w d \quad \text{instead of} \quad V_c = 1.9\sqrt{f_c'}\, b_w d + 2500\rho_w \frac{V_u d}{M_u}$$

Therefore, use $x_3 = 35.59$ in. in Fig. 6.11 (as in Fig. 6.10 of Ex. 6.1) as being accurate enough for all practical purposes. Proportion the spacing of the stirrups as in Fig. 6.11b.

The shear diagram showing all these details is given in Fig. 6.11. It can be seen that this refined solution reduced the shearing force taken by the stirrups at the d critical section by the difference between the 49,585 lb of Ex. 6.1 and the 60,600 lb of Ex. 6.2, as shown in the darkly shaded portion. This difference is taken by the plain concrete in the web. There is a small saving in the number of stirrups that can be justified only if the beam section designed by the refined method is extensively repetitively used in a multifloor multispan building.

Note in these shear problems that if concentrated loads act on the beam close to the midspan or, in the case of reversible loads, Fig. 6.12, almost constant stirrup spacing throughout the span becomes necessary. The spacing to be used would be that required at the critical section at distance d from the face of the support. Superposition of the shear diagram for a concentrated live load over that of the distributed load due to self-weight or otherwise gives the total shear force for stirrup spacing determination.

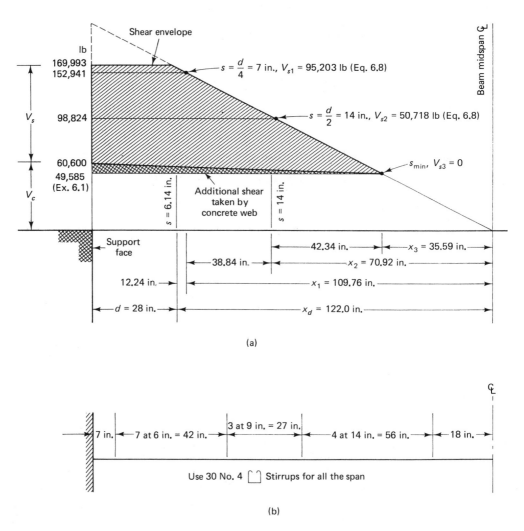

(a)

(b)

Figure 6.11   Stirrups arrangement for Ex. 6.2: (a) shear envelope and stirrup design segments; (b) vertical stirrups spacing.

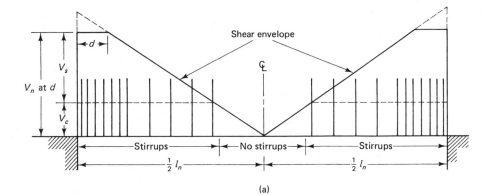

(a)

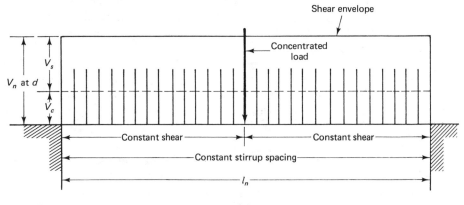

(b)

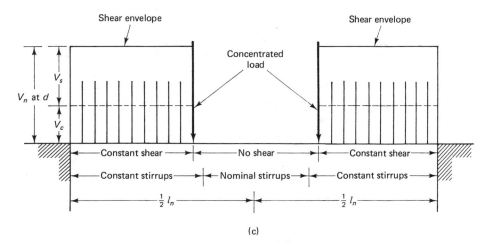

(c)

Figure 6.12 Schematic stirrups distribution: (a) stirrups spacing for uniformly distributed load on beam; (b) stirrups spacing for centrally loaded beam; (c) stirrups spacing for third-point loaded beam.

6.9 DEEP BEAMS

Deep beams are structural elements loaded as beams but having a large depth/thickness ratio and a shear span/depth ratio not exceeding 2 to 2.5., where the shear span is the clear span of the beam for distributed load. Floor slabs under horizontal loads, wall slabs under vertical loads, short span beams carrying heavy loads, and some shear walls are examples of this type of structural element.

Because of the geometry of deep beams, they behave as two-dimensional rather than one-dimensional members and are subjected to a two-dimensional state of stress. As a result, plane sections before bending do not necessarily remain plane after bending. The resulting strain distribution is no longer considered as linear, and shear deformations that are neglected in normal beams become significant compared to pure flexure. Consequently, the stress block becomes nonlinear even at the elastic stage. At the limit state of ultimate load, the compressive stress distribution in the concrete would no longer follow the same parabolic shape or intensity as that shown in Fig. 5.2c for a normal beam.

Figure 6.13 illustrates the linearity of the stress distribution at midspan prior to cracking in a normal beam where the effective span/depth ratio exceeds a value of $3\frac{1}{2}$ to 4. In contrast, Fig. 6.14a shows the nonlinearity of stress at midspan corresponding to the nonlinear strain under discussion. It can also be recognized that the magnitude of the maximum tensile stress at the bottom fiber far exceeds the magnitude of the maximum compressive stress. The stress trajectories in Fig. 6.14b and c confirm this observation. Note the steepness and concentration of the principal tensile stress trajectories at midspan and the concentration of the compressive stress trajectories at the support for both cases of loading of the beam at top or bottom.

The concrete cracks in a direction perpendicular to the tensile principal stress trajectories. As the load increases, the cracks widen and propagate, and more cracks open. Hence less and less concrete remains to resist the indeterminate state of stress. Because the shear span is small, the compressive stresses in the support region affect the magnitude and direction of the principal tensile stresses such that they become less inclined and lower in value.

In many cases the cracks would almost be vertical or follow the direction of the compression trajectories, with the beam almost shearing off from the support in a total shear failure. Hence, in the case of deep beams, horizontal reinforcement is needed throughout the height of the beams, in addition to the vertical shear reinforcement along the span. From Fig. 6.14b and c and the steep gradient of the tensile stress

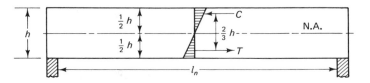

Figure 6.13   Elastic distribution in normal beams ($l_n/h \geq 3\frac{1}{2}$ to 5).

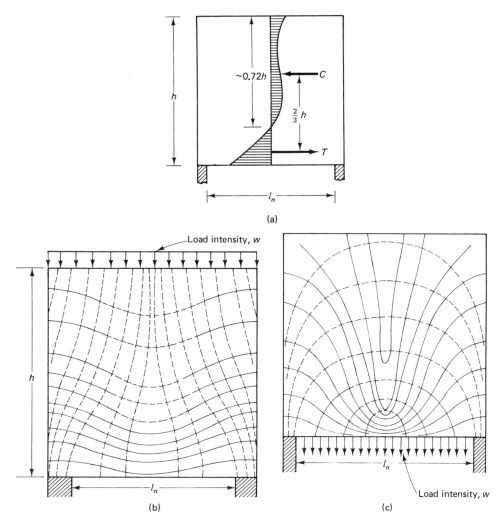

Figure 6.14 Elastic stress distribution in deep beams: (a) deep beam ($l_n/h \leq$ 1.0); (b) principal stress trajectories in deep beams loaded on top; (c) principal stress trajectories in deep beams loaded at bottom.

trajectories at the lower fibers, a concentration of horizontal reinforcing bars is required to resist the high tensile stresses at the lower regions of the deep beam.

Additionally, the high depth/span ratio of the beam should provide an increased resistance to the external shear load due to a higher compressive arch action. Consequently, it should be expected that the nominal resisting shear force V_c for deep beams will considerably exceed the V_c value for normal beams.

In summary, shear in deep beams is a major consideration in their design. The magnitude and spacing of both the vertical and horizontal shear reinforcement differ considerably from those used in normal beams, as well as the expressions that have to be used for their design.

6.9.1 Design Criteria for Shear in Deep Beams Loaded at the Top

From the discussion in Section 6.9, it can be inferred that deep beams ($a/d < 2.5$ and $l_n/d < 5.0$) have a higher nominal shear resistance V_c than do normal beams. While the critical section for calculating the factored shear force V_u is taken at distance d from the face of the support in normal beams, the shear plane in the deep beam is considerably steeper in inclination and closer to the support. If x is the distance of the failure plane from the face of the support, l_n the clear span for uniformly distributed load, and a the shear arm or span for concentrated loads, the expression for distance is

$$\text{Uniform load:} \qquad x = 0.15l_n \tag{6.17a}$$

$$\text{Concentrated load:} \quad x = 0.50a \tag{6.17b}$$

In either case, the distance x should not exceed the effective depth d.
The factored shear force V_u has to satisfy the condition

$$V_u \leq \phi(8\sqrt{f_c'}\, b_w d) \qquad \text{for } l_n/d < 2.0 \tag{6.18a}$$

or

$$V_u \leq \phi\left[\frac{2}{3}\left(10 + \frac{l_n}{d}\right)\sqrt{f_c'}\, b_w d\right] \qquad \text{for } 2 \leq l_n/d \leq 5 \tag{6.18b}$$

If not, the section has to be enlarged. The strength reduction factor $\phi = 0.85$.
The nominal shear resisting force V_c of the plain concrete can be taken as

$$V_c = \left(3.5 - 2.5\frac{M_u}{V_u d}\right)\left(1.9\sqrt{f_c'} + 2500\rho_w \frac{V_u d}{M_u}\right)b_w d \leq 6\sqrt{f_c'}\, b_w d \tag{6.19a}$$

where $1.0 < 3.5 - 2.5(M_u/V_u d) \leq 2.5$. This factor is a multiplier of the basic equation for V_c in normal beams to account for the higher resisting capacity of deep beams. The ACI code allows this higher resisting capacity provided that some minor unsightly cracking is tolerated if V_u exceeds the first shear cracking load; otherwise, the designer can use

$$V_c = 2\sqrt{f_c'}\, b_w d \tag{6.19b}$$

When the factored shear V_u exceeds ϕV_c, shear reinforcement has to be provided such that $V_u \leq \phi(V_c + V_s)$, where V_s is the force resisted by the shear reinforcement:

$$V_s = \left(\frac{A_v}{s_v}\frac{1 + l_n/d}{12} + \frac{A_{vh}}{s_h}\frac{11 - l_n/d}{12}\right)f_y d \tag{6.20}$$

where A_v = total area of vertical reinforcement spaced at s_v in the horizontal direction at both faces of the beam

A_{vh} = total area of horizontal reinforcement spaced at s_h in the vertical direction at both faces of the beam

$$\text{Maximum } s_v \leq \frac{d}{5} \quad \text{or} \quad 18 \text{ in.}$$

$$\text{whichever is smaller} \qquad (6.21a)$$

$$\text{Maximum } s_h \leq \frac{d}{3} \quad \text{or} \quad 18 \text{ in.}$$

and

$$\text{Minimum } A_v = 0.0015bs_v$$

$$(6.21b)$$

$$\text{Minimum } A_{vh} = 0.0025bs_h$$

The shear reinforcement required at the critical section must be provided throughout the deep beams.

In the case of continuous deep beams, because of the large stiffness and negligible rotation of the beam section at the supports, the continuity factor at the first interior support has a value close to 1.0. Consequently, the same reinforcement for shear can be used in all spans for all practical purposes if all the spans are equal and similarly loaded.

6.9.2 Design Criteria for Flexure in Deep Beams

6.9.2.1 Simply Supported Beams

The ACI code does not specify a design procedure but requires a rigorous nonlinear analysis for the flexural analysis and design of deep beams. The simplified provisions presented in this section are based on the recommendations of the Euro-International Concrete Committee (CEB).

Figure 6.14a shows a schematic stress distribution in a homogeneous deep beam having a span/depth ratio $l_n/h \simeq 1.0$. It was experimentally observed that the moment lever arm does not change significantly even after initial cracking. Since the nominal resisting moment is

$$M_n = A_s f_y (\text{moment arm } jd) \qquad (6.22a)$$

the reinforcement area A_s for flexure is

$$A_s = \frac{M_u}{\phi f_y jd} \geq \frac{200bd}{fy} \qquad (6.22b)$$

The lever arm as recommended by CEB is

$$jd = 0.2(l + 2h) \qquad \text{for } 1 \leq l/h < 2 \qquad (6.23a)$$

and

$$jd = 0.6l \qquad \text{for } l/h < 1 \qquad (6.23b)$$

where l is the effective span measured center to center of supports or 1.15 clear span l_n, whichever is smaller. The tension reinforcement has to be placed in the lower segment of beam height such that the segment height is

$$y = 0.25h - 0.05l < 0.20h \qquad (6.24)$$

It should consist of closely spaced small-diameter bars well anchored into the supports.

6.9.2.2 Continuous Beams

Continuous deep beams can be treated in the same manner as simply supported deep beams, except that additional reinforcement has to be provided for the negative moment at the support. Figure 6.15 presents stress trajectories of the principal tensile and compressive stresses in a continuous deep beam. Comparing this diagram to Fig. 6.14b for the simply supported case, one can observe the similarity of the steepness of the tensile stress trajectories at midspan. At the continuous supports, the total section is in tension.

The concentration of the tensile stress trajectories at the support regions of the continuous deep beam necessitates a concentration of well-anchored horizontal shear reinforcement. The required total flexural reinforcement area

$$A_s = \frac{M_u}{\phi f_y jd} \geq \frac{200bd}{f_y}$$

as in Eq. 6.22b for the simply supported beam. The lever arm jd is, however, different and has a value

$$jd = 0.2(l + 1.5h) \qquad \text{for } 1 \leq l/h \leq 2.5 \qquad (6.25a)$$

$$jd = 0.5l \qquad\qquad \text{for } l/h < 1.0 \qquad\qquad (6.25b)$$

The distribution of the negative flexural rinforcement A_s in continuous beams should be such that the steel area A_{s1} should be placed in the top 20% of the beam depth, and

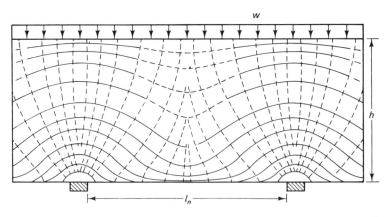

Figure 6.15   Tensile and compression trajectories in a continuous deep beam. Solid line: tension trajectories; dashed line: compression trajectories.

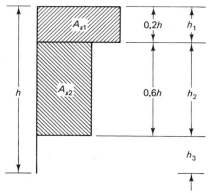

Figure 6.16   Distribution of horizontal flexural steel in continuous deep beams.

the balance steel area A_{s2} at the next 60% of the beam depth as shown in Fig. 6.16. The value of

$$A_{s1} = 0.5(l/h - 1)A_s \qquad (6.26a)$$

$$A_{s2} = A_s - A_{s1} \qquad (6.26b)$$

For cases where the ratio l/h has a value equal to or less than 1.0, use nominal steel for A_{s1} in the top 20% of the beam depth and provide the total A_s in the next 60% of the depth. In the lower h_3 zone the positive reinforcement coming from the beam span should pass through the support for anchorage and continuity.

6.9.3 Sequence of Deep Beams Design Steps for Shear

The following is a recommended procedure for the design of shear reinforcement in deep beams based on ACI requirements. The sequence of steps should essentially be similar to those in Section 6.7 for web reinforcement design in normal beams. Additionally, flexural reinforcement has to be provided to resist the stresses due to bending.

1. Check whether the beam can be classified as a deep beam, that is, $a/d < 2.5$ or $l_n/d < 5.0$ for a concentrated or a uniform load, respectively.
2. Determine the critical section distances x from the face of support: $x = 0.5a$ for concentrated load and $x = 0.15l_n$ for distributed load. Calculate the factored V_u at the critical section and check whether it is less than the maximum $\phi V_n = V_u$ permitted by Eqs. 6.18a and 6.18b; if not, enlarge the beam section.
3. Calculate the shear resisting capacity V_c of the plain concrete from Eq. 6.19.
4. Calculate V_s if $V_u > \phi V_c$ and choose s_v and s_h by assuming the size of shear reinforcement in both the horizontal and vertical directions.
5. Verify if the size and maximum spacing from step 4 satisfy Eqs. 6.21a and 6.21b; if not, revise and recheck using Eq. 6.20.

6. Select reasonable size and spacing of the shear reinforcement in both horizontal and vertical directions. Where possible, use welded wire fabric mats since they provide superior anchorage of the reinforcement to tied bar mats and are easier to handle and keep in position at both faces of the deep beam.

7. Design the flexural reinforcement as in Section 6.9.2 after determining the moment lever arm jd for the particular case of simply supported or continuous deep beams.

8. Distribute the flexural reinforcement in accordance with Eqs. 6.26a and 6.26b and Fig. 6.16 if the beam is continuous. If the beam is simply supported, concentrate the flexural horizontal longitudinal bars in the lower $(0.25h - 0.05l) \leq 0.20h$ part of the beam depth.

9. Sketch a detailed schematic of the distribution of both the shear and the flexural reinforcement.

6.9.4 Example 6.3: Design of Shear Reinforcement in Deep Beams

A simply supported beam having a clear span $l_n = 10$ ft (3.05 m) is subjected to a uniformly distributed live load of 86,000 lb/ft (1255.6 kN/m) on the top. The height h of the beam is 6 ft (1.83 m) and its thickness b is 20 in. (508 mm). The area of its horizontal tension steel is 8.0 in.^{2} (5161 mm^2), determined in Ex. 6.4. Given:

$$f'_c = 4000 \text{ psi (27.6 MPa)}$$
$$f_y = 60,000 \text{ psi (414 MPa)}$$

Design the shear reinforcement for this beam.

Solution

Check l_n/d evaluate factored shear force V_u (Step 1)

Assume that $d \simeq 0.9h = 0.9 \times 6 \times 12 \simeq 65$ in. (1651 mm).

$$\frac{l_n}{d} = \frac{10.0 \times 12}{65} = 1.85 < 5$$

Hence treat as a deep beam.

$$\text{Beam self-weight} = \frac{20 \times 72}{144} \times 150 = 1500 \text{ lb/ft (21.9 kN/m)}$$

$$\text{Total factored load} = 1.7 \times 86,000 + 1.4 \times 1500$$
$$= 148,300 \text{ lb/ft (2165.2 kN/m)}$$

$$\text{Distance of the critical section} = 0.15l_n = 0.15 \times 10.0$$
$$= 1.5 \text{ ft} = 18 \text{ in. (457.2 mm)}$$

The factored shear force V_u at the critical section is

$$V_u = \frac{148,300 \times 10}{2} - 148,300 \times \frac{18}{12} = 519,050 \text{ lb } (2308.7 \text{ kN})$$

Nominal shear strength V_n and resisting capacity V_c
(Steps 2 and 3)

$$\phi V_n = \phi(8\sqrt{f_c'}\, b_w d) = 0.85(8\sqrt{4000} \times 20 \times 65)$$

$$\doteq 559,091 \text{ lb } (2486.8 \text{ kN}) > 519,050 \qquad \text{O.K.}$$

$$M_u = \frac{148,300 \times 10 \text{ ft}}{2} \times 1.5 - \frac{148,300 \times (1.5)^2}{2}$$

$$= 945,412.5 \text{ ft-lb} = 11,344,950 \text{ in.-lb}$$

$$\frac{M_u}{V_u d} = \frac{11,344,950}{519,050 \times 65} = 0.3363$$

$$3.5 - 2.5\frac{M_u}{V_u d} = 3.5 - 2.5 \times 0.3363 = 2.66 > 2.5 \qquad \text{use 2.5}$$

$$\rho_w = \frac{8.0}{20 \times 65} = 0.0062$$

$$\frac{V_u d}{M_u} = 2.97$$

From Eq. 6.19:

$$V_c = 2.5\left(1.9\sqrt{f_c'} + 2500\,\rho_w \frac{V_u d}{M_u}\right) b_w d$$

$$= 2.5(1.9\sqrt{4000} + 2500 \times 0.0062 \times 2.97) \times 20 \times 65 = 540,155 \text{ lb}$$

$$6\sqrt{f_c'}\, b_w d = 493,315 \text{ lb} < 540,155 \text{ lb}$$

Hence $V_c = 493,315$ lb (2144.3 kN) controls.

Shear reinforcement (steps 4 and 5)

Assume No. 3 (9.52 mm diameter) bars placed both horizontally and vertically on both faces of the beam.

$$A_v = 2 \times 0.11 = 0.22 \text{ in.}^2(141.9 \text{ mm}^2) = A_{vh}$$

$$\phi V_s = V_u - \phi V_c$$

or

$$V_s = \frac{V_u}{\phi} - V_c = \frac{519,050}{0.85} - 493,315 = 117,332 \text{ lb } (521.9 \text{ kN})$$

$$V_s = \left(\frac{A_v}{s_v}\frac{1 + l_n/d}{12} + \frac{A_{vh}}{s_h}\frac{11 - l_n/d}{12}\right)f_y d$$

Assume that $s_v = s_h = s$ (similar spacing in both the vertical and horizontal directions). Hence

$$117,332 = \left(\frac{0.22}{s} \frac{1 + 120/65}{12} + \frac{0.22}{s} \frac{11 - 120/65}{12} \right) 60,000 \times 65$$

$$s = 7.31 \text{ in. (185.7 mm)}$$

The maximum permissible spacing of vertical bars $s_v = d/5$ or 18 in., whichever is smaller.

$$s_v = \frac{65}{5} = 13 \text{ in.} \qquad \text{hence } s_v = 7.31 \text{ in. controls}$$

The maximum permissible spacing of horizontal bar $s_h = d/3$ or 18 in., whichever is smaller.

$$s_h = \frac{65}{3} = 21.7 \text{ in.} \qquad \text{hence } s_h = 7.31 \text{ in. controls}$$

Use spacing $s_v = s_h = 7$ in. (177.8 mm).

Check for minimum steel:

Minimum $A_v = 0.0015 b s_v = 0.0015 \times 20 \times 7 = 0.21$ in.$^2 < 0.22$ in.^{2} O.K.

Minimum $A_{vh} = 0.0025 b s_h = 0.0025 \times 20 \times 7 = 0.35$ in.$^2 > 0.22$ in.^{2}

Hence No. 3 bars are not adequate for horizontal steel. No. 4 bars on both faces $= 2 \times 0.20$ in.$^2 = 0.40$ in.^{2}. Use vertical No. 3 bars at 7 in. center to center (9.53 mm diameter at 177.8 mm center to center) and horizontal No. 4 bars at 7 in. center to center (12.70 mm diameter at 177.8 mm center to center), Fig. 6.17. Use of No. 4 bars instead of No. 3 bars in Eq. 6.20 for V_s would give a higher value of the force V_s that the shear reinforcement is resisting.

A better reinforcing system would be to use welded wire fabric in deep beams. For a comparable reinforcing area needed, use size D20 welded wire fabric (0.5 in. = 12.7 mm diameter) spaced at $s_h = 6$ in. (152 mm) center to center in the vertical direction and $s_v = 8$ in. (203 mm) center to center in the horizontal direction.

6.9.6 Example 6.4: Flexural Steel in Deep Beams

Design the flexural reinforcement for the beam in Ex. 6.3.

Solution

$l_n = 120$ in. (3048 mm) and $h = 72$ in. (1828 mm). Since the width of the supports is not given, assume that $l = 1.15 l_n = 138$ in. (3505 mm). The external factored load $U = 148,300$ lb/ft.

External factored moment $M_u = \dfrac{w_u l_n^2}{8} = \dfrac{148,300(10.0)^2}{8} = 1,853,750$ ft-lb

$$= 22,245,000 \text{ in.-lb (2513.7 kNm)}$$

$$\frac{l}{h} = \frac{138}{72} = 1.92 < 2$$

$$jd = 0.2(138 + 2 \times 72) = 56.4 \text{ in.}$$

$$A_s = \frac{22,245,000}{0.9 \times 56.4 \times 60,000} = 7.30 \text{ in.}^2 (4708.5 \text{ mm}^2)$$

Use four No. 9 horizontal bars on each face, area = 8.00 in.^{2}. The height over which A_s is to be distributed above the lower beam face is

$$0.25h - 0.05l = 0.25 \times 72 - 0.05 \times 138 = 11.1 \text{ in.}$$

$$\text{Spacing of flexural steel} = \frac{11.1}{3} = 3.7 \text{ in.}$$

Space four No. 9 bars at 3.5 in. center to center vertical spacing on each face of the deep beam to be well anchored into the supports (28.6-mm-diameter bars at 76.2 mm spacing). Figure 6.17a and b give, respectively, a sectional elevation and a horizontal cross section showing the details of the vertical and horizontal shear reinforcement as well as the flexural reinforcement concentrated at the lower 10.5 in. of the deep beam.

6.9.6 Example 6.5: Reinforcement Design for Continuous Deep Beams

Design the reinforcement necessary for an interior span of a continuous beam over several supports if the loading and the properties of the beam are the same as those of Ex. 6.3.

Solution

Shear reinforcement

Since the deep beam has a large stiffness, the shear continuity factor for the first interior support is assumed to equal 1.0. Hence use the same vertical and horizontal shear reinforcement as in Ex. 6.3. Use size D20 welded wire fabric (0.5 in. = 12.7 mm diameter) spaced at 6 in. (152 mm) center to center in the horizontal direction and 8 in. (203 mm) center to center in the vertical direction, Fig. 6.18.

Flexural Reinforcement

Assume that $d = 65$. in. from Ex. 6.3. The approximate positive factored moment at midspan is

$$+M_u = \frac{w_u l_n^2}{16} = \frac{148,300(10)^2}{16} = 926,875 \text{ ft-lb}$$
$$= 11,122,500 \text{ in.-lb } (1257 \text{ kNm})$$

$$+M_n = \frac{M_u}{\phi} = 12,358,333 \text{ in.-lb } (1397 \text{ kNm})$$

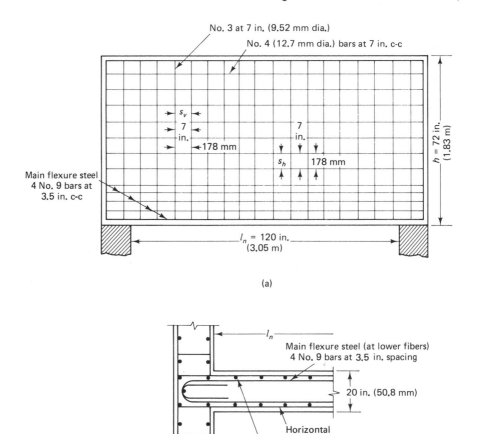

Figure 6.17 Reinforcement for a simply supported deep beam (Ex. 6.4): (a) sectional elevation of beam; (b) cross-sectional plan of beam at support.

lever arm $jd = 56.4$ in. (Ex. 6.3)

$$+A_s = \frac{M_n}{f_y jd} = \frac{12,358,333}{60,000 \times 56.4} = 3.65 \text{ in.}^2 < \frac{200bd}{f_y}$$

$$= \frac{200(20 \text{ in.})(0.9 \times 72 \text{ in.})}{60,000}$$

$$= 4.32 \text{ in.}^2$$

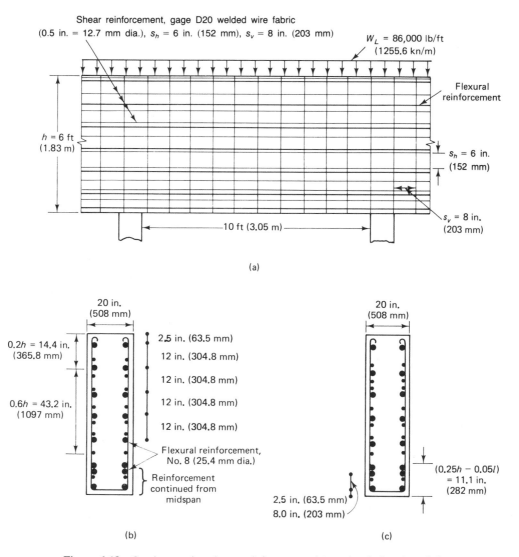

Figure 6.18 Continuous deep beam reinforcement: (a) sectional elevation of the beam; (b) section over support; (c) section at midspan.

Use three No. 8 bars on each face (three 25.4 mm diameter on each face), area = 4.74 in.^{2} (3057 mm^2). Continue the reinforcement over all the beam span into the support as in Fig. 6.18.

The maximum negative factored moment at an interior span is

$$-M_u = \frac{w_u l_n^2}{12} = \frac{148,300(10)^2}{12} = 14,830,000 \text{ in.-lb (1675.8 kN-m)}$$

lever arm $jd = 0.2(l + 1.5h) = 0.2(138 + 1.5 \times 72) = 49.2$ in.

The negative nominal moment of resistance is

$$-M_n = \frac{M_u}{\phi} = \frac{14,830,000}{0.9} = 16,477,778 \text{ in.-lb}$$

$$\text{total negative steel } A_s = \frac{16,477,778}{60,000 \times 49.2} = 5.58 \text{ in.}^2 \ (3599 \text{ mm}^2)$$

The reinforcement area A_{s1} to be provided for the upper zone is

$$0.5\left(\frac{l}{h} - 1\right)A_s = 0.5\left(\frac{138}{72} - 1\right)5.58 = 2.56 \text{ in.}^2$$

$$h_1 = 0.2 \times 72 = 14.4 \text{ in.}$$

$$A_{s2} = 5.58 - 2.56 = 3.02 \text{ in.}^2 \text{ over } h_2 = 72.0 - 14.4 - 14.4 = 43.2 \text{ in.}$$

Using No. 8 bars (25.4 mm diameter):

Zone h_1: Two No. 8 bars on each face (3.14 in.^{2} > 2.56 in.^{2})
Zone h_2: Three No. 8 bars on each face (4.74 in.^{2} > 3.02 in.^{2})

Figure 6.18 shows in elevation and cross section the arrangement of reinforcement for this beam.

6.10 BRACKETS OR CORBELS

Brackets or corbels are short-haunched cantilevers that project from the inner face of columns to support heavy concentrated loads or beam reactions. They are very important structural elements for supporting precast beams, gantry girders, and any other forms of precast structural systems. Precast and prestressed concrete is becoming increasingly dominant, and larger spans are being built, resulting in heavier shear loads at supports. Hence the design of brackets and corbels has become increasingly important. The safety of the total structure could depend on the sound design and construction of the supporting element, in this case the corbel, necessitating a detailed discussion of this subject.

In brackets or corbels, the ratio of the shear arm or span to the corbel depth is often less than 1.0. Such a small ratio changes the state of stress of a member into a two-dimensional one, as discussed in the case of deep beams. Shear deformations would hence affect their nonlinear stress behavior in the elastic state and beyond, and the shear strength becomes a major factor. They differ from deep beams in the existence of potentially large horizontal forces transmitted from the supported beam to the corbel or bracket. These horizontal forces result from long-term shrinkage and creep deformation of the supported beam, which in most cases is anchored to the bracket.

The cracks are usually mostly vertical or steeply inclined pure shear cracks. They often start from the point of application of the concentrated load and propagate

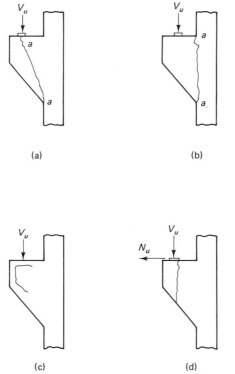

Figure 6.19 Failure patterns: (a) diagonal shear; (b) shear friction; (c) anchorage splitting; (d) vertical splitting.

toward the bottom reentrant corner junction of the bracket to the column face as in Fig. 6.19a, or start at the upper reentrant corner of the bracket or corbel and proceed almost vertically through the corbel toward its lower fibers, as shown in Fig. 6.19b. Other failure patterns in such elements are shown in Fig. 6.19c and d. They can also develop through a combination of the ones illustrated. Bearing failure can also occur by crushing of the concrete under the concentrated load-bearing plate, if the bearing area is not adequately proportioned.

As will be noticed in the subsequent discussion, detailing of the corbel or bracket reinforcement is of major importance. Failure of the element can be attributed in many cases to incorrect detailing that does not realize full anchorage development of the reinforcing bars.

6.10.1 Shear Friction Hypothesis for Shear Transfer in Corbels

Corbels cast at different times than the main supporting columns can have a potential shear crack at the interface between the two concretes through which shear transfer has to develop. As discussed in the case of deep beams, the smaller the ratio a/d, the larger the tendency for pure shear to occur through essentially vertical planes. This

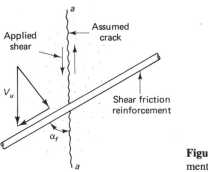

Figure 6.20   Shear-friction reinforcement at crack.

behavior is more accentuated in the case of corbels with a potential interface crack between two dissimilar concretes.

The shear friction approach in this case is recommended by the ACI, as shown in Fig. 6.19b. An assumption is made of an already cracked vertical plane (a–a in Fig. 6.20) along which the corbel is considered to slide as it reaches its limit state of failure. A coefficient of friction μ is used to transform the horizontal resisting forces of the well-anchored closed ties into a vertical nominal resisting force larger than the external factored shear load. Hence the nominal vertical resisting shear force

$$V_n = A_{vf}f_y\mu \tag{6.27a}$$

to give

$$A_{vf} = \frac{V_n}{f_y\mu} \tag{6.27b}$$

where A_{vf} is the total area of the horizontal, anchored closed shear ties.

The external factored vertical shear has to be $V_u \leq \phi V_n$, where for normal concrete,

$$V_n \leq 0.2f_c'b_wd \tag{6.28a}$$

or

$$V_n \leq 800b_wd \tag{6.28b}$$

whichever is smaller. The required effective depth d of the corbel can be determined from Eq. 6.28a or 6.28b, whichever gives a larger value.

For all-lightweight or sand-lightweight concretes, the shear strength V_n should not be taken greater than $(0.2 - 0.07\ a/d)f_c'b_wd$ nor $(800 - 280\ a/d)\ b_wd$ in pounds.

If the shear friction reinforcement is inclined to the shear plane such that the shear force produces some tension in the shear friction steel,

$$V_n = A_{vf}f_y(\mu \sin \alpha_f + \cos \alpha_f) \tag{6.28c}$$

where α_f is the angle between the shear friction reinforcement and the shear plane. The reinforcement area becomes

$$A_{vf} = \frac{V_n}{f_y(\mu \sin \alpha_f + \cos \alpha_f)} \tag{6.28d}$$

The assumption is made that all the shear resistance is due to the resistance at the crack interface between the corbel and the column. The ACI coefficient of friction μ has the following values:

Concrete cast monolithically	1.4λ
Concrete placed against hardened roughened concrete	1.0λ
Concrete placed against unroughened hardened concrete	0.6λ
Concrete anchored to structural steel	0.7λ

where $\lambda = 1.0$ for normalweight concrete, 0.85 for sand-lightweight concrete, and 0.75 for all lightweight concrete.

High values of the friction coefficient μ are used so that the calculated shear strength values are in agreement with experiments. If considerably higher strength concretes are used in the corbels, such as polymer-modified concretes, to interface with the normal concrete of the supporting columns, higher μ values could logically be used for such cases as those listed above. Work by the author in Ref. 6.13 substantiates the use of higher values.

Part of the horizontal steel A_{vf} is incorporated in the top tension tie and the remainder of A_{vf} is distributed along the depth of the corbel as in Fig. 6.21. Evaluation of the top horizontal primary reinforcement layer A_s will be discussed in the next section.

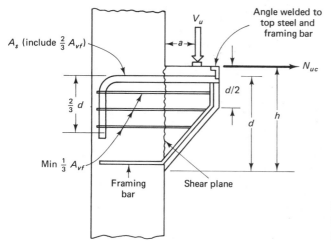

Figure 6.21 Reinforcement schematic for corbel design by shear friction hypothesis.

6.10.2 Horizontal External Force Effect

When the corbel or bracket is cast monolithically with the supporting column or wall and is subjected to a large horizontal tensile force N_{uc} produced by the beam supported by the corbel, a modified approach is used, often termed the *strut theory approach*. In all cases, the horizontal factored force N_{uc} cannot exceed the vertical factored shear V_u. As seen in Fig. 6.22, reinforcing steel A_n has to be provided to resist the force N_{uc}.

$$A_n = \frac{N_{uc}}{\phi f_y} \tag{6.29}$$

and

$$A_f = \frac{V_u a + N_{uc}(h - d)}{\phi f_y jd} \tag{6.30}$$

Reinforcement A_f also has to be provided to resist the bending moments caused by V_u and N_{uc}.

The value of N_{uc} considered in the design should not be less than $0.20V_u$. The flexural steel area A_f can be approximately obtained by the usual expression for the limit state at failure of beams, that is,

$$A_f = \frac{M_u}{\phi f_y jd} \tag{6.31}$$

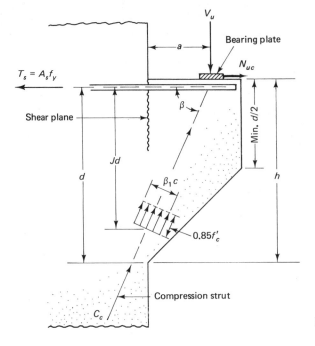

Figure 6.22 Compression strut in corbel.

where $M_u = V_u a + N_{uc}(h - d)$. The axis of such an assumed section lies along a compression strut inclined at an angle β to the tension tie A_s, as shown in Fig. 6.22. The volume C_c of the compressive block is

$$C_c = 0.85f_c'\beta_1 cb = \frac{T_s}{\cos\beta} = \frac{A_s f_y}{\cos\beta} = \frac{V_u}{\sin\beta} \qquad (6.32a)$$

for which the depth $\beta_1 c$ of the block is obtained perpendicular to the *direction* of the compressive strut,

$$\beta_1 c = \frac{A_s f_y}{0.85f_c' b \cos\beta} \qquad (6.32b)$$

The effective depth d minus the $\beta_1 c/2 \cos\beta$ in the vertical direction gives the lever arm jd between the force T_s and the horizontal component of C_c in Fig. 6.22. Therefore,

$$jd = d - \frac{\beta_1 c}{2\cos\beta} \qquad (6.32c)$$

If jd is substituted in Eq. 6.31,

$$A_f = \frac{M_u}{\phi f_y (d - \beta_1 c/2 \cos\beta)} \qquad (6.33)$$

To eliminate several trials and adjustments, the lever arm jd from Eq. 6.32c can be approximated for all practical purposes in most cases as

$$jd \simeq 0.85d \qquad (6.34a)$$

so that

$$A_f = \frac{M_u}{0.85\phi f_y d} \qquad (6.34b)$$

The area A_s of the primary tension reinforcement (tension tie) can now be calculated and placed as shown in Fig. 6.23.

$$A_s \geq \tfrac{2}{3}A_{vf} + A_n \qquad (6.35)$$

or

$$A_s \geq A_f + A_n \qquad (6.36)$$

whichever is larger:

$$\rho = \frac{A_s}{bd} \geq 0.04\frac{f_c'}{f_y}$$

If A_h is assumed to be the total area of the closed stirrups or ties parallel to A_s,

$$A_h \geq 0.5(A_s - A_n) \qquad (6.37)$$

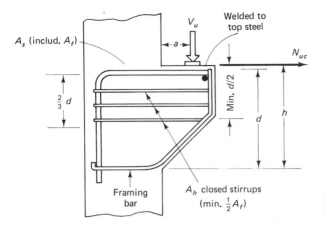

Figure 6.23 Reinforcement schematic for corbel design by strut theory.

The bearing area under the external load V_u on the bracket should not project beyond the straight portion of the primary tension bars A_s, nor should it project beyond the interior face of the transverse welded anchor bar shown in Fig. 6.23.

6.10.3 Sequence of Corbel Design Steps

As discussed in the preceding section, a horizontal factored force N_{uc}, a vertical factored force V_u, and a bending moment $[V_u a + N_{uc}(h - d)]$ basically act on the corbel. To prevent failure, the corbel has to be designed to resist these three parameters simultaneously by one of the following two methods, depending on the type of corbel construction sequence, that is, whether the corbel is cast monolithically with the column or not:

(a) For monolithically cast corbel with the supporting column, by evaluating the steel area A_h of the closed stirrups which are placed below the primary steel ties A_s. Part of A_h is due to the steel area A_n from Eq. 6.29 resisting the horizontal force N_{uc}.

(b) Calculating the steel area A_{vf} by the shear friction hypothesis if the corbel and the column are *not* cast simultaneously, using part of A_{vf} along the depth of the corbel stem and incorporating the balance in the area A_s of the primary top steel reinforcing layer.

The primary tension steel area A_s is the major component of both methods a and b. Calculations of A_s depend on whether Eq. 6.35 or 6.36 governs. If Eq. 6.35 controls, $A_s = \frac{2}{3}A_{vf} + A_n$ is used and the remaining $\frac{1}{3}A_{vf}$ is distributed over a depth $\frac{2}{3}d$ adjacent to A_s.

If Eq. 6.36 controls, $A_s = A_f + A_n$ with the addition of $\frac{1}{2}A_f$ provided as closed stirrups parallel to A_s and distributed within $\frac{2}{3}d$ vertical distance adjacent to A_s.

In both cases, the primary tension reinforcement plus the closed stirrups automatically yield the total amount of reinforcement needed for either type of corbel. Since the mechanism of failure is highly indeterminate and randomness can be expected in the propagation action of the shear crack, it is sometimes advisable to choose the larger calculated value of the primary top steel area A_s in the corbel regardless of whether the corbel element is cast simultaneously with the supporting column.

The horizontal closed stirrups are also a major element in reinforcing the corbel, as seen from the foregoing discussions. Occasionally, additional inclined closed stirrups are also used.

The following sequence of steps is proposed for the design of the corbel:

1. Calculate the factored vertical force V_u and the nominal resisting force V_n of the section such that $V_n \geq V_u/\phi$, where $\phi = 0.85$ for all calculations. V_u/ϕ should be $\leq 0.20 f'_c b_w d$ or $\leq 800 b_w d$. If not, the concrete section at the support should be enlarged.
2. Calculate $A_{vf} = V_n/f_y \mu$ for resisting the shear friction force and use in the subsequent calculation of the primary tension top steel A_s.
3. Calculate the flexural steel area A_f and the direct tension steel area A_n, where

$$A_f = \frac{V_u a + N_{uc}(h - d)}{\phi f_y j_d} \quad \text{and} \quad A_n = \frac{N_{uc}}{\phi f_y}$$

4. Calculate the primary steel area:
   (a) $A_s = \frac{2}{3}A_{vf} + A_n$
   (b) $A_s = A_f + A_n$
   and select whichever is larger. If case (a) controls, the remaining $\frac{1}{3}A_{vf}$ has to be provided as closed stirrups parallel to A_s and distributed with a $\frac{2}{3}d$ distance adjacent to A_s, as in Fig. 6.21.

   If case (b) controls, use in addition $\frac{1}{2}A_f$ as closed stirrups distributed within a distance $\frac{2}{3}d$ adjacent to A_s, as in Fig. 6.23.

$$A_h \geq 0.5(A_s - A_n)$$

and

$$\rho = \frac{A_s}{bd} \geq 0.04\frac{f'_c}{f_y}$$

or

$$\text{minimum } A_s = 0.04\frac{f'_c}{f_y}bd$$

5. Select the size and spacing of the corbel reinforcement with special attention to the detailing arrangements, as many corbel failures are due to incorrect detailing.

6.10.4 Example 6.5: Design of a Bracket or Corbel

Design a corbel to support a factored vertical load V_u = 90,000 lb (180 kN) acting at a distance a = 5 in. (127 mm) from the face of the column. It has a width b = 10 in. (254 mm), a total thickness h = 18 in. (457.2 mm), and an effective depth d = 14 in. (355.6 mm). Given:

> f'_c = 5000 psi (34.5 MPa), normalweight concrete
> f_y = 60,000 psi (414 MPa)

Assume the corbel to be either cast after the supporting column was constructed, or both cast simultaneously. Neglect the weight of the corbel.

Solution

Step 1

$$V_n \geq \frac{V_u}{\phi} = \frac{90,000}{0.85} = 105,882 \text{ lb}$$

$$0.2 f'_c b_w d = 0.2 \times 5000 \times 10 \times 14 = 140,000 \text{ lb} > V_n$$

$$800 b_w d = 800 \times 10 \times 14 = 112,000 \text{ lb} > V_n \qquad \text{O.K.}$$

Step 2

(a) Monolithic construction; normal-weight concrete $\mu = 1.4\lambda$:

$$A_{vf} = \frac{V_u}{\phi f_y \mu} = \frac{105,882}{60,000 \times 1.4} = 1.261 \text{ in.}^2 \text{ (813.3 mm}^2)$$

(b) Nonmonolithic construction; $\mu = 1.0\lambda$:

$$A_{vf} = \frac{105,882}{60,000 \times 1.0} = 1.765 \text{ in.}^2 \text{ (1138.2 mm}^2)$$

Choose the larger A_{vf} = 1.765 in.^{2} as controlling.

Step 3

Since no value of the horizontal external force N_{uc} transmitted from the superimposed beam is given, use

minimum $N_{uc} = 0.20 V_u = 0.2 \times 90,000 = 18,000$ lb

$$A_f = \frac{M_u}{\phi f_y jd} = \frac{V_u a + N_{uc}(h - d)}{\phi f_y jd} \qquad \text{where } jd \approx 0.85d$$

$$= \frac{90,000 \times 5 + 18,000(18 - 14)}{0.90 \times 60,000(0.85 \times 14)} = 0.812 \text{ in.}^2 \text{ (523.9 mm}^2)$$

$$A_n = \frac{N_{uc}}{\phi f_y} = \frac{18,000}{0.85 \times 60,000} = 0.353 \text{ in.}^2 \text{ (277.6 mm}^2)$$

Step 4

Check the controlling area of primary steel A_s:

(a) $A_s = (\frac{2}{3}A_{vf} + A_n) = \frac{2}{3} \times 1.765 + 0.353 = = 1.529$ in.^{2}
(b) $A_s = A_f + A_n = 0.812 + 0.353 = 1.165$ in.^{2}

$$\text{minimum } A_s = 0.04\frac{f'_c}{f_y}bd = 0.04 \times \frac{5,000}{60,000} \times 10 \times 14 = 0.47 \text{ in.}^2$$

$$< 1.529 \qquad \text{O.K.}$$

Provide $A_s = 1.529$ in.^{2} (986.2 mm^2).
Horizontal closed stirrups:
Since case (a) controls,

$$\tfrac{1}{3}A_{vf} = \tfrac{1}{3} \times 1.765 = 0.588 \text{ in.}^2$$

$$A_h = 0.5(A_s - A_n) = 0.5(1.529 - 0.353) = 0.588 \text{ in.}^2$$

Use as the larger of the two values $\frac{1}{3}A_{vf}$ and A_h.

Step 5

Select bar sizes:

(a) Required $A_s = 1.529$ in.^{2}; use three No. 7 bars $= 1.80$ in.^{2} (three bars of diameter 22.2 mm $= 1161$ mm^2).
(b) Required $A_h = 0.588$ in.^{2}; use three No. 3 closed stirrups $= 2 \times 3 \times 0.11 = 0.66$ in.^{2} spread over $\frac{2}{3}d = 9.33$ in. vertical distance. Hence use three No. 3 closed stirrups at 3 in. center to center. Also use three framing size No. 3 bars and one welded No. 3 anchor bar.

Details of the bracket reinforcement are shown in Fig. 6.24. The bearing area under the load has to be checked and the bearing pad designed such that the bearing stress at the factored load V_u should not exceed $\phi(0.85f'_cA_1)$, where A_1 is the pad area.

$$V_u = 90,000 \text{ lb} = 0.70(0.85 \times 5000)$$

$$A_1 = \frac{90,000}{0.70 \times 0.85 \times 5000} = 30.25 \text{ in.}^2 \text{ (19,516 mm}^2\text{)}$$

Use a plate $5\frac{1}{2}$ in. $\times 5\frac{1}{2}$ in. Its thickness has to be designed based on the manner in which V_u is applied.

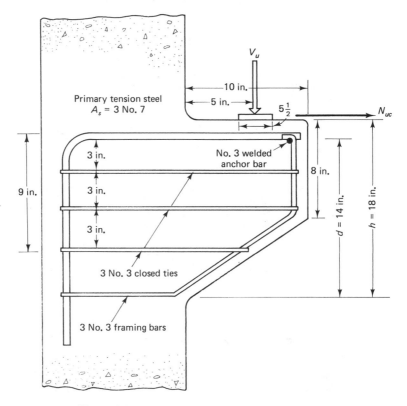

Figure 6.24 Corbel reinforcement details (Ex. 6.5).

SELECTED REFERENCES

6.1   ACI-ASCE Committee 426, "The Shear Strength of Reinforced Concrete Members," *Journal of the Structural Division, ASCE,* Vol. 99, No. ST6, June 1973, 1091–1187.

6.2   Taylor, H. P. J., "The Fundamental Behavior of Reinforced Concrete Beams in Bending and Shear," *Special Publication SP-42,* Vol. 1, American Concrete Institute, Detroit, 1974, pp. 43–77.

6.3   Zsutty, T. C., "Beam Shear Strength Prediction by Analysis of Existing Data," *Journal of the American Concrete Institute,* Proc. Vol. 65, November 1968, pp. 943–951.

6.4   Mattock, A. H., "Diagonal Tension Cracking in Concrete Beams with Axial Forces," *Journal of the Structural Division, ASCE,* Vol. 95, No. ST9, September 1969, pp. 1887–1900.

6.5   Laupa, A., Seiss, C. P., and Newmark, N. M., "Strength in Shear of Reinforced Concrete Beams," *University of Illinois Bulletin No. 428,* Vol. 52, March 1955, 73 pp.

6.6 Moody, K. G., Viest, I. M., Elstner, R. C. and Hognestad, E., "Shear Strength of Reinforced Concrete Beams," *Journal of the American Concrete Institute,* Proc. Vols. 51-15, 51-21, 51-28, and 51-34, February 1954, pp. 317–332, 417–434, 525–539, 697–732, respectively.

6.7 Comité Euro-International du Béton (CEB), "International Recommendations for the Design and Construction of Concrete Structures," June 1970, 80 pp.; and *"CEB-FIP" Model Code For Concrete Structures,* Vol. 2, Paris, April 1978, 345 pp.

6.8 Park, R., and Paulay, T., *Reinforced Concrete Structures,* Wiley, New York, 1975, 768 pp.

6.9 Raths, C. H., and Kriz, L. B., "Connections in Precast Concrete Structures— Strength of Corbels," *Journal of the Prestressed Concrete Institute,* Proc. Vol. 10, No. 1, February 1965, pp. 16–47.

6.10 Leonhardt, F., "Über die Kunst des Bewehrens von Stahlbetontragwerken," *Beton-und Stahlbetonbau,* Vol. 60, No. 8, pp. 181–192; No. 9, pp. 212–220.

6.11 ACI-ASCE Committee 426, "Suggested Revisions to Shear Provisions for Building Codes," ACI 426 IR-77, American Concrete Institute, Detroit, 1979, 84 pp.

6.12 Mattock, A. H., Chen, K. C., and Soongswang, K., "The Behavior of Reinforced Concrete Corbels," *Journal of the Prestressed Concrete Institute,* Vol. 21, No. 2, April 1976, pp. 52–77.

6.13 Nawy, E. G., and Ukadike, M. M., "Shear Transfer in Concrete and Polymer Modified Concrete Members Subjected to Shear Load," *Journal of the American Society for Testing and Materials, Proceedings,* Philadelphia, March 1983, pp. 83–97.

6.14 ACI Committee 340, *Strength Design Handbook,* Vol. 1: *Beams, Slabs, Brackets, Footings and Pile Caps,* Special Publication SP-17(81), American Concrete Institute, Detroit, 1981, 508 pp.

PROBLEMS FOR SOLUTION

6.1 A simply supported beam has a clear span $l_n = 22$ ft (6.70 m) and is subjected to an external uniform service dead load $w_D = 1200$ lb per ft (17.5 kN/m) and live load $w_L = 900$ lb per ft (13.1 kN/m). Determine the maximum factored vertical shear V_u at the critical section. Also determine the nominal shear resistance V_c by both the short method and by the more refined method of taking the contribution of the flexural steel into account. Design the size and spacing of the diagonal tension reinforcement. Given:

$$b_w = 12 \text{ in. (304.8 mm)}$$
$$d = 17 \text{ in. (431.8 mm)}$$
$$h = 20 \text{ in. (508.0 mm)}$$
$$A_s = 6.0 \text{ in.}^2 \text{ (3780 mm}^2)$$

$f'_c = 4000$ psi (27.6 MPa), normalweight concrete
$f_y = 60,000$ psi (413.7 MPa)

Assume that no torsion exists.

6.2   Solve Problem 6.1 assuming that the beam is made of sand-lightweight concrete and that it is subjected to an axial service compressive load of 2500 lb acting at its plastic centroid.

6.3   A cantilever beam is subjected to a concentrated service live load of 25,000 lb. (111.2 kN) acting at a distance of 3 ft 6 in. (1.07 m) from the wall support. Its cross section is 10 in. × 20 in. with an effective depth $d = 17$ in. (431.8 mm). Design the stirrups needed. Given:

$f'_c = 3000$ psi (20.7 MPa), normalweight
$f_y = 40,000$ psi (275.8 MPa)

6.4   The first interior span of a continuous beam has a clear span $l_n = 18$ ft (5.49 m) and is subjected to an intensity of external uniform service live load $w_L = 1800$ lb per linear foot (26.3 kN/m), and a service dead load $w_d = 2200$ lb per linear foot (32.1 kN/m) not including its self-weight. Design the section for flexure and diagonal tension, including the size and spacing of the stirrups, assuming that the beam width $b_w = 15$ in. (381.0 mm). Assume that the beam is not subjected to torsion and that all spans are equal. Given:

$f'_c = 5000$ psi (34.47 MPa), normalweight concrete
$f_y = 60,000$ psi (413.7 MPa)

6.5   A continuous beam has two equal spans $l_n = 18$ ft. (5.49 m) and is subjected to an external service dead load w_D of 350 lb per ft. (5.1 kN/m) and a service live load w_L of 900 lb per ft. (13.2 kN/m). In addition an external service concentrated dead load P_D of 20,000 lb and an external service concentrated live load P_L of 28,500 lb (12.8 kN) are applied to one midspan only. Design the diagonal tension reinforcement necessary. Given:

$f'_c = 5000$ psi (34.47 MPa), normalweight concrete
$f_y = 60,000$ psi (413.7 MPa)

optional

6.6   Design the vertical stirrups for a beam having the shear diagram shown in Fig. 6.25 assuming that $V_c = 2\sqrt{f'_c}\,b_w d$. Given:

$b_w = 14$ in. (355.6 mm)
$d_w = 20$ in.
$V_{u1} = 75,000$ lb (333.6 kN)
$V_{u2} = 60,000$ lb (266.9 kN)

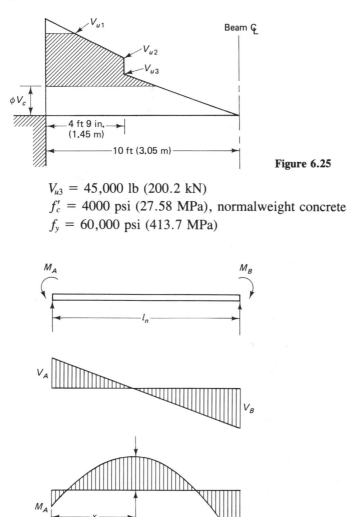

Figure 6.25

$V_{u3} = 45,000$ lb (200.2 kN)
$f'_c = 4000$ psi (27.58 MPa), normalweight concrete
$f_y = 60,000$ psi (413.7 MPa)

Figure 6.26

6.7   Calculate the nominal shear strength V_c of the plain concrete in the web of the continuous normalweight concrete beam shown in Fig. 6.26 using the more refined expression for evaluating the shear. Given:

$$\rho_w = 0.025$$

$$\frac{l_n}{d} = 16$$

$$\frac{x}{l_n} = 0.45$$

$$M_0 = -\frac{w_u l_n^2}{8} = 120,000 \text{ ft-lb (162.7 kNm)}$$

$M_x = 55,000$ ft-lb (74.6 kNm)

$f_c' = 5000$ psi (34.47 MPa), lightweight concrete

$f_y = 60,000$ psi (413.7 MPa)

Also compute the intensity of factored load w_u per foot to which this span is subjected.

6.8 A simply supported deep beam has a clear span $l_n = 10$ ft (3.1 m) and an effective center-to-center span $l = 11$ ft 6 in. (3.5 m). The total depth of the beam is $h = 8$ ft 10 in. (2.7 m). It is subjected to a uniform factored load on the top fibers of intensity $w = 120,000$ lb/ft (1601.8 kN/m), including its self-weight. Design the beam for flexure and shear. Given:

$f_c' = 4500$ psi (31.03 MPa), normalweight concrete

$f_y = 60,000$ psi (413.7 MPa)

Assume the beam to be loaded only in its plane and that wind and earthquakes are not a consideration.

6.9 Solve Problem 6.8 if the same beam was continuous over three spans and was subjected to the same intensity of load.

6.10 Design a bracket to support a concentrated factored load $V_u = 125,000$ lb (556.0 kN) acting at a lever arm $a = 4$ in. (101.6 mm) from the column face, horizontal factored force $N_{uc} = 40,000$ lb (177.9 kN). Given:

$b = 14$ in. (355.6 mm)

$f_c' = 5000$ psi (34.47 MPa), normalweight concrete

$f_y = 60,000$ psi (413.7 MPa)

Assume that the bracket was cast after the supporting column cured and that the column surface at the bracket location was not roughened before casting the bracket. Detail the reinforcing arrangements for the bracket.

6.11 Solve Problem 6.10 if the structural system was made from monolithic sand-lightweight concrete in which the corbel or bracket is cast simultaneously with the supporting column.

7

Torsion

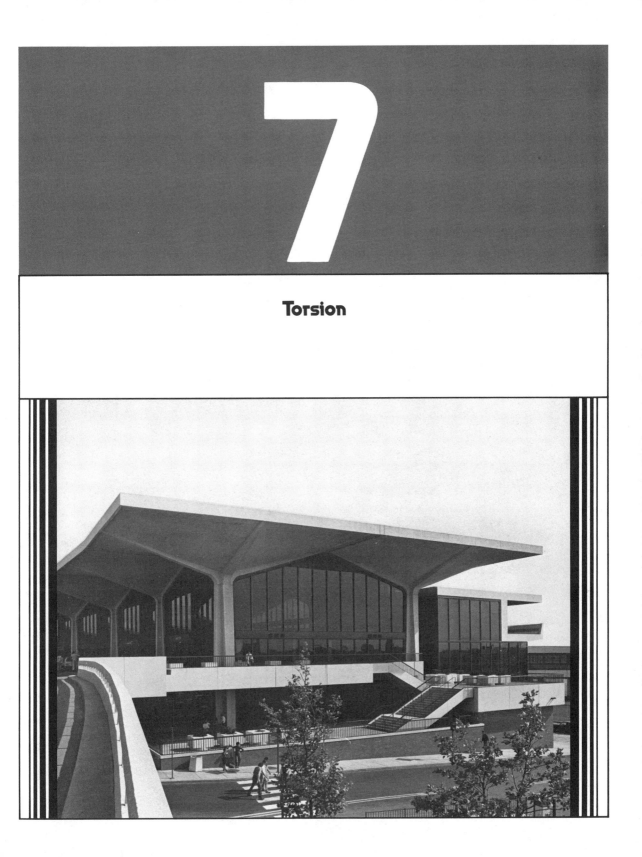

7.1 INTRODUCTION

Torsion occurs in monolithic concrete construction primarily where the load acts at a distance from the longitudinal axis of the structural member. An end beam in a floor panel, a spandrel beam receiving load from one side, a canopy or a bus-stand roof projecting from a monolithic beam on columns, peripheral beams surrounding a floor opening, or a helical staircase are all examples of structural elements subjected to twisting moments. These moments occasionally cause excessive shearing stresses. As a result, severe cracking can develop well beyond the allowable serviceability limits unless special torsional reinforcement is provided. Photos 40 and 41 illustrate the extent of cracking at failure of a beam in torsion. They show the curvilinear plane of twist caused by the imposed torsional moments. In actual spandrel beams of a structural system, the extent of damage due to torsion is usually not as severe, as seen in photos 42 and 43. This is due to the redistribution of stresses in the structure. However, loss of integrity due to torsional distress should always be avoided by proper design of the necessary torsional reinforcement.

An introduction to the subject of torsional stress distribution has to start with the basic elastic behavior of simple sections, such as circular or rectangular sections. Most concrete beams subjected to twist are components of rectangles. They are usually flanged sections such as T beams and L beams. Although circular sections are rarely a consideration in normal concrete construction, a brief discussion of torsion in

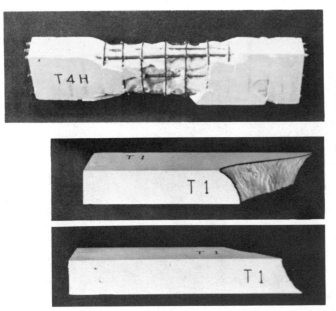

Photo 40 Reinforced plaster beam at failure in pure torsion. (Rutgers tests: Law, Nawy, et al.)

(a)

(b)

Photo 41 Plain mortar beam in pure torsion: (a) top view; (b) bottom view. (Rutgers tests: Law, Nawy, et al.)

Photo 42   Reinforced concrete beams in torsion—testing setup. (Courtesy of Thomas T. C. Hsu.)

Photo 43   Close-up of torsional cracking of beams in the preceding photograph. (Courtesy of Thomas T. C. Hsu.)

circular sections serves as a good introduction to the torsional behavior of other types of sections.

Shear stress is equal to shear strain times the shear modulus at the elastic level in circular sections. As in the case of flexure, the stress is proportional to its distance from the neutral axis (i.e., the center of the circular section) and is maximum at the extreme fibers. If r is the radius of the element, $J = \pi r^4/2$, its polar moment of inertia, and v_{te} the elastic shearing stress due to an elastic twisting moment T_e,

$$v_{te} = \frac{T_e r}{J} \qquad \text{(a)}$$

When deformation takes place in the circular shaft, the axis of the circular cylinder is assumed to remain straight. All radii in a cross section also remain straight (i.e., without warping) and rotate through the same angle about the axis. As the circular element starts to behave plastically, the stress in the plastic outer ring becomes constant while the stress in the inner core remains elastic, as shown in Fig. 7.1. As the whole cross section becomes plastic, $b = 0$ and the shear stress

$$v_{tf} = \frac{3}{4} \frac{T_p r}{J} \qquad \text{(b)}$$

where v_{tf} is the nonlinear shear stress due to an ultimate twisting moment T_p, where the subscript f denotes failure.

In rectangular sections, the torsional problem is considerably more complicated. The originally plane cross sections undergo warping due to the applied torsional moment. This moment produces axial as well as circumferential shear stresses with zero values at the corners of the section and the centroid of the rectangle, and maximum values on the periphery at the middle of the sides, as seen in Fig. 7.2. The maximum torsional shearing stress would occur at midpoints A and B of the larger dimension of the cross section. These complications plus the fact that reinforced

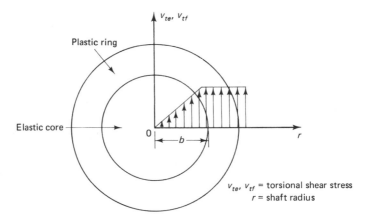

Figure 7.1   Torsional stress distribution through circular section.

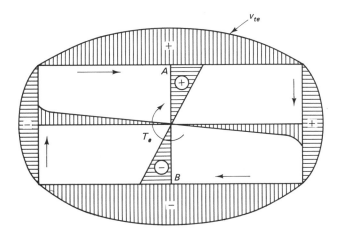

Figure 7.2 Pure torsion stress distribution in a rectangular section.

concrete sections are neither homogeneous nor isotropic make it difficult to develop exact mathematical formulations based on physical models such as Eqs. (a) and (b) for circular sections.

For over 60 years, the torsional analysis of concrete members has been based on either (1) the classical theory of elasticity developed through mathematical formulations coupled with membrane analogy verifications (St.-Venant's), or (2) the theory of plasticity represented by the sand-heap analogy (Nadai's). Both theories were applied essentially to the state of pure torsion. But it was found experimentally that the elastic theory is not entirely satisfactory for the accurate prediction of the state of stress in concrete in pure torsion. The behavior of concrete was found to be better represented by the plastic approach. Consequently, almost all developments in torsion as applied to concrete and reinforced concrete have been in the latter direction.

7.2 PURE TORSION IN PLAIN CONCRETE ELEMENTS

7.2.1 Torsion in Elastic Materials

St.-Venant presented in 1853 his solution to the elastic torsional problem with warping due to pure torsion which develops in noncircular sections. Prandtl in 1903 demonstrated the physical significance of the mathematical formulations by his membrane analogy model. The model establishes particular relationships between the deflected surface of the loaded membrane and the distribution of torsional stresses in a bar subjected to twisting moments. Figure 7.3 shows the membrane analogy behavior for rectangular as well as L-shaped forms.

For small deformations, it can be proved that the differential equation of the deflected membrane surface has the same form as the equation that determines the stress distribution over the cross section of the bar subjected to twisting moments. Similarly, it can be demonstrated that (1) the tangent to a contour line at any point of

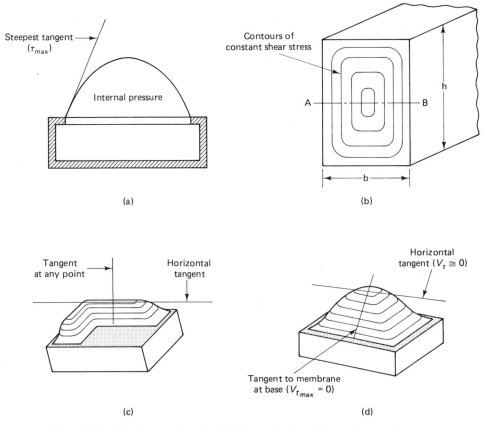

Figure 7.3 Membrane analogy in elastic pure torsion: (a) membrane under pressure; (b) contours in a real beam or in a membrane; (c) L section; (d) rectangular section.

a deflected membrane gives the direction of the shearing stress at the corresponding cross section of the actual membrane subjected to twist; (2) the maximum slope of the membrane at any point is proportional to the magnitude of shear stress τ at the corresponding point in the actual member; (3) the twisting moment to which the actual member is subjected is proportional to *twice* the volume under the deflected membrane.

It can be seen from Figs. 7.2 and 7.3b that the torsional shearing stress is inversely proportional to the distance between the contour lines. The closer the lines, the higher the stress, leading to the previously stated conclusion that the maximum torsional shearing stress occurs at the middle of the longer side of the rectangle. From the membrane analogy, this maximum stress has to be proportional to the steepest slope of the tangents at points A and B.

If δ = maximum displacement of the membrane from the tangent at point A, then from basic principles of mechanics and St.-Venant's theory,

$$\delta = b^2 G\theta \tag{7.1a}$$

where G is the shear modulus and θ is the angle of twist. But $v_{t(max)}$ is proportional to the slope of tangent; hence

$$v_{t(max)} = k_1 bG\theta \tag{7.1b}$$

where the k's are constants. The corresponding torsional moment T_e is proportional to *twice* the volume under the membrane, or

$$T_e \propto 2(\tfrac{2}{3}\,\delta bh) = k_2\,\delta bh$$

or

$$T_e = k_3 b^3 hG\theta \tag{7.1c}$$

From Eqs. 7.1b and 7.1c,

$$v_{t(max)} = \frac{T_e b}{kb^3 h} \simeq \frac{T_e b}{J_1} \tag{7.1d}$$

The denominator $kb^3 h$ in Eq. 7.1d represents the polar moment of inertia J_1 of the section. Comparing Eq. 7.1d to Eq. (a) for the circular section shows the similarity of the two expressions except that the factor k in the equation for the rectangular section takes into account the shear strains due to warping. Equation 7.1d can be further simplified to give

$$v_{t(max)} = \frac{T_e}{kb^2 h} \tag{7.2}$$

It can also be written to give the stress at planes inside the section, such as an inner concentric rectangle of dimensions x and y, where x is the shorter side, so that

$$v_{t(max)} = \frac{T_e}{kx^2 y} \tag{7.3}$$

It is important to note in using the membrane analogy approach that the torsional shear stress changes from one point to another along the same axis as AB in Fig. 7.3, because of the changing slope of the analogous membrane, rendering the torsional shear stress calculations lengthy.

7.2.2 Torsion in Plastic Materials

As indicated earlier, the plastic sand-heap analogy provides a better representation of the behavior of brittle elements such as concrete beams subjected to pure torsion. The torsional moment is also proportional to *twice* the volume under the heap and the maximum torsional shearing stress is proportional to the slope of the sand heap. Figure 7.4 is a two- and three-dimensional illustration of the sand heap. The torsional moment T_p in Fig. 7.4d is proportional to twice the volume of the rectangular heap shown in parts (b) and (c). It can also be recognized that the slope of the sand-heap sides as a

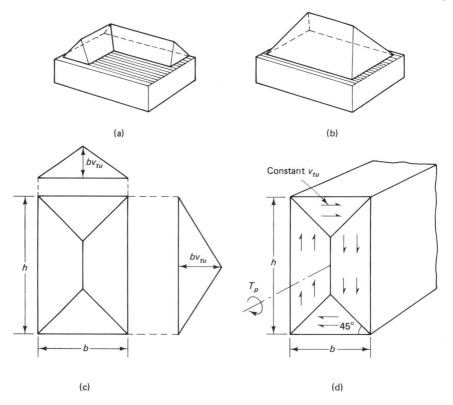

Figure 7.4 Sand-heap analogy in plastic pure torsion: (a) sand-heap L section; (b) sand-heap rectangular section; (c) plan of rectangular section; (d) torsional shear stress.

measure of the torsional shearing stress is *constant* in the sand-heap analogy approach, whereas it is continuously variable in the membrane analogy approach. This characteristic of the sand heap considerably simplifies the solutions.

7.2.3 Sand-Heap Analogy Applied to L Beams

Most concrete elements subjected to torsion are flanged sections, most commonly L beams comprising the external wall beams of a structural floor. The L beam in Fig. 7.5 is chosen in applying the plastic sand-heap approach to evaluate its torsional moment capacity and shear stresses to which it is subjected.

The sand heap is broken into three volumes:

V_1 = pyramid representing a square cross-sectional shape = $y_1 b_w^2 / 3$

V_2 = tent portion of the web representing a rectangular cross-sectional
    shape = $y_1 b_w (h - b_w)/2$

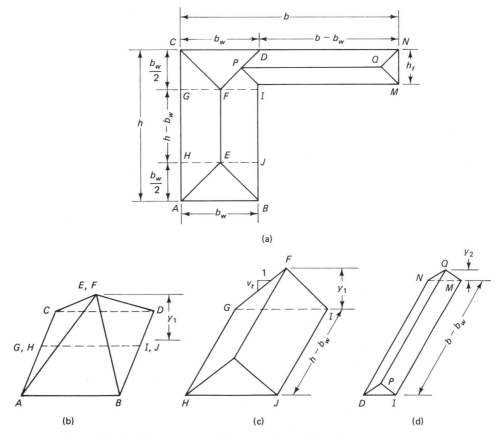

Figure 7.5   Sand-heap analogy of flanged section: (a) sand heap on L-shaped cross section; (b) composite pyramid from web (V_1); (c) tent segment from web (V_2); (d) transformed tent of beam flange (V_3).

V_3 = tent representing the flange of the beam, transferring part *PDI* to
$$NQM = y_2 h_f(b - b_w)/2$$

Torsional moment is proportional to twice the volume of the sand heaps; hence

$$T_p \simeq 2\left[\frac{y_1 b_w^2}{3} + \frac{y_1 b_w(h - b_w)}{2} + \frac{y_2 h_f(b - b_w)}{2}\right] \tag{7.4}$$

Also, torsional shear stress is proportional to the slope of the sand heaps; hence

$$y_1 = \frac{v_t b_w}{2} \tag{7.5a}$$

$$y_2 = \frac{v_t h_f}{2} \tag{7.5b}$$

Substituting y_1 and y_2 from Eqs. 7.5a and 7.5b into Eq. 7.4 gives us

$$v_{t(max)} = \frac{T_p}{(b_w^2/6)(3h - b_w) + (h_f^2/2)(b - b_w)} \tag{7.6}$$

If both the numerator and denominator of Eq. 7.6 are divided by $(b_w h)^2$ and the terms rearranged, we have

$$v_{t(max)} = \frac{T_p h / (b_w h)^2}{[\frac{1}{6}(3 - b_w/h)] + \frac{1}{2}(h_f/b_w)2(b/h - b_w/h)} \tag{7.7a}$$

If one assumes that C_t is the denominator in Eq. 7.7a and $J_E = C_t/(b_w h)^2$, Eq. 7.7a becomes

$$v_{t(max)} = \frac{T_p h}{J_E} \tag{7.7b}$$

where J_E is the equivalent polar moment of inertia and a function of the shape of the beam cross section. Note that Eq. 7.7b is similar in format to Eq. 7.1d from the membrane analogy except for the different values of the denominators J and J_E. Equation 7.7a can be readily applied to rectangular sections by setting $h_f = 0$.

It must also be recognized that concrete is not a perfectly plastic material; hence the actual torsional strength of the plain concrete section has a value lying between the membrane analogy and the sand-heap analogy values.

Equation 7.7b can be rewritten designating $T_p = T_c$ as the nominal torsional resistance of the plain concrete and $v_{t(max)} = v_{tc}$ using ACI terminology, so that

$$T_c = k_2 b^2 h v_{tc} \tag{7.8a}$$

$$T_c = k_2 x^2 y v_{tc} \tag{7.8b}$$

where x is the smaller dimension of the rectangular section.

Extensive work by Hsu and confirmed by others has established that k_2 can be taken as $\frac{1}{3}$. This value originated from research in the skew-bending theory of plain concrete. It was also established that $6\sqrt{f_c'}$ can be considered as a limiting value of the pure torsional strength of a member without torsional reinforcement. Using a reduction factor of 2.5 for the first cracking torsional load $v_{tc} = 2.4\sqrt{f_c'}$ and using $k_2 = \frac{1}{3}$ in Eq. 7.8, results in

$$T_c = 0.8\sqrt{f_c'}\, x^2 y \tag{7.9a}$$

where x is the shorter side of the rectangular section. The high reduction factor of 2.5 is used to offset any effect of bending moments that might be present.

If the cross section is a T or L section, the area can be broken into component rectangles as in Fig. 7.6, such that

$$T_c = 0.8\sqrt{f_c'} \sum x^2 y \tag{7.9b}$$

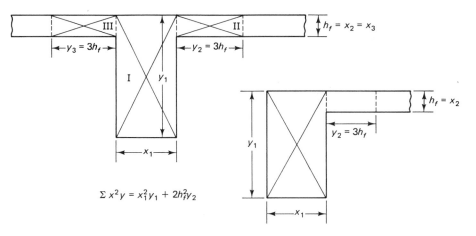

$$\Sigma\, x^2 y = x_1^2 y_1 + 2h_f^2 y_2$$

Figure 7.6 Component rectangles for T_c calculation.

7.3 TORSION IN REINFORCED CONCRETE ELEMENTS

Torsion rarely occurs in concrete structures without being accompanied by bending and shear. The foregoing should give a sufficient background on the contribution of the plain concrete in the section toward resisting *part* of the combined stresses resulting from torsional, axial, shear, or flexural forces. The capacity of the plain concrete to resist torsion when in combination with other loads could, in many cases, be lower than when it resists the same factored external twisting moments alone. Consequently, torsional reinforcement has to be provided to resist the excess torque.

Inclusion of longitudinal and transverse reinforcement to resist part of the torsional moments introduces a new element in the set of forces and moments in the section. If

T_n = required total nominal torsional resistance of the section including the reinforcement

T_c = nominal torsional resistance of the plain concrete

T_s = torsional resistance of the reinforcement

then

$$T_n = T_c + T_s \tag{7.10a}$$

or

$$T_s = T_n - T_c \tag{7.10b}$$

In order to study the contribution of the longitudinal steel bars and the closed transverse bars so that T_s can be evaluated, one has to analyze the system of forces acting on the warped cross sections of the structural element at the limit state of failure. Basically, two approaches are presently acceptable:

1. The skew-bending theory, which is based on the plane deformation approach to plane sections subjected to bending and torsion.
2. The truss analogy theory and its extension as compression field theory. It applies to the torsional stirrups a modified truss analogy comparable to that used for the design of shear stirrups.

7.3.1 Skew-Bending Theory

This theory considers in detail the internal deformational behavior of the series of transverse warped surfaces along the beam. Initially proposed by Lessig, it had subsequent contributions from Collins, Hsu, Zia, Gesund, Mattock, and Elfgren among the several researchers in this field. T. C. C. Hsu made a major contribution experimentally to the development of the skew-bending theory as it presently stands. In his recent book (Ref. 7.12), Hsu details the development of the theory of torsion as applied to concrete structures and how the skew-bending theory formed the basis of the present ACI code provisions on torsion. The complexity of the torsional problem can thus permit in this textbook only the brief discussion that follows.

The failure surface of the normal beam cross section subjected to bending moment M_u remains plane after bending, as in Fig. 7.7a. If a twisting moment T_u is

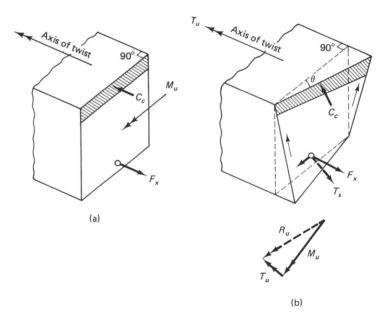

Figure 7.7   Skew bending due to torsion: (a) bending before twist; (b) bending and torsion.

also applied exceeding the capacity of the section, cracks develop on three sides of the beam cross section and compressive stresses on portions of the fourth side along the beam. As torsional loading proceeds to the limit state at failure, a skewed failure surface results due to the combined torsional moment T_u and bending moment M_u. The neutral axis of the skewed surface and the shaded area in Fig. 7.7b denoting the compression zone would no longer be straight but subtend a varying angle θ with the original plane cross sections.

Prior to cracking, neither the longitudinal bars nor the closed stirrups have any appreciable contribution to the torsional stiffness of the section. At the postcracking stage of loading, the stiffness of the section is reduced, but its torsional resistance is considerably increased, depending on the amount and distribution of *both* the longitudinal bars and the transverse *closed* ties. It has to be emphasized that little additional torsional strength can be achieved beyond the capacity of the plain concrete in the beam unless both longitudinal torsion bars and transverse ties are used.

The skew-bending theory idealizes the compression zone by considering it to be of uniform depth. It assumes the cracks on the remaining three faces of the cross section to be uniformly spread, with the steel ties (stirrups) at those faces carrying the tensile forces at the cracks and the longitudinal bars resisting shear through dowel action with the concrete. Figure 7.8a shows the forces acting on the skewly bent plane. The polygon in Fig. 7.8b gives the shear resistance F_c of the concrete, the force T_l of the active longitudinal steel bars in the compression zone, and the normal compressive block force C_c.

The torsional moment T_c of the resisting shearing force F_c generated by the shaded compressive block area in Fig. 7.8a is thus

$$T_c = \frac{F_c}{\cos 45°} \times \text{ its arm about forces } F_v \text{ in Fig. 7.8a}$$

or

$$T_c = \sqrt{2}\, F_c(0.8x) \tag{7.11a}$$

where x is the shorter side of the beam. Extensive tests (Refs. 7.9 and 7.12) to evaluate F_c in terms of internal stress in concrete, $k_1\sqrt{f'_c}$, and the geometrical torsional constants of the section, $k_2 x^2 y$, led to the expression

$$T_c = \frac{2.4}{\sqrt{x}}\, x^2 y \sqrt{f'_c} \tag{7.11b}$$

The dowel forces F_x and F_y are assumed to be proportional to the cross-sectional areas of these bars. If a ratio is established between the proportion of torsional resistance given by the dowel forces F_x and F_y and the torsional resistance of the hoop forces F_v, torsional moments would be the summations

$$\sum F_v(\tfrac{1}{2}x_1),\ \sum F_x(\tfrac{1}{2}y_0),\ \sum F_y(\tfrac{1}{2}x_0),\ \sum T_l(\tfrac{1}{2}x_0)$$

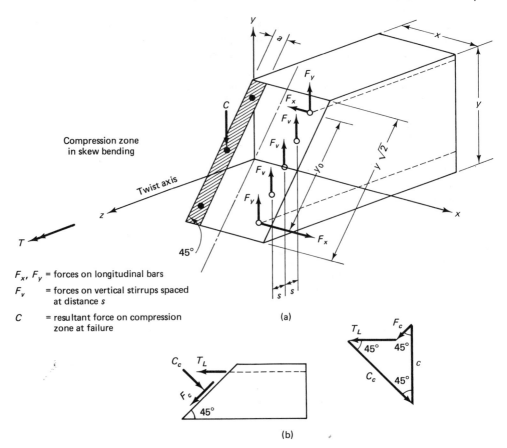

F_x, F_y = forces on longitudinal bars
F_v = forces on vertical stirrups spaced at distance s
C = resultant force on compression zone at failure

(a)

(b)

Figure 7.8 Forces on the skewly bent planes: (a) all forces acting on skew plane at failure; (b) vector forces on compression zone.

The dimensions x_1 and y_1 are, respectively, the shorter and the longer center-to-center dimensions of the closed rectangular stirrups, and the dimensions x_0 and y_0 are the corresponding center-to-center dimensions of the longitudinal bars at the corners of the stirrups. The resulting expression for the torsional strength, T_s, provided by the hoops and the longitudinal steel in the rectangular section is

$$T_s = \alpha_1 \frac{x_1 y_1 A_t f_y}{s} \tag{7.12}$$

where $\alpha_1 = 0.66 + 0.33 y_1/x_1$, so that the total nominal torsional moment of resistance is $T_n = T_c + T_s$ or

$$T_n = \frac{2.4}{\sqrt{x}} x^2 y \sqrt{f'_c} + \left(0.66 + 0.33\frac{y_1}{x_1}\right)\frac{x_1 y_1 A_t f_y}{s} \tag{7.13}$$

7.3.2 Space Truss Analogy Theory

This theory was originally developed by Ramsch and extended later by Lampert and Collins, with additional work by Hsu, Thurliman, Elfgren, and others. Further refinement was introduced by Collins and Mitchell (Ref. 7.11) as a compression field theory. The space truss analogy is an extension of the model used in the design of the shear-resisting stirrups, in which the diagonal tension cracks, once they start to develop, are resisted by the stirrups. Because of the nonplanar shape of the cross sections due to the twisting moment, a space truss composed of the stirrups is used as the diagonal tension members, and the idealized concrete strips at 45° between the cracks are used as the compression members, as shown in Fig. 7.9.

It is assumed in this theory that the concrete beam behaves in torsion similar to a thin-walled box with a constant shear flow in the wall cross section, producing a constant torsional moment. The use of hollow-walled sections rather than solid sections proved to give essentially the same ultimate torsional moment, provided that the walls are not too thin. Such a conclusion is borne out by tests which have shown that the torsional strength of the solid sections is composed of the resistance of the closed stirrup cage, consisting of the longitudinal bars and transverse stirrups, and the

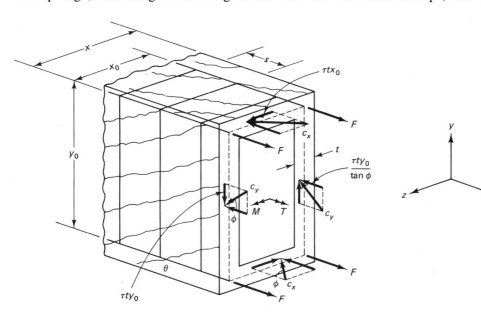

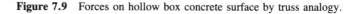

F = tensile force in each longitudinal bar
c_x = inclined compressive force on horizontal side
c_y = inclined compressive force on vertical side
τt = shear flow force per unit length of wall

Figure 7.9   Forces on hollow box concrete surface by truss analogy.

idealized concrete inclined compression struts in the plane of the cage wall. The compression struts are the inclined concrete strips between the cracks in Fig. 7.9.

The CEB-FIP code is based on the space truss model. In this code, the effective wall thickness of the hollow beam is taken as $\frac{1}{6}D_0$ where D_0 is the diameter of the circle inscribed in the rectangle connecting the corner longitudinal bars, namely, $D_0 = x_0$ in Fig. 7.9. In summary, the absence of the core does not affect the strength of such members in torsion—hence the acceptability of the space truss analogy approach based on hollow sections.

If the shear flow in the walls of the box section is τt, where τ is the shear stress, and F is the tensile force in each longitudinal bar at the corner, the force equilibrium equation would be

$$4F = 2\frac{\tau t x_0}{\tan \phi} + 2\frac{\tau t y_0}{\tan \phi} \tag{7.14}$$

and the moments due to the shear flow forces would be

$$T_n = \tau t\, y_0 x_0 + \tau t x_0 y_0 \tag{7.15}$$

If A_t is the area of the stirrup cross section, and f_y is the yield strength of the stirrup spaced at a distance s, then

$$A_t f_y = \tau t s \tan \phi \tag{7.16a}$$

Also, if A_l is the total area of the four longitudinal bars at the corners,

$$F = \tfrac{1}{4}A_l f_y \tag{7.16b}$$

Solving Eqs. (7.14), (7.15), and (7.16a) leads to

$$T_n = 2x_0 y_0\sqrt{\frac{A_l f_y A_t f_y}{2s(x_0 + y_0)}} \tag{7.17}$$

For the case of equal volumes of longitudinal steel and transverse stirrups (i.e. $\phi = 45°$), the torsional moment of resistance T_n at failure would be

$$T_n = 2\frac{A_t f_y}{s}x_0 y_0 \tag{7.18}$$

Note the format similarity of Eq. 7.12, developed by the skew-bending theory, and Eq. 7.18, developed by the space truss analogy theory.

7.4 BEHAVIOR OF CONCRETE UNDER COMBINED TORSION, SHEAR, AND BENDING

7.4.1 Combined Torsion and Shear

Thus far the discussion has presented the internal resisting mechanism and the accompanying forces, moments, and stresses in the plain concrete and in the reinforcement

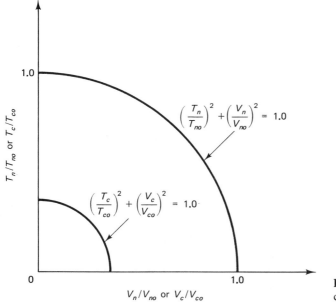

Figure 7.10 Interaction diagram for combined torsion and shear.

when a one-dimensional structural element is subjected to twisting moments. When external torsion is accompanied by external shear, the same section is subjected to higher shearing stresses because of the combined effect of the two loading types as they interact with each other. A beam's resistance to combined torsion and shear is less than its resistance to either of these two parameters when acting alone. Consequently, an interaction relationship becomes necessary in a similar manner as that developed for combined bending and axial load discussed in Chapter 9. Figure 7.10 represents the following nondimensional interaction expression relating torsion to shear:

1. Member without web steel:

$$\left(\frac{T_c}{T_{c0}}\right)^2 + \left(\frac{V_c}{V_{c0}}\right)^2 \leq 1.0 \qquad (7.19a)$$

T_c and V_c are the nominal external torsion and shear when acting *simultaneously*. T_{c0} and V_{c0} are the nominal values for torsion and shear when each acts alone.

2. Members reinforced for combined torsion and shear:

$$\left(\frac{T_n}{T_{n0}}\right)^2 + \left(\frac{V_n}{V_{n0}}\right)^2 \leq 1.0 \qquad (7.19b)$$

T_n and V_n represent the nominal torsional and shear strengths to resist T_u and V_u when acting *simultaneously*. $T_{n0} = T_c + T_s$ represents the nominal torsional resistance of the reinforced web when pure torsion *alone* acts on the section; $V_{n0} = V_c + V_s$ represents the nominal shear resistance of the reinforced web when shear *alone* acts on the section. Equation 7.19a can be rewritten using the approximate values of T_c from Eq. 7.11b and V_c from Eq. 6.9 for the nonreinforced web:

$$\left(\frac{T_c}{0.8\sqrt{f_c'}\ \Sigma\ x^2y}\right)^2 + \left(\frac{V_c}{2\sqrt{f_c'}b_wd}\right)^2 \leq 1.0 \qquad (7.20)$$

In the case of the reinforced web subjected to combined torsion and shear, an upper limit on T_{n0} and V_{n0} has to be placed to ensure yielding of the web reinforcement at the limit state of failure. Based on test results,

$$T_{n0} \leq 12\sqrt{f_c'}\ \frac{\Sigma x^2y}{3} \qquad \text{and} \qquad V_{n0} \leq 10\sqrt{f_c'}\ b_wd$$

Consequently, Eq. 7.19b becomes

$$\left(\frac{T_n}{4\sqrt{f_c'}\ \Sigma\ x^2y}\right)^2 + \left(\frac{V_n}{10\sqrt{f_c'}b_wd}\right)^2 \leq 1.0 \qquad (7.21)$$

It can be seen by comparing Eqs. 7.20 and 7.21 that $T_n = 5T_c$. The ACI code simplifies the procedure by requiring that

$$T_s \leq 4T_c \qquad (7.22)$$

Otherwise, the beam cross section has to be enlarged.

7.4.2 Combined Torsion and Bending

When bending acts simultaneously with torsion, the bending capacity of the section is drastically reduced. As a result, cracking due to the torsional shear stress is generated at low load levels. Figure 7.11c shows the vector resultant R_u for combined bending and torsional moments which cause warping of the section, as shown in Fig. 7.7b.

In a manner similar to the case of combined torsion and shear, an interaction relationship is established relating torsion to bending when they act *simultaneously*. It has to be assumed that the section is reinforced with both compression and tension steel.

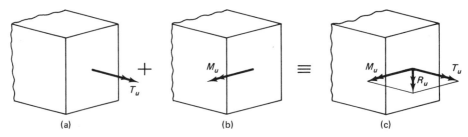

Figure 7.11 Schematic vector representation of combined torsion and bending: (a) torsion; (b) bending; (c) combined bending and torsion.

Two cases can develop for which the following interaction expressions are applicable:

1. When the tension steel yields in the tension zone,

$$\left(\frac{T_n}{T_{n0}}\right)^2 = r\left(1 - \frac{M_n}{M_{n0}}\right) \tag{7.23a}$$

2. When the tension yielding occurs in the flexural compression zone,

$$\left(\frac{T_n}{T_{n0}}\right)^2 = 1 + r\frac{M_n}{M_{n0}} \tag{7.23b}$$

where T_n = nominal torsional moment strength equivalent to T_u/ϕ
T_{n0} = nominal torsional resistance of the reinforced web when pure torsion acts alone
M_n = nominal bending moment strength M_u/ϕ
M_{n0} = nominal bending strength when bending alone acts
$r = \dfrac{A_s f_y}{A_s' f_y}$

Figure 7.12 shows the interaction diagram for combined torsion and bending for force ratio $r = A_s f_y/A_s' f_y$ between the values 1.0 and 3.0.

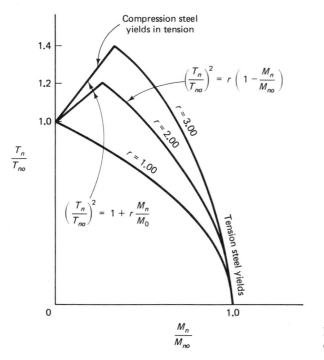

Figure 7.12 Interaction diagram for combined torsion and bending.

7.4.3 Combined Bending, Shear, and Torsion

A combination of these three parameters results in an interaction three-dimensional surface. The scope of the book limits the feasibility of an in-depth discussion. The applicable expression results from superimposing the effect of the combined torsion and shear on the effect of the combined bending and torsion from the two interaction cases in Sections 7.4.1 and 7.4.2. The ACI code requires (1) calculation of the transverse web steel for shear, adding it to the separately calculated transverse web steel for torsion; and (2) calculation of the longitudinal steel for torsion, adding it to the tension steel required for bending, but distributing it symmetrically on all faces of the cross section.

7.5 DESIGN OF REINFORCED CONCRETE BEAMS SUBJECTED TO COMBINED TORSION, BENDING, AND SHEAR

7.5.1 Torsional Behavior of Structures

The torsional moment acting on a particular structural component such as a spandrel beam can be calculated using normal structural analysis procedures. Design of the particular component needs to be based on the limit state at failure. Therefore, the nonlinear behavior of a structural system after torsional cracking must be identified in one of the following two conditions: (1) no redistribution of torsional stresses to other members after cracking, and (2) redistribution of torsional stresses and moments after cracking to effect deformation compatibility between intersecting members.

Stress resultants due to torsion in statically determinate beams can be evaluated from equilibrium conditions alone. Such conditions require a design for the full factored external torsional moment, as no redistribution of torsional stresses is possible. This state is often termed *equilibrium torsion*. An edge beam supporting a cantilever canopy as in Fig. 7.13 is such an example.

The edge beam has to be designed to resist the *total* external factored twisting moment due to the cantilever slab; otherwise, the structure will collapse. Failure would be caused by the beam not satisfying conditions of equilibrium of forces and moments resulting from the large external torque.

In statically indeterminate systems, stiffness assumptions, compatibility of strains at the joints, and redistribution of stresses may affect the stress resultants, leading to a reduction of the resulting torsional shearing stresses. A reduction is permitted in the value of the factored moment used in the design of the member if part of this moment can be redistributed to the intersecting members. The ACI code allows a maximum factored torsional moment at the critical section d from the face of the supports:

$$T_u = \phi\left(4\sqrt{f_c'}\,\frac{\Sigma x^2 y}{3}\right) \qquad (7.24)$$

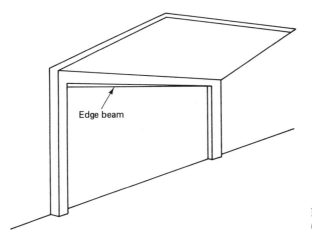

Figure 7.13 No redistribution torsion (equilibrium torsion).

Neglect of the full effect of the total external torsion in this case does not, in effect, lead to failure of the structure but may result in excessive cracking if $\phi(4\sqrt{f'_c}\ \Sigma x^2 y/3)$ is considerably smaller in value than the actual factored torque. An example of compatibility torsion can be seen in Fig. 7.14.

Beams B_2 apply twisting moments T_u at sections 1 and 2 of spandrel beam AB in Fig. 7.14b. The magnitudes of relative stiffnesses of beam AB and transverse beams B_2 determine the magnitudes of rotation at intersecting joints 1 and 2. Because of continuity and two-way action, the end moments for beams B_2 at their intersections with spandrel beam AB will not be fully transferred as twisting moments to the column supports at A and B. They would be greatly reduced as moment redistribution results in transfer of most of the end bending moments from ends 1 and 2 to ends 3 and 4 as well as the midspan of beams B_2. T_u at each spandrel beam supports A and B and at the critical section a distance d from these supports is determined from Eq. 7.24.

$$T_u = \phi\left(4\sqrt{f'_c}\frac{\Sigma x^2 y}{3}\right)$$

If the actual factored torque due to beams B_2 is less than that given by Eq. 7.24, the beam has to be designed for the lesser torsional value. Torsional moments are neglected, however, if

$$T_u < \phi\left(0.5\sqrt{f'_c}\ \Sigma\ x^2 y\right) \tag{7.25}$$

When the factored torsional moment T_u exceeds $\phi(0.5\sqrt{f'_c}\ \Sigma\ x^2 y)$ the ACI code requires that the plain concrete web in sections be designed for

$$V_c = \frac{2\sqrt{f'_c}b_w d}{\sqrt{1 + [2.5C_t(T_u/V_u)]^2}} \tag{7.26a}$$

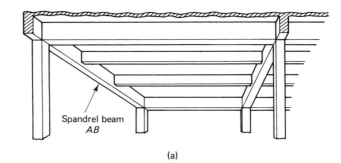

(a)

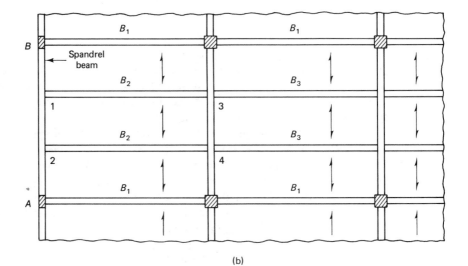

(b)

Figure 7.14 Torsion redistribution (compatibility): (a) isometric view of one end panel; (b) plan of a typical one-way floor system.

and

$$T_c = \frac{0.8\sqrt{f'_c}\,\Sigma x^2 y}{\sqrt{1 + (0.4V_u/C_t T_u)^2}} \qquad (7.26b)$$

Equations 7.26a and 7.26b are derived from Eq. 7.20 assuming that the ratio of the torsional moment to the shear force remains constant throughout the loading history. When the contribution of torsion reinforcement is taken into account, the ACI limits the torsional force T_s resisted by the steel to a value not exceeding $4T_c$, as in Eq. 7.22.

7.5.2 Torsional Web Reinforcement

As indicated in Section 7.3.1, meaningful additional torsional strength due to the addition of torsional reinforcement can be achieved only by using *both* stirrups and longitudinal bars. Ideally, equal volumes of steel in both the closed stirrups and the longitudinal bars should be used so that both participate equally in resisting the twisting moments. This principle is the basis of the ACI expressions for proportioning the torsional web steel. If s is the spacing of the stirrups, A_l is the total cross-sectional area of the longitudinal bars, and A_t is the cross section of one stirrup leg, where the dimensions of the stirrup are x_1 in the short direction and y_1 in the long direction, then

$$2A_t(x_1 + y_1) = A_l s \qquad (7.27a)$$

so that

$$2A_t = \frac{A_l s}{x_1 + y_1} \qquad (7.27b)$$

Hence the total torsional web steel, including both the closed stirrups and the longitudinal bars for Eqs. 7.27a and 7.27b, becomes

$$A_{\text{total}} = 2A_t + \frac{A_l s}{x_1 + y_1} \qquad (7.28a)$$

But, from Eq. 7.12,

$$A_t = \frac{T_s s}{\alpha_1 x_1 y_1 f_y} \qquad (7.28b)$$

where $\alpha_1 = 0.66 + 0.33 y_1/x_1 \le 1.5$ and T_s is the torsional resisting moment of the torsional web steel. If T_c is the nominal torsional resistance of the plain concrete in web,

$$T_s = T_n - T_c \qquad (7.29)$$

From Eq. 7.27b and using the ACI expression for A_t for combined torsion and shear, where

$$2A_t = \frac{200 \times s}{f_y}\frac{T_u}{T_u + V_u/3C_t}$$

the longitudinal torsional reinforcement can be expressed as

$$A_l = \left(\frac{400 \times s}{f_y}\frac{T_u}{T_u + V_u/3C_t} - 2A_t\right)\frac{x_1 \times y_1}{s} \qquad (7.30)$$

where $C_t = b_w d/\Sigma x^2 y$. The term $2A_t$ in Eq. 7.30 cannot be less than $50\, b_w s/f_y$ since this value is the minimum $2A_t$ for the torsional stirrups to be effective. A thorough discussion and detailed derivation of Eq. 7.30 is presented in Ref. 7.12.

A reduction in stirrups can be compensated by an increase in longitudinal steel as long as the total torsional steel *volume* remains the same. If the spacing s of the stirrups is small such that $2A_t$ is considerably larger than the minimum value $50b_w s/f_y$, it is not uncommon that A_l from Eq. 7.30 gives a negative value so that the minimum A_l from Eq. 7.27a for equal volumes of stirrups and longitudinal bars is invoked; that is,

$$A_l = 2A_t \frac{x_1 + y_1}{s} \tag{7.31}$$

The total area A_{vt} of the closed stirrups for combined torsion and shear becomes

$$A_{vt} = 2A_t + A_v \geq \frac{50b_w s}{f_y} \tag{7.32}$$

7.5.3 Design Procedure for Combined Torsion and Shear

The following is a summary of the recommended sequence of design sterps. A flowchart describing the sequence of operations in graphical form is shown in Fig. 7.15.

1. Classify whether the applied torsion is equilibrium or compatibility torsion. Determine the critical section and calculate the factored torsional moment T_u. The critical section is taken at a distance d from the face of the support. If T_u is less than $\phi(0.5\sqrt{f_c'}\ \Sigma\ x^2 y)$, torsional effects can be neglected.
2. Calculate the nominal torsional resistance T_c of the plain concrete web:

$$T_c = \frac{0.8\sqrt{f_c'}\ \Sigma\ x^2 y}{\sqrt{1 + (0.4V_u/C_t T_u)^2}}$$

where $C_t = b_w d/\Sigma\ x^2 y$. Members subjected to significant axial tension may be designed for a T_c value which is multiplied by $(1 + N_u/500Ag)$, where N_u is negative for tension.

Check if T_u exceeds ϕT_c. If it does not, disregard torsional effect. If it does, calculate the value T_s of that portion of the twisting moment to be resisted by the steel reinforcement. For equilibrium torsion

$$T_s = T_n - T_c$$

For compatibility torsion

$$T_s = \frac{4}{3}\sqrt{f_c'}\ \Sigma\ x^2 y - T_c \quad \text{or} \quad T_s = T_n - T_c$$

whichever is less. The value of T_n has to be at least equivalent to T_u/ϕ. If $T_s > 4T_c$, enlarge the section.

Select the closed stirrups to be used as transverse reinforcement. A minimum No. 3 (9.5 mm diameter) bar size can be used. If s = constant spacing of

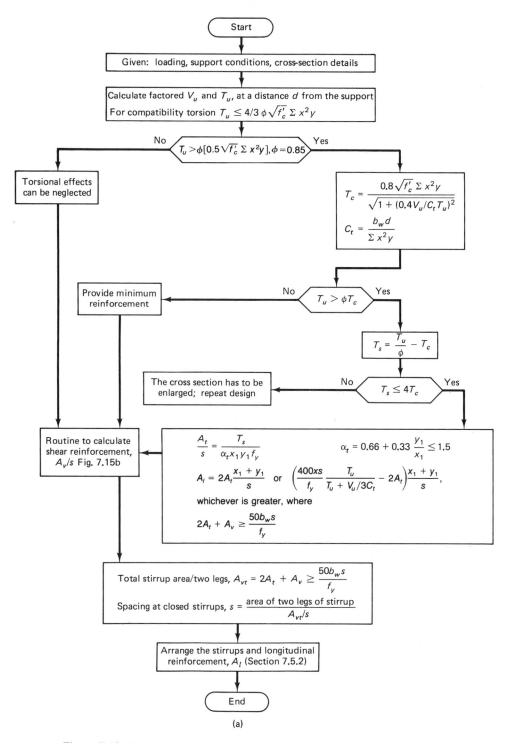

Figure 7.15 Flowchart for the design reinforcement for combined shear and torsion: (a) torsional web steel; (b) shear web steel.

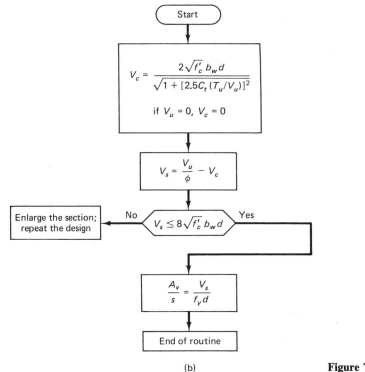

(b)   **Figure 7.15** (*cont.*)

the stirrups, calculate the area of torsion stirrup per *one* leg per unit spacing:

$$\frac{A_t}{s} = \frac{T_s}{\alpha_t x_1 y_1 f_y}$$

3. Calculate the required shear reinforcement A_v per unit spacing in a transverse section. V_u is the factored external shear force at the critical section, V_c is the nominal shear resistance of the concrete in the web, and V_s is the shearing force to be resisted by the stirrups:

$$\frac{A_v}{s} = \frac{V_s}{f_y d}$$

where $V_s = V_n - V_c$ and

$$V_c = \frac{2\sqrt{f_c'}\, b_w d}{\sqrt{1 + [2.5 C_t (T_u/V_u)]^2}}$$

The value of V_n has to be at least equal to V_u/ϕ.

4. Obtain the total A_{vt}, the area of closed stirrups for torsion and shear, and design the stirrups such that

$$A_{vt} = 2A_t + A_v \geq \frac{50b_w s}{f_y}$$

5. Calculate the area of longitudinal reinforcement A_l required for torsion where

$$A_l = 2A_t \frac{x_1 + y_1}{s}$$

or

$$A_l = \left(\frac{400xs}{f_y} \frac{T_u}{T_u + V_u/3C_t} - 2A_t \right) \frac{x_1 + y_1}{s}$$

whichever is larger. A_l calculated using the second expression need not exceed

$$A_l = \left(\frac{400xs}{f_y} \frac{T_u}{T_u + V_u/3C_t} - \frac{50b_w s}{f_y} \right) \frac{x_1 + y_1}{s}$$

6. Arrange the reinforcement using the following guidelines:
    (a) Spacing s of closed stirrups should be less than $(x_1 + y_1)/4$ or 12 in.
    (b) Longitudinal bars should be equally spaced around the perimeter of the closed stirrups. The distance between the bars should be less than 12 in. and at least one longitudinal bar should be placed in each corner.
    (c) Design yield strength of torsion reinforcement should not exceed 60,000 psi.
    (d) Stirrups used for torsion reinforcement have to be anchored through a distance d from the extreme compression fibers. Closed ties with hooks at ends achieve this effect.
    (e) Torsional reinforcement should be provided at least a distance $(d + b)$ beyond the point theoretically required, in order to cover any potential excessive shearing stresses.

7.5.4 Example 7.1: Design of Web Reinforcement for Combined Torsion and Shear in a T-Beam Section

A T-beam cross section has the geometrical dimensions shown in Fig. 7.16. A factored external shear force acts at the critical section, having a value $V_u = 15,000$ lb (67.5 kN). It is subjected to the following torques: (a) equilibrium factored external torsional moment $T_u = 500,000$ in.-lb (57.15 kN-m), (b) compatibility factored $T_u = 75,000$ in.-lb (8.47 kN-m), and (c) compatibility factored $T_u = 300,000$ in.-lb. Given:

   Bending reinforcement $A_s = 3.4$ in.^{2} (2193 mm^2)
   $f_c' = 4000$ (27.58 MPa), normalweight concrete
   $f_y = 60,000$ (413.7 MPa)

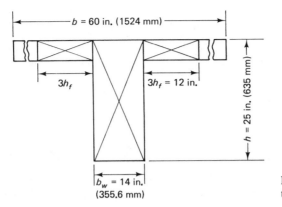

Figure 7.16   Component rectangles of the T beam.

Design the web reinforcement needed for this section.

Solution

(a) Equilibrium torsion:

Factored torsional moment (Step 1)

Given equilibrium torsional moment = 500,000 in.-lb (57.15 kN-m). The total torsional moment must be provided for in the design. From Fig. 7.16

$$\Sigma x^2y = 14^2 \times 25 + 4^2 \times 3 \times 4 + 4^2 \times 3 \times 4 = 5284 \text{ in.}^3$$

$$\phi(0.5\sqrt{f_c'}\,\Sigma x^2y) = 0.85 \times 0.5 \times \sqrt{4000} \times 5284 = 142,030 \text{ in.-lb} < T_u$$

Therefore, stirrups should be provided.

Torsion closed stirrup design (Step 2)

$$T_n = \frac{T_u}{\phi} = \frac{500,000}{0.85} = 588,235 \text{ in.-lb (66.47 kN-m)}$$

$$T_c = \frac{0.8\sqrt{f_c'}\,\Sigma x^2y}{\sqrt{1 + (0.4V_u/C_tT_u)^2}}$$

Assume an effective cover of 2.5 in. and $d = 25.0 - 2.5 = 22.5$ in.

$$C_t = \frac{b_wd}{\Sigma x^2y} = \frac{14 \times 22.5}{5284} = 0.0596$$

$$T_c = \frac{0.8\sqrt{4000} \times 5284}{\sqrt{1 + \left(\dfrac{0.4 \times 15,000}{0.0596 \times 500,000}\right)^2}} = 262,092 \text{ in.-lb (29.61 kN-m)}$$

Also assume that both T_c and V_c are constant for all practical purposes to the midspan of this beam.

$$T_s = T_n - T_c = 588,235 - 262,092 = 326,143 \text{ in.-lb (36.85 kN-m)}$$

Assume $1\frac{1}{2}$ in. clear cover and No. 4 closed stirrups.

$$x_1 = 14 - 2(1.5 + 0.25) = 10.5 \text{ in.}$$

$$y_1 = 25 - 2(1.5 + 0.25) = 21.5 \text{ in.}$$

$$\alpha_t = 0.66 + 0.33 \times \frac{21.5}{10.5} = 1.34 < 1.5$$

Use $\alpha_t = 1.34$.

$$\frac{A_t}{s} = \frac{T_s}{f_y \alpha_t x_1 y_1} = \frac{326{,}143}{60{,}000 \times 1.34 \times 10.5 \times 21.5}$$

$$= 0.0180 \text{ in.}^2/\text{in. spacing/one leg}$$

Shear stirrup design (Step 3)

$$V_c = \frac{2\sqrt{f_c'}\, b_w d}{\sqrt{1 + [2.5 C_t (T_u/V_u)]^2}} = \frac{2\sqrt{4000} \times 14 \times 22.5}{\sqrt{1 + (2.5 \times 0.0596 \times 500{,}000/15{,}000)^2}}$$

$$= 7865 \text{ lb (35.39 kN)}$$

$$V_s = V_n - V_c = \frac{15{,}000}{0.85} - 7865 = 9782 \text{ lb (44.2 kN)}$$

$$\frac{A_v}{s} = \frac{V_s}{f_y d} = \frac{9782}{60{,}000 \times 22.5} = 0.0072 \text{ in.}^2/\text{in. spacing/two legs}$$

Combined closed stirrups for torsion and shear (Step 4)

$$\frac{A_{vt}}{s} = \frac{2A_t}{s} + \frac{A_v}{s} = 2 \times 0.0180 + 0.0072 = 0.0432 \text{ in.}^2/\text{in. /two legs}$$

Try No. 3 (9.5 mm diameter) closed stirrups. The area for two legs = 0.22 in.^{2} (142 mm^2).

$$s = \frac{\text{area of stirrup cross section}}{\text{required } A_{vt}/s} = \frac{0.22}{0.0432} = 5.09 \text{ in.}$$

maximum allowable spacing, $s_{max} = \dfrac{x_1 + y_1}{4} = \dfrac{10.5 + 21.5}{4} = 8 \text{ in.} > 5.09 \text{ in. O.K.}$

Use No. 3 closed stirrups at 5 in. (127 mm) center to center.

$$\text{minimum stirrup area required} = A_v + 2A_t = \frac{50 b_w s}{f_y} = \frac{50 \times 14 \times 5}{60{,}000} = 0.0583 \text{ in.}^2$$

$$\text{area provided} = 0.22 \text{ in.}^2 > 0.0583 \text{ in.}^2 \quad \text{O.K.}$$

Longitudinal torsion steel design (Step 5)

$$A_l = 2A_t \frac{x_1 + y_1}{s} = 2 \times 0.018(10.5 + 21.5) = 1.152 \text{ in.}^2$$

Also

$$A_l = \left(\frac{400xs}{f_y} \frac{T_u}{T_u + V_u/3C_t} - 2A_t \right) \frac{x_1 + y_1}{5}$$

(or substituting $50b_w s/f_y$ for $2A_t$ whichever controls).

$$\frac{50b_w s}{f_y} = 0.0583 < 2A_t = 2 \times 0.0180 \times 5 = 0.18 \text{ in.}^2$$

Use $2A_t = 0.18$ in.^{2}. Hence

$$A_l = \left(\frac{400 \times 14 \times 5}{60,000} \frac{500,000}{500,000 + \dfrac{15,000}{3 \times 0.0596}} - 0.18 \right) \frac{10.5 + 21.5}{5}$$

$$= 1.41 \text{ in.}^2 > 1.152 \text{ in.}^2$$

Therefore, use $A_l = 1.41$ in.^{2}.

Distribution of torsion longitudinal bars

Torsional $A_l = 1.41$ in.^{2}. Assume that $\frac{1}{4}A_l$ goes to the top corners and $\frac{1}{4}A_l$ goes to the bottom corners of the stirrups, to be added to the flexural bars. The balance, $\frac{1}{2}A_l$, would thus be distributed equally to the vertical faces of the beam cross section at a spacing not to exceed 12 in. center to center.

$$\text{midspan } \Sigma A_s = \frac{A_l}{4} + A_s = \frac{1.41}{4} + 3.4 = 3.75 \text{ in.}^2$$

Provide five No. 8 (25.4 mm diameter) bars at the bottom. Provide two No. 4 (12.7 mm diameter) bars with an area of 0.40 in.^{2} at the top. The required area of $A_l/4$ is 0.35 in.^{2}. The area of steel needed for each vertical face = 0.35 in.^{2}. Provide two No. 4 (12.7 mm diameter) bars on each side. Figure 7.17 shows the geometry of the cross section.

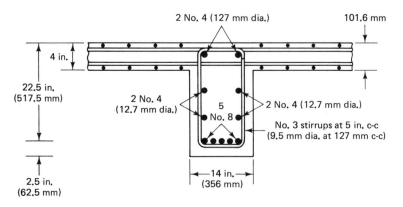

Figure 7.17   Web reinforcement details, Ex. 7.1(a).

Solution

(b) Compatibility torsion:

Factored torsional moment (Step 1)

Given T_u = 75,000 in.-lb (8.47 kN/m). Using the results of case (a), one gets

$$\phi(0.5 \sqrt{f_c'} \, \Sigma x^2 y) = 0.85 \times 0.5 \times \sqrt{4000} \times 5284 \text{ in.}^3$$

$$= 142{,}030.5 \text{ in.-lb (16.04 kN-m)} > T_u$$

$$= 75{,}000 \text{ in.-lb}$$

Hence torsion effects can be neglected, and only design for shear is needed.

Solution

(c) Compatibility torsion:

Factored torsional moment (Step 1)

Given that T_u = 300,000 in.-lb (33.9 kN-m) is greater than $\phi(0.5 \sqrt{f_c'} \, \Sigma x^2 y)$. Hence stirrups should be provided. Since this is a compatibility torsion, the section can be designed for a torsional moment of $\phi(4 \sqrt{f_c'} \, \Sigma x^2 y / 3)$ if the external torsion exceeds this value.

$$\phi\left(4 \sqrt{f_c'} \, \frac{\Sigma x^2 y}{3}\right) = 378{,}748 \text{ in.-lb} > \text{given } T_u = 300{,}000 \text{ in.-lb}$$

Hence the section should be designed for T_u = 300,000 in.-lb.

Torsion closed stirrup design (Step 2)

Using Eq. 7.26b,

$$T_c = 253{,}467 \text{ in.-lb (28.64 kN-m)}$$

$$T_s = T_n - T_c = \frac{300{,}000}{0.85} - 253{,}467 = 99{,}474 \text{ in.-lb (11.24 kN-m)}$$

$$\frac{A_t}{s} = \frac{99{,}474}{60{,}000 \times 1.34 \times 10.5 \times 21.5} = 0.0055 \text{ in.}^2/\text{in. /one leg.}$$

Shear stirrup design (Step 3)

$$V_c = \frac{2\sqrt{4000} \times 14 \times 22.5}{\sqrt{1 + (2.5 \times 0.0596 \times 300{,}000/15{,}000)^2}} = 12{,}676 \text{ lb (57.04 kN)}$$

$$V_s = V_n - V_c = \frac{15{,}000}{0.85} - 12{,}676 = 4971 \text{ lb (22.40 kN)}$$

$$\frac{A_v}{s} = \frac{4971}{60{,}000 \times 22.5} = 0.0037 \text{ in.}^2/\text{in. /two legs}$$

Combined closed stirrups for torsion and shear (Step 4)

$$\frac{A_{vt}}{s} = \frac{2A_t}{s} + \frac{A_v}{s} = 2 \times 0.0055 + 0.0037 = 0.0147 \text{ in.}^2/\text{in. /two legs}$$

Trying No. 3 ties with an area $= 2 \times 0.11 = 0.22$ in.^{2} (9.5 mm diameter, $A_s = $ 142 mm^2) yields

$$s = \frac{\text{area of tie cross section } A_s}{\text{required } A_{vt}/s} = \frac{0.22}{0.0147} = 14.96 \text{ in.}$$

$$\text{maximum permissible spacing } s_{max} = \frac{x_1 + y_1}{4} = \frac{10.5 + 21.5}{4} = 8 \text{ in.} < 16.06 \text{ in.}$$

Hence provide No. 3 (9.5 mm diameter) closed stirrups at 8 in. center to center (203.2 mm center to center).

$$\text{minimum stirrup area required} = \frac{50 \times 14 \times 8}{60,000} = 0.0933 \text{ in.}^2$$

$$\text{area provided} = 0.22 \text{ in.}^2 > 0.0933 \text{ in.}^2 \quad \text{O.K.}$$

Longitudinal torsion steel design (Step 5)

$$A_l = 2A_t \frac{x_1 + y_1}{s} = 0.01(10.5 + 21.5) = 0.32 \text{ in.}^2$$

$$\frac{50b_w s}{f_y} = 0.093 > 2A_t = 0.011 \times 8 = 0.088 \text{ in.}^2$$

Hence, alternatively,

$$A_l = \left(\frac{400 \times 8 \times 14}{60,000} \frac{300,000}{300,000 + \dfrac{15,000}{3 \times 0.0596}} - 0.093 \right) \frac{10.5 + 21.5}{8}$$

$$= 1.96 \text{ in.}^2$$

Therefore, A_l to be provided $= 1.96$ in.^{2}.

Distribution of torsion longitudinal bars

Torsional $A_l = 1.96$ in.^{2}, so $A_l/4 = 0.49$ in.^{2}. Using the same logic as that followed in case (a), provide five No. 8 (25.4 mm diameter) bars at the bottom face. The area required, $A_s + A_l/4 = 3.89$ in.^{2}; the area provided $= 3.95$ in.^{2}. The required area at the top corners and at each vertical face $= A_l/4 = 0.49$ in.^{2}. Provide two No. 5 bars (15.9 mm diameter) at the top and at each of two vertical sides, giving 0.62 in.^{2} in each area. Figure 7.18 shows the geometry of the section reinforcement.

7.5.5 Example 7.2: Equilibrium Torsion Web Steel Design

A normalweight concrete canopy slab on continuous beams spans 24 ft (7.32 m) on several supports, as shown in Fig. 7.19. It carries a uniform service live load of 30 psf (1.44 kPa). Design the interior span spandrel beam $A1$–$A2$ for diagonal tension and torsion. Assume no wind or earthquake and neglect creep and shrinkage effects. Given:

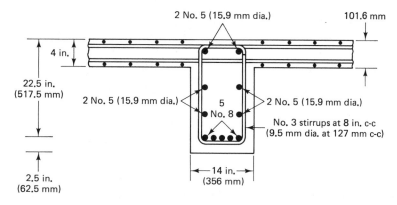

Figure 7.18   Web reinforcement details, Ex. 7.1(c).

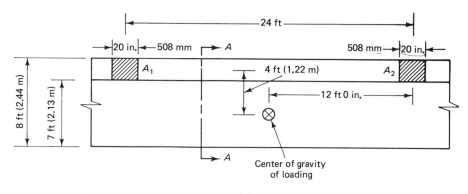

(a)

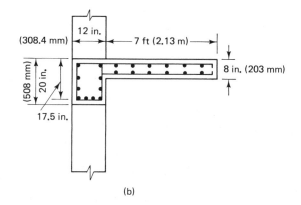

(b)

Figure 7.19   Plan and sectional elevation, Ex. 7.2: (a) plan; (b) section *A–A*.

f'_c = 4000 psi (27.6 MPa)

f_y = 60,000 psi (413.7 MPa)

Exterior columns = 12 in. × 20 in. (304.8 × 508 mm)

Midspan A_s = 1.50 in.^{2} (967.74 mm^2)

Support A_s = 2.4 in.^{2} (1548 mm^2)

Support A'_s = 0.8 in.^{2} (516.13 mm^2)

Solution

Factored torsional moment (Step 1)

Beam A 1–A 2 is a case of nonredistribution torsion because the torsional resistance of the beam is required to maintain equilibrium. Hence the section has to be designed to resist the total external factored torsional moment.

$$\text{service dead load of the cantilever slab} = \frac{8.0}{12} \times 150 = 100.0 \text{ psf (5.08 kPa)}$$

service live load = 30 psf (1.44 kPa)

factored load U = 1.4 × 100.0 + 1.7 × 30 = 191 psf (9.1 kPa)

total load on the cantilever slab = 191 × 24 × 7 = 32,088 lb (144.4 kN)

This load acts at center of gravity of loading shown in Fig. 7.19a, having a moment arm = 4.0 ft (1.22 m). Hence the maximum factored moment at the center line of the support = $\frac{1}{2}$(32,088 × 4) = 64,176 ft-lb.

Note that the reaction at the supports is half of the total torque acting on the slab, as shown in Fig. 7.20, because the center of gravity of the twisting moment is midway between the supports. Since the load is uniformly distributed, the torsional moment variation will be linear along the span. Figure 7.21a shows the torsional envelope for this beam with the T_c value assumed constant along the span. The factored torsional moment at the critical section d (17.5 in.) from the face of the support is

$$T_u = 64,176.0 \left(\frac{12 - \dfrac{10 + 17.5}{12}}{12} \right) = 51,920.0 \text{ ft-lb}$$

$$= 623,040 \text{ in.-lb (70.4 kN-m)}$$

$$T_n = \frac{51,920}{0.85} = 61,082 \text{ ft-lb (82.83 kN-m)}$$

Shear force distribution: Since the beam is to be designed for combined shear and torsion, the distribution of the shear force along the span needs to be determined. Using

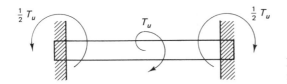

Figure 7.20   Distribution of torsional moment.

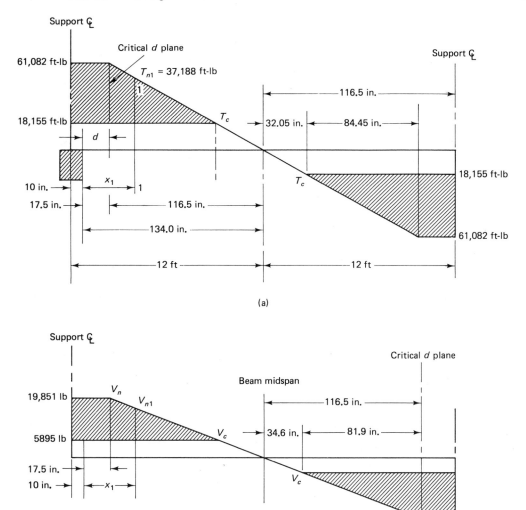

Figure 7.21   (a) Torsion and (b) shear strength envelopes for beam A_1–A_2, Ex. 7.2.

the factored load U, the shear force at the centerline of the support due to the load from the slab $= (\frac{1}{2})(191 \times 8 \times 24) = 18{,}336$ lb. The reaction due to the self-weight of the web is

$$1.4\left[\frac{12 \times (20-8)}{144} \times 150\right] \times \frac{24}{2} = 2520 \text{ lb}$$

Therefore, the total factored shear at the center line of support $= 18{,}336 + 2520 = 20{,}856$ lb. The factored shear $V_u \le \phi V_n$ at the critical section is

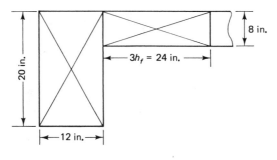

Figure 7.22   Component rectangles.

$$V_u = 20,856 \left(\frac{12 - \dfrac{10 + 17.5}{12}}{12} \right) = 16,873 \text{ lb (77.05 kN)}$$

$$V_n = \frac{16,873}{0.85} = 19,851 \text{ lb (88.30 kN)}$$

Note that a continuity factor of 1.15 would have had to be used in calculating V_u if the *first* interior support section was being considered. Figure 7.21b shows the factored shear envelope for the beam $A\,1$–$A\,2$ with the V_c value assumed constant along the span. From Fig. 7.22:

$$\Sigma x^2 y = 12^2 \times 20 + 8^2 \times 24 = 4416 \text{ in.}^3$$

$$\phi(0.5 \sqrt{f'_c}\, \Sigma x^2 y) = 0.85 \times 0.5 \sqrt{4000} \times 4416 = 118,699 \text{ in.-lb} < T_u$$

Hence torsional effects should be considered in the design.

Torsion closed stirrups design (Step 2)

At the critical section,

$$T_n = \frac{T_u}{\phi} = \frac{623,040}{0.85} = 732,988 \text{ in.-lb} = 61,082 \text{ ft-lb (82.82 kN-m)}$$

$$T_c = \frac{0.8 \sqrt{f'_c}\, \Sigma x^2 y}{\sqrt{1 + (0.4 V_u / C_t T_u)^2}}$$

$$C_t = \frac{b_w d}{\Sigma x^2 y} = \frac{12 \times 17.5}{4416} = 0.0476$$

$$T_c = \frac{0.8 \sqrt{4000} \times 4416}{\sqrt{1 + \left(\dfrac{0.4 \times 16,873}{0.0476 \times 623,040} \right)^2}} = 217,863 \text{ in.-lb} = 18,155 \text{ ft-lb (24.62 kN-m)}$$

$$T_s = T_n - T_c = 732,988 - 217,863 = 515,125 \text{ in.-lb} = 42,927 \text{ ft-lb (58.21 kN-m)}$$

$$T_s (515,125 \text{ in.-lb}) < 4T_c (871,452 \text{ in.-lb}) \qquad \text{O.K.}$$

Assume $1\frac{1}{2}$ in. clear cover and No. 4 closed stirrups.

$$x_1 = 12 - 2(1.5 + 0.25) = 8.5 \text{ in.}$$

$$y_1 = 20 - 2(1.5 + 0.25) = 16.5 \text{ in.}$$

$$\alpha_t = 0.66 + 0.33 \times \frac{16.5}{8.5} = 1.30 < 1.5$$

Use $\alpha_t = 1.30$.

$$\frac{A_t}{s} = \frac{T_s}{f_y \alpha_t x_1 y_1} = \frac{515,125}{60,000 \times 1.3 \times 8.5 \times 16.5} = 0.0471 \text{ in.}^2/\text{in. spacing/leg}$$

Shear stirrup design (Step 3)

$$V_c = \frac{2\sqrt{f_c'} \, b_w d}{\sqrt{1 + [2.5C_t(T_u/V_u)]^2}} = \frac{2\sqrt{4000} \times 12 \times 17.5}{\sqrt{1 + (2.5 \times 0.0476 \times 623,040/16,873)^2}}$$

$$= 5895 \text{ lb} \ (26.21 \text{ kN})$$

$$V_s = V_n - V_c = \frac{16,873}{0.85} - 5895 = 13,956 \text{ lb} \ (62.07 \text{ kN})$$

$$\frac{A_v}{s} = \frac{V_s}{f_y d} = \frac{13,956}{60,000 \times 17.5} = 0.0133 \text{ in.}^2/\text{in. spacing/two legs}$$

Combined closed stirrups for torsion and shear (Step 4)

$$\frac{A_{vt}}{s} = \frac{2A_t}{s} + \frac{A_v}{s} = 2 \times 0.0471 + 0.0133 = 0.1075 \text{ in.}^2/\text{in./two legs}$$

$$\frac{50b_w}{f_y} = 0.01 < \frac{A_v + 2A_t}{s} \qquad \text{O.K.}$$

Try No. 4 closed stirrups, area $= 2 \times 0.2 = 0.4 \text{ in.}^2$.

$$s = \frac{\text{area of the cross section}}{\text{required } A_{vt}/s} = \frac{0.40}{0.1075} = 3.72 \text{ in. c-c} \ (94.54 \text{ mm c-c})$$

Maximum allowable spacing

$$s_{\max} = \frac{x_1 + y_1}{4} = \frac{8.5 + 16.5}{4} = 6.25 \text{ in.} > 3.72 \text{ in.} \qquad \text{O.K.}$$

Therefore, provide No. 4 (12.7 mm diameter) closed stirrups at 3.5 in. center to center (89 mm center to center) at the critical section up to the face of the support.

Increasing stirrup spacing: Since the torsional moment and the shear force decrease toward the center of the beam, the spacing of stirrups could be increased as one moves toward the midspan. If No. 4 closed stirrups are provided at the maximum allowable spacing of 6.25 in. center to center, the torsional capacity T_s of the stirrups would be

$$T_s = \frac{\alpha_t x_1 y_1 f_y A_t}{s} = \frac{1.30 \times 8.5 \times 16.5 \times 60,000 \times 0.20}{6.25}$$

$$= 350,064 \text{ in.-lb}$$

V_c and T_c values are assumed constant throughout the span in the calculation of T_s for practical considerations. Nominal torsional resistance T_{n1} due to $T_c + T_{s1}$ corresponding to No. 4 stirrups for torsion only at maximum $s = 6.25$ in. is $T_n = 217{,}863 + 350{,}064 = 567{,}927$ in.-lb $= 47{,}327$ ft-lb. The distance x_1, corresponding to T_{n1} from support face to plane 1–1 in Fig. 7.21a, is

$$x_1 = 17.5 + \left(116.5 - \frac{47{,}327}{61{,}082} \times 116.5\right) = 43.73 \text{ in.}$$

Increase the original $s = 3.5$ in. at a plane $x_1 = 43.73$ from the face of the support, say at 45 in. from the critical d section.

Design of the stirrups:

$$T_{n1} \text{ at } x_1 = \frac{116.5 - 45}{116.5} \times 61{,}082 = 37{,}488 \text{ ft-lb} = 449{,}857 \text{ in.-lb}$$

$$V_{n1} \text{ at } x_1 = \frac{116.5 - 45}{116.5} \times 19{,}851 = 12{,}183 \text{ lb}$$

$$T_{s2} = 449{,}857 - 217{,}863 = 231{,}994 \text{ in.-lb}$$

$$V_{s2} = 12{,}183 - 5895 = 6288 \text{ lb}$$

$$\frac{A_t}{s} = \frac{231{,}994}{60{,}000 \times 1.3 \times 8.5 \times 16.5} = 0.0212 \text{ in.}^2/\text{in./leg}$$

$$\frac{A_v}{s} = \frac{V_s}{f_y d} = \frac{6288}{60{,}000 \times 17.5} = 0.0060 \text{ in.}^2/\text{in./two legs}$$

$$\frac{A_{vt}}{s} = 0.0060 + 2 \times 0.0212 = 0.0484 \text{ in.}^2/\text{in./two legs}$$

For No. 4 closed stirrups,

$$s = \frac{0.20 \times 2}{0.0484} = 8.25 \text{ in. c-c} > 6.25 \text{ in. allowed}$$

Therefore, change stirrups No. 4 spacing to 6.25 in. center to center starting about 45 in. away from the critical d section toward and up to the midspan. Stirrups have to be used up to a distance $d + b = 17.5 + 12.0 = 29.5$ in., beyond which they are theoretically not required. From Fig. 7.21b it can be seen that they should be used up to the midspan section. Figure 7.23 shows schematically the spacing of the closed stirrups.

Longitudinal torsional steel design (Step 5)

$$A_l = \frac{2A_t(x_1 + y_1)}{s} = 2 \times 0.0471 (8.5 + 16.5) = 2.36 \text{ in.}^2$$

Also

$$A_l = \left(\frac{400xs}{f_y} \frac{T_u}{T_u + V_u/3C_t} - 2A_t\right)\frac{x_1 + y_1}{s}$$

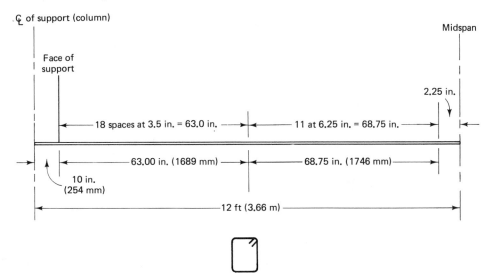

Use 60 No. 4 stirrups for the whole span

Figure 7.23   Stirrups arrangement for Ex. 7.2.

Substituting $50b_w s/f_y$ for $2A_t$ if it is larger than $2A_t$, we have

$$\frac{50b_w s}{f_y} = \frac{50 \times 12 \times 3.5}{60,000} = 0.035 < 2A_t = 2 \times 0.0471 \times 3.5 = 0.33 \text{ in.}^2$$

Hence use $2A_t = 0.33$ in.^{2}.

Alternatively,

$$A_l = \left(\frac{400 \times 3.5 \times 12}{60,000} \frac{623,040}{623,040 + 16,873/3 \times 0.0476} - 0.33 \right) \frac{8.5 + 16.5}{3.5}$$

$$= -0.676 \text{ in.}^2$$

This does not control, since the volume of longitudinal steel A_l has to be equal to the minimum volume of the transverse closed stirrups. Therefore, $A_l = 2.36$ in.^{2} (1523 mm^2) controls. Although refinement of design can permit reducing the number of longitudinal torsional bars as one approaches the midspan because of the drop in the value of T_u, use the same A_l up to the midspan for practical considerations.

To distribute A_l evenly on all faces of the beam, use $\frac{1}{4}A_l$ at each vertical face with $\frac{1}{4}A_l$ at the top corners and $\frac{1}{4}A_l$ at the bottom corners to be added to the flexural reinforcement. $A_l/4 = 2.36/4 = 0.59$ in.^{2} (381 mm^2). Use two No. 5 = 0.62 in.^{2} (12.7 mm diameter) on each vertical side for both the support and midspan sections.

Support section:

$$A_s = \frac{A_l}{4} + A_s = 0.59 + 2.4 = 2.99 \text{ in.}^2$$

Use four No. 8 bars = 3.16 in.^{2} (25.4 mm diameter).

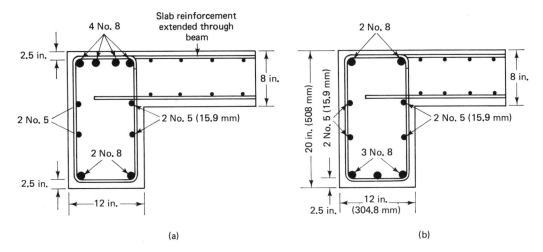

Figure 7.24 Web reinforcement details: (a) support section; (b) midspan section.

$$A'_s = \frac{A_l}{4} + A'_s = 0.59 + 0.8 = 1.39 \text{ in.}^2$$

Use two No. 8 bars = 1.58 in.^{2}.

Midspan section:

$$A_s = \frac{A_l}{4} + A_s = 0.59 + 1.50 = 2.09 \text{ in.}^2$$

Use three No. 8 bars = 2.37 in.^{2}.

Since the torque is less as the midspan is approached, two of the top No. 8 longitudinal bars can be cut off prior to reaching the midspan section. Figure 7.24a and b give the reinforcing details of the sections at the support and midspan respectively.

7.5.6 Example 7.3: Compatibility Torsion Web Steel Design

A parking-garage floor system of one-way slabs on beams is shown in Fig. 7.25. Typical panel dimensions are 12 ft 6 in. × 50 ft (3.81 m × 15.24 m) on centers. Design the exterior spandrel beam A_1–B_1 for combined torsion and shear, assuming that the sections are adequately designed for bending. Given:

Service live load = 50 psf (2.4 kPa)
Slab thickness   = 5 in. (127 mm)
f'_c = 4000 psi (27.58 MPa), normalweight concrete
f_y = 60,000 psi (413.7 MPa)
Height floor to floor = 10 ft
Exterior columns = 14 in. × 24 in. (356 mm × 610 mm)
Interior columns = 24 in. × 24 in. (610 mm × 610 mm)

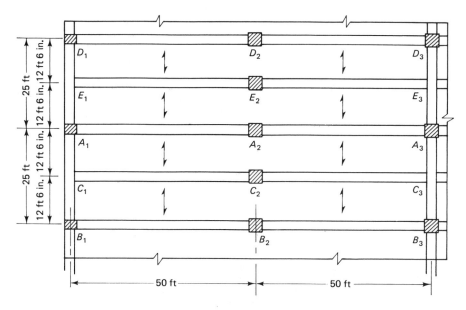

Figure 7.25 Plan of floor systems.

All beams = 14 in. × 30 in. (356 mm × 762 mm)

Required flexural reinforcement for beam A_1-B_1:

    Midspan $A_s = 1.69$ in.^{2}

    Support $A_s = 2.16$ in.^{2}

    Support $A_s' = 0.90$ in.^{2}

Solution

Factored torsional moment (Steps 1, 2, and 3)

1. Beam A_1-B_1 is a case of compatibility torsion because it is part of a continuous floor system where redistribution of moments takes place. The torsional moment due to C_2-C_1 at intersection C_1 is redistributed in directions C_1-C_2 due to the flexibility and rotation of the beam section at C_1 compared to its rigidity at A_1 and B_1. Hence the maximum factored torsional value to be applied to the section at each of the two ends (Fig. 7.26a) is

$$T_u = \phi\left(4\sqrt{f_c'}\,\frac{\Sigma x^2 y}{3}\right)$$

$$\Sigma x^2 y = 14^2 \times 30 + 5^2 \times 15 = 6255 \text{ in.}^3$$

$$T_u = 0.85 \times 4\sqrt{4000} \times \frac{6255}{3}$$

$$= 448,348 \text{ in.-lb} = 37,362.3 \text{ ft-lb}$$

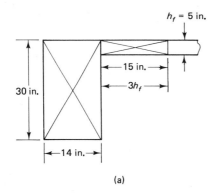

(a)

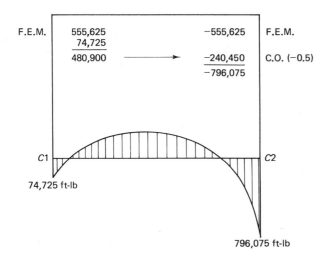

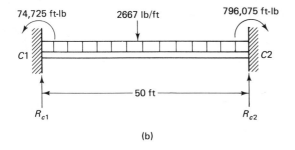

(b)

Figure 7.26 (a) Component rectangles;
(b) bending moments for beam C1–C2.

2. *Fixed end moments in beam C_1–C_2:*

$$\text{service dead load} = \left[\frac{5.0}{12} \times 12.5 + \frac{(30 - 5) \times 14}{144}\right] 150 = 1146 \text{ lb/ft (16.7 kN/m)}$$

service live load $= 50 \times 12.5 = 625$ lb/ft (9.1 kN/m)

factored load $U = 1.4 \times 1146 + 1.7 \times 625 = 2667$ lb/ft (38.9 kNm)

$$\text{fixed end moment} = \frac{w_u l^2}{12} = \frac{2667(50)^2}{12} = 555,625 \text{ ft-lb}$$

The factored torque in compatibility torsion that beam C_2–C_1 applies at connection C_1 is

$$T_u = 2 \times 37,362.3 = 74,725 \text{ ft-lb}$$

This value is less than the factored end moment $w_u l^2 / 12$ at end C_1. Hence the torsional moment to be used at midspan of A_1–B_1 is $T_u = 74,725$ ft-lb. Perform the moment distribution shown in Fig. 7.26b to determine the reaction R_{c1} and R_{c2}.

3. *Beam reaction at C_1 and the resulting shear in beam A_1–B_1:*

$$\sum M_{c2} = 0 \quad \text{or} \quad 50R_{c1} + 796,075 - 74725 - \frac{2667(50)^2}{2} = 0$$

$$R_{c1} = \frac{-796,075 + 74,725 + 3,333,750}{50} = 52,248 \text{ lb}$$

$$\text{factored self-weight of } A_1\text{–}B_1 = 1.4\left(\frac{14 \times 30}{144}\right)150 = 613 \text{ lb/ft}$$

distance of critical section in A_1–B_1 from column center line

$$= d + \frac{14}{2} = (30 - 2.5) + \frac{14}{2} = 34.5 \text{ in.}$$

$$V_u = \frac{52,248}{2} + 613\left(12.5 - \frac{34.5}{12}\right) = 32,024 \text{ lb}$$

Beams A_1–B_1 would be subjected to the torsion and shear envelopes shown in Fig. 7.27.

Torsion closed stirrup design (Step 2)

$$T_n = \frac{37,362.3 \times 12}{0.85} = 527,468 \text{ in.-lb (59.60 kN-m)}$$

$$T_c = \frac{0.8\sqrt{f_c'}\,\Sigma x^2 y}{\sqrt{1 + (0.4V_u/C_t T_u)^2}}$$

$$C_t = \frac{b_w d}{\Sigma x^2 y} = \frac{14 \times 27.5}{6255} = 0.0616$$

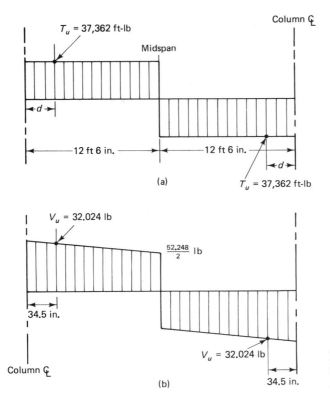

Figure 7.27 (a) Torsion and (b) shear factored force envelopes for beam A1-B1, Ex. 7.3.

$$T_c = \frac{0.8\sqrt{4000} \times 6255}{\sqrt{1 + \left(\dfrac{0.4 \times 32,024}{0.0616 \times 448,348}\right)^2}} = 287,103 \text{ in.-lb} \,(32.44 \text{ kN-m})$$

$$T_s = 527,468 - 287,103 = 240,365 \text{ in.-lb} \,(27.16 \text{ kN-m})$$

to be resisted by the closed ties. Assume $1\frac{1}{2}$ in. clear cover and No. 4 closed stirrups.

$$x_1 = 14 - 2(1.5 + 0.25) = 10.5 \text{ in.}$$

$$y_1 = 30 - 2(1.5 + 0.25) = 26.5 \text{ in.}$$

$$\alpha_t = 0.66 + 0.33 \times \frac{26.5}{10.5} = 1.493 < 1.5 \qquad \text{use } \alpha_t = 1.493$$

$$\frac{A_t}{s} = \frac{T_s}{f_y \alpha_t x_1 y_1} = \frac{240,365}{60,000 \times 1.493 \times 10.5 \times 26.5} = 0.0096 \text{ in.}^2/\text{in. spacing/leg}$$

Shear stirrup design (Step 3)

$$V_c = \frac{2\sqrt{f_c'}b_w d}{1 + [2.5C_t(T_u/V_u)]^2} = \frac{2\sqrt{4,000} \times 14 \times 27.5}{\sqrt{1 + \sqrt{(2.5 \times 0.0616 \times 448,348/32,024)^2}}}$$

$$V_c = 20,490 \text{ lb } (91.14 \text{ kN})$$

$$V_s = V_n - V_c = \frac{32,024}{0.85} - 20,490 = 17,185 \text{ lb } (76.44 \text{ kN})$$

$$\frac{A_v}{s} = \frac{V_s}{f_y d} = \frac{17,185}{60,000 \times 27.5} = 0.0104 \text{ in.}^2/\text{in. spacing/two legs}$$

Combined closed stirrups for torsion and shear (Step 4)

$$\frac{A_{vt}}{s} = \frac{2A_t}{s} + \frac{A_v}{s} = 2 \times 0.0096 + 0.0104 = 0.0296 \text{ in.}^2/\text{in./two legs}$$

Try No. 3 ties:

$$2 \times 0.11 = 0.22 \text{ in.}^2 (9.5 \text{ mm diameter, } A_s = 142 \text{ mm}^2)$$

$$s = \frac{\text{area of tie cross section } A_s}{\text{required } A_{vt}/s} = \frac{0.22}{0.0296} = 7.43 \text{ in. c-c}$$

$$\text{maximum allowable spacing } s_{\max} = \frac{x_1 + y_1}{4} = \frac{10.5 + 26.5}{4} = 9.25 \text{ in.} > 7.43 \text{ in.}$$

Check stirrup spacing needed at midspan (Step 5)

$$\text{midspan } V_c = \frac{2\sqrt{4000} \times 14 \times 27.5}{\sqrt{1 + (2.5 \times 0.0616 \times 448,348/26,124)^2}} = 17,233 \text{ lb}$$

$$V_s = \frac{52,248}{2 \times 0.85} - 17,233 = 13,501 \text{ lb}$$

$$\frac{A_v}{s} = \frac{13,501}{60,000 \times 27.5} = 0.0082 \text{ in.}^2/\text{in./two legs}$$

$$\text{midspan } T_c = \frac{0.8\sqrt{4000} \times 6,255}{\sqrt{1 + \left(\dfrac{0.4 \times 26,124}{0.0616 \times 448,348}\right)^2}} = 296,002 \text{ in.-lb}$$

$$T_s = 527,468 - 296,002 = 231,466 \text{ in.-lb}$$

$$\frac{A_t}{s} = \frac{231,466}{60,000 \times 1.493 \times 10.5 \times 26.5} = 0.0093 \text{ in.}^2/\text{in./one leg}$$

$$\frac{A_{vt}}{s} = 2 \times 0.0093 + 0.0082 = 0.0266 \text{ in.}^2/\text{in./two legs}$$

$$s = \frac{0.22}{0.0266} = 8.27 \text{ in. c-c}$$

Use No. 3 closed stirrups at $7\frac{1}{4}$ in. center to center (184 mm spacing).

$$\text{minimum stirrups requirement} = A_v + 2A_t = \frac{50b_w s}{f_y} = \frac{50 \times 14 \times 7.25}{60,000}$$

$$\text{minimum stirrups requirement} = 0.0846 \text{ in.}^2$$

$$\text{area provided} = 0.22 \text{ in.}^2 > 0.0846 \text{ in.}^2 \quad \text{O.K.}$$

Longitudinal torsion steel design

$$A_l = 2A_t \frac{x_1 + y_1}{s} = 2 \times 0.0096(10.5 + 26.5) = 0.71 \text{ in.}^2$$

Also,

$$A_l = \left(\frac{400xs}{f_y} \frac{T_u}{T_u + V_u/3C_t} - 2A_t \right) \frac{x_1 + y_1}{s}$$

(or substituting $50b_w s / f_y$ for $2A_t$, whichever controls)

$$\frac{50b_w s}{f_y} = 0.0846 < 2A_t = 2 \times 0.0096 \times 7.25 = 0.1392 \text{ in.}^2$$

Hence

$$A_l = \left(\frac{400 \times 14 \times 7.25}{60,000} \frac{448,348}{448,348 + \dfrac{32,024}{3 \times 0.0616}} - 0.1392 \right) \frac{10.5 + 26.5}{7.25}$$

$$= 1.78 \text{ in.}^2 (1148 \text{ mm}^2)$$

Use $A_l = 1.78 \text{ in.}^2$ (1148 mm²). Bars have to be placed around the perimeter of the web, spaced at not more than 12 in. and with a bar in each corner of the closed stirrup or tie. Combine the corner bars with the longitudinal flexural reinforcement.

Distribution of torsion longitudinal bars

Torsional $A_l = 1.78 \text{ in.}^2$ (use same for midspan and support sections). Assume that $\frac{1}{4}A_l$ goes to the top corners and $\frac{1}{4}A_l$ would thus be distributed equally to the vertical faces of the beam cross section at a spacing not to exceed 12 in. center to center.

$$\text{midspan} \sum A_s = \frac{A_l}{4} + A_s = \frac{1.78}{4} + 1.69 = 2.14 \text{ in.}^2$$

Five No. 6 bars $= 2.20 \text{ in.}^2$, three bars to continue up to the support (five bars 19.1 mm diameter):

$$\sum A_s = \frac{A_l}{4} + A_s = \frac{1.78}{4} + 2.16 = 2.61 \text{ in.}^2$$

Six No. 6 bars $= 2.64 \text{ in.}^2$ (six bars 19.1 mm diameter):

$$A_s \text{ at each vertical face} = \frac{A_l}{4} = \frac{1.78}{4} = 0.45 \text{ in.}^2 \text{ (three No. 4 bars} = 0.60 \text{ in.}^2\text{)}$$

Use three No. 4 bars on each face (three bars 12.7 mm diameter). Some of the longitudinal bars are cut off before reaching the midspan, as in Ex. 7.2. Figure 7.28 shows the geometry of the spandrel beam cross section at both the midspan and the support.

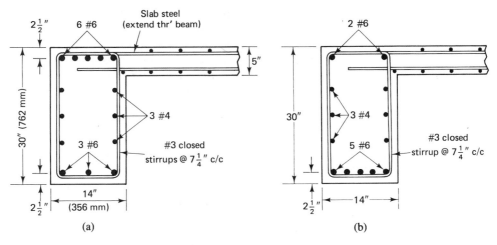

Figure 7.28   Web reinforcement details: (a) support section; (b) midspan section.

SELECTED REFERENCES

7.1   Timoshenko, S., *Strength of Materials,* Part II: *Advanced Theory,* D. Van Nostrand, New York, 1952, 501 pp.

7.2   Nadai, A., *Plasticity: A Mechanics of the Plastic State of Matter,* McGraw-Hill, New York, 1931, 349 pp.

7.3   Cowan, H. J., "Design of Beams Subject to Torsion Related to the New Australian Code," *Journal of the American Concrete Institute,* Proc. Vol. 56, January 1960, pp. 591–618.

7.4   Gesund, H., Schnette, F. J., Buchanan, G. R., and Gray, G. A., "Ultimate Strength in Combined Bending and Torsion of Concrete Beams Containing Both Longitudinal and Transverse Reinforcement," *Journal of the American Concrete Institute,* Proc. Vol. 61, December 1964, pp. 1509–1521.

7.5   Lessig, N. N., "Determination of Carrying Capacity of Reinforced Concrete Elements with Rectangular Cross-section Subjected to Flexure with Torsion," *Zhelezonbeton,* 1959, pp. 5–28.

7.6   Zia, P., "Tension Theories for Concrete Members," *Special Publication SP 18-4,* American Concrete Institute, Detroit, 1968, pp. 103–132.

7.7   Hsu, T. T. C., "Ultimate Torque of Reinforced Concrete Members," *Journal of the Structural Division, ASCE,* Vol. 94, No. ST2, February 1968, pp. 485–510.

7.8   Rangan, B. V., and Hall, A. J., "Strength of Rectangular Prestressed Concrete Beams in Combined Torsion, Bending and Shear," *Journal of the American Concrete Institute,* Proc. Vol. 70, April 1973, 270–279.

7.9   Wang, C. K., and Salmon, C. G., *Reinforced Concrete Design,* 3rd ed. Harper & Row, New York, 1979, 918 pp.

7.10 Thurliman, B., "Torsional Strength of Reinforced and Prestressed Concrete Beams—CEB Approach, U.S. and European Practices," *Special Publication,* American Concrete Institute, Detroit, 1979, pp. 117–143.

7.11 Collins, M. P., and Mitchell, D., "Shear and Torsion Design of Prestressed and Non-prestressed Concrete Beams," *Journal of the Prestressed Concrete Institute,* Proc. Vol. 25, No. 5, September–October 1980, pp. 32–100.

7.12 Hsu, T. T. C., *Torsion of Reinforced Concrete,* Van Nostrand Reinhold, New York, 1983, 510 pp.

PROBLEMS FOR SOLUTION

7.1 Calculate the torsional capacity T_c for the sections shown in Fig. 7.29. Given:

$V_u/T_u = 0.05$
$f'_c = 4000$ psi (27.6 MPa), normalweight concrete

7.2 A cantilever beam is subjected to a concentrated service live load of 20,000 lb (90 kN) acting at a distance of 3 ft 6 in. (1.07 m) from the wall support. In addition, the beam has to resist an equilibrium factored torsion $T_u = 300,000$ in.-lb (33.89 kN/m). The beam cross section is 12 in. × 24 in. (304.8 mm × 609.6 mm) with an effective depth of 22.5 in. (571.5 mm). Design the stirrups and the additional longitudinal steel needed. Given:

$f'_c = 3500$ psi
$f_y = 60,000$ psi
$A_s = 4.0$ in.^{2} (2580.64 mm^2)

7.3 The first interior span of a four-span continuous beam has a clear span $l_n = 18$ ft (5.49 m). The beam is subjected to a uniform external service dead load $w_D = 1700$ plf (24 pkN/m) and a service live load $w_L = 2200$ plf (32.1 kN/m). Design the section for flexure, diagonal tension, and torsion. Select the size and spacing of the closed stirrups and extra longitudinal steel that might be needed for torsion. Assume that the beam width $b_w = 15$ in. (381.0 mm) and that redistribution of torsional stresses is possible such that the external torque T_u can be assumed as $\phi(4\sqrt{f'_c} \Sigma x^2 y/3)$. Given:

$f'_c = 5000$ psi (34.47 MPa), normalweight concrete
$f_y = 60,000$ psi (413.7 MPa)

(Note that this problem is similar to Problem 6.4 except that torsional moment is added.)

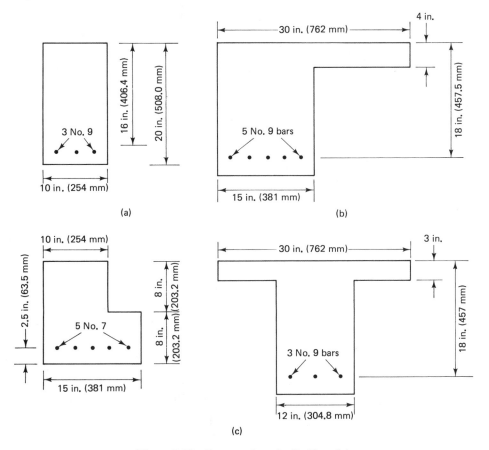

Figure 7.29 Cross sections for Problem 7.1.

7.4 A continuous beam has the shear and torsion envelopes shown in Fig. 7.30. The beam dimensions are $b_w = 14$ in. (355.6 mm) and $d = 25$ in. (63.5 mm). It is subjected to factored shear forces $V_{u1} = 75,000$ lb (333.6 kN), $V_{u2} = 60,000$ lb, and $V_{u3} = 45,000$ lb. Design the beam for torsion and shear and detail the web reinforcement. Given:

$f'_c = 4000$ psi (27.58 MPa), lightweight concrete
$f_y = 60,000$ psi (413.7 MPa)

The required reinforcement is as follows:

Midspan $A_s = 3.0$ in.^{2}
Support $A_s = 3.6$ in.^{2}, $A'_s = 0.7$ in.^{2}

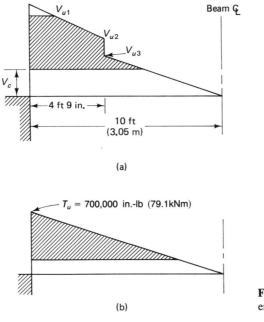

(a)

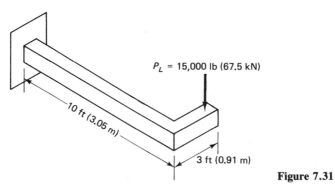

(b)

Figure 7.30   (a) Shear and (b) torsion envelopes.

Figure 7.31

7.5   Design the rectangular beam shown in Fig. 7.31 for bending, shear, and torsion. Assume that the beam width $b = 12$ in. (305 mm). Given:

$$f'_c = 4000 \text{ psi (27.58 MPa)}$$
$$f_y = 60,000 \text{ psi (413.8 MPa)}$$

7.6   An exterior spandrel beam A_1–B_1, part of the monolithic floor system shown in Fig. 7.32, has a center-to-center span of 36 ft and a slab thickness $h_f = 6$ in. (152.4 mm) on beams 15 in. × 36 in. in cross section. It is subject to a service

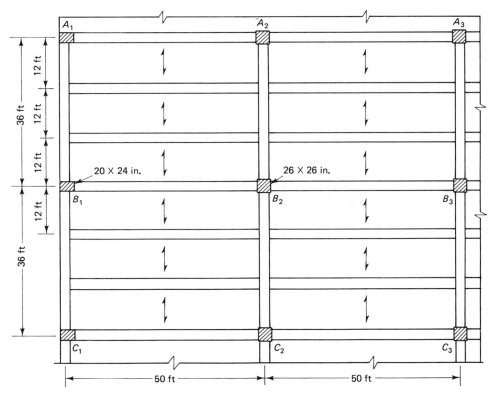

Figure 7.32

live load = 50 psf (2.4 kPa). Design the shear and torsion reinforcement necessary to resist the external factored loads. Given:

f'_c = 4000 psi (27.58 MPa), normalweight concrete
f_y = 60,000 psi (413.7 MPa)

Assume that the required flexural reinforcement for beam A_1–B_1 is:

Midspan A_s = 2.09 in.^{2}
Support A_s = 3 in.^{2}, A'_s = 1.6 in.^{2}

8

Serviceability of Beams & One-Way Slabs

8.1 INTRODUCTION

Serviceability of a structure is determined by its deflection, cracking, extent of corrosion of its reinforcement, and surface deterioration of its concrete. Surface deterioration can be minimized by proper control of mixing, placing, and curing of the concrete. If the surface is exposed to potentially damaging chemicals, such as in a chemical factory or a sewage plant, a special type of cement with appropriate additives should be used in the concrete mix. Use of adequate cover as recommended in Chapters 4 and 5, proper quality control of the materials and the application of proper crack control and deflection control criteria to the design can minimize and in most cases eliminate these problems.

This chapter deals with the evaluation of deflection and cracking behavior of beams and one-way slabs in some detail. It is intended to give the designer adequate basic background on the effect of cracking on the stiffness of the member, the short-term and long-term deflection performance, and the manner in which the cracked concrete beam element can still perform adequately and aesthetically without loss of reliability in its performance. Deflection of two-way action slabs and plates is given in Chapter 11 with numerical examples of deflection calculations for short-term as well as long-term loading.

8.2 SIGNIFICANCE OF DEFLECTION OBSERVATION

The working stress method of design and analysis used prior to the 1970s limited the stress in concrete to about 45% of its compressive strength, and the stress in the steel to less than 50% of its yield strength. Elastic analysis was applied to the design of structural frames as well as reinforced concrete sections. The structural elements were proportioned to carry the highest service-level moment along the span of the member, with redistribution of moment effect often largely neglected. As a result, heavier sections with higher reserve strength resulted as compared to those obtained by the current ultimate strength approach.

Higher-strength concretes having f'_c values in excess of 12,000 psi (82.74 MPa) and higher-strength steels are being used in strength design and expanding knowledge of the properties of the materials has resulted in lower values of load factors and reduced reserve strength. Hence more slender and efficient members are specified with deflection becoming a more pronounced controlling criteria.

Beams and slabs are rarely built as isolated members, but a monolithic part of an integrated system. Excessive deflection of a floor slab may cause dislocations in the partitions it supports. Excessive deflection of a beam can damage a partition below, and excessive deflection of a lintel beam above a window opening could crack the glass panels. In the case of open floors or roofs such as top garage floors, ponding of water can result. For these reasons, deflection control criteria are necessary such as those given in Table 11.3.

8.3 DEFLECTION BEHAVIOR OF BEAMS

The load–deflection relationship of a reinforced concrete beam is basically trilinear, as idealized in Fig. 8.1. It is composed of three regions prior to rupture:

Region I: Precracking stage where a structural member is crack-free (Fig. 8.2)

Region II: Postcracking stage where the structural member develops acceptable controlled cracking both in distribution and width

Region III: Postserviceability cracking stage where the stress in the tension reinforcement reaches the limit state of yielding

8.3.1 Precracking Stage: Region I

The precracking segment of the load–deflection curve is essentially a straight line defining full elastic behavior. The maximum tensile stress in the beam in this region

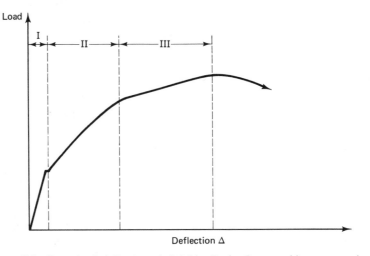

Figure 8.1 Beam load–deflection relationship. Region I, precracking stage; region II, postcracking stage; region III, postserviceability stage (steel yields).

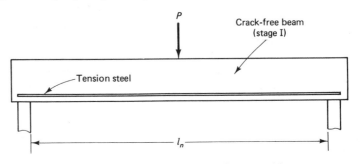

Figure 8.2 Centrally loaded beam at the precracking stage.

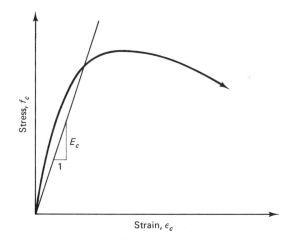

Figure 8.3 Stress–strain diagram of concrete.

is less than its tensile strength in flexure, namely, less than the modulus of rupture f_r of concrete. The flexural stiffness EI of the beam can be estimated using Young's modulus E_c of concrete and the moment of inertia of the uncracked reinforced concrete cross section. The load-deflection behavior is dependent on the stress-strain relationship of the concrete as a significant factor. A typical stress-strain diagram of concrete is shown in Figure 8.3.

The value of E_c can be estimated using the ACI empirical expression given in Chapter 3:

$$E_c = 33 w_c^{1.5} \sqrt{f_c'}$$

or

$$E_c = 57{,}000 \sqrt{f_c'} \qquad \text{for normalweight concrete}$$

An accurate estimation of the moment of inertia I necessitates consideration of the contribution of the steel reinforcement A_s. This can be done by replacing the steel area by an equivalent concrete area $(E_s/E_c)A_s$ since the value of Young's modulus E_s of the reinforcement is higher than E_c. One can transform the steel area to an equivalent concrete area, calculate the center of gravity of the transformed section, and obtain the transformed moment of inertia I_{gt}.

Example 8.1 presents a typical calculation of I_{gt} for a transformed rectangular section. Most designers, however, use a gross moment of inertia I_g based on the uncracked concrete section, disregarding the additional stiffness contributed by the steel reinforcement as insignificant.

The precracking region stops at the initiation of the first flexural crack when the concrete stress reaches its modulus of rupture strength f_r. Similarly to the direct tensile splitting strength, the modulus of rupture of concrete is proportional to the square root of its compressive strength. For design purposes, the value of the modulus for normal weight concrete may be taken as

$$f_r = 7.5 \sqrt{f_c'} \tag{8.1}$$

If lightweight concrete is used, the value of f_r from Eq. 8.1 is multiplied by 0.75 for all lightweight concrete and by 0.85 for sand-lightweight concrete.

If the distance of the extreme tension fiber from the center of gravity of the section is y_t and the cracking moment is M_{cr},

$$M_{cr} = \frac{I_g f_r}{y_t} \tag{8.2}$$

For a rectangular section

$$y_t = \frac{h}{2} \tag{8.3}$$

where h is the total thickness of the beam. Equation 8.2 is derived from the classical bending equation $\sigma = Mc/I$ for elastic and homogeneous materials.

Calculations of deflection for this region are not important since very few reinforced concrete beams remain uncracked under actual loading. However, mathematical knowledge of the variation in stiffness properties is important since segments of the beam along the span in the actual structure can remain uncracked.

8.3.1.1 Example 8.1: Alternative Methods of Cracking Moment Evaluation

Calculate the cracking moment M_{cr} for the beam cross section shown in Fig. 8.4, using both (a) transformed and (b) gross cross-section alternatives in the solution. Given:

$f'_c = 4000$ psi (27.6 MPa)
$f_y = 60,000$ psi (414 MPa)
$E_s = 29 \times 10^6$ psi (200,000 MPa), normalweight concrete

Reinforcement: four No. 9 bars (four bars 28.6 mm diameter) placed in two bundles.

Solution

(a) *Transformed section solution:* Depth of center-of-gravity axis, $\bar{y}$, can be obtained using the first moment of area, namely,

$$\left[bh + \left(\frac{E_s}{E_c} - 1 \right)A_s \right]\bar{y} = bh\frac{h}{2} + \left(\frac{E_s}{E_c} - 1 \right)A_s d$$

Note that $(E_s/E_c) - 1$ is used instead of E_s/E_c to account for the concrete displaced by the reinforcing bars.

It is customary to denote $n = E_s/E_c$ as the modular ratio. Taking moments about the top extreme fibers of the section,

$$\bar{y} = \frac{(bh^2/2) + (n - 1)A_s d}{bh + (n - 1)A_s}$$

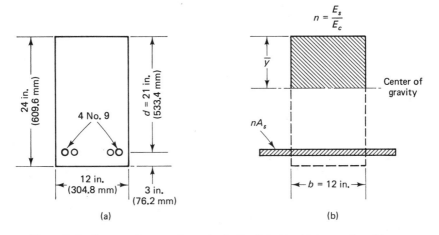

Figure 8.4   Cross-section transformation in Ex. 8.1: (a) midspan section; (b) transformed section.

For normalweight 4000-psi concrete,

$$E_c = 57,000\sqrt{4000}$$

$$= 3.6 \times 10^6 \text{ psi } (24.8 \times 10^6 \text{ MPa})$$

$$n = \frac{29 \times 10^6}{3.6 \times 10^6} = 8.1$$

$$\bar{y} = \frac{\dfrac{12 \times (24)^2}{2} + (8.1 - 1)4.0 \times 21}{12 \times 24 + (8.1 - 1)4.0} = 12.8 \text{ in. } (325.1 \text{ mm})$$

If the moment of inertia of steel reinforcement about its own axis is neglected as insignificant,

$$\text{transformed section } I_{gt} = \frac{bh^3}{12} + bh(12.8 - 12.0)^2 + (n - 1)A_s(d - \bar{y})^2$$

or

$$I_{gt} = \frac{12 \times 24^3}{12} + 12 \times 24 \times 0.8^2 + 7.1 \times 4.0(21 - 12.8)^2$$

$$= 15,918 \text{ in.}^4 (66.22 \times 10^8 \text{ mm}^4)$$

The distance of the center of gravity of the transformed section from the lower extreme fibers is

$$y_t = 24 - 12.8 = 11.2 \text{ in. } (284.5 \text{ mm})$$

$$f_r = 7.5\sqrt{4000} = 474.3 \text{ psi } (3.27 \text{ MPa})$$

$$M_{cr} = \frac{I_g f_r}{y_t} = \frac{15,918 \times 474.3}{11.2} = 674,100 \text{ in.-lb } (76.17 \text{ kN-m})$$

(b) *Gross-section solution:*

$$\bar{y} = \frac{h}{2} = 12 \text{ in.}$$

$$\text{cross section } I_g = \frac{bh^3}{12} = \frac{12 \times 24^3}{12} = 13,824 \text{ in.}^4$$

$$y_t = 12 \text{ in. (304.8 mm)}$$

$$f_r = 474.3 \text{ psi}$$

$$M_{cr} = \frac{13,824 \times 474.3}{12} = 546,394 \text{ in.-lb (61.74 kN-m)}$$

There is a difference of about 15% in the value of I_g and 19% in the value of M_{cr}. Even though this percentage difference in the values of the I_g and M_{cr} obtained by the two methods seems somewhat high, such a difference in the deflection calculation values is not of real significance and in most cases does not justify using the transformed-section method for evaluating M_{cr}.

8.3.2 Postcracking Service Load Stage: Region II

The precracking region ends at the initiation of the first crack and moves into region II of the load–deflection diagram in Fig. 8.1. Most beams lie in this region at service loads. A beam undergoes varying degrees of cracking along the span corresponding to the stress and deflection levels at each section. Hence cracks are wider and deeper at midspan, whereas only narrow minor cracks develop near the supports in a simple beam.

When flexural cracking develops, the contribution of the concrete in the tension zone reduces substantially. Hence the flexural rigidity of the section is reduced, making the load–deflection curve less steep in this region than in the precracking stage segment. As the magnitude of cracking increases, stiffness continues to decrease, reaching a lower-bound value corresponding to the reduced moment of inertia of the cracked section. At this limit state of service load cracking, the contribution of tension-zone concrete to the stiffness is neglected. The moment of inertia of the cracked section designated as I_{cr} can be calculated from the basic principles of mechanics.

Strain and stress distributions across the depth of a typical cracked rectangular concrete section are shown in Fig. 8.5. The following assumptions are made with respect to deflection computation based on extensive testing verification: (1) the strain distribution across the depth is assumed to be linear; (2) concrete does not resist any tension; (3) both concrete and steel are within the elastic limit; and (4) strain distribution is similar to that assumed for strength design, but the magnitudes of strains, stresses, and stress distribution are different.

To calculate the moment of inertia, the value of the neutral axis depth, c, should be determined from horizontal force equilibrium.

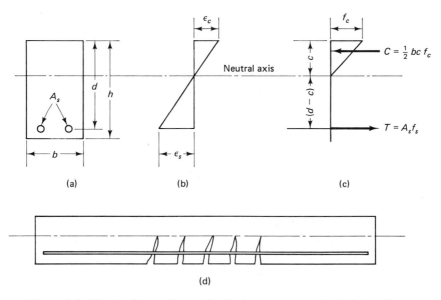

Figure 8.5 Elastic strain and stress distributions across a cracked reinforced concrete section: (a) cross section; (b) strain; (c) elastic stress and force; (d) cracked beam prior to failure in flexure.

$$A_s f_s = bc\frac{f_c}{2} \tag{8.4a}$$

Since the steel stress $f_s = E_s\epsilon_s$ and concrete stress, $f_c = E_c\epsilon_c$, Eq. 8.4a can be rewritten as

$$A_s E_s \epsilon_s = \frac{bc}{2} E_c \epsilon_c \tag{8.4b}$$

From similar triangles in Fig. 8.5b,

$$\frac{\epsilon_c}{c} = \frac{\epsilon_s}{d - c} \tag{8.5a}$$

or

$$\epsilon_s = \epsilon_c\left(\frac{d}{c} - 1\right) \tag{8.5b}$$

From Eqs. 8.4b and 8.5b,

$$A_s E_s \epsilon_c\left(\frac{d}{c} - 1\right) = \frac{bc}{2} E_c \epsilon_c \tag{8.6a}$$

or

$$\frac{A_s E_s}{E_c}\left(\frac{d}{c} - 1\right) = \frac{bc}{2} \tag{8.6b}$$

Replacing the modular ratio E_s/E_c by n, Eq. 8.6b can be rewritten as

$$\frac{bc^2}{2} + nA_s c - nA_s d = 0 \tag{8.6c}$$

The value of c can be obtained by solving the quadratic equation 8.6c. The moment of inertia I_{cr} can be obtained from

$$I_{cr} = \frac{bc^3}{3} + nA_s(d - c)^2 \tag{8.7}$$

where the term $bc^3/3$ in Eq. 8.7 denotes the moment of inertia of the *compressive* area bc about the neutral axis, namely, the base of the compression rectangle, neglecting the section area in tension below the neutral axis. The reinforcing area is multiplied by n to transform it to its equivalent in concrete for contribution to the section stiffness. The moment of inertia of the steel about its own axis is disregarded as negligible.

Only part of the beam cross section is cracked in the case under discussion. As seen from Fig. 8.5d, the uncracked segments below the neutral axis along the beam span possess some degree of stiffness which contributes to the overall beam rigidity. The actual stiffness of the beam lies between $E_c I_g$ and $E_c I_{cr}$, depending on such other factors as (1) extent of cracking, (2) distribution of loading, and (3) contribution of the concrete, as seen in Fig. 8.5d between the cracks. Generally, as the load approaches the steel yield load level, the stiffness value approaches $E_c I_{cr}$.

Branson developed simplified expressions for calculating the effective stiffness $E_c I_e$ for design. The Branson equation, verified as applicable to most cases of reinforced and prestressed beams and universally adopted for deflection calculations, defines the effective moment of inertia as

$$I_e = \left(\frac{M_{cr}}{M_a}\right)^3 I_g + \left[1 - \left(\frac{M_{cr}}{M_a}\right)^3\right] I_{cr} \le I_g \tag{8.8a}$$

Equation 8.8a is also written in the form

$$I_e = I_{cr} + \left(\frac{M_{cr}}{M_a}\right)^3 (I_g - I_{cr}) \le I_g \tag{8.8b}$$

The effective moment of inertia I_e as shown in Eq. 8.8b depends on the maximum moment M_a along the span in relation to the cracking moment capacity M_{cr} of the section.

8.3.2.1 Example 8.2: Effective Moment of Inertia of Cracked Beam Sections

Calculate the moment of inertia I_{cr} and the effective moment of inertia I_e of the beam cross section in Ex. 8.1 if the external maximum service load moment is 2,000,000 in.-lb (226 kN-m). Given (Ex. 8.1):

$b = 12$ in.
$d = 21$ in.
$h = 24$ in.
$A_s = 4.0$ in.^{2}
$f_c' = 4000$ psi
$f_y = 60,000$ psi
$E_s = 29 \times 10^6$ psi
$E_c = 3.6 \times 10^6$ psi
$n = 8.1$

Solution

From Eq. 8.6c,

$$\frac{12c^2}{2} + 8.1 \times 4.0c - 8.1 \times 4.0 \times 21 = 0$$

Hence neutral axis depth $c = 8.3$ in. (210.8 mm). From Eq. 8.7,

$$I_{cr} = \frac{12.0 \times 8.3^3}{3} + 8.1 \times 4.0(21.0 - 8.3)^2 = 7513 \text{ in.}^4(31.25 \times 10^8 \text{ mm}^4)$$

Using the I_{gt} and M_{cr} values of Ex. 8.1, which include the effect of the transformed steel area,

$$I_e = 7513 + \left(\frac{674,100}{2,000,000}\right)^3 (15,918 - 7513)$$

$$= 7835 \text{ in.}^4(32.59 \times 10^8 \text{ mm}^4) < I_g \qquad \text{as expected}$$

If the gross-cross-section values for I_g and M_{cr} are used without including the effect of transformed A_s, the effective moment of inertia becomes

$$I_e = 7513 + \left(\frac{546,394}{2,000,000}\right)^3 (13,824 - 7513)$$

$$= 7642 \text{ in.}^4(31.79 \times 10^8 \text{ mm}^4) < I_g$$

Comparison of the two values of effective I_e calculated by the two methods (7835 in.^{4} versus 7642 in.^{4}) shows an insignificant difference. Hence use of the gross-section properties in Eq. 8.8 is, in most cases, adequate particularly when one considers the variability in the loads and the randomness in the properties of concrete.

8.3.3 Postserviceability Cracking Stage and Limit State of Deflection Behavior at Failure: Region III

The load–deflection diagram of Fig. 8.1 is considerably flatter in region III than in the preceding regions. This is due to substantial loss in stiffness of the section because of extensive cracking and considerable widening of the stabilized cracks throughout the span. As the load continues to increase, the strain ϵ_s in the steel bars at the tension side continues to increase beyond the yield strain ϵ_y with no additional stress. The beam is considered at this stage to have structurally failed by initial yielding of the tension steel. It continues to deflect without additional loading, the cracks continue to open, and the neutral axis continues to rise toward the outer compression fibers. Finally, a secondary compression failure develops, leading to total crushing of the concrete in the maximum moment region followed by rupture.

The increase in the beam load level between first yielding of the tension reinforcement in a simple beam and the rupture load level varies between 4 and 10%. The deflection value before rupture, however, can be several times that at the steel yield level, depending on the beam span/depth ratio, the steel percentage, the type of loading, and the degree of confinement of the beam section. An ultimate deflection value 8 to 12 times the first yield deflection has frequently been observed in tests.

Postyield deflection and limit deflection at failure are not of major significance in design, hence are not being discussed in any detail in this text. It is important, however, to recognize the reserve deflection capacity as a measure of ductility in structures in earthquake zones and in other areas where the probability of overload is high.

8.4 LONG-TERM DEFLECTION

Time-dependent factors magnify the magnitude of deflection with time. Consequently, the design engineer has to evaluate immediate as well as long-term deflection in order to ensure that their values satisfy the maximum permissible criteria for the particular structure and its particular use.

Time-dependent effects are caused by the superimposed creep, shrinkage, and temperature strains. These additional strains induce a change in the distribution of stresses in the concrete and the steel, resulting in an increase in the curvature of the structural element for the same external load.

The calculation of creep and shrinkage strains at a given time is a complex process as discussed in Chapter 3. One has to consider how these time-dependent concrete strains affect the stress in the steel and the curvature of the concrete element. In addition, consideration has to be given to the effect of progressive cracking on the change in stiffness factors, considerably complicating the analysis and design process. Consequently, an empirical approach to evaluate deflection under sustained loading is, in many cases, more practical.

The additional deflection under sustained loading and long-term shrinkage in accordance with the ACI procedure can be calculated using a multiplying factor:

$$\lambda = \frac{T}{1 + 50\rho'} \qquad (8.9)$$

where ρ' is the compression reinforcement ratio calculated at midspan for simple and

Photo 45   Deflected simply supported beam prior to failure. (Tests by Nawy et al.)

Photo 46   Deflected continuous prestressed beam prior to failure. (Tests by Nawy, Potyondy, et al.)

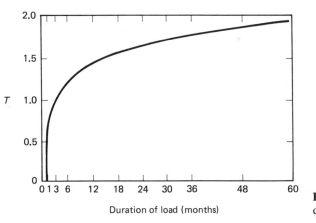

Figure 8.6 Multipliers for long-term deflection.

continuous beams and T is a factor which is taken as 1.0 for loading time duration of 3 months, 1.2 for 6 months, 1.4 for 12 months, and 2.0 for 5 years or more.

If the instantaneous deflection is Δ_i, the additional time-dependent deflection becomes $\lambda \Delta_i$ and the total long-term deflection would be $(1 + \lambda)\Delta_i$. Since live loads are not present at all times, only part of the live load in addition to the more permanent dead load are considered as the sustained load. Figure 8.6 gives the relationship between the load duration in months and the multiplier T in Eq. 8.6. It is seen from this plot that the maximum multiplier value $T = 2.0$ represents a nominal limiting time-dependent factor for 5 years' duration of loading. In effect, the expression for the long-term factor λ in Eq. 8.9 has similar characteristics as the stiffness EI of a section in that it is a function of the material property T and the section property $(1 + 50\rho')$.

The total long-term deflection is

$$\Delta_{LT} = \Delta_L + \lambda_\infty \Delta_D + \lambda_t \Delta_{LS} \qquad (8.10)$$

where Δ_L = immediate live-load deflection
Δ_D = immediate dead-load deflection
Δ_{LS} = sustained live-load deflection (a percentage of the immediate Δ_L determined by expected duration of sustained load)
λ_∞ = time-dependent multiplier for infinite duration of sustained load
λ_t = time-dependent multiplier for limited load duration

The value of the multiplier λ is the same for normal or lightweight concrete.

8.5 PERMISSIBLE DEFLECTIONS IN BEAMS AND ONE-WAY SLABS

Permissible deflections in a structural system are governed primarily by the amount that can be sustained by the interacting components of a structure without loss of aesthetic appearance and without detriment to the deflecting member. The level of acceptability of deflection values is a function of such factors as the type of building, the use or nonuse of partitions, the presence of plastered ceilings, or the sensitivity of equipment or vehicular systems that are being supported by the floor. Since deflection limitations have to be placed at service load levels, structures designed conservatively for low concrete and steel stresses would normally have no deflection problems. Present-day structures, however, are designed by ultimate load procedures efficiently utilizing high-strength concretes and steels. More slender members resulting from such designs would have to be better controlled for serviceability deflection performance, immediate and long-term.

8.5.1 Empirical Method of Minimum Thickness Evaluation for Deflection Control

The ACI code recommends in Table 8.1 minimum thickness for beams as a function of the span length, where no deflection computations are necessary if the member is not supporting or attached to construction likely to be damaged by large deflections. Other deflections would have to be calculated and controlled as in Table 8.2. If the total beam thickness is less than required by the table, the designer should verify the deflection serviceability performance of the beam through detailed computations of the immediate and long-term deflections.

Same as ACI 9.5A

TABLE 8.1   MINIMUM THICKNESS OF BEAMS AND ONE-WAY SLABS UNLESS DEFLECTIONS ARE COMPUTED[a]

Member[b]	Minimum thickness, h			
	Simply supported	One end continuous	Both ends continuous	Cantilever
Solid one-way slabs	$l/20$	$l/24$	$l/28$	$l/10$
Beams or ribbed one-way slabs	$l/16$	$l/18.5$	$l/21$	$l/8$

[a]Clear span length l is in inches. Values given should be used directly for members with normalweight concrete ($w_c = 145$ pcf) and grade 60 reinforcement. For other conditions, the values should be modified as follows:

1. For structural lightweight concrete having unit weights in the range 90 to 120 lb/ft^3, the values should be multiplied by $(1.65 - 0.005w_c)$ but not less than 1.09, where w_c is the unit weight in lb/ft^3.
2. For f_y other than 60,000 psi, the values should be multiplied by $(0.4 + f_y/100,000)$.

[b]Members not supporting or attached to partitions or other construction likely to be damaged by large deflections.

TABLE 8.2 MINIMUM PERMISSIBLE RATIOS OF SPAN (ℓ) TO DEFLECTION (Δ)
(ℓ = longer span)

Type of member	Deflection, Δ, to be considered	$(l/\Delta)_{min}$
Flat roofs not supporting and not attached to nonstructural elements likely to be damaged by large deflections	Immediate deflection due to live load L	180[a]
Floors not supporting and not attached to nonstructural elements likely to be damaged by large deflections	Immediate deflection due to live load L	360
Roof or floor construction supporting or attached to nonstructural elements likely to be damaged by large deflections	That part of total deflection occurring after attachment of nonstructural elements; sum of long-term deflection due to all sustained loads (dead load plus any sustained portion of live load) and immediate deflection due to any additional live load[b]	480[c]
Roof or floor construction supporting or attached to nonstructural elements not likely to be damaged by large deflections		240[c]

[a]Limit not intended to safeguard against ponding. Ponding should be checked by suitable calculations of deflection, including added deflections due to ponded water, and considering long-term effects of all sustained loads, camber, construction tolerances, and reliability of provisions for drainage.

[b]Long-term deflection has to be determined but may be reduced by the amount of deflection calculated to occur before attachment of nonstructural elements. This reduction is made on the basis of accepted engineering data relating to time–deflection characteristics of members similar to those being considered.

[c]Ratio limit may be lower if adequate measures are taken to prevent damage to supported or attached elements, but should not be lower than tolerance of nonstructural elements.

8.5.2 Permissible Limits of Calculated Deflection

The ACI code requires that the calculated deflection for a beam or one-way slab has to satisfy the serviceability requirement of minimum permissible deflection for the various structural conditions listed in Table 8.2 of Section 8.5.1 if Table 8.1 is *not* used. However, long-term effects cause measurable increases in deflection with time and result sometimes in excessive overstress in the steel and concrete. Hence it is always advisable to calculate the total time-dependent deflection Δ_{LT} in Eq. 8.10 and design the beam size based on the permissible span/deflection ratios of Table 8.2.

8.6 COMPUTATION OF DEFLECTIONS

Deflection of structural members is a function of the span length, support, or end conditions, such as simple support or restraint due to continuity, the type of loading, such as concentrated or distributed load, and the flexural stiffness EI of the member.

The general expression for the maximum deflection Δ_{max} in an elastic member can be expressed from basic principles of mechanics as

$$\Delta_{max} = K\frac{Wl_n^3}{48EI_c} \tag{8.11}$$

where W = total load on the span
$\quad l_n$ = clear span length
$\quad E$ = modulus of concrete
$\quad I_c$ = moment of inertia of the section
$\quad K$ = a factor depending on the degree of fixity of the support

Equation 8.11 can also be written in terms of moment such that the deflection at any point in a beam is

$$\Delta = k\frac{ML^2}{E_c I_e} \tag{8.12}$$

where k = a factor depending on support fixity and load conditions
$\quad M$ = moment acting on the section
$\quad I_e$ = effective moment of inertia

Table 8.3 gives the maximum elastic deflection values in terms of the gravity load for typical beams loaded with uniform or concentrated load.

TABLE 8.3 MAXIMUM DEFLECTION EXPRESSIONS FOR MOST COMMON LOAD AND SUPPORT CONDITIONS

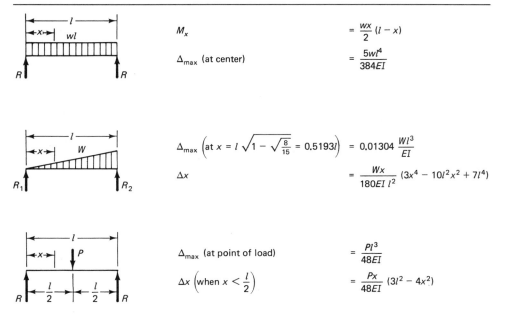

M_x	$= \dfrac{wx}{2}(l-x)$
Δ_{max} (at center)	$= \dfrac{5wl^4}{384EI}$

$\Delta_{max}\left(\text{at } x = l\sqrt{1-\sqrt{\tfrac{8}{15}}} = 0.5193l\right)$	$= 0.01304\,\dfrac{Wl^3}{EI}$
Δx	$= \dfrac{Wx}{180EI\, l^2}(3x^4 - 10l^2 x^2 + 7l^4)$

Δ_{max} (at point of load)	$= \dfrac{Pl^3}{48EI}$
$\Delta x \left(\text{when } x < \dfrac{l}{2}\right)$	$= \dfrac{Px}{48EI}(3l^2 - 4x^2)$

TABLE 8.3 *(Cont.)*

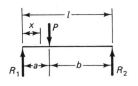

$$\Delta_{max}\left(\text{at } x = \sqrt{\frac{a(a + 2b)}{3}} \text{ when } a > b\right) = \frac{Pab(a + 2b)\ \sqrt{3a(a + 2b)}}{27EI\ l}$$

$$\Delta a \text{ (at point of load)} = \frac{Pa^2b^2}{3EI\ l}$$

$$\Delta x \text{ (when } x < a) = \frac{Pbx}{6EI\ l}\ (l^2 - b^2 - x^2)$$

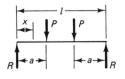

$$\Delta_{max} \text{ (at center)} = \frac{Pa}{24EI}\ (3l^2 - 4a^2)$$

$$\Delta x \text{ (when } x < a) = \frac{Px}{6EI}\ (3la - 3a^2 - x^2)$$

$$\Delta x \text{ (when } x > a \text{ and } < (l - a)) = \frac{Pa}{6EI}\ (3lx - 3x^2 - a^2)$$

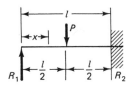

$$\Delta_{max}\left(\text{at } x = l\ \sqrt{\frac{1}{5}} = 0.4472l\right) = \frac{Pl^3}{48EI\ \sqrt{5}} = 0.009317\ \frac{Pl^3}{EI}$$

$$\Delta x \text{ (at point of load)} = \frac{7Pl^3}{768EI}$$

$$\Delta x \left(\text{when } x < \frac{l}{2}\right) = \frac{Px}{96EI}\ (3l^2 - 5x^2)$$

$$\Delta x \left(\text{when } x > \frac{l}{2}\right) = \frac{P}{96EI}\ (x - l)^2\ (11x - 2l)$$

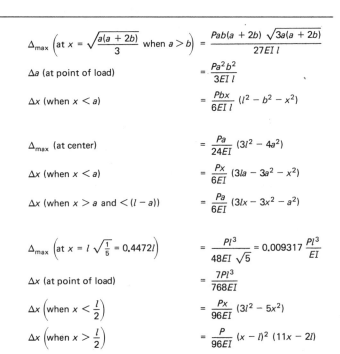

$$\Delta_{max} \text{ (at free end)} = \frac{wl^4}{8EI}$$

$$\Delta x = \frac{w}{24EI}\ (x^4 - 4l^3x + 3l^4)$$

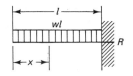

$$\Delta_{max} \text{ (at center)} = \frac{wl^4}{384EI}$$

$$\Delta x = \frac{wx^2}{24EI}\ (l - x)^2$$

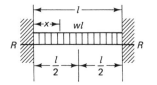

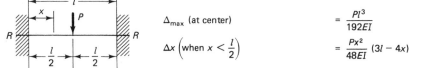

$$\Delta_{max} \text{ (at center)} = \frac{Pl^3}{192EI}$$

$$\Delta x \left(\text{when } x < \frac{l}{2}\right) = \frac{Px^2}{48EI}\ (3l - 4x)$$

TABLE 8.3 *(Cont.)*

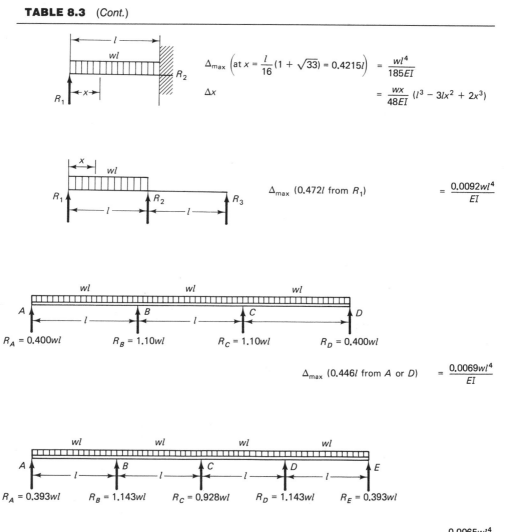

$$\Delta_{max}\left(\text{at } x = \frac{l}{16}(1 + \sqrt{33}) = 0.4215l\right) = \frac{wl^4}{185EI}$$

$$\Delta x = \frac{wx}{48EI}(l^3 - 3lx^2 + 2x^3)$$

$$\Delta_{max} (0.472l \text{ from } R_1) = \frac{0.0092wl^4}{EI}$$

$R_A = 0.400wl$ $R_B = 1.10wl$ $R_C = 1.10wl$ $R_D = 0.400wl$

$$\Delta_{max} (0.446l \text{ from } A \text{ or } D) = \frac{0.0069wl^4}{EI}$$

$R_A = 0.393wl$ $R_B = 1.143wl$ $R_C = 0.928wl$ $R_D = 1.143wl$ $R_E = 0.393wl$

$$\Delta_{max} (0.440l \text{ from } A \text{ or } E) = \frac{0.0065wl^4}{EI}$$

8.6.1 Example 8.3: Deflection Behavior of a Uniformly Loaded Simple Span Beam

A simply supported uniformly loaded beam has a clear span $l_n = 27$ ft (8.23 m), a width $b = 10$ in. (254 mm), and a total depth $h = 16$ in. (406 mm), $d = 13.0$ in. (330 mm), and $A_s = 1.32$ in.^{2} (851.6 mm^2). It is subjected to a service dead-load moment $M_D = 215,000$ in.-lb (24.3 kN-m), and a service live-load moment $M_L = 250,000$ in.-lb (28.3 kN-m). Determine if the beam satisfies the various deflection criteria for short-term and long-term loading. Assume that 60% of the live load is continuously applied for 24 months. Given:

f'_c = 5000 psi (34.5 MPa), normalweight concrete
f_y = 60,000 psi (413.7 MPa)

Solution

$$E_c = 33w^{1.5}\sqrt{5000} = 4.29 \times 10^6 \text{ psi (29,580 MPa)}$$

$$E_s = 29 \times 10^6 \text{ psi (200,000 MPa)}$$

$$\text{modular ratio } n = \frac{E_s}{E_c} = \frac{29.0 \times 10^6}{4.29 \times 10^6} = 6.76$$

$$f_r = 7.5\sqrt{f'_c} = 7.5\sqrt{5000} = 530.3 \text{ psi (3.66 MPa)}$$

Minimum required depth

From Table 8.1,

$$h_{\min} = \frac{l_n}{16} = \frac{\overset{27.0}{\cancel{28.0}} \times 12}{16} = 21.0 \text{ in. (533.4 mm)} > \text{actual } h = 16.0 \text{ in.}$$

Hence deflection calculations have to be made.

Effective moment of inertia I_e

$$I_g = \frac{bh^3}{12} = \frac{10(16)^3}{12} = 3413.3 \text{ in.}^4$$

$$y_t = \frac{16.0}{2} = 8.0 \text{ in.}$$

$$M_{cr} = \frac{f_r I_g}{y_t} = \frac{530.3 \times 3413.3}{8.0} = 226,259 \text{ in.-lb}$$

Depth of neutral axis c:

$$d = 16.0 - 3.0 = 13.0 \text{ in.} \qquad A_s = 1.32 \text{ in.}^2$$

$$\frac{10c^2}{2} = nA_s(d - c)$$

or $5c^2 = 6.76 \times 1.32(13.0 - c)$, to get $c = 4.03$ in.

$$I_{cr} = \frac{10c^3}{3} + 6.76 \times 1.32(13.0 - c)^2 = \frac{10(4.03)^3}{3} + 8.923(13.0 - 4.03)^2$$

$$= 936.1 \text{ in.}^4$$

Dead load:

$$\frac{M_{cr}}{M_a} = \frac{226,259}{215,000} = 1.05 > 1.0$$

Use $M_{cr} = M_a$ and $I_e = I_g$ since the dead-load moment is smaller than the cracking moment (the beam will not crack at the dead-load level).

Dead load + 60% live load:

$$\frac{M_{cr}}{M_a} = \frac{226,259}{215,000 + 0.6 \times 250,000} = 0.62$$

Dead load + live load:

$$\frac{M_{cr}}{M_a} = \frac{226,259}{215,000 + 250,000} = 0.49$$

$$I_e = \left(\frac{M_{cr}}{M_a}\right)^3 I_g + \left[1 - \left(\frac{M_{cr}}{M_a}\right)^3\right] I_{cr}$$

Dead load:

$$I_e = 3413.3 \text{ in.}^4$$

Dead load + 0.6 live load:

$$I_e = 0.24 \times 3413.3 + 0.76 \times 936.1 = 1530.6 \text{ in.}^4$$

Dead load + live load:

$$I_e = 0.12 \times 3413.3 + 0.88 \times 936.1 = 1233.4 \text{ in.}^4$$

Short-term deflection

$$\Delta = \frac{5wl^4}{384EI} = \frac{5Ml_n^2}{48EI} = \frac{5(27.0 \times 12)^2 M}{48 \times 4.29 \times 10^6 I} = 0.0025\frac{M}{I} \text{ in.}$$

Immediate live-load deflection:

$$\Delta_L = \frac{0.0025(215,000 + 250,000)}{1233.4} - \frac{0.0025(215,000)}{3413.3}$$

$$= 0.943 - 0.158 = 0.785 \text{ in.}$$

Immediate dead-load deflection:

$$\Delta_D = \frac{0.0025 \times 215,000}{3413.3} = 0.158 \text{ in.}$$

Immediate 60% live-load deflection:

$$\Delta_{LS} = 0.0025\left(\frac{215,000 + 250,000 \times 0.6}{1530.6} - \frac{215,000}{3413.3}\right)$$

$$= 0.597 - 0.158 = 0.439 \text{ in.}$$

Long-term deflection

From Eq. 8.10,

$$\Delta_{LT} = \Delta_L + \lambda_\infty \Delta_D + \lambda_t \Delta_{LS}$$

$$\lambda = \frac{T}{1 + 50\rho'} \qquad \text{where } \rho' = 0 \text{ for singly reinforced beam}$$

$$T \text{ for 5 years or more} = 2.0 \qquad \lambda_\infty = \frac{2.0}{1 + 0} = 2.0$$

$$T \text{ for 24 months} = 1.65 \qquad \lambda_t = \frac{1.65}{1} = 1.65$$

$$\Delta_{LT} = 0.785 + 2.0 \times 0.158 + 1.65 \times 0.439 = 1.825 \text{ in.}$$

Deflection requirements (Table 8.2)

$$\frac{l_n}{180} = \frac{27 \times 12}{180} = 1.80 \text{ in.} > \Delta_L$$

$$\frac{l_n}{360} = 0.90 \text{ in.} \qquad\qquad > \Delta_L$$

$$\frac{l_n}{480} = 0.68 \text{ in.} \qquad\qquad < \Delta_{LT}$$

$$\frac{l_n}{240} = 1.35 \text{ in.} \qquad\qquad < \Delta_{LT}$$

Hence the use of this beam is limited to floors or roofs not supporting or attached to nonstructural elements such as partitions.

8.7 DEFLECTION OF CONTINUOUS BEAMS

Similar to problem due 4/29/86

As discussed in Chapter 5, a continuous reinforced concrete beam would have a flanged section at midspan, and sometimes a doubly reinforced section at the support if the reinforcement at the bottom fibers of the support section are adequately tied and anchored. Consequently, it is necessary to be able to find the effective moment of inertia I_e of T sections and of doubly reinforced sections. A simple procedure is to use the weighted-average section properties as required by the ACI code:

Look at Table 9.5C

1. Beams with both ends continuous:

$$\text{average } I_e = 0.70 I_m + 0.15(I_{e1} + I_{e2}) \tag{8.13}$$

2. Beams with one end continuous

$$\text{average } I_e = 0.85 I_m + 0.15(I_{ec}) \tag{8.14}$$

where I_m = midspan section I_e
        I_{e1}, I_{e2} = I_e for the respective beam ends and
        I_{ec} = I_e of continuous end

It is seen from Eqs. 8.13 and 8.14 that the controlling moment of inertia for deflection evaluation is the midspan-section effective moment of inertia.

Moment envelopes have to be used to calculate the positive and negative values of I_e. If the continuous beam is subjected to a single heavy concentrated load, only the midspan effective moment of inertia I_e is to be used.

8.7.1 Deflection of T Beams

The most common nonrectangular sections are the flanged T and L beams. The same principles used for deflection computations of rectangular sections can be applied to the nonrectangular ones. The contribution of the compressive resisting force can be obtained using the appropriate concrete area, as explained below.

As in the case of rectangular beams, the contribution of steel to the moment of inertia of the uncracked section is disregarded. The cross section of the beam in Fig. 8.7a is divided into two areas for the purpose of calculating I_g.

$$\text{depth of center of gravity } \bar{y} = \frac{A_1 y_1 + A_2 y_2}{A_1 + A_2} \tag{8.15a}$$

$$y_t = h - \bar{y} \tag{8.15b}$$

The gross moment of inertia, I_g, for the two rectangles is

Design
Mwk

$$I_g = \frac{bh_f^3}{12} + bh_f\left(\bar{y} - \frac{h_f}{2}\right)^2 + \frac{b_w(h - h_f)^3}{12} + b_w(h - h_f)\left(y_t - \frac{h - h_f}{2}\right)^2 \tag{8.16}$$

For the cracked section, the depth c of the neutral axis is calculated from the horizontal force equilibrium, as in Fig. 8.7b and c. If the depth of neutral axis falls within the flange thickness, the beam behaves as a rectangular section having a width b of the flange and an effective depth d.

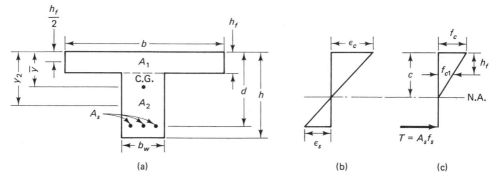

Figure 8.7 Stress and strain distribution across depth of flanged sections: (a) geometry; (b) strains; (c) stresses.

Photo 47   Deflected simply supported beam at failure. (Tests by Nawy et al.)

When the depth c of neutral axis falls below the flange thickness h_f, the appropriate areas of concrete in the flange and the web of the section and corresponding stresses are applied in the calculation of the compression force. The average stress in the flange area bh_f would be $(f_c + f_{c1})/2$, where f_{c1} is the stress at the bottom of the flange. Using similar triangles yields

$$f_{c1} = f_c \frac{c - h_f}{c} \tag{8.17}$$

The average stress in compression for the web area, $b_w(c - h_f)$, would be $f_{c1}/2$ based on the triangular distribution of stress. Hence the force equilibrium equation can be written as

$$A_s f_y = bh_f \frac{f_c + f_{c1}}{2} + b_w(c - h_f)\frac{f_{c1}}{2} \tag{8.18a}$$

Using Eqs. 8.17 and 8.18a,

$$2A_s E_s \epsilon_s = bh_f E_c \epsilon_c \left(1 + \frac{c - h_f}{c}\right) + b_w(c - h_f)E_c \epsilon_c \frac{c - h_f}{c} \tag{8.18b}$$

Expressing ϵ_s in terms of ϵ_c and using modular ratio n gives us

$$2nA_s \frac{d - c}{c} = bh_f \frac{2c - h_f}{c} + b_w(c - h_f)\frac{c - h_f}{c} \tag{8.18c}$$

or

$$b_w(c - h_f)^2 - 2nA_s(d - c) + bh_f(2c - h_f) = 0 \tag{8.18d}$$

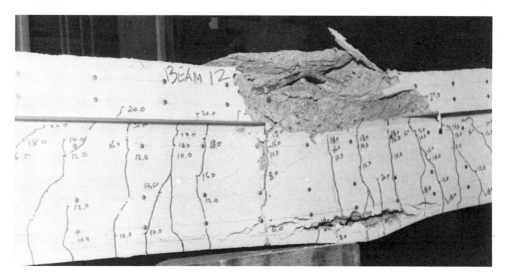

Photo 48   Flexural stabilized cracks at failure. (Tests by Nawy et al.)

The quadratic equation 8.18a has to be solved to obtain c. Once c is known, the moment of inertia I_{cr} of the cracked section can be calculated using the following expression: T-section

Design
Hwk

$$I_{cr} = \frac{1}{3}b_w(c - h_f)^3 + \frac{1}{12}bh_f^3 + bh_f\left(c - \frac{h_f}{2}\right)^2 + nA_s(d - c)^2 \qquad (8.19)$$

The effective moment of inertia I_e and deflection Δ can be computed as in the case of rectangular sections using Eqs. 8.8a and 8.8b. In the case of L sections, expressions for I_{cr} such as those of Eq. 8.19 can be developed in a similar manner as for T sections.

8.7.2 Deflection of Beams with Compression Steel

Beams with compression reinforcement can be treated similarly to singly reinforced sections except that the contribution of the compression reinforcement to the stiffness of the beam should be considered because of its high stiffening effect. For the moment of inertia of the uncracked section, I_g can be used with sufficient accuracy. The contribution of the compression steel A_s' to the cracked moment of inertia I_{cr} has to be included. Also, Eq. 8.6c has to be modified for calculating the neutral-axis depth c of the beam. If the compressive force $A_s'f_s'$ of the steel is added to the compressive force of the concrete, Eq. 8.4a as seen from Fig. 8.8 becomes

$$A_sf_s = bc\frac{f_c}{2} - A_s'f_c\frac{c - d'}{c} + A_s'f_s' \qquad (8.20a)$$

where d' is the effective cover of compression reinforcement.

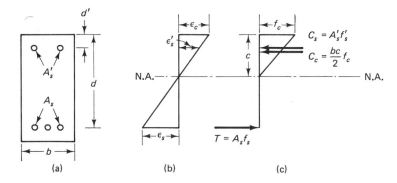

Figure 8.8 Stress and strain distribution at service load in doubly reinforced beam: (a) geometry; (b) strains; (c) stresses.

As in the case of singly reinforced concrete beams (Eqs. 8.4 to 8.6), Eq. 8.20a can be written in the form

$$\frac{bc^2}{2} + [nA_s + (n-1)A'_s]c - nA_s d - (n-1)A'_s d' = 0 \qquad (8.20b)$$

The moment of inertia I_{cr} of the cracked section can therefore be expressed as

$$I_{cr} = \frac{bc^3}{3} + nA_s(d-c)^2 + (n-1)A'_s(c-d')^2 \qquad (8.21)$$

The procedure for calculating the effective moment of inertia I_e and the deflection Δ is the same as in the case of singly reinforced beams.

8.7.3 Deflection Bending Moments in Continuous Beams

The flexural moment envelope has to be constructed for the total continuous beam span in order to evaluate the effective moment of inertia I_e. The usual methods of structural analysis are followed in finding the continuity moments at supports and the positive midspan moments for the various spans. Once these moments are determined, the immediate central postelastic (i.e., postcracking) deflection can be evaluated.

As in the case of simply supported beams, the deflection Δ can be written either in terms of load as in Eq. 8.11 or in terms of moment as in Eq. 8.12. If an interior span AB subjected to a uniform load is isolated as in Fig. 8.9, the midspan deflection Δ_c is

$$\Delta_c = \frac{5l^2}{48EI}[M_m + 0.1(M_a + M_b)] \qquad (8.22)$$

where M_a, M_b = negative service load bending moments
M_0 = simple span service load static moment
M_m = midspan moment

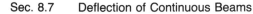

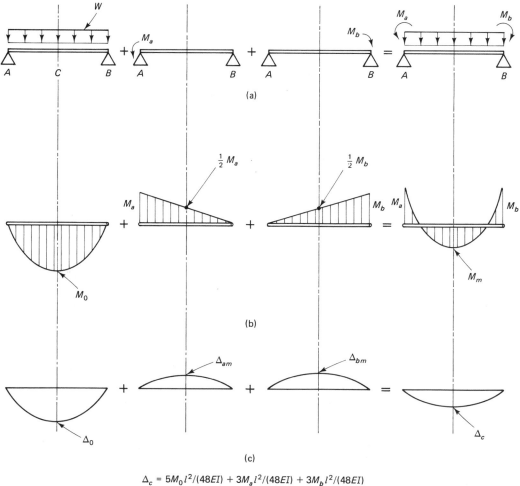

$$\Delta_c = 5M_0l^2/(48EI) + 3M_al^2/(48EI) + 3M_bl^2/(48EI)$$

$$= \frac{5l^2}{48EI}\,[M_m + 0.1\,(M_a + M_b)]$$

Figure 8.9 Deflection bending moments in continuous beams: (a) loads; (b) moments; (c) deflections, using superposition.

As the exterior span is subjected to the largest positive and negative moments, deflection calculations control for this span in most cases.

8.7.4 Example 8.4: Deflection of a Continuous Four-Span Beam

A reinforced concrete beam supporting a 4-in. (101.6-mm) slab is continuous over four equal spans $l = 36$ ft (10.97 m) as shown in Fig. 8.10. It is subjected to a uniformly distributed load $w_D = 700$ lb/ft (10.22 kN/m), including its self-weight and a service

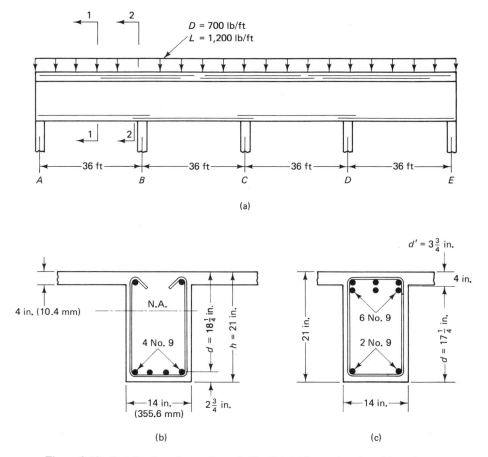

Figure 8.10 Details of continuous beam in Ex. 8.4: (a) beam elevation; (b) section 1–1; (c) section 2–2.

live load $w_L = 1200$ lb/ft (17.52 kN/m). The beam has the dimensions $b = 14$ in. (355.6 mm), $d = 18.25$ in. (463.6 mm) at midspan, and a total thickness $h = 21.0$ in. (533.4 mm). The first interior span is reinforced with four No. 9 bars at midspan (28.6 mm diameter) at the bottom fibers and six No. 9 bars at the top fibers of the support section.

Calculate the maximum deflection of the continuous beam, determine what code deflection criteria it meets, and what limitations, if any, have to be placed on its use. Given:

$f'_c = 4000$ psi (27.8 MPa), normalweight concrete

$f_y = 60,000$ psi (413.7 MPa)

50% of the live load is sustained 36 months on the structure

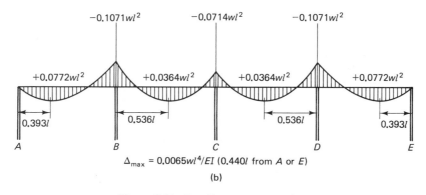

$$\Delta_{max} = 0.0065wl^4/EI \; (0.440l \text{ from } A \text{ or } E)$$

(b)

Figure 8.11   Bending moment envelope.

Solution

Minimum depth requirement

From Table 8.1,

$$\text{minimum } h = \frac{l}{18.5} = \frac{36.0 \times 12}{18.5} = 23.35 \text{ in.}$$

$$\text{actual } h = 21.0 \text{ in.} < 23.35 \text{ in.}$$

Deflection calculations have to be made.

Material properties and bending moment envelope

$$E_c = 57,000 \sqrt{f_c'} = 57,000 \sqrt{4000} = 3.6 \times 10^6 \text{ psi (24,822 MPa)}$$

$$E_s = 29 \times 10^6 \text{ psi (200,000 MPa)}$$

$$\text{modular ratio } n = \frac{E_s}{E_c} = \frac{29 \times 10^6}{3.6 \times 10^6} = 8.1$$

$$\text{modulus of rupture } f_r = 7.5 \sqrt{f_c'} = 7.5 \sqrt{4000} = 474.3 \text{ psi (3.27 MPa)}$$

From bending moment analysis, the bending moment diagram for the beam is shown in Fig. 8.11. For deflection, the largest moments are in end spans *AB* and *ED*.

positive moment $= 0.0772wl^2$

$$+M_D = 0.0772 \times 700(36.0)^2 \times 12 = 840,430 \text{ in.-lb}$$

$$+M_L = 0.0772 \times 1200(36.0)^2 \times 12 = 1,440,737 \text{ in.-lb}$$

$$+(M_D + M_L) = 0.0772 \times 1900(36.0)^2 \times 12 = 2,281,167 \text{ in.-lb}$$

negative moment $= 0.1071wl^2$

$$-M_D = 0.1071 \times 700(36.0)^2 \times 12 = 1,165,933 \text{ in.-lb}$$

$$-M_L = 0.1071 \times 1200(36.0)^2 \times 12 = 1,998,743 \text{ in.-lb}$$

$$-(M_D + M_L) = 0.1071 \times 1900(36.0)^2 \times 12 = 3,164,676 \text{ in.-lb}$$

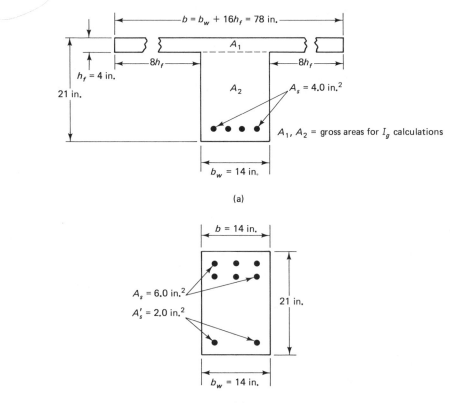

Figure 8.12 Gross moment of inertia I_g cross sections in Ex. 8.4: (a) midspan section; (b) support section.

Effective moment of inertia I_e

Figure 8.12 shows the theoretical midspan and support cross sections to be used for calculations of the gross moment of inertia I_g.

1. *Midspan section:*

width of T-beam flange $= b_w + 16h_f = 14.0 + 16 \times 4.0 = 78$ in. (1981 mm)

Depth from compression flange to the elastic centroid from Eq. 8.15a:

$$\bar{y} = \frac{A_1 y_1 + A_2 y_2}{A_1 + A_2}$$

$$= \frac{78(4 \times 2) + 14 \times (21 - 4) \times 12.5}{78 \times 4 + 14 \times 17} = 6.54 \text{ in.}$$

$$y_t = h - \bar{y} = 21.0 - 6.54 = 14.46 \text{ in.}$$

From Eq. 8.16,

$$I_g = \frac{78(4)^3}{12} + 78 \times 4\left(6.54 - \frac{4}{2}\right)^2 + \frac{14(21 - 4)^3}{12}$$

$$+ 14(21 - 4)\left(14.46 - \frac{21 - 4}{2}\right)^2$$

$$= 21{,}033 \text{ in.}^4$$

$$M_{cr} = \frac{f_r I_g}{y_t} = \frac{474.3 \times 21{,}033}{14.46} = 689{,}900 \text{ in.-lb}$$

Depth of neutral axis:

$$A_s = \text{four No. 9 bars} = 4.0 \text{ in.}^2$$

From Eq. 8.18d:

$$14(c - 4.0)^2 - 2 \times 8.1 \times 4.0(18.25 - c) + 78 \times 4(2c - 4.0)$$

or $c^2 + 41.17c - 157.0$ to give $c = 3.5$ in. Hence the neutral axis is inside the flange and the section is analyzed as a rectangular section.

From Eq. 8.6c, for rectangular sections,

$$\frac{78c^2}{2} + 8.1 \times 4 \times c - 8.1 \times 4 \times 18.25 = 0$$

Therefore, $c = 3.5$ in.

$$I_{cr} = \frac{78.0(3.5)^3}{3} + 8.1 \times 4(18.25 - 3.5)^2 = 8163.8 \text{ in.}^4$$

Ratio M_{cr}/M_a:

$$D \text{ ratio} = \frac{689{,}900}{840{,}430} = 0.821$$

$$D + 50\% \text{ } L \text{ ratio} = \frac{689{,}900}{840{,}430 + 0.5 \times 1{,}440{,}737} = 0.442$$

$$D + L \text{ ratio} = \frac{689{,}900}{2{,}281{,}167} = 0.302$$

Effective moment of inertia for midspan section:

$$I_e = \left(\frac{M_{cr}}{M_a}\right)^3 I_g + \left[1 - \left(\frac{M_{cr}}{M_a}\right)^3 I_{cr}\right]$$

I_e for dead load $= 0.5534 \times 21{,}033 + 0.4466 \times 8163.8 = 15{,}286 \text{ in.}^4$

I_e for $D + 0.5L = 0.0864 \times 21{,}033 + 0.9136 \times 8163.8 = 9276 \text{ in.}^4$

I_e for $D + L = 0.0275 \times 21{,}033 + 0.9725 \times 8163.8 = 8518 \text{ in.}^4$

2. *Support section:*

$$I_g = \frac{bh^3}{12} = \frac{14(21)^3}{12} = 10{,}804.5 \text{ in.}^4$$

$$y_t = \frac{21.0}{2} = 10.5 \text{ in.}$$

$$M_{cr} = \frac{f_r I_g}{y_t} = \frac{474.3 \times 10,804.5}{10.5} = 488,055 \text{ in.-lb}$$

Depth of neutral axis:

$$A_s = \text{six No. 9} = 6.0 \text{ in.}^2 \ (3870 \text{ mm}^2)$$

$$A_s' = \text{two No. 9} = 2.0 \text{ in.}^2 \ (1290 \text{ mm}^2)$$

$$d = 21.0 - 3.75 = 17.25 \text{ in. } (438.2 \text{ mm})$$

From Eq. 8.20b,

$$\frac{14c^2}{2} + [8.1 \times 6.0 + (8.1 - 1)2.0]c - 8.1 \times 6.0$$

$$\times 17.25 - (8.1 - 1) \times 2.0 \times 2.75 = 0$$

or $c^2 + 8.97c - 125.34 = 0$, to give $c = 7.58$ in.

From Eq. 8.21, the cracking moment of inertia is

$$I_{cr} = \frac{bc^3}{3} + nA_s(d - c)^2 + (n - 1)A_s'(c - d')^2$$

$$= \frac{14(7.58)^3}{3} + 8.1 \times 6.0(17.25 - 7.58)^2 + (8.1 - 1)2.0(7.58 - 2.75)^2$$

$$= 6908.2 \text{ in.}^4$$

Ratio M_{cr}/M_a:

$$D \text{ ratio} = \frac{488,055}{1,165,933} = 0.42$$

$$D + 50\% \ L \text{ ratio} = \frac{488,055}{1,165,933 + 0.5 \times 1,998,743} = 0.225$$

$$D + L = \frac{488,055}{3,164,676} = 0.15$$

Effective moment of inertia for support section:

I_e for dead load $= 0.0741 \times 10,804.5 + 0.9259 \times 6908.2 = 7196.9$ in.^{4}

I_e for $D + 0.5L = 0.0122 \times 10,804.5 + 0.9878 \times 6908.2 = 6955.7$ in.^{4}

I_e for $D + L = 0.0034 \times 10,845.5 + 0.9966 \times 6908.2$ in.$^4 = 6921.6$ in.^{4}

Average effective I_e for continuous span

From Eq. 8.14,

average $I_e = 0.85I_m + 0.15I_{ec}$

dead load: $I_e = 0.85 \times 15{,}286 + 0.15 \times 7196.9 = 14{,}073$ in.^{4}

$D + 0.5L$: $I_e = 0.85 \times 9276 + 0.15 \times 6955.7 = 8928$ in.^{4}

$D + L$: $I_e = 0.85 \times 8518 + 0.15 \times 6921.6 = 8278$ in.^{4}

Short-term deflection

From Table 8.3, the maximum deflection for span AB or DE is

$$\Delta = \frac{0.0065wl^4}{EI}$$

l assumed $\simeq l_n$ for all practical purposes

$$\Delta = \frac{0.0065(36.0 \times 12)^4}{36 \times 10^6} \times \frac{w}{I_e} \times \frac{1}{12} = 5.240 \frac{w}{I_e} \text{ in.}$$

Immediate live-load deflection:

$$\Delta_L = \frac{5.240(1900)}{8278} - \frac{5.240(700)}{14{,}073} = 1.20 - 0.26 = 0.94 \text{ in.}$$

Immediate dead-load deflection:

$$\Delta_D = \frac{5.240(700)}{14{,}073} = 0.26 \text{ in.}$$

Immediate 50% live-load deflection:

$$\Delta_{LS} = \frac{5.240(1300)}{8928} - \frac{5.240(700)}{14{,}073} = 0.76 - 0.26 = 0.50 \text{ in.}$$

Long-term deflection

$$\rho' = \frac{A_s'}{bd} = 0 \qquad \text{at midspan in this case}$$

From Eq. 8.9,

$$\text{multiplier } \lambda = \frac{T}{1 + 50\rho'}$$

From Fig. 8.6,

$$T = 1.75 \text{ for 36-month sustained load}$$

$$T = 2.0 \text{ for 5-year loading}$$

Therefore,

$$\lambda_\infty = 2.0 \qquad \text{and} \qquad \lambda_t = 1.75$$

From Eq. 8.10, the total sustained load deflection is

$$\Delta_{LT} = \Delta_L + \lambda_\infty \Delta_D + \lambda_t \Delta_{LS}$$

or

$$\Delta_{LT} = 0.94 + 2.0 \times 0.26 + 1.75 \times 0.50 = 2.35 \text{ in. (59.6 mm)}$$

Deflection requirements (Table 8.2)

$$\frac{l}{180} = \frac{36 \times 12}{180} = 2.4 \text{ in.} > \Delta_L$$

$$\frac{l}{360} = 1.2 \text{ in.} > \Delta_L$$

$$\frac{l}{480} = 0.9 \text{ in.} < \Delta_{LT}$$

$$\frac{l}{240} = 1.8 \text{ in.} < \Delta_{LT}$$

Hence the continuous beam is limited to floors or roofs not supporting or attached to nonstructural elements such as partitions.

8.8 OPERATIONAL DEFLECTION CALCULATION PROCEDURE AND FLOWCHART

Deflection of structures affects their aesthetic appearance as well as their long-term serviceability. The following step-by-step procedure should be followed after the structural member is designed for flexure.

1. Compare the total design depth of the member with the minimum allowable value obtained from Table 8.1. If it is less than the allowable, proceed to perform a detailed calculation of short-term and long-term deflection. It is, however, always advisable to perform the detailed calculations regardless of the comparison with Table 8.1.
2. The detailed calculations should establish as a first step:
   (a) The gross moment of inertia I_g
   (b) The cracking moment M_{cr}, which is a function of the modulus of rupture of concrete
3. Calculate the depth c of the neutral axis of the *transformed* section. Find the cracking moment of inertia I_{cr}.
4. Find the effective moment of inertia I_e as follows:

$$I_e = \left(\frac{M_{cr}}{M_a}\right)^3 I_g + \left[1 - \left(\frac{M_{cr}}{M_a}\right)^3\right] I_{cr} \le I_g$$

or

$$I_e = I_{cr} + \left(\frac{M_{cr}}{M_a}\right)^3 (I_g - I_{cr}) \le I_g$$

The effective I_e has to be calculated for the following service load-level combinations:

(a) Dead load (D)

(b) Dead load + sustained portion of live load $(D + \alpha L$, where α is less than 1.0)

(c) Dead load + live load $(D + L)$

5. Calculate the immediate deflection based on I_e of the three combinations in step 4, using the elastic deflection expression in Table 8.3. If the beam is continuous over more than two supports, find the average I_e as follows:

    Both ends continuous:

$$\text{average } I_e = 0.70 I_m + 0.15(I_{e1} + I_{e2})$$

One end continuous:

$$\text{average } I_e = 0.85 I_m + 0.15 I_{ec}$$

6. Calculate the long-term deflection, finding first the time-dependent multiplier $\lambda = T/(1 + 50\rho')$ from values in Fig. 8.6. The total long-term deflection is

$$\Delta_{LT} = \Delta_L = \lambda_\infty \Delta_D + \lambda_t \Delta_{LS}$$

7. If $\Delta_{LT} <$ maximum permissible Δ in Table 8.2, limit the use of the structure to particular loading types or conditions, or enlarge the section. Figure 8.13 gives a flowchart of the operational sequence of deflection control checks that the designer engineer should use when deflection computations are necessary.

8.9 DEFLECTION CONTROL IN ONE-WAY SLABS

One-way slabs can be treated as rectangular beams of 12-in. (304.8-mm) width. As floor loads are specified as load intensity per unit area, such intensity on a one-way slab over a 1-ft width becomes pounds per linear foot. Reinforcement is chosen in terms of bar spacing instead of number of bars and the area of steel for a 12-in. width of slab can easily be calculated for the total number of bars in a 12-in.-width strip.

8.9.1 Example 8.5: Deflection Calculations for a Simply Supported One-Way Slab

A 5-in.-thick $(h = 127$ mm) one-way slab has a span of 12 ft (3.66 in.). It is subjected to a live load of 60 psf (2.88 kPa) in addition to its self-weight. Calculate the immediate and long-term deflections of this slab, assuming that 45% of the live load is sustained over a 24-month period. Determine what type of elements it should support. Given:

$f_c' = 3500$ psi (24.13 MPa)

$f_y = 60,000$ psi (413.7 MPa)

$E_s = 29 \times 10^6$ psi (200,000 MPa)

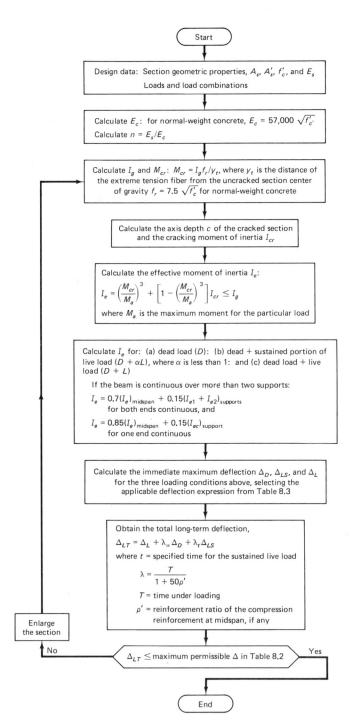

Figure 8.13 Deflection evaluation flowchart.

Steel reinforcement: No. 4 bars at 6 in. center-to-center spacing (12.7 mm diameter at 152.4 mm center to center)

Solution

Minimum depth requirement

From Table 8.1,

$$\text{minimum } h = \frac{l}{20} = \frac{12 \times 12}{20} = 7.20 \text{ in.}$$

$$\text{actual } h = 5 \text{ in.} < 7.20 \text{ in.}$$

Deflection calculations have to be made.

Material properties and bending moments

$$E_c = 57,000 \sqrt{f_c'} = 57,000 \sqrt{3500} = 3.37 \times 10^6 \text{ psi (23,256 MPa)}$$

$$E_s = 29 \times 10^6 \text{ psi (200,000 MPa)}$$

$$\text{modular ratio } n = \frac{E_s}{E_c} = \frac{29 \times 10^6}{3.37 \times 10^6} = 8.61$$

$$\text{modulus of rupture } f_r = 7.5 \sqrt{f_c'} = 7.5 \sqrt{3500} = 443.7 \text{ psi}$$

$$\text{gross moment of inertia } I_g = \frac{bh^3}{12} = \frac{12(5.0)^3}{12} = 125.0 \text{ in.}^4$$

$$\text{cracking moment } M_{cr} = \frac{f_r I_g}{y_t} = \frac{443.7 \times 125.0}{2.5} = 22,185 \text{ in.-lb}$$

$$\text{service load bending moment} = \frac{wl_n^2}{8} = \frac{w(12.0)^2}{8} \times 12 \text{ in.-lb} = 216w \text{ in.-lb}$$

Neutral-axis depth of transformed section

If c is the depth from the compression fibers to the neutral axis of the transformed section,

$$A_s = \text{No. 4 at 12 in.} = 0.40 \text{ in.}^2 \text{ per 12-in.-wide strip}$$

$$d = h - 0.75 - \frac{\phi}{2} = 5.0 - 0.75 - 0.25 = 4.0 \text{ in.}$$

From Eq. 8.6(c) for rectangular sections,

$$\frac{bc^2}{2} + nA_s c - nA_s d = 0$$

$$\frac{12c^2}{2} + 8.61 \times 0.40c - 8.61 \times 0.40 \times 4.0 = 0$$

or $c^2 + 0.574c - 2.296 = 0$, giving $c = 1.255$ in.

Effective moment of inertia

>   *Dead load:*

$$w_D = \text{self-weight of slab} = \frac{5}{12} \times 150 \text{ pcf} = 62.5 \text{ psf}$$

$$M_a = 216w = 216 \times 62.5 = 13,500 \text{ in.-lb} < M_{cr}$$

Hence the slab will not crack under dead load and $I_e = I_g = 125.0 \text{ in.}^4$.

>   *Dead load + 45% live load:*

$$M_a = 216(62.5 + 0.45 \times 60) = 19,332 \text{ in.-lb} < M_{cr}$$

Hence the slab will not crack under dead load and 45% sustained live load and $I_e = I_g = 125.0 \text{ in.}^4$.

>   *Dead + live load:*

$$M_a = 216(62.5 + 60.0) = 26,460 \text{ in.-lb} > M_{cr}$$

The section is cracked.

$$I_{cr} = \frac{bc^3}{3} + nA_s(d - c)^2 \qquad \text{from Eq. 8.7}$$

or

$$I_{cr} = \frac{12(1.255)^3}{3} + 8.61 \times 0.40(4.0 - 1.255)^2 = 33.86 \text{ in.}^4$$

$$\frac{M_{cr}}{M_a} = \frac{22,185}{26,460} = 0.838$$

$$I_e = \left(\frac{M_{cr}}{M_a}\right)^3 I_g + \left[1 - \left(\frac{M_{cr}}{M_a}\right)^3\right] I_{cr} = 0.59 \times 125.0 + 0.41 \times 33.86 = 87.63 \text{ in.}^4$$

Short-term deflection

>   From Table 8.3,

$$\Delta = \frac{5wl_n^2}{384E_cI_e} = \frac{5w(12.0 \times 12)^4}{384 \times 3.37 \times 10^6 I_e} \times \frac{1}{12} = \frac{0.1384}{I_e} w \text{ in.}$$

>   *Immediate live-load deflection:*

$$\Delta_L = \frac{0.1384(62.5 + 60.0)}{87.63} - \frac{0.1384(62.5)}{125.0} = 0.194 - 0.069 = 0.125 \text{ in. (3.2 mm)}$$

>   *Immediate dead-load deflection:*

$$\Delta_D = \frac{0.1384(62.5)}{125.0} = 0.069 \text{ in. (1.8 mm)}$$

>   *Immediate 45% LL deflection:*

$$\Delta_{LS} = \frac{0.1384(62.5 + 0.45 \times 60)}{125.0} - \frac{0.1384(62.5)}{125.0}$$

$$= 0.099 - 0.069 = 0.030 \text{ in. } (0.8 \text{ mm})$$

Long-term deflection

From Eq. 8.9, multiplier $\lambda = T/(1 + 50\rho')$. From Fig. 8.5, $T = 1.65$ for 24-month sustained load. Therefore,

$$\lambda_\infty = 2.0 \qquad \text{and} \qquad \lambda_t = 1.65$$

From Eq. 8.10, the total sustained load deflection is

$$\Delta_{LT} = \Delta_L + \lambda_\infty \Delta_D + \lambda_t \Delta_{LS}$$

or

$$\Delta_{LT} = 0.125 + 2.0 \times 0.069 + 1.65 \times 0.030 = 0.313 \text{ in. } (7.9 \text{ mm})$$

Deflection requirements (Table 8.2)

$$\frac{l}{180} = \frac{12 \times 12}{180} = 0.80 \text{ in. } > \Delta_L$$

$$\frac{l}{360} = 0.40 \text{ in. } > \Delta_L$$

$$\frac{l}{480} = 0.30 \text{ in. } \simeq \Delta_{LT}$$

$$\frac{l}{240} = 0.60 \text{ in. } > \Delta_{LT}$$

Therefore, the slab can support sensitive attached nonstructural elements which are otherwise damaged by large deflections.

8.10 FLEXURAL CRACKING IN BEAMS AND ONE-WAY SLABS

8.10.1 Fundamental Behavior

Concrete cracks at an early stage of its loading history because it is weak in tension. Consequently, it is necessary to study its cracking behavior and control the width of the flexural cracks. Cracking contributes to the corrosion of the reinforcement, surface deterioration, and its long-term detrimental effects.

Increased use of high-strength reinforcing steels having 60,000 to 100,000 psi (413.7 to 551.6 MPa) yield strength and with high stresses occurring at low load levels is becoming prevalent. Also, higher-strength concrete in excess of 9000 to 14,000 psi

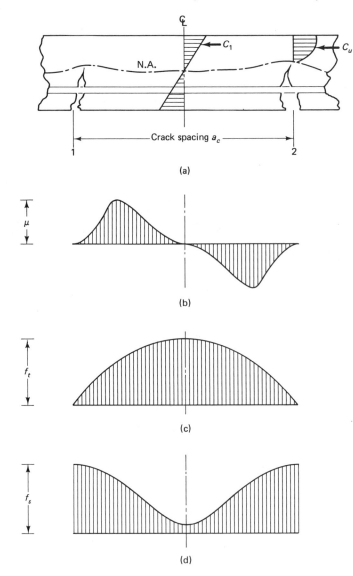

Figure 8.14 Longitudinal stress distribution between two adjacent cracks when cracks are fully developed: (a) crack development geometry; (b) ultimate bond stress μ; (c) longitudinal tensile stress f_t in the concrete; (d) longitudinal tensile stress f_s in the steel.

strength in compression (62.1 to 96.6 MPa) and optimal utilization of the material in the strength theories of analysis and design are possible today. Hence prediction and control of cracking and crack widths are essential for reliable serviceability performance under long-term loading.

Two types of stresses act on the tensile stretched zones of the concrete surrounding the tension reinforcement shown in Fig. 8.14a. They are longitudinal and lateral sets of stresses. As the longitudinal bending stress acts, the tensile zone undergoes a lateral contraction before cracking, resulting in lateral compression between the concrete and the reinforcing bars or wires. At the moment that a flexural crack starts to develop, this biaxial lateral compression has to disappear at the crack because the longitudinal tension in the concrete becomes zero at the crack location.

The longitudinal bond stress gradually reaches its peak at the crack. This causes the tensile stress f_t in the concrete at that location suddenly to reach its maximum value. The concrete can no longer withstand any tension because of the high stress concentration at the moment of incipient fracture, and it splits, as seen in Fig. 8.14a.

The stress in the concrete is dynamically transferred to the reinforcing steel (Fig. 8.14d). At the transfer of stress, the tensile stress in the concrete at the cracked section is relieved, becoming zero at the crack (Fig. 8.14c). Laterally, the neutral-axis position rises at the cracked section in order to maintain equilibrium at that section.

The distance a_c between two adjacent cracks is the stabilized crack spacing, namely the distance between two cracks when they continue to widen under load as principal cracks while other previously formed cracks between them close due to redistribution of stress. In other words, cracks stabilize when no new cracks form in the structural member. A schematic plot of the crack width versus crack spacing is given in Fig. 8.15. It illustrates in the almost horizontal plateau of the diagram the load at which the crack spacing becomes stabilized.

The width of each of the two cracks would be a function of the difference in elongation between the reinforcing bars and the surrounding stretched concrete over a length a_c. From a practical viewpoint, the elongation of the concrete and the shrinkage strain can be neglected as insignificant. Hence

$$\text{crack width } w = \alpha_c^\beta \, \epsilon_c^\gamma \qquad (8.23)$$

The value of γ partly depends on whether the reinforced concrete member is one- or two-dimensional, while α and β are experimental nonlinearity constants.

It has been proven that a_c varies as $(1/k_1\mu')$, $k_2 f_t'$ and $d_b/k_3\rho_t$, where μ is the bond stress, f_t' is the tensile strength of the concrete, d_b is the diameter of the steel bar,

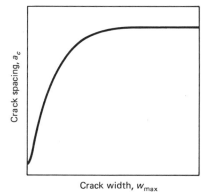

Figure 8.15 Schematic variation of crack width versus crack spacing.

$\rho_t = A_s/A_t$ is the ratio of the steel area at the tension side of the section, and A_t is the area of concrete in tension; k_1, k_2, and k_3 are constants.

8.10.2 Crack-Width Evaluation

While Eq. 8.23 is the basic mathematical model for the evaluation of the maximum crack width, the large number of variables involved, the randomness of cracking behavior, and the large degree of scatter require extensive idealization and simplification. One simplification based on a statistical study of test data of several investigators is the Gergely–Lutz expression,

$$w_{max} = 0.076\beta f_s \sqrt[3]{d_c A} \tag{8.24}$$

where w_{max} = crack width in units of 0.001 in. (0.0254 mm)

$\beta = (h - c)/(d - c)$ = depth factor average value = 1.20

d_c = thickness of cover to the center of the first layer of bars (in.)

f_s = maximum stress (ksi) in the steel at service load level with $0.6f_y$ to be used if no computations are available

A = area of concrete in tension divided by the number of bars (in.^{2}) = bt/γ_{bc}, where γ_{bc} is defined as the number of bars at the tension side

It is to be noted that allowance of $f_s = 0.6f_y$ in lieu of actual steel stress computations is applicable only to normal structures. Special precautions have to be taken for structures exposed to very aggressive climates such as chemical factories or offshore structures. Additionally, the depth of the concrete area in tension in reinforced concrete is determined by having the center of gravity of the bars as the centroid of the concrete area in tension. Hence, for a single layer of bars, the depth t of the concrete area in tension equals $2d_c$. The shaded area in Fig. 8.16 gives the total concrete area in tension.

8.10.3 Example 8.6: Maximum Crack Width in a Reinforced Concrete Beam

Calculate the maximum crack width for a rectangular simply supported beam which has the cross section shown in Fig. 8.16. The beam span is 30 ft (9.14 m). It carries a working uniform load of 1000 lb/ft, including its own weight (14.6 kN/m). Given:

f'_c = 5000 psi, normalweight concrete (34.47 MPa)

f_y = 60,000 psi (413.7 MPa)

E_s = 29 × 10^6 psi (200,000 MPa)

Solution

Alternative using the actual steel stress

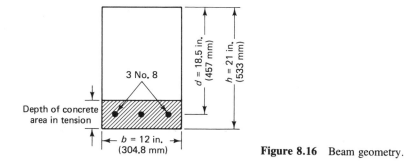

Figure 8.16 Beam geometry.

gross moment of inertia $I_g = \dfrac{bh^3}{12} = \dfrac{12(21)^3}{12} = 9261.0$ in.^{4}

modulus of rupture $f_r = 7.5\sqrt{f_c'} = 7.5\sqrt{5000} = 530.3$ psi (2.66 MPa)

cracking moment $M_{cr} = \dfrac{I_g}{y_t}f_r = \dfrac{9261.0 \times 530.3}{10.5} = 467{,}725$ in.-lb

maximum beam moment $M_a = \dfrac{wl_n^2}{8} = \dfrac{1000(30.0)^2}{8} = 112{,}500$ lb-ft

$$= 1{,}350{,}000 \text{ in.-lb}$$

$\dfrac{bc^2}{2} + nA_sc - nA_sd = 0$

$A_s = 2.37$ in.^{2} (1528.7 mm^2)

$E_c = 57{,}000\sqrt{5000} = 4.03 \times 10^6$ psi (27,797 MPa)

$n = \dfrac{E_s}{E_c} = \dfrac{29 \times 10^6}{4.03 \times 10^6} = 7.20$

$6c^2 + 7.2 \times 2.37c - 7.2 \times 2.37 \times 18.5 = 0$ to give $c = 5.97$ in. (149.2 mm)

From Eq. 8.7, the cracked moment of inertia is

$$I_{cr} = \dfrac{bc^3}{3} + nA_s(d - c)^2$$

$$= \dfrac{12(5.97)^3}{3} + 7.2 \times 2.37(18.5 - 5.97)^2 = 3530.2 \text{ in.}^4$$

steel stress $f_s = \dfrac{M_a}{I_{cr}}(d - c)n$

$$= \dfrac{1{,}350{,}000}{3530.2}(18.5 - 5.97) \times 7.2$$

$$= 34{,}500 \text{ psi (238.05 MPa)} < 36{,}000 \text{ psi}\qquad \text{O.K.}$$

steel stress $f_s = 34.5$ ksi   to be used in Eq. 8.24

$$\beta = \frac{h - c}{d - c} = \frac{21.0 - 5.97}{18.5 - 5.97} = 1.20$$

$$A = \frac{bt}{\text{no. of bars}} = \frac{b(2d_c)}{\gamma_{bc}} = \frac{12.0(2 \times 2.5)}{3} = 20 \text{ in.}^2$$

$$w_{max} = 0.076\beta f_s \sqrt[3]{d_c A} \times 10^{-3}$$

$$= 0.076 \times 1.20 \times 34.5\sqrt[3]{2.5 \times 20.0} \times 10^{-3}$$

$$= 0.0116 \text{ in. } (0.29 \text{ mm})$$

Alternative using $f_s = 0.6f_y$

$\beta = 1.20$ for beams

$f_s = 0.6f_y = 0.6 \times 60.0 = 36.0$ ksi

$$w_{max} = 0.076 \times 1.20 \times 36.0\sqrt[3]{3.0 \times 2.4} \times 10^{-3} = 0.0137 \text{ in. } (\approx0.35 \text{ mm})$$

The preceding solution alternative is usually unnecessary due to its length and rigor. It is presented to illustrate the computation of the actual value of the stress in the main longitudinal steel at service load levels. Such computations might be necessary for crack-width evaluation where low service stress levels have to be used in such flexural designs as in the case of water retaining and sanitary engineering structures. The stress $f_s \leq 0.6f_y$ gives a load factor of 1.67 for the limit state at failure.

8.10.4 Crack-Width Evaluation for Beams Reinforced with Bundled Bars

The bond stress between the reinforcing bars and the surrounding concrete is a major parameter affecting flexural crack spacing and hence crack width. The contact area of bundled bars is less than that of the isolated bars if they act independently. Using the perimetric reduction deduced from Fig. 8.17, the cracking equation becomes

$$w_{max} = 0.076\beta f_s \sqrt[3]{d_c' A'} \qquad (8.25)$$

where w_{max} is the crack width in units of 0.001 in. and $A' = bt/\gamma_{bc}'$ with the factor for γ_{bc}' shown in Fig. 8.17a. d_c' is the depth of cover to the center of gravity of the bundle. The steps for calculation of w_{max} are identical to those for beams reinforced for nonbundled bars.

8.10.5 Example 8.7: Maximum Crack Width in a Beam Reinforced with Bundled Bars

Find the maximum flexural crack width for a reinforced concrete beam which has the cross-sectional geometry shown in Fig. 8.18. Given:

$f_y = 60,000$ psi

$f_s = 0.6f_y = 36,000$ psi

A_s = two bundles of three No. 8 bars each (25.4 mm diameter)

size of stirrups = No. 4 (12.7 mm diameter)

Solution

d'_c = center of gravity of the three bars from the outer tension fibers

$$= (1.5 + 0.5) + \frac{2 \times 0.5 + 1 \times 1.5}{3} = 2.83 \text{ in.}$$

t = depth of the concrete area in tension

$$= 2 \times 2.83 = 5.66 \text{ in.}$$

γ_{bc} = number of bars if all are of the same diameter, or the total steel area divided by the area of the largest bar if more than one size is used

$$= 6 \text{ in this case}$$

$$\gamma'_{bc} = 0.650\gamma_{bc} = 0.650 \times 6 = 3.9$$

$$A = \frac{bt}{\gamma'_{bc}} = \frac{10 \times 5.66}{3.9} = 14.51 \text{ in.}^2$$

$$w_{max} = 0.076 \times 1.20 \times 36.0\sqrt[3]{2.83 \times 14.51} \times 10^{-3} = 0.011 \text{ in. (0.29 mm)}$$

Two bars
$\gamma'_{bc} = 0.815\gamma_{bc}$

Three bars
$\gamma'_{bc} = 0.650\gamma_{bc}$

Four bars
$\gamma'_{bc} = 0.570\gamma_{bc}$

where $A' = \dfrac{bt}{\gamma'_{bc}}$ $Z = f_s \sqrt{d'_c A'}$

(a)

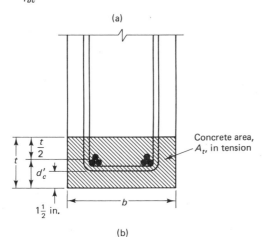

$\dfrac{t}{2}$

t

d'_c

$1\frac{1}{2}$ in.

b

Concrete area, A_t, in tension

(b)

Figure 8.17 Perimetric reduction factors for beams with bundled bars: (a) perimetric factors; (b) section geometry of the concrete area in tension.

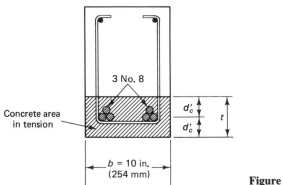

Figure 8.18 Beam geometry.

8.10.6 Z Factor for Crack-Control Check in Beams

Crack-control checks are necessary only when tension steel is used with yield strength f_y exceeding 40,000 psi (275.8 MPa). The ACI code, in order to reduce the length of calculations, recommends a Z factor where

$$Z = f_s \sqrt[3]{d_c A} \qquad \text{kips per inch} \qquad (8.26a)$$

for beams reinforced with nonbundled bars. For beams reinforced with bundled bars:

$$Z = f_s \sqrt[3]{d_c' A'} \qquad \text{kips per inch} \qquad (8.26b)$$

A Z value not exceeding 175 for interior exposure corresponds to w_{max} = 0.016 in. (40 mm) and 145 for exterior exposure corresponds to w_{max} = 0.013 in. (0.33 mm). It is to be emphasized that smaller reinforcing bars at closer spacing within the tensile zone give more even distribution of cracks, hence are preferable for crack control. Also, use of the Z-factor should be discouraged since it gives the designer *no* physical indication of the crack width magnitude.

8.11 PERMISSIBLE CRACK WIDTHS

The maximum crack width that a structural element should be permitted to develop depends on the particular function of the element and the environmental conditions to which the structure is liable to be subjected to. Table 8.4 from the ACI Committee 224 report on cracking serves as a reasonable guide on the permissible crack widths in concrete structures under the various environmental conditions encountered. Engineering judgment has to be exercised in the determination of the maximum crack width that can be tolerated.

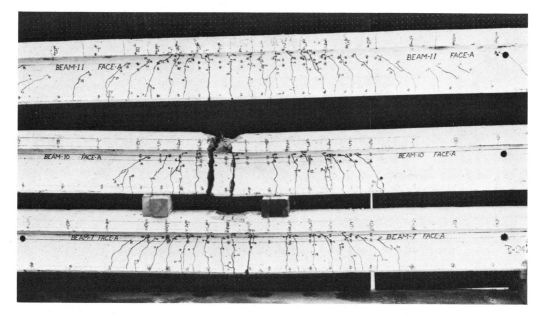

Photo 49   Typical flexural crack formation in beams.

TABLE 8.4   PERMISSIBLE CRACK WIDTHS

Exposure condition	Tolerable crack width	
	in.	mm
Dry air or protective membrane	0.016	0.41
Humidity, moist air, soil	0.012	0.30
Deicing chemicals	0.007	0.18
Seawater and seawater spray; wetting and drying	0.006	0.15
Water-retaining structures (excluding nonpressure pipes)	0.004	0.10

8.11.1 Example 8.8: Crack Control in a Rectangular Beam

Check whether the beam in Ex. 8.6 satisfies serviceability criteria for crack control for the following three environmental conditions using $f_s = 0.6f_y$: (a) interior exposure using Eq. 8.22, (b) interior exposure using the Z-factor approach of Section 8.10.6, and (c) exposure to deicing chemicals.

Solution

(a) The maximum permissible crack width = 0.016 in. The expected crack width = 0.0137 in. from the solution in Ex. 8.6. Hence the beam satisfies the crack control criteria.

(b) $z = f_s\sqrt[3]{d_cA} = 36.0\sqrt[3]{2.5 \times 20} = 132.63$ kips/in. < 175; O.K.

(c) The maximum permissible crack width from Table 8.4 is $w_{max} = 0.007$ in. The expected crack width = 0.0137 in. > 0.007 in. Hence the beam does not satisfy the crack control criteria and the reinforcement content at the tension side has to be redesigned and increased.

8.11.2 Example 8.9: Crack Control in a Beam Reinforced with Bundled Bars

Verify if the beam in Ex. 8.7 satisfies the crack control criteria for (a) aggressive chemical environment, (b) water-retaining structures, (c) exterior exposure, and (d) also find the maximum width of the beam based on a Z-factor value = 175 kips/in. for interior exposure.

Solution

The expected crack width = 0.011 in. (from Ex. 8.7).

(a) The maximum permissible crack width from Table 8.4 is $w_{max} = 0.007$ in. < 0.011 in. Hence the beam does not satisfy the crack control criteria.

(b) The permissible $w_{max} = 0.004$ in. < 0.011 in. The beam does not satisfy the crack control criteria.

(c) The permissible $w_{max} = 0.013$ in. (ACI code) $>$ expected $w_{max} = 0.011$ in. The beam satisfies the crack control criteria.

(d) $t = 5.66$ in. from Ex. 8.7:

$$Z = 175 = f_s\sqrt[3]{d'_cA} = 36.0\sqrt[3]{2.83A}$$

$$A = \left(\frac{175}{36.0}\right)^3 \times \frac{1}{2.83} = 40.6 \text{ in.}^2$$

But $A = bt/\gamma'_{bc}$, where $\gamma'_{bc} = 0.650\gamma_{bc} = 3.9$ from Ex. 8.7, or $40.6 = bt/3.90$.

$$b = \frac{40.6 \times 3.90}{5.66} = 27.98 \text{ in. (710.5 mm)}$$

Hence the maximum allowable width of the beam web is 27.98 in. to satisfy the crack control criteria for interior exposure.

SELECTED REFERENCES

8.1 ACI Committee 435, "Allowable Deflections," *Journal of the American Concrete Institute,* Proc. Vol. 65, No. 6, June 1968, p. 433.

8.2 Branson, D. E., "Design Procedures for Computing Deflections," *Journal of the American Concrete Institute,* Proc. Vol. 65, September 1968, pp. 730–742.

8.3 ACI Committee 435, "Variability of Deflections of Simply Supported Reinforced Concrete Beams," *Journal of the American Concrete Institute,* Proc. Vol. 69, January 1972, pp. 29–35.

8.4 Branson, D. E., *Deformation of Concrete Structures,* McGraw-Hill, New York, 1977.

8.5 Nawy, E. G., "Crack Control in Reinforced Concrete Structures," *Journal of the American Concrete Institute,* Proc. Vol. 65, October 1968, pp. 825–838.

8.6 Gergely, P., and Lutz, L. A., *Maximum Crack Width in Reinforced Concrete Flexural Members,* Special Publication SP-20, American Concrete Institute, Detroit, 1968.

8.7 Nawy, E. G., "Crack Control in Beams Reinforced with Bundled Bars," *Journal of the American Concrete Institute,* Proc. Vol. 69, October 1972, pp. 637–640.

8.8 ACI Committee 224, "Control of Cracking in Concrete Structures," *Journal of the American Concrete Institute,* Proc. Vol. 77, October 1980, pp. 35–76; Proc. Vol. 69, December 1972, pp. 717–753.

8.9 Nawy, E. G. and Blair, K. W., "Further Studies on Flexural Crack Control in Structural Slab Systems;" "Discussion" by ACI Code Committee 318; and "Authors' Closure," *Journal of the American Concrete Institute,* Proc. Vol. 70, January 1973, pp. 61–63.

PROBLEMS FOR SOLUTION

8.1 Calculate I_g and I_{cr} for cross sections (a) through (f) in Fig. 8.19. Given:

$f_c = 4000$ psi (27.58 MPa), normalweight concrete
$f_y = 60,000$ psi (413.7 MPa)
$E_s = 29 \times 10^6$ psi (200,000 MPa)

8.2 Calculate the maximum immediate and long-term deflection for a 6-in.-thick slab on simple supports spanning over 13 ft. The service dead and live loads are 70 psf (33.52 kPa) and 120 psf (57.46 kPa), respectively. The reinforcement consists of No. 5 bars (16 mm diameter) at 6 in. center to center (154.2 mm center to center). Also check which limitations, if any, need to be placed on its usage. Assume that 60% of the live load is sustained over a 30-month period. Given:

$f'_c = 4500$ psi (31.03 MPa)
$E_s = 29 \times 10^6$ psi (200,000 MPa)

8.3 Calculate the deflection due to dead load, and dead load plus live load, for the following cases in Problem 8.1: cross sections (a), (d), and (e). Use for service-load levels $0.2M_u$ as maximum dead-load moment and $0.35M_u$ as maximum

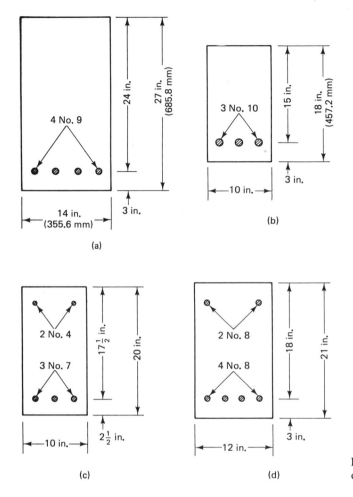

Figure 8.19 Beam cross sections for deflection calculations.

live-load moment. Assume that all beams are simply supported and have a span of 22 ft (6.71 m).

8.4 Repeat Problem 8.2 assuming the slab to be continuous over four supports. The top tension reinforcement at the support consists of No. 5 bars at 4 in. center to center and the compression reinforcement consists of No. 5 bars at 12-in. centers.

8.5 A beam supporting a 4-in. slab is continuous over four supports. The center-to-center spans are 26 ft with the end span resting on an outer wall. It has a web width $b_w = 12$ in. and a total thickness $h = 18$ in. and carrying a service live load $W_L = 6000$ lb/ft and a service dead load $W_D = 1800$ lb/ft, including its self-weight. The midspan tension reinforcement consists of $A_s = $ four No. 8 bars (28.6 mm) and the support reinforcement is comprised of $A_s = $ six No. 10 bars (32.3 mm) and $A_s' = $ three No. 8 bars. Calculate the maximum immediate and long-term deflections of this beam assuming that 55% of the sustained live load

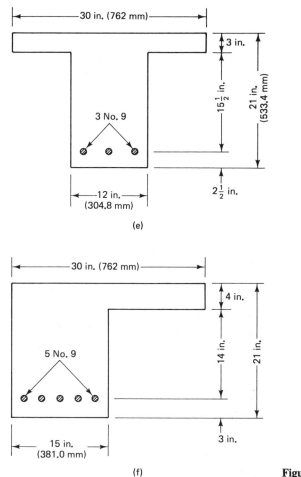

Figure 8.19 (*cont.*)

acts over a 24-month period. Also check what deflection serviceability criteria this beam satisfies and whether it can support attached partitions and other elements that can be damaged by large deflections. Given:

$$f'_c = 5000 \text{ psi}$$
$$f_y = 60,000 \text{ psi (413.7 MPa)}$$
$$E_s = 29 \times 10^6 \text{ psi (200,000 MPa)}$$

8.6 Calculate the maximum expected flexural crack width in the beam of Problem 8.5 and verify if it satisfies the serviceability criteria for crack control if it is subjected to (a) interior exposure, and (b) freeze–thaw and deicing cycles.

8.7 A rectangular beam under simple bending has the dimensions shown in Fig. 8.20. It is subjected to an aggressive chemical environment. Calculate the

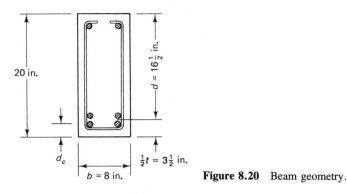

Figure 8.20 Beam geometry.

maximum expected flexural crack width and whether the beam satisfies the serviceability criteria for crack control. Given:

$$f'_c = 5000 \text{ psi } (34.47 \text{ MPa})$$
$$f_y = 60,000 \text{ psi } (413.7 \text{ MPa})$$
minimum clear cover $= 1\frac{1}{2}$ in. (38.1 mm)

8.8 Find the Z factor in Problem 8.7 and determine if it satisfies the serviceability criteria for crack control if it is subjected to exterior exposure conditions.

8.9 Find the maximum web of a beam reinforced with bundled bars to satisfy the crack-control criteria for interior exposure conditions. Given:

$$f'_c = 4000 \text{ psi } (27.6 \text{ MPa})$$
$$f_y = 60,000 \text{ psi } (413.7 \text{ MPa})$$
$A_s =$ two bundles of three No. 9 bars each (three bars 28.6 mm diameter each in bundle)

No. 4 stirrups used (13 mm diameter)

9

Combined Compression & Bending:
Columns

9.1 INTRODUCTION

Columns are vertical compression members of a structural frame intended to support the load-carrying beams. They transmit loads from the upper floors to the lower levels and then to the soil through the foundations. Since columns are compression elements, failure of one column in a critical location can cause the progressive collapse of the adjoining floors and the ultimate total collapse of the entire structure.

Structural column failure is of major significance in terms of economic as well as human loss. Thus extreme care needs to be taken in column design, with a higher reserve strength than in the case of beams and other horizontal structural elements, particularly since compression failure provides little visual warning.

As will be seen in subsequent sections, the ACI code requires a considerably lower strength reduction factor ϕ in the design of compression members than the ϕ factors in flexure, shear, or torsion. The discussion presented in Chapter 4 on the probability of failure and reliability of performance explains and justifies in more detail the reasons for the additional reserve strength needed in proportioning compression members.

The principles of stress and strain compatibility used in the analysis (design) of beams discussed in Chapter 5 are equally applicable to columns. A new factor is introduced, however: the addition of an external axial force to the bending moments acting on the critical section. Consequently, an adjustment becomes necessary to the force and moment equilibrium equations developed for beams to account for combined compression and bending.

The amount of reinforcement in the case of beams was controlled so as to have ductile failure behavior. In the case of columns, the axial load will occasionally dominate; hence compression failure behavior in cases of a large axial load/bending moment ratio cannot be avoided.

As the load on a column continues to increase, cracking becomes more intense along the height of the column at the transverse tie location. At the limit state of failure, the concrete cover in tied columns or the shell of concrete outside the spirals of spirally confined columns spalls and the longitudinal bars become exposed. Additional load leads to failure and local buckling of the individual longitudinal bars at the unsupported length between the ties. It is noted that at the limit state of failure, the concrete cover to the reinforcement spalls first after the bond is destroyed.

As in the case of beams, the strength of columns is evaluated on the basis of the following principles:

1. A linear strain distribution exists across the thickness of the column.
2. There is no slippage between the concrete and the steel (i.e., the strain in steel and in the adjoining concrete is the same).
3. The maximum allowable concrete strain at failure for the purpose of strength calculations = 0.003 in./in.

298

4.   The tensile resistance of the concrete is negligible and is disregarded in computations.

9.2 TYPES OF COLUMNS

Columns can be classified on the basis of the form and arrangement of reinforcement, the position of the load on the cross section, and the length of the column in relation to its lateral dimensions.

The form and arrangement of the reinforcement identify three types of columns, as shown in Fig. 9.1:

1.   Rectangular or square columns reinforced with longitudinal bars and lateral ties (Fig. 9.1a).
2.   Circular columns reinforced with longitudinal reinforcement and spiral reinforcement, or lateral ties (Fig. 9.1b).
3.   Composite columns where steel structural shapes are encased in concrete. The structural shapes could be placed inside the reinforcement cage, as shown in Fig. 9.1c.

Although tied columns are the most commonly used because of lower construction costs, spirally bound rectangular or circular columns are also used where increased ductility is needed, such as in earthquakes zones. The ability of the spiral column to sustain the maximum load at excessive deformations prevents the complete collapse of structure before total redistribution of moments and stresses is complete. Figure 9.2 shows the large increase in ductility (toughness) due to the effect of spiral binding.

Based on the position of the load on the cross section, columns can be classified as concentrically or eccentrically loaded, as shown in Fig. 9.3. Concentrically loaded columns (Fig. 9.3a) carry no moment. In practice, however, all columns have to be designed for some unforeseen or accidental eccentricity due to such causes as imperfections in the vertical alignment of formwork.

Eccentrically loaded columns (Fig. 9.3b and c) are subjected to moment in addition to the axial force. The moment can be converted to a load P and an eccentricity e, as shown in Fig. 9.3b and c. The moment can be uniaxial, as in the case of an exterior column in a multistory building frame, or when two adjacent panels are not similarly loaded, such as columns A and B in Fig. 9.4. A column is considered biaxially loaded when bending occurs about both the X and Y axes, such as in the case of corner column C of Fig. 9.4.

Failure of columns could occur as a result of material failure by initial yielding of the steel at the tension face or initial crushing of the concrete at the compression face, or by loss of lateral structural stability (i.e., through buckling).

If a column fails due to initial material failure, it is classified as a *short column*. As the length of the column increases, the probability that failure will occur by

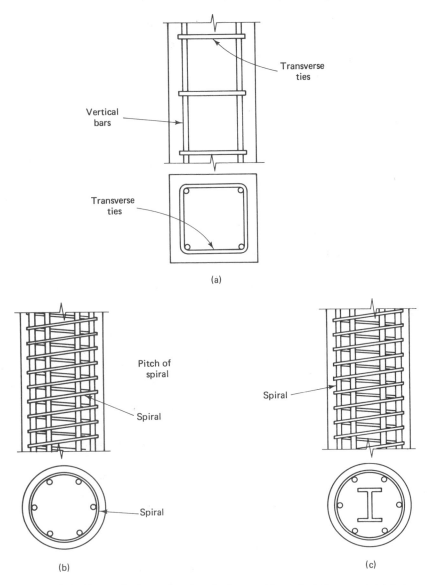

Figure 9.1 Types of columns based on the form and type of reinforcement: (a) tied column; (b) spiral column; (c) composite column.

buckling also increases. Therefore, the transition from the short column (material failure) to the long column (failure due to buckling) is defined by using the ratio of the effective length kl_u to the radius of gyration r. The height, l_u, is the unsupported length of the column and k is a factor that depends on end conditions of the column and whether it is braced or unbraced. For example, in the case of unbraced columns,

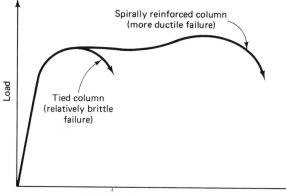

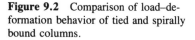

Figure 9.2 Comparison of load–deformation behavior of tied and spirally bound columns.

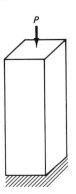

(a)

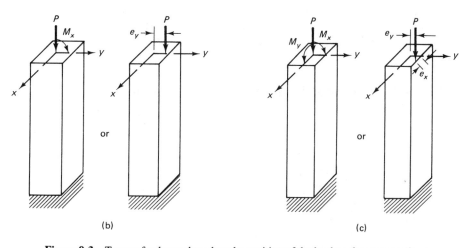

(b)   (c)

Figure 9.3 Types of columns based on the position of the load on the cross section: (a) concentrically loaded column; (b) axial load plus uniaxial moment; (c) axial load plus biaxial moment.

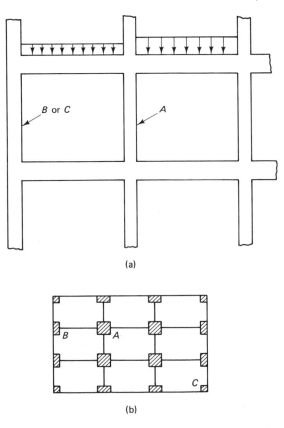

(a)

(b)

Figure 9.4 Bending of columns: (a) frame elevation; (b) framing plan. *A*, interior column under nonsymmetrical load-uniaxial bending; *B*, exterior column, uniaxial bending; *C*, exterior corner column, biaxial bending.

if kl_u/r is less than or equal to 22, such a column is classified as a short column, in accordance with the ACI load criteria. Otherwise, it is defined as a long or a slender column. The ratio kl_u/r is called the *slenderness ratio*.

9.3 STRENGTH OF SHORT CONCENTRICALLY LOADED COLUMNS

Consider a column of gross cross-sectional area A_g with width b and total depth h, reinforced with a total area of steel A_{st} on all faces of the column. The net cross-sectional area of the concrete is $A_g - A_{st}$.

Figure 9.5 presents the stress history in the concrete and the steel as the column load is increased. Both the steel and the concrete behave elastically at first. At a strain of approximately 0.002 in./in. to 0.003 in./in., the concrete reaches its maximum strength f_c'. Theoretically, the maximum load that the column can take occurs when the stress in the concrete reaches f_c'. Further increase is possible if strain hardening occurs in the steel at about 0.003 in./in. strain levels.

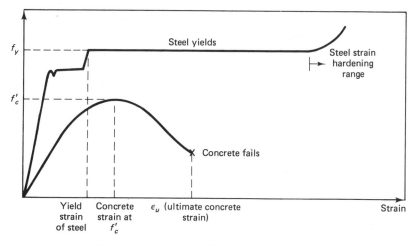

Figure 9.5   Stress–strain behavior of concrete and steel (concentric load).

Therefore, the maximum concentric load capacity of the column can be obtained by adding the contribution of the concrete, which is $(A_g - A_{st})0.85f'_c$, and the contribution of the steel, which is $A_{st}f_y$, where A_g is the total gross area of the concrete section and A_{st} is the total steel area $= A_s + A'_s$. The value of $0.85f'_c$ instead of f'_c is used in the calculation since it is found that the maximum attainable strength in the actual structure approximates $0.85f'_c$. Thus the nominal concentric load capacity, P_0, can be expressed as

$$P_0 = 0.85f'_c(A_g - A_{st}) + A_{st}f_y \qquad (9.1)$$

It should be noted that concentric load causes uniform compression throughout the cross section. Consequently, at failure the strain and stress will be uniform across the cross section, as shown in Fig. 9.6.

It is highly improbable to attain zero eccentricity in actual structures. Eccentricities could easily develop because of factors such as slight inaccuracies in the layout of columns and unsymmetric loading due to the difference in thickness of the slabs in adjacent spans or imperfections in the alignment, as indicated earlier. Hence a minimum eccentricity of 10% of the thickness of the column in the direction perpendicular to its axis of bending is considered as an acceptable assumption for columns with ties and 5% for spirally reinforced columns.

To reduce the calculations necessary for analysis and design for minimum eccentricity, the ACI code specifies a reduction of 20% in the axial load for tied columns and a 15% reduction for spiral columns. Using these factors, the maximum nominal axial load capacity of columns cannot be taken greater than

Tied

Column

$$P_{n(max)} = 0.8[0.85f'_c(A_g - A_{st}) + A_{st}f_y] \qquad (9.2)$$

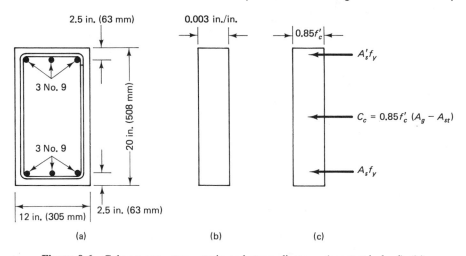

Figure 9.6 Column geometry—strain and stress diagrams (concentric load): (a) cross section; (b) concrete strain; (c) stress (forces).

for tied reinforced columns and

Spirally Reinforced $$P_{n(max)} = 0.85[0.85f'_c(A_g - A_{st}) + A_{st}f_y]$$ (9.3)

for spirally reinforced columns.

These nominal loads should be reduced further using strength reduction factors ϕ, as explained in later sections. Normally, for design purposes, $A_g - A_{st}$ can be assumed to be equal to A_g without great loss in accuracy.

9.3.1 Example 9.1: Analysis of an Axially Loaded Short Rectangular Tied Column

A short tied column is subjected to axial load only. It has the geometry shown in Fig. 9.6a and is reinforced with three No. 9 bars (28.6 mm diameter) on each of the two faces parallel to the x axis of bending. Calculate the maximum nominal axial load strength $P_{n(max)}$. Given:

$$f'_c = 4000 \text{ psi (27.6 MPa)}$$
$$f_y = 60,000 \text{ psi (414 MPa)}$$

Solution

$A_s = A'_s = 3$ in.^{2}. Therefore, $A_{st} = 6$ in.^{2}. Using Eq. 9.2 yields

$$P_{n(max)} = 0.8\{0.85 \times 4000[(12 \times 20) - 6] + 6 \times 60,000\}$$

$$= 924,480 \text{ lb (4110 kN)}$$

If $A_g - A_{st}$ is taken as equal to A_g, it results in

$$P_{n(\max)} = 0.8(0.85 \times 4000 \times 12 \times 20 + 6 \times 60,000)$$

$$= 940,800 \text{ lb } (4180 \text{ kN})$$

Note from Fig. 9.6b and c that the entire concrete cross section is subjected to a uniform stress of $0.85f_c'$ and a uniform strain of 0.003 in./in.

9.3.2 Example 9.2: Analysis of an Axially Loaded Short Circular Column

A 20-in.-diameter short spirally reinforced circular column is symmetrically reinforced with six No. 8 bars, as shown in Fig. 9.7. Calculate the strength $P_{n(\max)}$ of this column if subjected to axial load only. Given:

$$f_c' = 4000 \text{ psi } (27.6 \text{ MPa})$$
$$f_y = 60,000 \text{ psi } (414 \text{ MPa})$$

Solution

$$A_{st} = 4.74 \text{ in.}^2$$

$$A_g = \frac{\pi}{4}(20)^2 = 314 \text{ in.}^2 \qquad \frac{\pi d^2}{4}$$

Using Eq. 9.3 yields

$$P_{n(\max)} = 0.85[0.85 \times 4000(314 - 4.74) + 4.74 \times 60,000] \quad \leftarrow \text{ spirally reinf.}$$

$$= 1,135,501 \text{ lb } (5065 \text{ kN})$$

or assuming that $A_g - A_{st} \simeq A_g$,

$$P_{n(\max)} = 0.85[0.85 \times 4000 \times 314 + 4.74 \times 60,000]$$

$$= 1,149,200 \text{ lb } (5062 \text{ kN})$$

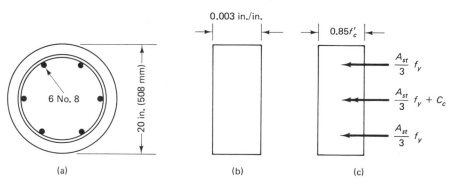

Figure 9.7   Column geometry—strain and stress diagrams (concentric load): (a) cross section; (b) concrete strain; (c) stress (forces).

9.4 STRENGTH OF ECCENTRICALLY LOADED COLUMNS: AXIAL LOAD AND BENDING

9.4.1 Behavior of Eccentrically Loaded Short Columns

The same principles concerning the stress distribution and the equivalent rectangular stress block applied to beams are equally applicable to columns. Figure 9.8 shows a typical rectangular column cross section with strain, stress, and force distribution diagrams. The diagram differs from Fig. 5.4 in the introduction of an additional longitudinal nominal force P_n at the limit failure state acting at an eccentricity e from the plastic (geometric) centroid of the section. The depth of neutral axis primarily determines the strength of the column.

The equilibrium expressions for forces and moments from Fig. 9.8 can be expressed as follows for nonslender columns.

$$\text{nominal axial resisting force } P_n \text{ at failure} = C_c + C_s - T_s \tag{9.4}$$

Nominal resisting moment M_n, which is equal to $P_n e$, can be obtained by writing the moment equilibrium equation about the plastic centroid. For columns with symmetrical reinforcement, the plastic centroid is the same as the geometric centroid.

$$M_n = P_n e = C_c\left(\bar{y} - \frac{a}{2}\right) + C_s(\bar{y} - d') + T_s(d - \bar{y}) \tag{9.5}$$

Since

$$C_c = 0.85 f_c' ba$$

$$C_s = A_s' f_s'$$

$$T_s = A_s f_s$$

Equations 9.4 and 9.5 can be rewritten as

$$P_n = 0.85 f_c' ba + A_s' f_s' - A_s f_s \tag{9.6}$$

$$M_n = P_n e = 0.85 f_c' ba\left(\bar{y} - \frac{a}{2}\right) + A_s' f_s'(\bar{y} - d') - A_s f_s(d - \bar{y}) \tag{9.7}$$

In Eqs. 9.6 and 9.7, the depth of the neutral axis c is assumed to be less than the effective depth d of the section and the steel at the tension face is in actual tension. Such a condition changes if the eccentricity e of the axial force P_n is very small. For such small eccentricities where the total cross section is in compression, contribution of the tension steel should be added to the contribution of concrete and compression steel. The term $A_s f_s$ in Eqs. 9.6 and 9.7 in such a case would have a positive sign since all the steel is in compression. It is also assumed that $(ba - A_s') \simeq ba$; that is, the volume of concrete displaced by compression steel is negligible.

If a computer is used for the analysis (design), more refined solutions can be obtained with minimum effort. Hence the area of concrete displaced with the steel is

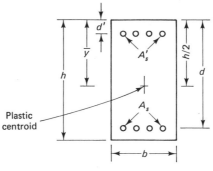

Cross section

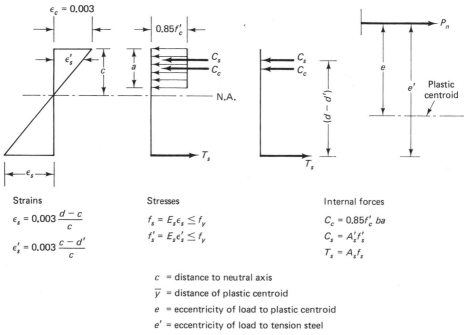

Strains	Stresses	Internal forces
$\epsilon_s = 0.003\dfrac{d-c}{c}$	$f_s = E_s\epsilon_s \leq f_y$	$C_c = 0.85f'_c\, ba$
$\epsilon'_s = 0.003\dfrac{c-d'}{c}$	$f'_s = E_s\epsilon'_s \leq f_y$	$C_s = A'_s f'_s$
		$T_s = A_s f_s$

c = distance to neutral axis
$\overline{y}$ = distance of plastic centroid
e = eccentricity of load to plastic centroid
e' = eccentricity of load to tension steel
d' = effective cover of compression steel

Figure 9.8   Stresses and forces in columns.

considered in the computer-aided solutions presented in Chapter 13. The examples in this chapter include for comparison purposes those more refined results from the computer solutions. It can be observed that the error in neglecting the contribution of the displaced concrete is not significant.

It should be noted that the axial force P_n cannot exceed the maximum axial load strength $P_{n(\max)}$, calculated using Eq. 9.2. Depending on the magnitude of the eccentricity e, the compression steel A'_s or the tension steel A_s will reach the yield strength,

f_y. Stress f'_s reaches f_y when failure occurs by crushing of the concrete. If failure develops by yielding of the tension steel, f_s should be replaced by f_y. When the magnitude of f'_s or f_s is less than f_y, the actual stresses can be calculated using the following equations obtained from similar triangles in the strain distribution across the depth of the section (Fig. 9.8).

$$f'_s = E_s \epsilon'_s = E_s \frac{0.003(c - d')}{c} \leq f_y \tag{9.8}$$

$$f_s = E_s \epsilon_s = E_s \frac{0.003(d - c)}{c} \leq f_y \tag{9.9}$$

9.4.2 Basic Column Equations 9.6 and 9.7 and Trial-and-Adjustment Procedure for Analysis (Design) of Columns

Equations 9.6 and 9.7 determine the nominal axial load P_n that can be safely applied at an eccentricity e for any eccentrically loaded column. If one examines these two expressions, the following unknowns can be identified:

1. Depth of the equivalent stress block, a
2. Stress in compression steel, f'_s
3. Stress in tension steel, f_s
4. P_n for the given e, or vice versa

The stresses f'_s and f_s can be expressed in terms of the depth of neutral axis c as in Eqs. 9.8 and 9.9 and thus in terms of a. The two remaining unknowns, a and P_n, can be solved using Eqs. 9.6 and 9.7. However, combining Eqs. 9.6 to 9.9 leads to a cubical equation in terms of the neutral-axis depth c. One has also to check whether the steel stresses are less than the yield strength f_y. Hence the following trial-and-adjustment procedure is suggested for a general case of analysis (design).

For a given section geometry and eccentricity e, assume a value for the distance c down to the neutral axis. This value is a measure of the compression block depth a since $a = \beta_1 c$. Using the assumed value of c, calculate the axial load P_n using Eq. 9.6 and $a = \beta_1 c$. Calculate the stresses f'_s and f_s in compression and tension steel, respectively, using Eqs. 9.8 and 9.9. Also, calculate the eccentricity corresponding to the calculated load P_n using Eq. 9.7. This calculated eccentricity should match the given eccentricity e. If not, repeat the steps until a convergence is accomplished. If the calculated eccentricity is larger than the given eccentricity, this indicates that the assumed value for c and the corresponding depth a of the compression block is less than the actual depth. In such a case, try another cycle, assuming a larger value of c.

This trial-and-adjustment process converges rapidly and becomes exceedingly simpler if a computer program is used, as explained in Chapter 13. This discussion pertains to a general case. Simplifying assumptions can be made in most cases to shorten the iteration process.

9.5 MODES OF MATERIAL FAILURE IN COLUMNS

Based on the magnitude of strain in the steel reinforcement at the tension side (Fig. 9.8), the section is subjected to one of two initial conditions of failure as follows:

1. Tension failure by initial yielding of steel at the tension side
2. Compression failure by initial crushing of the concrete at the compression side

The balanced condition occurs when failure develops simultaneously in tension and in compression.

If P_n is the axial load and P_{nb} is the axial load corresponding to the balanced condition, then

$$P_n < P_{nb} \qquad \text{tension failure}$$

$$P_n = P_{nb} \qquad \text{balanced failure}$$

$$P_n > P_{nb} \qquad \text{compression failure}$$

In all these cases the strain-compatibility relationship must be maintained.

9.5.1 Balanced Failure in Rectangular Column Sections

As the eccentricity decreases, a gradual transition takes place from a primary tension failure to a primary compression failure. The balanced failure condition is reached when the tension steel reaches its yield strain ϵ_y at precisely the same load level as the concrete reaches its ultimate strain ϵ_c (0.003 in./in.) and starts crushing.

From similar triangles, an expression for the depth of neutral axis c_b at balanced condition can be written as (Fig. 9.8)

$$\frac{c_b}{d} = \frac{0.003}{0.003 + f_y/E_s} \tag{9.10a}$$

or using $E_s = 29 \times 10^6$ psi,

$$c_b = d\frac{87,000}{87,000 + f_y} \tag{9.10b}$$

$$a_b = \beta_1 c_b = \beta_1 d\frac{87,000}{87,000 + f_y} \tag{9.11}$$

The axial load corresponding to balanced condition P_{nb} and the corresponding eccentricity e_b can be determined by using this a_b in Eqs. 9.6 and 9.7.

$$P_{nb} = 0.85f'_c ba_b + A'_s f'_s - A_s f_y \tag{9.12}$$

$$M_{nb} = P_{nb} e_b = 0.85f'_c ba_b\left(\bar{y} - \frac{a_b}{2}\right) + A'_s f'_s(\bar{y} - d') + A_s f_y(d - \bar{y}) \tag{9.13}$$

where

$$f'_s = 0.003E_s \frac{c_b - d'}{c_b} \leq f_y \qquad (9.14)$$

and $\bar{y}$ is the distance from the compression fibers to the plastic or geometric centroid. Note that since a_b and f'_s are known, both P_{nb} and e_b can be calculated without going through any trial runs. If $A'_s = A_s$, then $\bar{y} = 0.5h$.

9.5.2 Example 9.3: Analysis of a Column Subjected to Balanced Failure

Calculate the nominal balanced load, P_{nb}, in Ex. 9.1 and the corresponding eccentricity, e_b, for the balanced failure condition if the column shown in Fig. 9.9 is subjected to combined bending and axial load. Given:

$b = 12$ in.
$d = 17.5$ in.
$h = 20$ in.
$d' = 2.5$ in.
$A_s = A'_s = 3.0$ in.^{2}
$f'_c = 4000$ psi
$f_y = 60,000$ psi

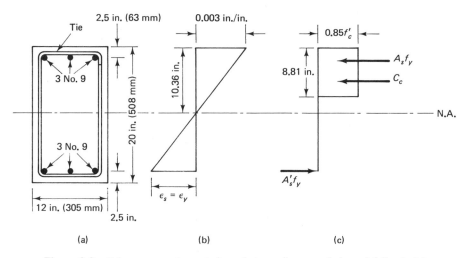

Figure 9.9 Column geometry—strain and stress diagrams (balanced failure): (a) cross section; (b) balanced strains; (c) stress.

Solution

Using Eq. 9.10b gives

$$c_b = 17.5\left(\frac{87,000}{87,000 + 60,000}\right) = 10.36 \text{ in.}$$

$$a_b = \beta_1 c_b = 0.85 \times 10.36 = 8.81 \text{ in.}$$

$$f'_s = 0.003 E_s \frac{c_b - d'}{c_b} \leq f_y$$

$$= (0.003)(29 \times 10^6)\left(\frac{10.36 - 2.5}{10.36}\right) = 66,006 \text{ psi} > f_y$$

Therefore,

$$f'_s = f_y = 60,000 \text{ psi}$$

Using Eq. 9.12, we have

$$P_{nb} = 0.85 \times 4000 \times 12 \times 8.81 + 3 \times 60,000 - 3 \times 60,000$$

$$= 359,448 \text{ lb}$$

Using Eq. 9.13 and $\bar{y} = h/2 = 10$ in. yields

$$M_{nb} = 0.85 \times 4000 \times 12 \times 8.81\left(10 - \frac{8.81}{2}\right) + 3 \times 60,000(10 - 2.5) + 3$$

$$\times 60,000(17.5 - 10)$$

$$= 4,711,111 \text{ in.-lb (532 kN-m)}$$

$$e_b = \frac{M_{nb}}{P_{nb}} = \frac{4,711,111}{359,448} = 13.1 \text{ in. (333 mm)}$$

(If displaced concrete is taken into account in the calculation, $P_{nb} = 348,986$ lb and $e_b = 13.3$ in.)

9.5.3 Tension Failure in Rectangular Column Sections

The initial limit state of failure in cases of large eccentricity occurs by yielding of steel at the tension side. The transition from compression failure to tension failure takes place at $e = e_b$. If e is larger than e_b or $P_n < P_{nb}$, the failure will be in tension through initial yielding of the tensile reinforcement. Equations 9.6 and 9.7 are applicable in the analysis (design) by substituting the yield strength f_y for the stress f_s in the tension reinforcement. The stress f'_s in the compression reinforcement may or may not be the yield strength and the actual stress f'_s should be calculated using Eq. 9.8.

Symmetrical reinforcement is usually used such that $A'_s = A_s$ in order to prevent the possible interchange of the compression reinforcement with the tension reinforcement during bar cage placement. Symmetry of reinforcement is also often necessary where the possibility exists of stress reversal due to change in wind direction.

Photo 51 Eccentrically loaded column at limit state of failure. (Tests by Nawy et al.)

If the compression steel is assumed to have yielded, and $A_s = A_s'$, Eqs. 9.6 and 9.7 can be rewritten as

$$P_n = 0.85f_c'ba \tag{9.15}$$

$$M_n = P_ne = 0.85f_c'ba\left(\bar{y} - \frac{a}{2}\right) + A_s'f_y(\bar{y} - d') + A_sf_y(d - \bar{y}) \tag{9.16a}$$

or

$$M_n = P_ne = 0.85f_c'ba\left(\frac{h}{2} - \frac{a}{2}\right) + A_sf_y(d - d') \tag{9.16b}$$

In Eq. 9.16b, the plastic (geometric) centroid is replaced by $h/2$ for symmetrical reinforcement and A_s' is replaced by A_s.

Additionally, Eqs. 9.15 and 9.16b can be combined to obtain a single equation for P_n. Replacing $0.85f_c'ba$ in Eq. 9.16b by Eq. 9.15 gives

$$P_ne = P_n\left(\frac{h}{2} - \frac{a}{2}\right) + A_sf_y(d - d') \tag{9.16c}$$

Since $a = P_n/0.85f'_c b$ from Eq. 9.15,

$$P_n e = P_n\left(\frac{h}{2} - \frac{P_n}{1.7f'_c b}\right) + A_s f_y(d - d') \tag{9.16d}$$

$$\frac{P_n^2}{1.7f'_c b} - P_n\left(\frac{h}{2} - e\right) - A_s f_y(d - d') = 0 \tag{9.16e}$$

If

$$\rho = \rho' = \frac{A_s}{bd} \tag{9.16f}$$

$$P_n = 0.85f'_c b\left[\left(\frac{h}{2} - e\right) + \sqrt{\left(\frac{h}{2} - e\right)^2 + \frac{2A_s f_y(d - d')}{0.85f'_c b}}\right] \tag{9.17}$$

and if

$$m = \frac{f_y}{0.85f'_c} \tag{9.18}$$

then Eq. 9.16e can be rewritten as

$$P_n = 0.85f'_c bd\left[\frac{h - 2e}{2d} + \sqrt{\left(\frac{h - 2e}{2d}\right)^2 + 2m\rho\left(1 - \frac{d'}{d}\right)}\right] \tag{9.19}$$

Replacing the eccentricity e (distance between the plastic centroid and the load) with e' (distance between the tension steel and the load), Eq. 9.19 can also be rewritten as

$$P_n = 0.85f'_c bd\left[\left(1 - \frac{e'}{d}\right) + \sqrt{\left(1 - \frac{e'}{d}\right)^2 + 2m\rho\left(1 - \frac{d'}{d}\right)}\right] \tag{9.20}$$

Note that $e' = [e + (d - h/2)]$ in Fig. 9.8 and

$$\frac{h - 2e}{2d} = 1 - \frac{e'}{d}$$

For nonstandard cases where the reinforcement is not symmetrical (i.e., if ρ is not equal to ρ') and if the concrete displaced by compression steel is taken into consideration (i.e., in Eqs. 9.15 and 9.16a), the compressive force contribution of concrete, C_c, is changed from $0.85f'_c ba$ to $0.85f'_c(ba - A'_s)$, then Eq. 9.19 changes to

$$P_n = 0.85f'_c bd\left[\rho'(m - 1) - \rho m + \left(1 - \frac{e'}{d}\right)\right]$$
$$+ \sqrt{\left(1 - \frac{e'}{d}\right)^2 + 2\left[\frac{e'}{d}(\rho m - \rho'm + \rho') + \rho'(m - 1)\left(1 - \frac{d'}{d}\right)\right]} \tag{9.21}$$

where e' is the distance between the axial force P_n and the tension steel (or eccentricity to tension steel).

$$\rho = \frac{A_s}{bd} \tag{9.22a}$$

$$\rho' = \frac{A'_s}{bd} \tag{9.22b}$$

Equations 9.20 and 9.21 are valid only if the compression steel yields. Otherwise, Eqs. 9.6, 9.7, and 9.8 should be used for obtaining P_n. Examples 9.4 and 9.5 illustrate the design process of a column controlled by initial tension failure.

9.5.4 Example 9.4: Analysis of a Column Controlled by Tension Failure; Stress in Compression Steel Equals Yield Strength

Calculate the nominal axial load strength P_n of the section in Ex. 9.1 (see Fig. 9.10) if the load acts at an eccentricity $e = 14$ in. (356 mm). Given:

$b = 12$ in.
$d = 17.5$ in.
$h = 20$ in.
$d' = 2.5$ in.
$A_s = A'_s = 3$ in.^{2}
$f'_c = 4000$ psi
$f_y = 60,000$ psi

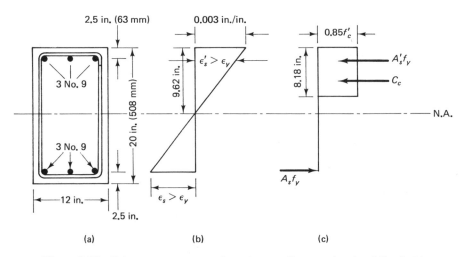

Figure 9.10 Column geometry—strain and stress diagrams (tension failure): (a) cross section; (b) strains; (c) stresses.

Solution

Using the results of Ex. 9.3, $e_b = 13.1$ in. $< e = 14$ in. Therefore, failure will occur by initial yielding of tension steel.

$$\rho = \rho' = \frac{A_s}{bd} = \frac{3}{12 \times 17.5} = 0.0143$$

$$m = \frac{60,000}{0.85 \times 4000} = 17.65$$

$$\frac{h - 2e}{2d} \quad \text{or} \quad 1 - \frac{e'}{d} = \frac{20 - 2 \times 14}{2 \times 17.5} = -0.2286$$

$$1 - \frac{d'}{d} = 1 - \frac{2.5}{17.5} = 0.8571$$

Using Eq. 9.19 or 9.20, we have

$$P_n = 0.85 \times 4000 \times 12 \times 17.5$$
$$[-0.2286 + \sqrt{(-0.2286)^2 + 2 \times 17.65 \times 0.0143 \times 0.8571}] = 333,979 \text{ lb}$$

$$a = \frac{P_n}{0.85 f_c' b} = \frac{333,979}{0.85 \times 4000 \times 12} = 8.19 \text{ in.}$$

$$c = \frac{8.19}{0.85} = 9.63 \text{ in.}$$

$$f_s' = 0.003 \times 29 \times 10^6 \left(\frac{9.63 - 2.5}{9.63} \right)$$

$$= 64,414 \text{ psi} > f_y \qquad \text{therefore, } f_s' = f_y \qquad \text{O.K.}$$

$$P_n = 333,438 \text{ lb } (1500.47 \text{ kN}) \text{ at } e = 14 \text{ in. } (356.6 \text{ mm})$$

(If the area of displaced concrete is taken into account, $P_n = 328,970$ lb.) If f_s' is less than f_y, a trial-and-adjustment procedure has to be used for the analysis.

It must be emphasized that in each analysis (design) problem, the balanced P_{nb}, M_{nb}, and hence e_b have to be evaluated to verify whether the appropriate expressions for tension failure or compression failure are applied in the solution.

9.5.5 Example 9.5: Analysis of a Column Controlled by Tension Failure; Stress in Compression Steel Less Than Yield Strength

A short rectangular reinforced concrete column is 12 in. $\times$ 15 in. (305 mm $\times$ 381 mm) as shown in Fig. 9.11 and is subjected to a load eccentricity $e = 12$ in. (305 mm). Calculate the safe nominal load strength P_n and the nominal moment strength M_n of the column section. Given:

$f_c' = 4000$ psi (27.6 MPa), normalweight concrete
$f_y = 60,000$ psi (414 MPa)

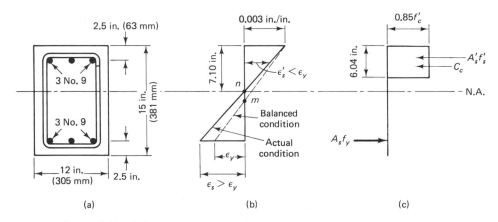

Figure 9.11 Column geometry—strain and stress diagrams (tension failure $f'_s < f_y$): (a) cross section; (b) strains; (c) stresses.

Eccentricity $e = 12$ in. (305 mm), and three No. 9 bars (28.7 mm diameter) for each of the compression and tension reinforcement.

Solution

To determine whether the failure is by crushing of concrete or yielding of steel, P_{nb} and e_b have to be calculated first.

$$A_s = A'_s = 3 \text{ in.}^2 (19.4 \text{ cm}^2)$$

$$d = 12.5 \text{ in.}$$

$$c_b = \left(\frac{87,000}{87,000 + 60,000}\right)12.5 = 7.4 \text{ in.}$$

$$a_b = 0.85 \times 7.4 = 6.3 \text{ in.}$$

$$f'_s = 29 \times 10^6 \times 0.003\left(\frac{7.4 - 2.5}{7.4}\right) = 57,609 \text{ psi} < f_y$$

Hence the compression steel did not yield.

$$P_{nb} = 0.85 \times 4000 \times 12 \times 6.3 + 3 \times 57,609 - 3 \times 60,000 = 249,867 \text{ lb}$$

$$M_{nb} = 0.85 \times 4000 \times 12 \times 6.3\left(\frac{15}{2} - \frac{6.3}{2}\right) + 3 \times 57,609\left(\frac{15}{2} - 2.5\right)$$

$$+ 3 \times 60,000\left(12.5 - \frac{15}{2}\right) = 2,882,259 \text{ in.-lb}$$

$$e_b = \frac{M_{nb}}{P_{nb}} = \frac{2,882,259}{249,867} = 11.5 \text{ in.}$$

The specified eccentricity $e = 12$ in. is $> e_b$. Hence failure will occur by initial yielding of steel. Point m on the vertical axis of the strain diagram in Fig. 9.11b denotes the position of the neutral axis at the balanced condition. For this condition, calculations showed that the strain in the compression steel is less than its yield strain. As the neutral-axis position rises to point n in the case of initial failure by yielding of the tension steel, the strain ϵ_s' in the compression steel will be less than that of the balanced condition, hence less than ϵ_y. Therefore, the trial-and-adjustment method should be used for the calculation of P_n. Since $c_b = 7.4$ in., assume a slightly smaller depth to the neutral axis for initial tension failure. Try $c = 7.10$ in.

$$a = 0.85 \times 7.10 = 6.04 \text{ in.}$$

$$f_s' = 29 \times 10^6 \times 0.003 \left(\frac{7.10 - 2.5}{7.10} \right) = 56{,}366 \text{ psi}$$

Since failure is by yielding of tension steel, the stress in the tension steel is equal to the yield strength, or

$$f_s = f_y = 60{,}000 \text{ psi}$$

$$P_n = 0.85 \times 4000 \times 12 \times 6.04 + 3 \times 56{,}366 - 3 \times 60{,}000$$

$$= 235{,}531 \text{ lb}$$

$$M_n = 0.85 \times 4000 \times 12 \times 6.04(7.5 - 3.02) + 3 \times 56{,}366 \times 5 + 3$$

$$\times 60{,}000 \times 5$$

$$= 2{,}849{,}508 \text{ in.-lb (322 kN-m)}$$

$$e = \frac{M_n}{P_n} = 12.10 \text{ in. (305 mm)}$$

Therefore,

$$P_n \simeq 236{,}000 \text{ lb (1050 kN)}$$

(If displaced concrete is taken into account in the calculations, $P_n = 235{,}050$ lb.)

9.5.6 Compression Failure in Rectangular Column Sections

For initial crushing of the concrete, the eccentricity e has to be less than the balanced eccentricity e_b and the stress in the tensile reinforcement below yield, that is, $f_s < f_y$.

The analysis (design) process necessitates applying the basic equilibrium equations 9.6 and 9.7, using the trial-and-adjustment procedure and ensuring strain-compatibility checks at all stages. The procedure is summarized in Section 9.4.2 and the following example illustrates its use in the analysis and design of reinforced concrete columns.

9.5.7 Example 9.6: Analysis of a Column Controlled by Compression Failure; Trial-and-Adjustment Procedure

Calculate the nominal load P_n of the section in Ex. 9.1 (see Fig. 9.12) if the column is subjected to a load eccentricity $e = 10$ in. (254 mm). Given:

$b = 12$ in. (305 mm)
$d = 17.5$ in. (445 mm)
$h = 20$ in. (508 mm)
$d' = 2.5$ in.
$A_s = A'_s = 3.0$ in.^{2} (1940 mm^2)
$f'_c = 4000$ psi (27.6 MPa)
$f_y = 60,000$ psi (414 MPa)

Solution

Using the results of Ex. 9.3, eccentricity for balanced failure, $e_b = 13.1$ in., which is larger than the given eccentricity of 10 in. Therefore, failure will occur by initial crushing of concrete at the compression face.

Trial 1

Assume that

$$c = 11.0 \text{ in. (254 mm)}$$

$$a = \beta_1 c = 0.85 \times 11.0 = 9.35 \text{ in.}$$

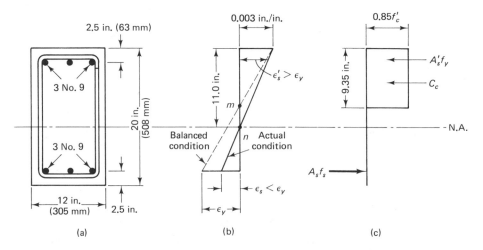

Figure 9.12 Column geometry—strain and stress diagrams (compression failure): (a) cross section; (b) strains; (c) stresses.

Photo 52 Compression side of eccentrically loaded column at failure. (Tests by Nawy et al.)

Using Eq. 9.8,

$$f'_s = 29 \times 10^6 \times 0.003\left(\frac{11.0 - 2.5}{11.0}\right) = 67,227 \text{ psi} > f_y$$

Therefore,

$$f'_s = f_y = 60,000 \text{ psi}$$

Using Eq. 9.9,

$$f_s = 29 \times 10^6 \times 0.003\left(\frac{17.5 - 11.0}{11.0}\right) = 51,409 \text{ psi}$$

Using Eq. 9.6,

$$P_n = 0.85 \times 4000 \times 12 \times 9.35 + 3 \times 60,000 - 3 \times 51,409$$
$$= 407,253 \text{ lb}$$

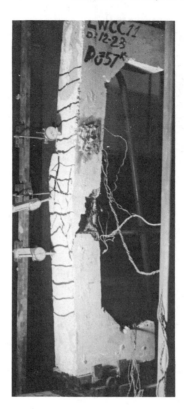

Photo 53 Tension side of eccentrically loaded column at rupture and cover spalling. (Tests by Nawy et al.)

Using Eq. 9.7,

$$M_n = 0.85 \times 4000 \times 12 \times 9.35\left(10 - \frac{9.35}{2}\right) + 3 \times 60,000(10 - 2.5)$$
$$+ 3 \times 51,409(17.5 - 10) = 4,538,083 \text{ in.-lb}$$

$$e = \frac{M_n}{P_n} = 11.14 \text{ in.} > 10 \text{ in.}$$

Therefore, an axial load of 407,253 lb can be applied at an eccentricity of 11.14 in.

Trial 2

Assume that $c = 11.5$ in.; hence a $= 0.85 \times 11.5 = 9.77$ in.

$$f'_s = 60,000 \text{ psi}$$
$$f_s = 45,391 \text{ psi}$$
$$P_n = 442,647 \text{ lb}$$
$$M_n = 4,410,218 \text{ in.-lb}$$
$$e = 9.96 \text{ in.} \simeq \text{ given eccentricity of 10 in.}$$

Therefore, for $e = 10$ in. (254 mm), P_n can be assumed to be 442,647 lb. (If displaced concrete is taken into account in the calculations, $P_n = 432,445$ lb.)

9.5.8 General Case of Columns Reinforced on All Faces: Exact Solution

When columns are reinforced with bars on all faces and those where the reinforcement in the parallel faces is nonsymmetrical, solutions have to be based on using first principles. Equations 9.6 and 9.7 have to be adjusted for this purpose and the trial-and-adjustment procedure adhered to. Strain-compatibility checks for strain in each reinforcing bar layer have to be performed at all load levels.

Figure 9.13 illustrates the case of a column reinforced on all four faces. Assume that

G_{sc} = center of gravity of the steel compressive force
G_{st} = center of gravity of the steel tensile force
F_{sc} = resultant steel compressive force = $\Sigma A'_s f_{sc}$
F_{st} = resultant steel tensile force = $\Sigma A_s f_{st}$

Equilibrium of the internal and external forces and moments requires that

$$P_n = 0.85 f'_c b \beta_1 c + F_{sc} - F_{st} \tag{9.23}$$

$$P_n e = 0.85 f'_c b \beta_1 c \left(\frac{h}{2} - \frac{1}{2} \beta_1 c \right) + F_{sc} y_{sc} + F_{st} y_{st} \tag{9.24}$$

Trial and adjustment is applied assuming a neutral-axis depth c and consequently a depth a of the equivalent rectangular block. The strain values in each bar layer are

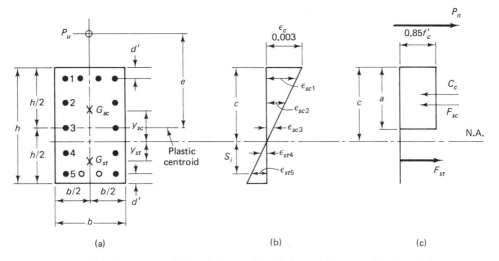

Figure 9.13   Column reinforced with steel on all faces: (a) cross section; (b) strain; (c) forces.

determined by the linear strain distribution in Fig. 9.13b to ensure strain compatibility. The stress in each reinforcing bar is obtained using the expression

$$f_{si} = E_s \epsilon_{si} = E_s \epsilon_c \frac{s_i}{c} = 87,000 \frac{c - s_i}{c} \tag{9.25}$$

where f_{si} has to be $\leq f_y$.

Find P_n corresponding to the assumed c in Eq. 9.23. Substitute into Eq. 9.24 the P_n value thus obtained with the parameter c as the unknown. If the resulting c is not close to the assumed value, proceed to another trial. The nominal resisting load P_n of the section would be the one corresponding to the trial depth c of the last trial cycle.

It is advisable in many instances also to add steel to the column faces which are perpendicular to the axis of bending such that their area does not exceed 25% of the area of the main steel.

9.6 WHITNEY'S APPROXIMATE SOLUTION IN LIEU OF EXACT SOLUTIONS

Empirical expressions proposed by Whitney can rapidly be used in lieu of the trial-and-adjustment method, although with some loss in accuracy.

9.6.1 Rectangular Concrete Columns

These expressions are presented particularly for circular columns since longhand trial-and-adjustment procedures for their analysis or design can be time consuming. Strain-compatibility checks for the reinforcement require that the strain in the bars be evaluated at each level across the depth of the section; hence the use of an empirical one-step expression becomes useful for a quick analysis. The use of hand-held computers and personal computers reduces or preempts the need for the use of the Whitney empirical approach in the case of designers familiar with computer use. Chapter 13 applies the exact method of analysis and design of circular and rectangular columns with strain compatibility checks at all stages through the use of hand-held computer programs developed for this purpose. The student and the design engineer can modify or develop other programs to suit the particular personal computer utilizing the detailed flowcharts in Chapter 13.

Whitney's solution is based on the following assumptions.

1. Reinforcement is symmetrically placed in single layers parallel to the axis of bending in rectangular sections.
2. Compression steel has yielded.
3. Concrete displaced by the compression steel is negligible compared to the total concrete area in compression; hence no correction is made for the concrete displaced by the compression steel.

4. For the purpose of calculating the contribution C_c of the concrete, the depth of the stress block is assumed to be $0.54d$, corresponding to an average value of a for balanced conditions in rectangular sections.

5. The interaction curve in the compression zone is a straight line, as shown in Fig. 9.17.

For most cases, Whitney's method leads to a conservative solution except when the factored load P_u has a value higher than the balanced load P_{ub}, as in Ex. 9.7(b), and the external eccentricity e is very small. Otherwise, the method leads to a non-conservative solution as illustrated in Ex. 9.7(b) and as seen in the shaded area BST of Fig. 9.17.

If compression controls, the equation can be written as

$$P_n = \frac{A_s' f_y}{[e/(d - d')] + 0.5} + \frac{bhf_c'}{(3he/d^2) + 1.18} \tag{9.26}$$

The following example illustrates the use of this equation.

9.6.2 Example 9.7: Analysis of a Column Controlled by Compression Failure; Whitney's Equation

Calculate the nominal strength load P_n for the section in Ex. 9.6 using Whitney's equation if the load eccentricity is (a) $e = 6$ in. (152.4 mm), and (b) $e = 10$ in. (254 mm).

Solution

(a) $e = 6$ in.:

$$P_n = \frac{3 \times 60,000}{[6/(17.5 - 2.5)] + 0.5} + \frac{12 \times 20 \times 4000}{[(3 \times 20 \times 6)/17.5^2] + 1.18}$$

$$= 607,555 \text{ lb } (2734.0 \text{ kN})$$

Exact solution, using trial and adjustment and including the displaced concrete, gives $P_n = 608,458$ lb (2738.0 kN). The approximate solution is conservative.

(b) $e = 10$ in.: Using Eq. 9.26,

$$P_n = \frac{3 \times 60,000}{[10/(17.5 - 2.5)] + 0.5} + \frac{12 \times 20 \times 4000}{[(3 \times 20 \times 10)/17.5^2] + 1.18}$$

$$= 460,098 \text{ lb } (2070.4 \text{ kN})$$

The exact solution, using the trial-and-adjustment procedure and including the effect of the displaced concrete, gives $P_n = 433,138$ lb (1960 kN), showing that the approximate solution is not always conservative, as discussed above.

9.6.3 Circular Concrete Columns

As in the case of rectangular columns, force and moment equilibrium equations can be used to solve for the unknown nominal axial load P_n for any given eccentricity. The equilibrium equations are similar to Eqs. 9.6 and 9.7 except that (1) the shape of the area under compressive stress will be a segment of a circle, and (2) reinforcing bars are not grouped together parallel to the compression and tension sides. Therefore, the force and stress in each bar should be considered separately. The area and the center of gravity of the segment of a circle in compression should be calculated using the appropriate mathematical expressions. This accurate approach can be easily adopted if one chooses to use handheld or desktop computers (see Chapter 13). The following simplified empirical Whitney's approach can be used for longhand calculations.

9.6.4 Empirical Method of Analysis of Circular Columns

Transform the circular column to an idealized equivalent rectangular column as shown in Ex. 9.8 and Fig. 9.14. For compression failure, the equivalent rectangular column would have (1) the thickness in the direction of bending equal to $0.8h$, where h is the outside diameter of the circular column (Fig. 9.14b); (2) the width of the idealized rectangular column to be obtained from the same gross area A_g of the circular columns such that $b = A_g/0.8h$; and (3) the total area of reinforcement A_{st} to be equally divided in two parallel layers and placed at a distance of $2D_s/3$ in the direction of bending, where D_s is the diameter of the cage measured center to center of the outer vertical bars. For tension failure, use the actual column for evaluating C_c, but place 40% of the steel A_{st} in parallel at a distance of $0.75D_s$, as shown in Fig. 9.14. The equivalent column method provides satisfactory results for most cases.

Once the dimensions of the equivalent rectangular column are established, the analysis (design) can be made as for rectangular columns. The equations for tension and compression failure can also be expressed in terms of the dimensions of the circular column, as follows:

For tension failure:

$$P_n = 0.85f_c'h^2\left[\sqrt{\left(\frac{0.85e}{h} - 0.38\right)^2 + \frac{\rho_g mD_s}{2.5h}} - \left(\frac{0.85e}{h} - 0.38\right)\right] \quad (9.27)$$

For compression failure:

$$P_n = \frac{A_{st}f_y}{(3e/D_s) + 1.00} + \frac{A_g f_c'}{[9.6he/(0.8h + 0.67D_s)^2] + 1.18} \quad (9.28)$$

where h = diameter of section

D_s = diameter of the reinforcement cage center to center of the outer vertical bars

e = eccentricity to the plastic centroid of section

$\rho_g = \dfrac{A_{st}}{A_g} = \dfrac{\text{gross steel area}}{\text{gross concrete area}}$

$m = \dfrac{f_y}{0.85f_c'}$

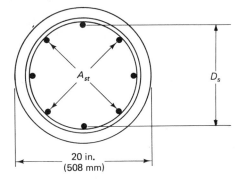

(a)

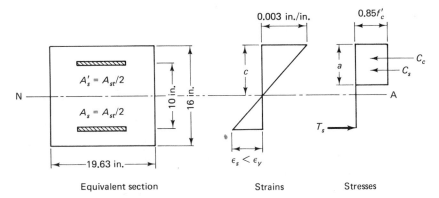

(b)

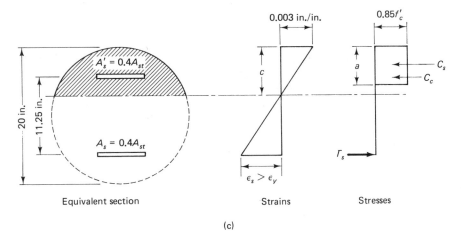

(c)

Figure 9.14 Equivalent column section: (a) given circular section (A_{st}, total reinforcement area); (b) equivalent rectangular section (compression failure); (c) equivalent column (tension failure).

9.6.5 Example 9.8: Calculation of Equivalent Rectangular Cross Section for a Circular Column

Obtain an equivalent rectangular cross section for the circular column shown in Fig. 9.14a.

Solution

$$\text{thickness of the rectangular section} = 0.8 \times 20 = 16 \text{ in.}$$

$$\text{width of the rectangular section} = (\pi/4)\frac{(20)^2}{16} = 19.63 \text{ in.}$$

$$d - d' = \frac{2}{3} \times 15 = 10 \text{ in.}$$

$$A_s = A'_s = \frac{A_{st}}{2}$$

9.6.6 Example 9.9: Analysis of a Circular Column

A circular column 20 in. (508 mm) in diameter is reinforced with six No. 8 equally spaced bars. Calculate (a) the load and eccentricity for the balanced failure condition, (b) the load P_n for $e = 16.0$ in. (406 mm), and (c) the load P_n for $e = 5.0$ in. (127 mm). Assume the column to be nonslender (short) and spirally reinforced. Given:

$$f'_c = 4000 \text{ psi (27.6 MPa)}$$
$$f_y = 60,000 \text{ psi (414 MPa)}$$

Solution

For case (a), an equivalent rectangular column shown in Fig. 9.15 is used for the analysis. Using the results of Ex. 9.8 for the equivalent column: $h = 16$ in., $d = 13$ in., $d' = 3$ in., $b = 19.63$ in., and $A'_s = A_s = 3 \times 0.79 = 2.37$ in.^{2}.

(a) *Balanced failure:*

$$c_b = 13\left(\frac{87,000}{87,000 + 60,000}\right) = 7.69 \text{ in.}$$

$$a_b = 0.85 \times 7.69 = 6.54 \text{ in.}$$

$$f'_s = 0.003 \times 29 \times 10^6\left(\frac{7.69 - 3}{7.69}\right) = 53,060 \text{ psi}$$

$$f_y = 60,000 \text{ psi}$$

$$P_{nb} = 0.85 \times 4000 \times 19.63 \times 6.54 + 2.37 \times 53,060 - 2.37 \times 60,000$$

$$= 420,045 \text{ lb}$$

If the actual circular cross section is used in the analysis instead of the equivalent

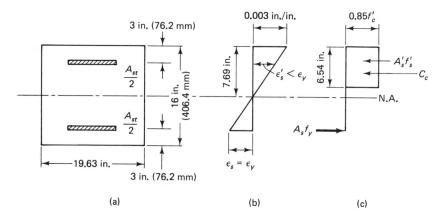

Figure 9.15 Column geometry—strain and stress diagrams (balanced failure): (a) equivalent section; (b) strains; (c) stresses.

section and an exact strain-compatibility analysis made, $P_{nb} = 454,330$ lb and $e_b = 7.18$ in. The exact solution is presented in Chapter 13 using the handheld computer program.

$$M_{nb} = 0.85 \times 4000 \times 19.63 \times 6.54 \left(8.0 - \frac{6.54}{2}\right) + 2.37 \times 53,060$$
$$\times 5 + 2.37 \times 60,000 \times 5$$
$$= 3,404,371 \text{ in.-lb}$$

$$e_b = \frac{M_{nb}}{P_{nb}} = \frac{3,404,371}{420,045} = 8.10 \text{ in. (205.9 mm)}$$

For cases (b) and (c), Whitney's formulas involving the actual dimensions of the circular column can be used directly.

(b) *Large eccentricity:* For $e = 16$ in. $> e_b$, tension failure controls. Assuming an effective cover of 2.5 in. to the center of the longitudinal reinforcement,

$$D_s = 20 - 2 \times 2.5 = 15 \text{ in.}$$

$$\rho_g = \frac{2 \times 2.37}{314} = 0.015$$

$$m = \frac{60,000}{0.85 \times 4000} = 17.65$$

Using Eq. 9.27 gives

$$P_n = 0.85 \times 4000 \times 400 \left[\sqrt{\left(\frac{0.85 \times 16}{20} - 0.38\right)^2 + \frac{0.015 \times 17.65 \times 15}{2.5 \times 20}} \right.$$
$$\left. - \left(\frac{0.85 \times 16}{20} - 0.38\right) \right]$$
$$= 151,793 \text{ lb (675 kN)}$$

$$\phi P_n = 0.75 \times 151,793 = 113,845 \text{ lb}$$

[Using strain compatibility (Chapter 13), $P_n = 173,940$ lb.]

(c) *Small eccentricity:* For $e = 5.0$ in. $< e_b$; compression failure controls. Using Eq. 9.28, we have

$$\text{total steel area } A_{st} = A_s + A'_s = 2 \times 2.37 = 4.74 \text{ in.}^2 \text{ (3057.3 mm}^2)$$

$$\text{gross concrete area } A_g = \frac{\pi(20.0)^2}{4} = 314.2 \text{ in.}^2 \text{ (2025.3 mm}^2)$$

$$P_n = \frac{4.74 \times 60,000}{\dfrac{3 \times 5.0}{15} + 1} + \frac{314.2 \times 4000}{\dfrac{9.6 \times 20 \times 5}{(0.8 \times 20 + 0.67 \times 15)^2} + 1.18}$$

$$= 626,577 \text{ lb (2780 kN)}$$

[Using strain compatibility (Chapter 13), $P_n = 621,653$ lb, indicating that the Whitney solution is in this case not conservative.]

9.7 COLUMN STRENGTH REDUCTION FACTOR ϕ

For members subject to flexure and relatively small axial loads, failure is initiated by yielding of the tension reinforcement and takes place in an increasingly ductile manner. Hence for small axial loads it is reasonable to permit an increase in the ϕ factor from that required for pure compression members. When the axial load vanishes, the member is subjected to pure flexure, and the strength reduction factor ϕ becomes 0.90. Figure 9.16a and b shows the zone in which the value of ϕ can be increased from 0.7 to 0.9 for tied columns and 0.75 to 0.9 for spiral columns. As the factored design compression load ϕP_n decreases beyond $0.1 A_g f'_c$, in Fig. 9.16a, the ϕ factor is increased from 0.7 to 0.9 for tied columns and 0.75 to 0.9 for spiral columns. For those cases where the value of P_{nb} is less than $0.1 A_g f'_c$, ϕ values are increased when the load $P_u < P_{ub}$ or $\phi P_n < \phi P_{nb}$, as in Fig. 9.16b.

The value $0.10 f'_c A_g$ is chosen by the ACI code as the design axial load value ϕP_n below which the ϕ factor could safely be increased for most compression members. In summary, if initial failure is in compression, the strength reduction factor ϕ is always 0.70 for tied columns and 0.75 for spirally reinforced columns.

The following expressions give variations in the value of ϕ for symmetrically reinforced compression members. The columns should have an effective depth not less than 70% of the total depth and the steel reinforcement should have a yield strength not exceeding 60,000 psi. For tied columns,

$$\phi = 0.90 - \frac{0.20\phi P_n}{0.1 f'_c A_g} \geq 0.70 \tag{9.29}$$

For spirally reinforced columns,

$$\phi = 0.90 - \frac{0.15\phi P_n}{0.1 f'_c A_g} \geq 0.75 \tag{9.30}$$

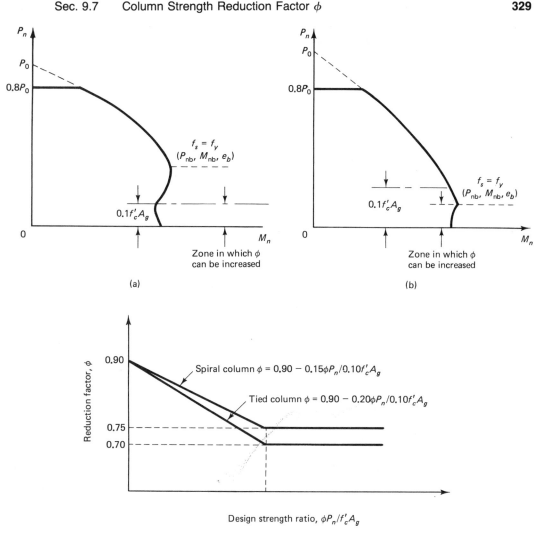

Figure 9.16   Controlling zones for modification of reduction factor ϕ in columns: (a) $0.1f'_c A_g < \phi P_{nb}$; (b) $0.1f'_c A_g > \phi P_{nb}$; (c) variation of ϕ for symmetrically reinforced compression members (P_n is the nominal axial strength at given eccentricity).

where

$$P_u = \phi P_n \qquad (9.31)$$

In both Eqs. 9.29 and 9.30, if ϕP_{nb} is less than $0.1f'_c A_g$, then ϕP_{nb} should be substituted for $0.1f'_c A_g$ in the denominator, using $0.7P_{nb}$ for tied and $0.75P_{nb}$ for spirally reinforced columns. Figure 9.16c presents the graphical representation of Eqs. 9.29 and 9.30.

9.7.1 Example 9.10: Calculation of Design Load Strength ϕP_n from Nominal Resisting Load P_n

Calculate the design loads P_u in Exs. 9.1 to 9.7 and 9.9 using the appropriate ϕ reduction factors.

Solution

Example 9.1:

$$P_{n(max)} = 924,480 \text{ lb, tied column}$$

$$\text{therefore, } \phi = 0.7$$

$$\phi P_{0(max)} = 0.7 \times 924,480 = 647,136 \text{ lb}$$

Example 9.2:

$$P_{n(max)} = 1,135,501 \text{ lb, spiral column}$$

$$\text{therefore, } \phi = 0.75$$

$$\phi P_{0(max)} = 0.75 \times 1,135,501 = 851,626 \text{ lb}$$

Example 9.3:

$$P_{nb} = 359,448 \text{ lb, tied column}$$

$$\text{therefore, } \phi = 0.7$$

$$\phi P_{nb} = 251,614 \text{ lb}$$

$$e_b = 13.1 \text{ in.}$$

Example 9.4:
$P_n = 333,438$ lb, tied column. Failure by yielding of steel at the tension side. Therefore, it has to be checked if the ϕ value is larger than 0.7.

$$0.1 A_g f'_c = 0.1 \times 12 \times 20 \times 4000 = 96,000 \text{ lb} < 0.7 P_n = 233,407 \text{ lb}$$

$$\text{therefore, } \phi = 0.7$$

$$\phi P_n = 0.7 \times 333,438 = 233,407 \text{ lb}$$

Example 9.5:

$$\phi = 0.7 \text{ (similar to previous case)}$$

$$\phi P_n = 0.7 \times 236,200 = 165,340 \text{ lb}$$

Example 9.6:
$P_n = 440,600$ lb, tied column. Failure is by crushing of concrete, axial force greater than balanced load. Therefore,

$$\phi = 0.7$$

$$\phi P_n = 0.7 \times 442,647 = 309,853 \text{ lb}$$

Example 9.7:
(a) P_n = 460,000 lb, tied column. Failure is by crushing of concrete. Therefore,

$$\phi = 0.7$$

$$\phi P_n = 0.7 \times 607,555 = 425,288 \text{ lb}$$

(b) P_n = 460,098 lb

$$\phi P_n = 0.7 \times 460,098 = 322,070 \text{ lb}$$

Example 9.9: Spiral column
(a) Balanced P_{ub} = 0.75 $\times$ P_{nb} = 0.75 $\times$ 420,045 = 315,034 lb.
(b) Tension failure, P_n = 151,793 lb

$0.1 f_c' A_g$ = 0.1 $\times$ 4000 $\times$ 314 = 125,600 lb $<$ ϕP_{nb} = 315,034 lb
ϕP_n = 0.75 $\times$ 151,793 = 113,845 lb
$f_c' A_g$ = 4000 $\times$ 314 = 1,256,000 lb

Therefore, using Eq. 9.30 and assuming that ϕ = 0.75, we have

$$\phi = 0.9 - \frac{1.5 \times 0.75 \times 151,793}{1,256,000}$$

$$= 0.764 > 0.75$$

Note: To obtain the exact answer, one should reiterate the calculation for ϕ until the answer converges. Therefore,

$$\phi P_n = 0.764 \times 151,793 = 115,970 \text{ lb}$$

(c) Compression failure:

$$\phi P_n = 0.75 P_n = 0.75 \times 626,577 = 469,933 \text{ lb}$$

9.8 LOAD–MOMENT STRENGTH INTERACTION DIAGRAMS (P–M DIAGRAMS) FOR COLUMNS CONTROLLED BY MATERIAL FAILURE

From the discussion in Sections 9.3 and 9.4 and the numerical examples presented, one can postulate that the capacity of reinforced concrete sections to resist combined axial and bending loads can be expressed by *P–M* interaction diagrams to relate the axial load to the bending moment in compression members. Figure 9.17 presents one such diagram. Notice that the approximation made in using the Whitney empirical approach is not always conservative, particularly when the factored P_u is close to the balanced case.

Each point on the curve represents one combination of nominal load strength P_n and nominal moment strength M_n corresponding to a particular neutral-axis location. The interaction diagram is separated into the tension control region and the compression control region by the balanced condition at point *B*. The following example illustrates the construction of the *P–M* diagram for a typical rectangular section.

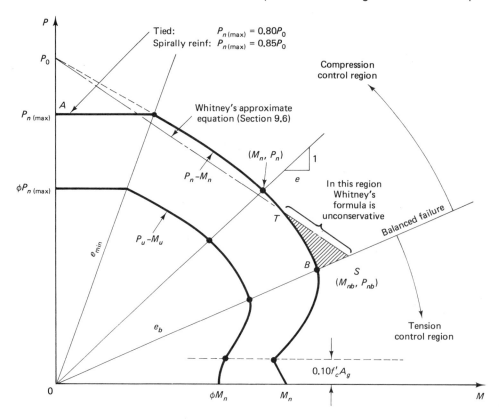

Figure 9.17   Typical load-moment strength (P–M) column interaction diagram.

9.8.1 Example 9.11: Construction of a Load–Moment Interaction Diagram

Construct a (P–M) diagram for a rectangular column (see Fig. 9.18) having the following geometry: width $b = 12$ in. (305 mm), thickness $h = 14$ in. (356 mm), reinforcement: four No. 11 bars (35.8 mm diameter). Given:

$$f'_c = 6000 \text{ psi (41.4 MPa)}$$
$$f_y = 60,000 \text{ psi (414 MPa)}$$
$$d' = 3.0 \text{ in. (76.2 mm)}$$

Solution

Concentric load

$$A_s = A'_s = 3.12 \text{ in.}^2$$

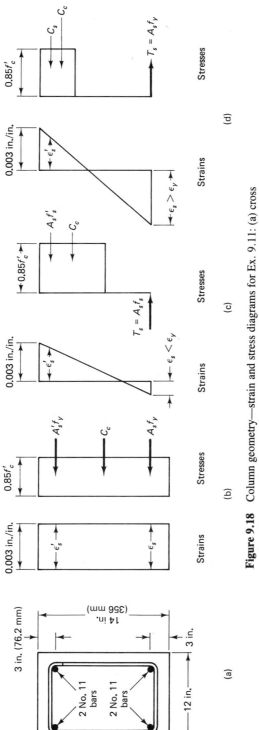

Figure 9.18 Column geometry—strain and stress diagrams for Ex. 9.11: (a) cross section; (b) concentric load (compression failure); (c) compression failure; (d) tension failure.

$$P_{n(max)} = 0.80(0.85f_c'A_g + A_{st}f_y)$$

$$= 0.80(0.85 \times 6000 \times 14 \times 12 + 2 \times 3.12 \times 60,000)$$

$$= 984,960 \text{ lb}$$

$$\phi P_{n(max)} = 0.7P_{n(max)} = 689,472 \text{ lb}$$

(If displaced concrete is taken into account in the calculations, $P_{n(max)} = 959,500$ lb and $\phi P_{n(max)} = 671,650$ lb.)

Balanced condition

$$d = 14.0 - 3.0 = 11.0 \text{ in.}$$

$$c_b = \frac{87,000}{87,000 + 60,000} \times 11.0 = 6.51 \text{ in.}$$

$$\epsilon_s' = 0.003 \times \frac{6.51 - 3.0}{6.51} = 0.0016 \text{ in./in.}$$

$$< \frac{f_y}{E_s} = \frac{60}{29,000} = 0.00207$$

$$f_s' = 29 \times 10^6 \times 0.0016 = 46,400 \text{ psi}$$

$$\beta_1 = 0.85 - \frac{0.05(6000 - 4000)}{1000} = 0.75 \text{ in.}$$

$$a_b = \beta_1 c_b = 0.75 \times 6.51 = 4.88 \text{ in.}$$

$$P_{nb} = 0.85f_c'ba_b + A_s'f_s' - A_sf_y$$

or

$$P_{nb} = 0.85 \times 6,000 \times 12 \times 4.88 + 3.12 \times 46,400 - 3.12 \times 60,000$$

$$= 298,656 + 144,768 - 187,200 = 256,224 \text{ lb}$$

$$M_{nb} = 0.85f_c'ba_b\left(\bar{y} - \frac{a}{2}\right) + A_s'f_s'(\bar{y} - d') + A_sf_y(d - \bar{y})$$

$\bar{y}$ = distance from extreme compression fibers to the plastic centroid

$$= \frac{h}{2} = \frac{14.0}{2} = 7 \text{ in.}$$

or

$$M_{nb} = 298,656\left(7.0 - \frac{4.88}{2}\right) + 144,768(7.0 - 3.0) + 187,200(11.0 - 7.0)$$

$$= 1,361,871 + 579,072 + 748,800 = 2,689,743 \text{ in.-lb}$$

$$e_b = \frac{M_{nb}}{P_{nb}} = \frac{2,689,743}{256,224} = 10.50 \text{ in.}$$

$$\phi P_{nb} = 0.7 P_{nb} = 179,357 \text{ lb} \qquad \phi M_{nb} = 1,882,820 \text{ in.-lb}$$

$$e_b = 10.5 \text{ in.}$$

(If displaced concrete is considered in the calculations, $\phi P_{nb} = 169,444$ lb and $e_b = 10.9$ in.)

Pure bending M_{n0}

Neglect the contribution of A_s' to the moment strength as insignificant when $P_u = 0$. Hence,

$$a = \frac{A_s f_y}{0.85 f_c' b} = \frac{3.12 \times 60,000}{0.85(6000)(12)} = \frac{187,200}{61,200} = 3.06 \text{ in.}$$

$$c = \frac{a}{\beta_1} = \frac{3.06}{0.75} = 4.08 \text{ in.} \qquad \epsilon_s' = 0.00079 \qquad f_s' = 22,910 \text{ psi}$$

$$M_{n0} = A_s f_y \left(d - \frac{a}{2} \right) = 187,200(11.0 - 1.53) = 1,772,784 \text{ in.-lb}$$

$$\phi M_{n0} = 0.9 \times 1,772,784 = 1,595,506 \text{ in.-lb}$$

For $c = 10$ in. $> c_b$: compression controls

$$\epsilon_s' = 0.003 \times \frac{10 - 3.0}{10} = 0.0021 \text{ in./in.}$$

$$\epsilon_y = \frac{f_y}{E_s} = \frac{60,000}{29 \times 10^6} = 0.0021 \simeq \epsilon_s' \qquad \text{therefore } f_s' = f_y$$

$$\epsilon_s = 0.003 \times \frac{11.0 - 10.0}{10} = 0.0003$$

$$f_s = 0.0003 \times 29 \times 10^6 = 8,700 \text{ psi}$$

$$a = \beta_1 c = 0.75 \times 10 = 7.5 \text{ in.}$$

$$C_c = 0.85(6000)(12)(7.5) = 459,000 \text{ lb}$$

$$C_s = 3.12(60,000) = 187,200 \text{ lb}$$

$$T_s = 3.12(8700) = 27,144 \text{ lb}$$

$$P_n = C_c + C_s - T_s$$

or

$$P_n = (459,000) + (187,200 - 27,144) = 619,056 \text{ lb}$$

$$M_n = C_c\left(\bar{y} - \frac{a}{2}\right) + C_s(\bar{y} - d') + T_s(d - \bar{y})$$

$$= 459,000(7.0 - 3.75) + 187,200(7.0 - 3.0) + 27,144(11.0 - 7.0)$$

$$= 1,491,750 + 748,800 + 108,576 = 2,349,126$$

$$\phi P_n = 0.7 P_n = 0.7 \times 619,056 = 433,339 \text{ lb}$$

(With displaced concrete taken into account, $\phi P_n = 422,200$ lb and $e = 3.79$ in.)

For $c = 4.2$ in. $< c_b$: tension control

$$a = \beta_1 c = 0.75 \times 4.2 = 3.15 \text{ in.}$$

$$\epsilon'_s = 0.003 \times \frac{4.2 - 3.0}{4.2} = 0.0009 < \epsilon_y$$

$$f'_s = 0.0009 \times 29 \times 10^6 = 24,857 \text{ psi}$$

$$f_s = f_y = 60,000 \text{ psi}$$

$$C_c = 0.85(6000)(12)(3.15) = 192,780 \text{ lb}$$

$$C_s = 3.12(24,857) = 77,554 \text{ lb}$$

$$T_s = 3.12(60,000) = 187,200 \text{ lb}$$

$$P_n = C_c + C_s - T_s = 192,780 + 77,554 - 187,200 = 83,134 \text{ lb}$$

$$M_n = C_c\left(\bar{y} - \frac{a}{2}\right) + C_s(\bar{y} - d') + T_s(d - \bar{y})$$

or

$$M_n = 192,780\left(7.0 - \frac{3.15}{2}\right) + 77,554(7.0 - 3.0) + 187,200(11.0 - 7.0)$$

$$= 1,044,868 + 310,216 + 748,800 = 2,103,884 \text{ in.-lb}$$

$$e = \frac{M_n}{P_n} = \frac{2,103,884}{83,134} = 25.31 \text{ in.}$$

Assuming that $\phi = 0.70$,

$$\phi P_n = 0.70 \times 83,134 = 58,194 \text{ lb}$$

$$0.1 A_g f'_c = 0.10(12.0 \times 14.0) \times 6000 = 100,800 \text{ lb}$$

$$\phi P_n < 0.1 A_g f'_c \qquad \text{hence } \phi > 0.70$$

Therefore,

$$\phi = 0.90 - \frac{0.2 \times 58{,}194}{0.1 \times 1{,}008{,}000} = 0.792$$

$$\phi P_n = 0.792 \times 83{,}134 = 65{,}842$$

Verify the first trial ϕ value:

$$\phi = 0.90 - \frac{0.2 \times 65{,}842}{0.1 \times 1{,}008{,}000} = 0.78$$

$$\phi P_n = 0.78 \times 83{,}134 = 64{,}845 \text{ lb}$$

(If displaced concrete is considered, $\phi P_n = 47{,}056$ lb and $e = 30.4$ in.)

For $0.10 f'_c A_g = \phi P_n$

Assume by trial and adjustment a value of $c = 4.85$ in.

$$a = \beta_1 c = 0.75 \times 4.85 = 3.64 \text{ in.}$$

$$\epsilon'_s = 0.003 \times \frac{4.85 - 3.0}{4.85} = 0.00114 < \epsilon_y$$

$$f'_s = 0.00114 \times 29 \times 10^6 = 33{,}060 \text{ psi}$$

$$C_c = 0.85 \times 6000 \times 12 \times 3.64 = 227{,}768 \text{ lb}$$

$$C_s = 3.12 \times 33{,}060 = 103{,}147 \text{ lb}$$

$$T_s = 3.12 \times 60{,}000 = 187{,}200 \text{ lb}$$

$$P_n = C_c + C_s - T_s = 227{,}768 + 103{,}147 - 187{,}200 = 143{,}715 \text{ lb}$$

$$\phi P_n = 0.70 \times 143{,}715 = 100{,}619 \text{ lb}$$

$$0.1 f'_c A_g = 100{,}800 \text{ lb} \cong \phi P_n \qquad \text{assumed } c \text{ value} \qquad \text{O.K.}$$

$$M_n = C_c \left(\bar{y} - \frac{a}{2} \right) + C_s (\bar{y} - d') + T_s (d - \bar{y}) = 227{,}768 \left(7.0 - \frac{3.64}{2} \right)$$

$$+ \; 103{,}147(7.0 - 3.0) + 187{,}200(11.0 - 7.0) = 2{,}341{,}226 \text{ lb}$$

$$e = \frac{M_n}{P_n} = \frac{2{,}341{,}226}{143{,}715} = 16.29 \text{ in.}$$

Additional points on the diagram are calculated by assigning other values for the neutral-axis depth c. The interaction diagram is presented in Fig. 9.19. Sets of charts of nondimensional interaction diagrams are available for various codes and units to facilitate speedy analysis and design in engineering offices. A typical chart from the ACI 340 SP-17(81) Handbook is shown in Fig. 9.20.

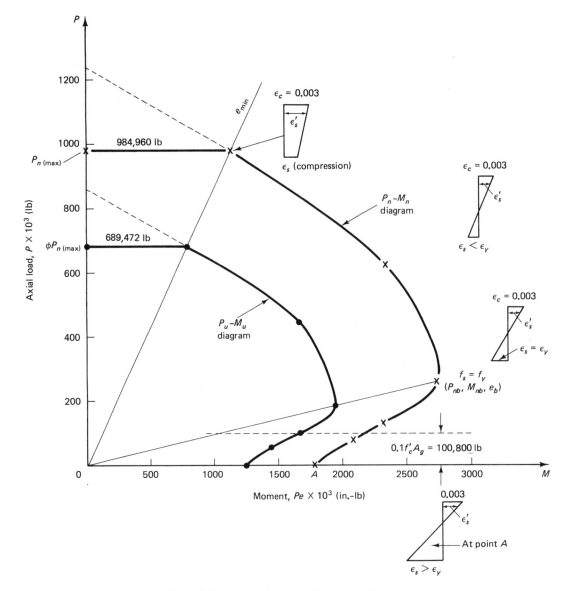

Figure 9.19   Interaction $P–M$ diagram for Ex. 9.11.

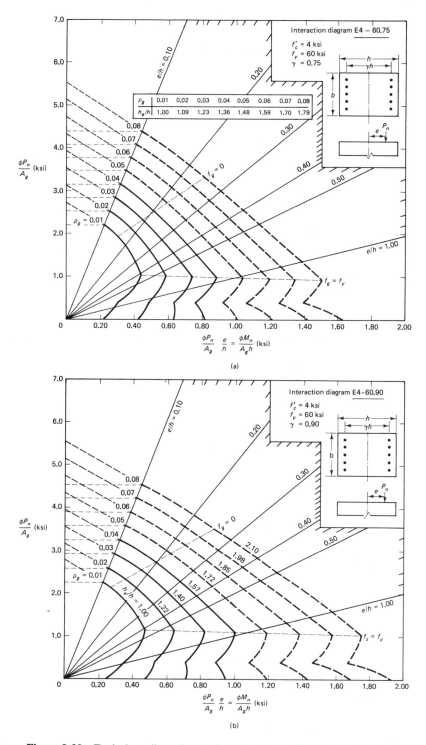

Figure 9.20 Typical nondimensional column interaction charts: (a) chart for small column sizes; (b) chart for larger column sizes.

9.9 PRACTICAL DESIGN CONSIDERATIONS

The following guidelines should be followed in the design and arrangement of reinforcement to arrive at a practical design.

9.9.1 Longitudinal or Main Reinforcement

Most columns are subjected to bending moment in addition to axial force. For this reason and to ensure some ductility, a mininum of 1% reinforcement should be provided in the columns. A reasonable reinforcement ratio is between 1.5 and 3.0%. Occasionally, in high-rise buildings where column loads are very large, 4% reinforcement is not unreasonable. Even though the code allows a maximum of 8% for longitudinal reinforcement in columns, it is not advisable to use more than 4% in order to avoid reinforcement congestion, especially at beam–column junctions.

A minimum of four longitudinal bars should be used in the case of tied columns. For spiral columns, at least six longitudinal bars should be used to provide hoop action in the spirals; see the ACI code for further discussion.

9.9.2 Lateral Reinforcement for Columns

Lateral Ties

Lateral reinforcement is required to prevent spalling of the concrete cover or local buckling of the longitudinal bars. The lateral reinforcement could be in the form of ties evenly distributed along the height of the column at specified intervals. Longitudinal bars spaced more than 6 in. apart should be supported by lateral ties, as shown in Fig. 9.21.

The following guidelines are to be followed for the selection of the size and spacing of ties.

1. The size of the tie should not be less than a No. 3 (9.5 mm) bar. If the longitudinal bar size is larger than No. 10 (32 mm), then No. 4 (12 mm) bars at least should be used as ties.
2. The vertical spacing of the ties must not exceed:
    (a) Forty-eight times the diameter of the tie
    (b) Sixteen times the diameter of the longitudinal bar
    (c) Least lateral dimension of the column

Figure 9.21 shows a typical arrangement of ties for four, six, and eight longitudinal bars in a column cross section.

Spirals

The other type of lateral reinforcement is spirals or helical lateral reinforcement as shown in Fig. 9.22. They are particularly useful in increasing ductility or member

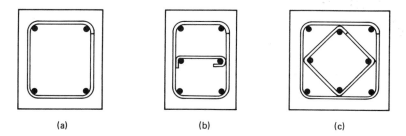

Figure 9.21   Typical ties arrangement for four, six, and eight longitudinal bars in a column: (a) one tie; (b) two ties; (c) two ties.

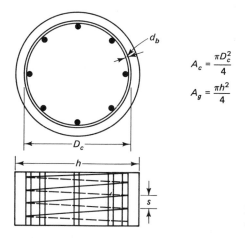

$$A_c = \frac{\pi D_c^2}{4}$$

$$A_g = \frac{\pi h^2}{4}$$

Figure 9.22   Helical or spiral reinforcement for columns.

toughness, hence are mandatory in high-earthquake-risk regions. Normally, concrete outside the confined core of the spirally reinforced column can totally spall under unusual and sudden lateral forces such as earthquake-induced forces. The columns have to be able to sustain most of the load even after the spalling of the cover in order to prevent the collapse of the building. Hence the spacing and size of spirals are designed to maintain most of the load-carrying capacity of the column, even under such severe load conditions.

Closely spaced spiral reinforcement increases the ultimate load capacity of columns. The spacing or pitch of the spiral is so chosen that the load capacity due to the confining spiral action compensates for the loss due to spalling of the concrete cover.

Equating the increase in strength due to confinement and the loss of capacity in spalling and incorporating a safety factor of 1.2, the following minimum spiral reinforcement ratio ρ_s is obtained:

$$\rho_s = 0.45\left(\frac{A_g}{A_c} - 1\right)\frac{f_c'}{f_{sy}} \qquad (9.32)$$

where $\rho_s = \dfrac{\text{volume of the spiral steel per one revolution}}{\text{volume of concrete core contained in one revolution}}$

$$A_c = \frac{\pi D_c^2}{4} \tag{9.33a}$$

$$A_g = \frac{\pi h^2}{4} \tag{9.33b}$$

h = diameter of the column
a_s = cross-sectional area of the spiral
d_b = nominal diameter of the spiral wire
D_c = diameter of the concrete core out-to-out of the spiral
f_{sy} = yield strength of the spiral reinforcement

To determine the pitch s of the spiral, calculate ρ_s using Eq. 9.32, choose a bar diameter d_b for the spiral, and calculate a_s; then obtain pitch s using Eq. 9.35b.

The spiral reinforcement ratio ρ_s, can be written as

$$\rho_s = \frac{a_s \pi (D_c - d_b)}{(\pi/4)D_c^2 s} \tag{9.34}$$

Therefore,

$$\text{pitch } s = \frac{a_s \pi (D_c - d_b)}{(\pi/4)D_c^2 \rho_s} \tag{9.35a}$$

or

$$s = \frac{4a_s(D_c - d_b)}{D_c^2 \rho_s} \tag{9.35b}$$

The spacing or pitch of spirals is limited to a range of 1 to 3 in. (25.4 to 76.2 mm) and the diameter should be at least $\frac{3}{8}$ in. (9.53 mm). The spirals should be well anchored by providing at least $1\frac{1}{2}$ extra turns when splicing of spirals rather than welding is used.

9.10 OPERATIONAL PROCEDURE FOR THE DESIGN OF NONSLENDER COLUMNS

The following steps can be used for the design of nonslender (short) columns where the behavior is controlled by material failure.

1. Evaluate the factored external axial load P_u and factored moment M_u. Calculate the eccentricity $e = M_u/P_u$.
2. Assume a cross section and the type of vertical reinforcement to be used. Fractional dimensions are to be avoided in selecting column sizes.
3. Assume a reinforcement ratio ρ between 1 and 4% and obtain the reinforcement area.

4. Calculate P_{nb} for the assumed section and determine the type of failure, whether by initial yielding of the steel or initial crushing of the concrete.
5. Check for the adequacy of the assumed section. If the section cannot support the factored load or it is oversized, hence uneconomical, revise the cross section and (or) the reinforcement and repeat steps 4 and 5.
6. Design the lateral reinforcement.

Figure 9.23 presents a flowchart for the sequence of calculations.

9.11 NUMERICAL EXAMPLES FOR ANALYSIS AND DESIGN OF NONSLENDER COLUMNS

9.11.1 Example 9.12: Design of a Column with Large Eccentricity; Initial Tension Failure

The tied reinforced concrete column in Fig. 9.24 is subjected to a service axial force due to dead load = 65,000 lb (289 kN) and a service axial force due to live load = 125,000 lb (556 kN). Eccentricity to the plastic centroid is e = 16 in. (406 mm).

Design the longitudinal and lateral reinforcement for this column, assuming a nonslender column with a total reinforcement ratio between 2 and 3%. Given:

f'_c = 4000 psi (27.6 MPa), normal-weight concrete
f_y = 60,000 psi (414 MPa)

Solution

Calculate the factored external load and moment (Step 1)

P_u = $1.4D + 1.7L$ = $1.4 \times 65,000 + 1.7 \times 125,000$ = 303,500 lb (1350 kN)

$P_u e$ = $303,500 \times 16$ = 4,856,000 in.-lb (549 kN-m)

Assume a section 20 in. × 20 in. and a total reinforcement ratio of 3% (Steps 2 and 3)

Assume that $\rho = \rho' = A_s/bd$ = 0.015 and d' = 2.5 in.

$$A_s = A'_s = 0.015 \times 20(20 - 2.5) = 5.25 \text{ in.}^2$$

Try five No. 9 bars—5.00 in.^{2} on each face (3225 mm^2).

$$\rho = \frac{5.00}{20 \times 17.5} = 0.0143$$

Check whether the given factored axial load P_u is greater than the balanced load, ϕP_{nb} (Step 4)

$$c_b = d\frac{87,000}{87,000 + f_y} = 17.5\left(\frac{87,000}{87,000 + 60,000}\right) = 10.4 \text{ in.}$$

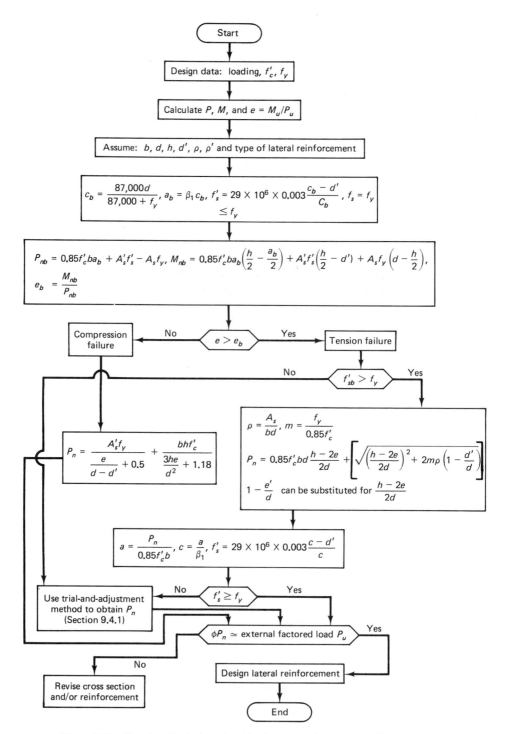

Figure 9.23 Flowchart for design of nonslender rectangular columns with bars on two faces only.

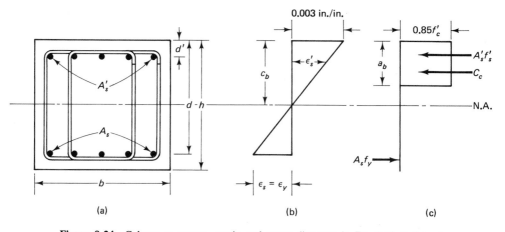

Figure 9.24 Column geometry—strain and stress diagrams in Ex. 9.12 (balanced failure): (a) cross section; (b) strains; (c) stresses.

$$a_b = \beta_1 c_b = 0.85 \times 10.4 = 8.82 \text{ in.}$$

$$f'_s = 0.003 \times 29 \times 10^6 \left(\frac{10.4 - 2.5}{10.4} \right)$$

$$= 66,086 \text{ psi} > f_y$$

Therefore, use $f'_s = f_y$. Using Eq. 9.6 gives

$$P_{nb} = 0.85 f'_c b a_b + A'_s f_y - A_s f_y$$

$$= 0.85 \times 4000 \times 20 \times 8.82 = 599,760 \text{ lb } (2670 \text{ kN})$$

$$\phi P_{nb} = 0.7 \times P_{nb} = 419,832 \text{ lb}$$

Since the given load $P_u = 303,500$ lb is less than ϕP_{nb}, the column will fail by initial yielding of the tension reinforcement.

Check the adequacy of the section (Step 5)

Using Eq. 9.19 or 9.20 yields

$$\rho = 0.0143 \qquad m = \frac{60,000}{0.85 \times 4000} = 17.65$$

$$\frac{h - 2e}{2d} \quad \text{or} \quad 1 - \frac{e'}{d} = \frac{20 - 32}{2 \times 17.5} = -0.34$$

$$1 - \frac{d'}{d} = 1 - \frac{2.5}{17.5} = 0.857$$

$$P_n = 0.85 \times 4000 \times 20 \times 17.5(-0.34 + \sqrt{0.12 + 2 \times 0.0143 \times 17.65 \times 0.857})$$

$$= 480,015 \text{ lb}$$

$$\phi P_n = 0.7 \times 480{,}015 = 336{,}011 \text{ lb (1512 kN)}$$

$$\phi P_n > 0.1 A_g f'_c \qquad \text{therefore, } \phi = 0.7 \qquad \text{O.K.}$$

Check if the compression steel stress $f'_s = f_y$:

$$a = \frac{480{,}015}{0.85 \times 4000 \times 20} = 7.06 \text{ in.}$$

$$c = \frac{a}{\beta_1} = 8.3 \text{ in.}$$

$$f'_s = 0.003 \times 29 \times 10^6 \left(\frac{8.3 - 2.5}{8.3}\right) = 60{,}795 \text{ psi} > f_y$$

therefore, $f'_s = f_y$ O.K.

An external load of 303,500 lb is less than 336,010 lb. Hence the design is satisfactory. (Using the more exact computer programs in Chapter 13, $\phi P_n = 327{,}964$ lb, which shows that using the results above can sometimes be unconservative.) Therefore, adopt a section 20 in. × 20 in. (508 mm × 508 mm) with five No. 9 bars on each side (5 bars 28.6 mm diameter) having $d = 17.5$ in. (445 mm).

Design of ties (Step 6)

Using No. 3 ties, spacing will be the minimum of:

(1) $16 \times \frac{9}{8} = 18$ in.

(2) $48 \times \frac{3}{8} = 18$ in.

(3)   Least dimension of 20 in.

Therefore, provide No. 3 ties at 18 in. center to center (9.53 mm diameter at 457 mm center to center).

9.11.2 Example 9.13: Design of a Column with Small Eccentricity; Initial Compression Failure

A nonslender column shown in Fig. 9.25 is subjected to a factored $P_u = 365{,}000$ lb (1620 kN) and a factored $M_u = 1{,}640{,}000$ in.-lb (185 kNm). Assume that the gross reinforcement ratio $\rho_g = 1.5$ to 2% and that the effective cover to the center of the longitudinal steel is $d' = 2\frac{1}{2}$ in. (63.5 mm). Design the column section and the necessary longitudinal and transverse reinforcement. Given:

$f'_c = 4500$ psi (31.03 MPa), normalweight concrete
$f_y = 60{,}000$ psi (414 MPa)

Solution

Calculation of factored design loads (Step 1)

$$P_u = 365{,}000 \text{ lb}$$

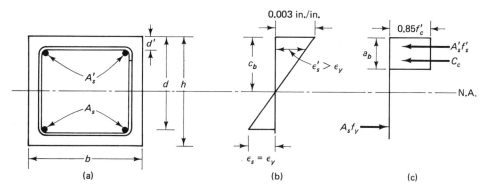

Figure 9.25 Column geometry—strain and stress diagrams in Ex. 9.13: (a) cross section; (b) strains (balanced case); (c) stresses.

$$e = \frac{1,640,000}{365,000} = 4.5 \text{ in. (114 mm)}$$

Assume a 15 in. × 15 in. (d = 12.5 in.) section (Steps 2 and 3)

Assume that the reinforcement ratio $\rho = \rho' = 0.01.$

$$A_s = A_s' \simeq 0.01 \times 15 \times 12.5 = 1.875 \text{ in.}^2$$

Provide two No. 9 bars on each side.

$$A_s = A_s' = 2.0 \text{ in.}^2 \text{ (1290 mm}^2\text{)}$$

Check whether the given load is less or greater than P_{ub} (Step 4)

$$d = 15.0 - 2.5 = 12.5 \text{ in. (317.5 mm)}$$

$$c_b = \frac{87,000}{87,000 + 60,000} \times 12.5 = 7.4 \text{ in.}$$

$$\beta_1 = 0.85 - 0.05 \left(\frac{4500 - 4000}{1000} \right) = 0.825$$

$$a = \beta_1 c = 0.825 \times 7.4 = 6.11 \text{ in.}$$

$$\epsilon_s' = 0.003 \times \frac{7.4 - 2.5}{7.4}$$

$$= 0.00199 \text{ in./in.} < \frac{f_y}{E_s}$$

$$f_s' = E_s \epsilon_s' = 29,000 \times 10^3 \times 0.00199 = 57,608 \text{ psi}$$

$$\phi P_{nb} = 0.7(0.85 f_c' b a_b + A_s' f_s' - A_s f_y)$$

$$= 0.7(0.85 \times 4500 \times 15 \times 6.11 + 2.0 \times 57,610 - 2 \times 60,000)$$

$$= 242,050 \text{ lb (1080 kN)}$$

$$\phi P_{nb} < P_u \qquad \text{compression failure controls}$$

Check the adequacy of the section (Step 5)

Using Eq. 9.26,

$$P_n = \frac{A_s' f_y}{\dfrac{e}{d-d'} + 0.5} + \frac{bhf_c'}{\dfrac{3he}{d^2} + 1.18}$$

$$= \frac{2 \times 60{,}000}{\dfrac{4.5}{12.5 - 2.5} + 0.5} + \frac{15 \times 15 \times 4500}{\dfrac{3 \times 15 \times 4.5}{12.5^2} + 1.18}$$

$$= 535{,}200 \text{ lb } (2380 \text{ kN})$$

$$= \phi P_n = 0.7 \times 535{,}200 = 374{,}000 \text{ lb} > 365{,}000 \text{ kips}$$

Therefore, the section is adequate to carry the load. Results from more exact computer solutions in Chapter 13 give $\phi P_n = 372{,}247$ lb. Use a column 15 in. × 15 in. ($d = 12.5$ in.) with two No. 9 bars on each face (381 mm × 381 mm size with two bars 28.6 mm diameter on each face).

Design of ties (Step 6)

Using No. 3 ties, the spacing will be the minimum of:

(1) $16 \times \frac{9}{8} = 18$ in.

(2) $48 \times \frac{3}{8} = 18$ in.

(3) 15 in.

Therefore, provide No. 3 ties at 15 in. (9.53 mm diameter at 381 mm center to center).

9.11.3 Example 9.14: Design of a Circular Spirally Reinforced Column

A spirally reinforced circular column is subjected to an external factored load $P_u = 110{,}000$ lb (489 kN) acting at an eccentricity to the plastic centroid of magnitude $e = 16$ in. (406 mm). Design the column cross section and the longitudinal and spiral reinforcement necessary, assuming a nonslender column with a total reinforcement ratio of about 2%. Given:

$f_c' = 4000$ psi (27.58 MPa), normalweight concrete
$f_y = 60{,}000$ psi (414 MPa)
$f_{sy} = 60{,}000$ psi (414 MPa)

Solution

Calculation of factored external loads (Step 1)

Given:

$P_u = 110{,}000$ lb
$e = 16$ in.

$$\text{required } P_n = \frac{P_u}{\phi} = \frac{110,000}{0.75} = 146,670 \text{ lb}$$

Try a 20-in. circular column with six No. 8 bars
(area = 4.74 in.²) (Steps 2 and 3)

Provide a clear cover of 1.5 in. and effective cover to the center of the bar of 2.5 in.

Check the adequacy of the section (Steps 4 and 5)

Using the results of Ex. 9.9(b), the available $P_n = 151,793$ lb with eccentricity $e = 16$ in. > required $P_n = 146,670$ or $\phi P_n = 113,845$ lb > $P_u = 110,000$ lb. Hence adopt the section. Use six No. 8 longitudinal bars.

Design the spiral reinforcement (Step 6)

Using Eq. 9.29,

$$\text{required } \rho_s = 0.45\left(\frac{A_g}{A_c} - 1\right)\frac{f_c'}{f_{sy}}$$

Using No. 3 spirals with a yield strength $f_y = 60,000$ psi:

clear concrete cover $d_c = 1.5$ in. (38.1 mm)

$$f_{sy} = 60,000 \text{ psi}$$

$$D_c = h - 2d_c = 20.0 - 2 \times 1.5 = 17.0 \text{ in. (431.8 mm)}$$

$$A_c = \frac{\pi(17.0)^2}{4} = 226.98 \text{ in.}^2$$

$$A_g = 314.0 \text{ in.}^2$$

$$\rho_s = 0.45\left(\frac{314.0}{226.98} - 1\right)\frac{4000}{60,000} = 0.0115$$

For No. 3 spirals, $a_s = 0.11$ in.² Using Eq. 9.35b,

$$\text{pitch } s = \frac{4a_s(D_c - d_b)}{D_c^2 \rho_s} = \frac{4 \times 0.11(17.0 - 0.375)}{(17.0)^2 \times 0.0115} = 2.20 \text{ in. (55.9 mm)}$$

Provide No. 3 spirals at $2\frac{1}{4}$ in. pitch (9.53 mm diameter spiral at 54.0 mm pitch).

9.12 LIMIT STATE AT BUCKLING FAILURE (SLENDER OR LONG COLUMNS)

Considerable literature exists on the behavior of columns subjected to stability considerations. If the column slenderness ratio exceeds the limits for short columns, the compression member will buckle prior to reaching its limit state of material failure. The strain in the compression face of the concrete at buckling load will be less than the 0.003 in./in. shown in Fig. 9.8. Such a column would be a slender member

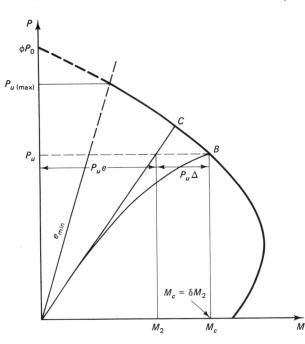

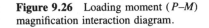

Figure 9.26 Loading moment (P–M) magnification interaction diagram.

subjected to combined axial load and bending, deforming laterally and developing additional moment due to the $P\Delta$ effect, where P is the axial load and Δ is the deflection of the column's buckled shape at the section being considered.

Consider a slender column subjected to axial load P_u at an eccentricity e. The buckling effect produces an additional moment of $P_u\Delta$. This moment reduces the load capacity from point C to point B in the interaction diagram, Fig. 9.26. The total moment ($P_u e + P_u\Delta$) is represented by point B in the diagram and the column can be designed for a larger or magnified moment M_c as a nonslender column.

In such a case, the load P_u is assumed to act at an eccentricity ($e + \Delta$) to produce moment M_c. The ratio M_c/M_2 is termed the magnification factor δ. If kl_u/r is the slenderness ratio, the lower limit for disregarding the slenderness ratio effect, namely, disregarding the need for stability analysis as specified by the ACI code is

$$\text{Braced frames:} \quad \frac{kl_u}{r} < 34 - 12\frac{M_1}{M_2} \tag{9.36a}$$

$$\text{Unbraced frames:} \quad \frac{kl_u}{r} < 22 \tag{9.36b}$$

where k is the column length factor, as shown in Fig. 9.27. M_1 and M_2 are the moments at the opposite ends of the compression member. M_2 is always larger than M_1 and the ratio M_1/M_2 is taken as positive for single curvature and negative for double curvature, as shown in Fig. 9.28a.

The effective length kl_u is used as the modified length of the column to account

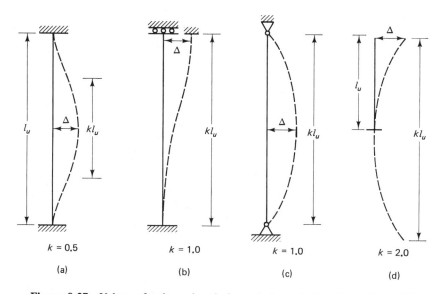

Figure 9.27 Values of column length factor k for typical end conditions: (a) fixed–fixed; (b) fixed–fixed with lateral motion; (c) pinned; (d) fixed–free.

for end restraints other than pinned. kl_u represents the length of an auxiliary pin-ended column which has an Euler buckling load equal to that of the column under consideration. Alternatively, it is the distance between the points of contraflexure of the member in its buckled form.

The value of the end restraint effective length factor k varies between 0.5 and 2.0.

Both column ends pinned, no lateral motion	$k = 1.0$
Both column ends fixed	$k = 0.5$
One end fixed, other end free	$k = 2.0$
Both column ends fixed, lateral motion exists	$k = 1.0$

Typical cases illustrating the buckled shape of the column for several end conditions and the corresponding length factors k are shown in Fig. 9.27.

For members in a structural frame, the end restraint lies between the hinged and fixed conditions. The actual k value can be determined from the Jackson and Moreland alignment charts in Fig. 9.28. In lieu of these charts, the following equations suggested in the ACI code commentary can also be used for calculating k.

1. *Braced compression members:*   An upper bound to the effective length factor may be taken as the smaller of the following two expressions:

$$k = 0.7 + 0.05(\psi_A + \psi_B) \le 1.0 \tag{9.37a}$$

$$k = 0.85 + 0.05\psi_{min} \le 1.0 \tag{9.37b}$$

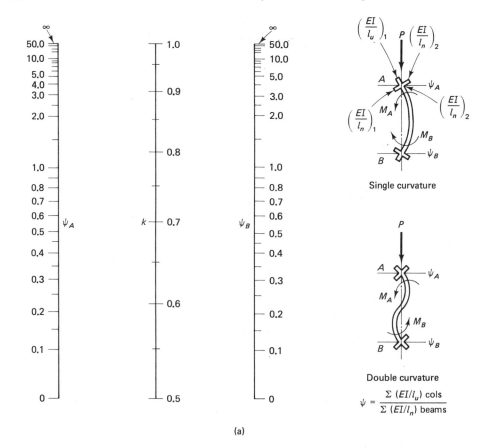

Figure 9.28 Effective length factor k for (a) braced and (b) unbraced frames.

where ψ_A and ψ_B are the values of ψ at the two ends of the column and $\psi_{\min}$ is the smaller of the two values. ψ is the ratio of the stiffness of all compression members to the stiffness of all flexural members in a plane at one end of the column. Or

$$\psi = \frac{\Sigma\, EI/l_u \text{ columns}}{\Sigma\, EI/l_n \text{ beams}} \tag{9.38}$$

where l_u is the unsupported length of the column; l_n is the clear beam span.
2. *Unbraced compression members restrained at both ends:*   The effective length may be taken as:
     For $\psi_m < 2$:

$$k = \frac{20 - \psi_m}{20}\sqrt{1 + \psi_m} \tag{9.39a}$$

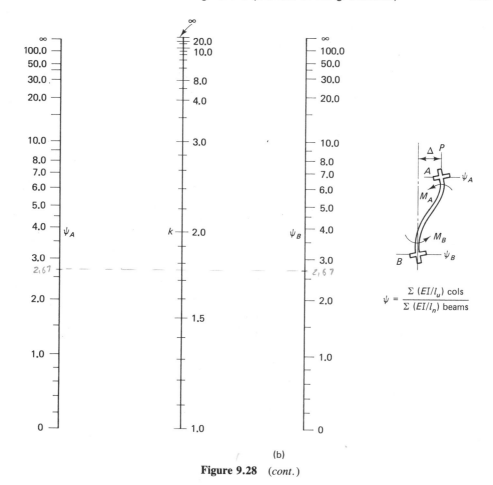

(b)

Figure 9.28   (*cont.*)

For $\psi_m \geq 2$:

$$k = 0.9\sqrt{1 + \psi_m}$$ (9.39b)

where ψ_m is the average of the ψ values at the two ends of the compression member.

3. *Unbraced compression members hinged at one end:*   The effective length factor may be taken as

$$k = 2.0 + 0.3\psi$$ (9.40)

where ψ is the value at the restrained end.

The radius of gyration $r = \sqrt{I_g/A_g}$ can be taken as $r = 0.3h$ for rectangular sections where h is the column dimension perpendicular to the axis of bending. For circular sections, the radius of gyration r is taken as $0.25h$.

If the value of kl_u/r is larger than that obtained from Eqs. 9.36a and 9.36b, two methods of stability analysis having general acceptance are recommended:

1. The *moment magnification method,* where the design of the member is based on a magnified moment:

$$M_c = \delta M_2 = \delta_b M_{2b} + \delta_s M_{2s} \qquad (9.41)$$

Lateral loads tend to increase or magnify the moments more than gravity loads. To account for the difference between the lateral and gravity loads, the magnifying δ factor is broken into two components, δ_b and δ_s, where δ_b is the magnification factor for predominant gravity load moment M_{2b}. The moment M_{2b} is defined as the larger factored end moment on a column due to loads that produce no appreciable side sway, that is, only gravity load moments.

δ_s is the magnification factor applied to the larger end moment M_{2s} due to loads that result in appreciable side sway, such as wind load moments. In the case of frames braced against side sway, braced with shear walls, the entire moment acting on the column is considered M_{2b} and δ_s is therefore assumed to be zero. Normally, if the lateral deflection of the building is less than span $l_n/1500$, the frame is considered as a braced frame.

2. A *second-order analysis,* taking into consideration the effect of deflections. It must be used in cases where $kl_u/r > 100$.

It is to be noted that all columns have to be designed for at least a minimum eccentricity of $(0.6 + 0.03h)$ in. That is, M_2 must be based on a minimum eccentricity of $(0.6 + 0.03h)$.

9.13 MOMENT MAGNIFICATION METHOD

The moment magnifiers δ_b and δ_s are dependent on the slenderness of the member, the stiffness of the entire frame, the restraint or applied moment at its ends, and the design cross section such that

$$\delta_b = \frac{C_m}{1 - P_u/\phi P_c} \geq 1 \qquad (9.42a)$$

$$\delta_s = \frac{1}{1 - \Sigma P_u/\phi \Sigma P_c} \geq 1 \qquad (9.42b)$$

where P_c = Euler buckling load = $\pi^2 EI/(kl_u)^2$
    kl_u = effective length (between points of inflection)
$\Sigma P_u, \Sigma P_c$ = summations for all columns in a story
    l_u = unsupported length of column
    C_m = a factor relating the actual moment diagram to an equivalent uniform moment diagram; for braced members subject to end loads only,

$$C_m = 0.6 + 0.4\frac{M_1}{M_2} \geq 0.4 \qquad \text{where } M_1 \leq M_2 \qquad (9.43)$$

and $M_1/M_2 > 0$ if no point of inflection exists between the column ends, Fig. 9.28a (single curvature). For other conditions, $C_m = 1.0$.

When the computed end eccentricities are less than $(0.6 + 0.03h)$ in., one may use the computed end moments to evaluate M_1/M_2 in Eq. 9.43. If computations show that there is essentially no moment at both ends of a compression member, the ratio M_1/M_2 shall be taken equal to 1.

An estimate of EI must include the effects of cracking and creep under long-term loading. For all members

$$EI = \frac{(E_c I_g/5) + E_s I_s}{1 + \beta_d} \qquad (9.44a)$$

For lightly reinforced members ($\rho_g \leq 3\%$), this may be simplified to

$$EI = \frac{E_c I_g/2.5}{1 + \beta_d} \qquad (9.44b)$$

where

$$\beta_d = \frac{\text{factored dead load moment}}{\text{factored total moment}} \leq 1$$

9.14 SECOND-ORDER ANALYSIS

This rigorous mathematical approach is needed if the slenderness ratio kl_u/r exceeds 100. The effect of deflection has to be taken into account and an appropriate reduced tangent modulus for concrete has to be used. The designer, with the aid of computers, can solve with limited efforts the set of simultaneous equations needed to determine the size of the reinforced concrete slender column. Charts can also be developed for the various eccentricity ratios and reinforcement and concrete strength combinations. It should be noted that the majority of columns in concrete building frames do not necessitate such an analysis since the slenderness ratio kl_u/r is, in most cases, below 100.

9.15 OPERATIONAL PROCEDURE AND FLOWCHART FOR THE DESIGN OF SLENDER (LONG) COLUMNS

1. Determine whether the frame has an appreciable side sway. If it does, use the magnification factors δ_b and δ_s. If the side sway is negligible, assume that $\delta_s = 0$. Assume a cross section. Calculate the eccentricity using the greater of the end moments and check whether it is more than the minimum allowable eccentricity, that is,

$$\frac{M_2}{P_u} \geq (0.6 + 0.03h) \text{ in.}$$

If the given eccentricity is less than the specified minimum, use the minimum value.

2. Calculate ψ_A and ψ_B using Eq. 9.38. Obtain k using Fig. 9.28 or Eqs. 9.39a and 9.39b. Calculate kl_u/r and determine whether the column is a short or long column. If the column is slender and kl_u/r is less than 100, calculate the magnified moment M_c. Using the M_c value obtained, calculate the equivalent eccentricity to be used if the column is to be designed as a short column. If kl_u/r is greater than 100, perform a second-order analysis.

3. Design the equivalent nonslender column. The flowchart, Fig. 9.29, presents the sequence of calculations. The necessary equations are provided in Section 9.11 and in the flowchart.

9.15.1 Example 9.15: Design of a Slender (Long) Column

A rectangular tied column is part of a 5 × 3 bays frame building subjected to uniaxial bending. Its clear height is $l_u = 18$ ft (5.55 m) and it is not braced against side sway. The factored external load $P_u = 726,000$ lb (3229 kN). The factored end moments are $M_1 = 550,000$ in.-lb (203.88 kN-m), $M_2 = 1,525,000$ in.-lb (172.25 kN-m). Design the column section and the reinforcement necessary for the following two conditions:

1. Consider gravity loads only assuming lateral side sway due to wind as negligible.
2. Consider side sway wind effects to cause a factored $P_u = 90,000$ lb (400.3 kN) and a factored $M_u = 1,220,000$ in.-lb. (137.9 kN-m)

Loads per floor of all columns at that level are:

$$\Sigma P_u = 15.5 \times 10^6 \text{ lb (68,944 kN)}$$
$$\Sigma P_c = 32.0 \times 10^6 \text{ lb (142,336 kN)}$$

Given:

$$\beta_d = 0.5, \ \psi_A = 2.0, \ \psi_B = 3.0$$
$$f'_c = 5000 \text{ psi (34.48 MPa)}$$
$$f_y = 60,000 \text{ psi (413.7 MPa)}$$
$$d' = 2.5 \text{ in. (63.5 mm)}$$

Solution

1. *Gravity Loading Only*

Check for side sway and minimum eccentricity (Step 1)

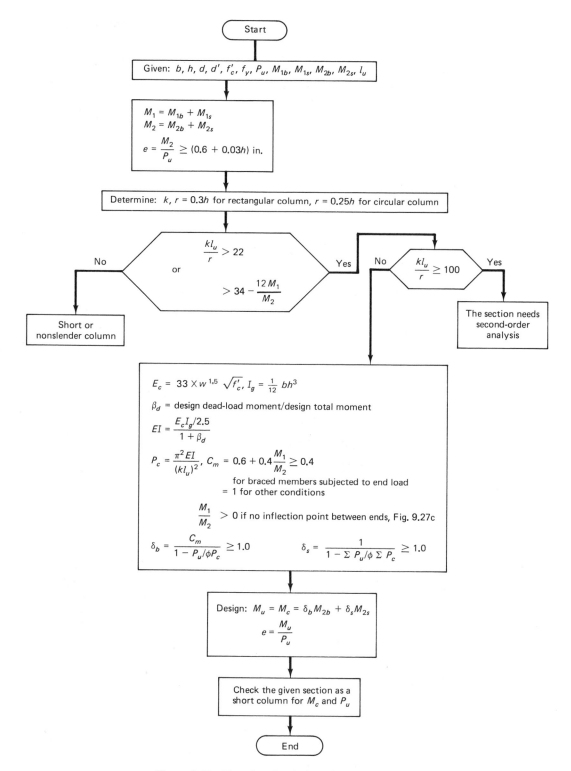

Figure 9.29 Flowchart for design of slender columns.

Since the frame has no appreciable side sway, the entire moment M_2 is taken as M_{2b} and magnification factor δ_s is taken as equal to zero. By trial and adjustment, a column section is assumed and analyzed. Try a section 21 in. × 21 in. (533.4 mm × 533.4 mm) as shown in Fig. 9.30a.

$$\text{actual eccentricity} = \frac{M_{2b}}{P_u} = \frac{1,525,000}{726,000} = 2.1 \text{ in. (52 mm)}$$

$$\text{minimum allowable eccentricity} = 0.6 + 0.03 \times 21 = 1.23 \text{ in.} < 2.1 \text{ in.}$$

Use $M_{2b} = 1,525,000$ in.-lb (172.26 kN-m).

Calculate the eccentricity to be used for equivalent short column (Step 2)

From the chart in Fig. 9.28b, $k = 1.7$.

$$\text{slenderness ratio } \frac{kl_u}{r} = \frac{1.7 \times 18 \times 12}{0.3 \times 21} = 58.29$$

As 58.29 is > 22 and < 100, use the moment magnification method.

$$E_c = 33w^{1.5} \sqrt{f_c'} = 33 \times 150^{1.5} \sqrt{5000}$$

$$= 4.29 \times 10^6 \text{ psi } (29.58 \times 10^6 \text{ kPa})$$

$$I_g = \frac{21(21)^3}{12} = 16,206.8 \text{ in.}^4$$

$$EI = \frac{E_c I_g/2.5}{1 + \beta_d} = \frac{4.29 \times 10^6 \times 16,206.8}{2.5} \times \frac{1}{1 + 0.5}$$

$$= 18.54 \times 10^9 \text{ lb-in.}^2$$

$$(kl_u)^2 = (1.7 \times 18 \times 12)^2 = 134.8 \times 10^3 \text{ in.}^2$$

Hence

$$P_c = \text{Euler buckling load} = \frac{\pi^2 EI}{(kl_u)^2}$$

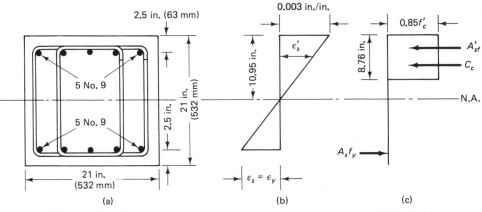

Figure 9.30 Column geometry—strain and stress diagrams (balanced failure): (a) cross section; (b) strains; (c) stresses.

$$= \frac{\pi^2 \times 18.54 \times 10^9}{134.8 \times 10^3} = 1.356 \times 10^6 \text{ lb} = 1356 \text{ kips (6032 kN)}$$

$C_m = 1.0$ for nonbraced column

$$\text{moment magnifier } \delta_b = \frac{C_m}{1 - P_u/\phi P_c} = \frac{1.0}{1 - \dfrac{726}{0.7 \times 1356}} = 4.253$$

$$\text{design moment } M_c = \delta_b M_{2b} = 4.253 \times 1{,}525{,}000$$

$$= 6{,}485{,}825 \text{ in.-lb (682.34 kN-m)}$$

Assume that the reduction factor $\phi = 0.70$.

$$\text{required } P_n = \frac{P_u}{\phi} = \frac{726{,}000}{0.7} = 1{,}037{,}143 \text{ lb (4612.8 kN)}$$

$$\text{required } M_n = \frac{6{,}485{,}825}{0.7} = 9{,}265{,}464 \text{ in.-lb (1957.22 kN-m)}$$

Hence design a nonslender column section for an axial load strength $P_n = 1{,}037{,}143$ lb and a moment strength $M_n = 9{,}265{,}464$ in.-lb.

$$e = \frac{9{,}265{,}464}{1{,}037{,}143} = 8.93 \text{ in. (227 mm)}$$

Design of an equivalent nonslender column (Step 3)

Analyze the assumed 21 in. $\times$ 21 in. square section. Assume that $\rho = p' \simeq 1.25\%$.

$$A_s = A_s' = 0.0125(21 \times 18.5) = 4.86 \text{ in.}^2$$

Provide five No. 9 bars (five 28 mm diameter) on each face; $A_s = A_s' = 5.0$ in.^{2} (3226 mm^2).

$$c_b = d \times \frac{87{,}000}{87{,}000 + f_y} = 18.5 \times \frac{87{,}000}{87{,}000 + 60{,}000} = 10.95 \text{ in. (278 mm)}$$

$$a_b = \beta_1 c_b = (0.85 - 0.05) \times 10.95 = 8.76 \text{ in. (222.5 mm)}$$

$$f_s' = 0.003 \times 29 \times 10^6 \left(\frac{10.95 - 2.5}{10.94} \right) = 67{,}137 \text{ psi} > f_y$$

therefore, $f_s' = f_y$

$$P_{nb} = 0.85 f_c' ba = 0.85 \times 5000 \times 21 \times 8.76$$

$$= 781{,}830 \text{ lb (3473.8 kN)}$$

$$M_{nb} = 9{,}584{,}800 \text{ in.-lb (2024.7 kN-m)}$$

$$e_b = 12.25 \text{ in.}$$

$$> e = 8.93 \text{ in.}$$

The failure will be in compression.

$$P_n = \cfrac{A'_s f_y}{\cfrac{e}{(d - d')} + 0.5} + \cfrac{bhf'_c}{\cfrac{3he}{d^2} + 1.18}$$

$$P_n = \cfrac{5.0 \times 60,000}{\cfrac{8.93}{18.5 - 2.5} + 0.5} + \cfrac{21 \times 21 \times 5000}{\cfrac{3 \times 21 \times 8.93}{(18.5)^2} + 1.18} = 1,064,383 \text{ lb } (4733 \text{ kN})$$

The actual nominal axial load strength P_n is larger than the required $P_n = 1,037,143$ lb, so the design is O.K. (Exact solution by Ch. 13 gives $P_n = 1,022,912$ lb)

Design of ties (Step 4)

Try No. 3 ties (9.52 mm diameter). The spacing must be at least $h = 21$ in. (533.4 mm).

16 diameter No. 9 bar = 18 in. (457.2 mm)

48 diameter No. 3 tie = 18 in. (457.2 mm)

Hence use No. 3 (9.52 mm diameter) closed ties spaced at 18 in. (455 mm) center to center.

Gravity and Wind Loading (side sway)

From part 1

$$P_c = 1,356,000 \text{ lb}$$

$$U = 0.75(1.4D + 1.7L + 1.7W)$$

also

$$U = 0.9D + 1.3W \text{ (did not control)}$$

Therefore,

$$P_u = 0.75(726,000 + 90,000) = 612,000 \text{ lb}$$

$$M_{2b} = 0.75 \times 1,525,000 = 1,143,750 \text{ in.-lb}$$

$$M_{2s} = 0.75 \times 1,220,000 = 915,000 \text{ in.-lb}$$

$$\delta_b = \cfrac{1.0}{1 - \cfrac{P_u}{\phi P_c}} = \cfrac{1.0}{1 - \cfrac{612,000}{0.7 \times 1,356,000}} = 2.81$$

$$\delta_s = \cfrac{1.0}{1 - \cfrac{\Sigma P_u}{\phi \Sigma P_c}} = \cfrac{1.0}{1 - \cfrac{15.5 \times 10^6}{0.7 \times 32.0 \times 10^6}} = 3.25$$

$$M_c = 2.81(1,143,750) + 3.25 \times 915,000 = 6,187,688 \text{ in.-lb}$$

$$\text{Required } P_n = \cfrac{612,000}{0.7} = 874,286 \text{ lb}$$

$$\text{Required } M_n = \frac{6,187,688}{0.7} = 8,839,554 \text{ in.-lb}$$

$$\text{Eccentricity } e = \frac{8,839,554}{874,286} = 10.11 \text{ in. } < e_b = 12.25 \text{ in., hence failure will}$$
be in compression.

Conditions for case (2) do not control since failure is still in compression and the required P_n is less than that for case (1).

Adopt the same section 21 in. $\times$ 21 in. with 5 No. 9 reinforcing bars on each of the two faces parallel to the neutral axis.

9.16 COMPRESSION MEMBERS IN BIAXIAL BENDING

9.16.1 Exact Method of Analysis

Columns in corners of buildings are compression members subjected to biaxial bending about both the x and the y axes as shown in Fig.9.31. Also, biaxial bending occurs due to imbalance of loads in adjacent spans and almost always in bridge piers. Such columns are subjected to moments M_{xx} about the x axis, creating a load eccentricity e_y, and a moment M_{yy} about the y axis creating a load eccentricity e_x. Thus the neutral axis is inclined at an angle θ to the horizontal.

The angle θ depends on the interaction of the bending moments about both axes and the magnitude of the load P_u. The compressive area in the column section can have one of the alternative shapes shown in Fig. 9.31c. Since such a column has to be designed from first principles, the trial-and-adjustment procedure has to be followed where compatibility of strain has to be maintained at all levels of the reinforcing bars. The process is similar to the one briefly outlined in Section 9.5.8 for columns with reinforcing bars on all faces. Additional computational effort is needed because of the position of the inclined neutral-axis plane and the four different possible forms of the concrete compression area.

Figure 9.32 shows the strain-distribution and forces on a biaxially loaded rectangular column cross section. G_c is the center of gravity of the concrete compression area, having coordinates x_c and y_c from the neutral axis in the directions x and y, respectively. G_{sc} is the resultant position of steel forces in the compression area having coordinates x_{sc} and y_{sc} from the neutral axis in the directions x and y, respectively. G_{st} is the resultant position of steel forces in the tension area having coordinates x_{st} and y_{st} from the neutral axis in directions x and y, respectively. From equilibrium of internal and external forces,

$$P_n = 0.85 f_c' A_c + F_{sc} - F_{st} \tag{9.45}$$

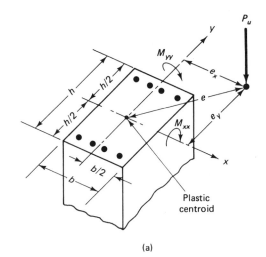

(a)

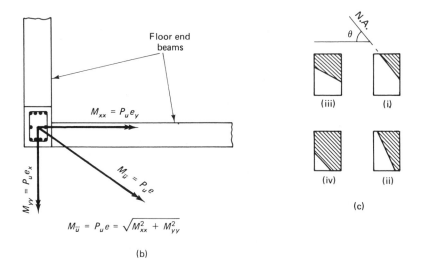

(b)

(c)

Figure 9.31 Corner column subjected to axial load: (a) biaxially stressed column cross section; (b) vector moments M_{xx} and M_{yy} in column plan.

where A_c = area of the compression zone covered by the rectangular stress block
F_{sc} = resultant steel compressive forces ($\Sigma A_s' f_{sc}$)
F_{st} = resultant steel tensile force ($\Sigma A_s f_{st}$)

From equilibrium of internal and external moments,

$$P_n e_x = 0.85 f_c' A_c x_c + F_{sc} x_{sc} + F_{st} x_{st} \qquad (9.46a)$$

$$P_n e_y = 0.85 f_c' A_c y_c + F_{sc} y_{sc} + F_{st} y_{st} \qquad (9.46b)$$

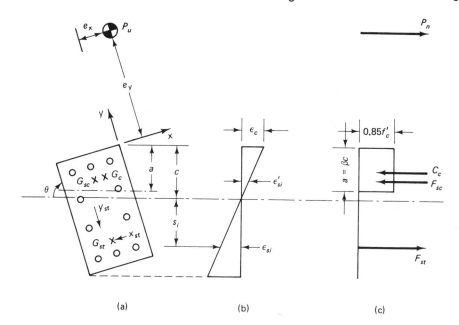

Figure 9.32 Strain-compatibility and forces in biaxially loaded rectangular columns: (a) cross section; (b) strain; (c) forces.

The position of the neutral axis has to be assumed in each trial and the stress calculated in *each* bar using

$$f_{si} = E_s \epsilon_{si} = E_c \epsilon_c \frac{s_i}{c} < f_y \qquad (9.47)$$

9.16.2 Load Contour Method

One method to give a rapid solution is to design the column for the vector sum of M_{xx} and M_{yy} and use a circular reinforcing cage in a square section for the corner column. However, such a procedure cannot be economically justified in most cases. Another design approach well proven by experimental verification is to transform the biaxial moments into an equivalent uniaxial moment and an equivalent uniaxial eccentricity. The section can then be designed for uniaxial bending as previously discussed in this chapter to resist the actual factored biaxial bending moments.

Such a method considers a failure surface instead of failure planes and is generally termed the *Bresler–Parme contour method*. This method involves cutting the three-dimensional failure surfaces in Fig. 9.33 at a constant value P_n to give an interaction plane relating M_{nx} and M_{ny}. In other words, the contour surface S can be

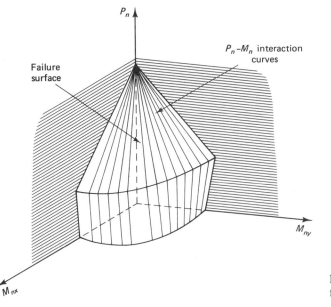

Figure 9.33 Failure interaction surface for biaxial column bending.

viewed as a curvilinear surface which includes a family of curves, termed the *load contours*.

The general nondimensional equation for the load contour at a constant load P_n may be expressed as follows:

$$\left(\frac{M_{nx}}{M_{ox}}\right)^{\alpha_1} + \left(\frac{M_{ny}}{M_{oy}}\right)^{\alpha_2} = 1.0 \tag{9.48}$$

where $M_{nx} = P_n e_y$ and $M_{ny} = P_n e_x$

$M_{ox} = M_{nx}$ at such an axial load P_n when M_{ny} or $e_x = 0$

$M_{oy} = M_{ny}$ at such an axial load P_n when M_{nx} or $e_y = 0$

The moments M_{ox} and M_{oy} are the *required* equivalent resisting moment strengths about the x and y axes, respectively.

α_1, α_2 = exponents depending on the cross-section geometry, steel percentage, and its location and material stresses f_c' and f_y

Equation 9.48 can be simplified using a common exponent and introducing a factor β for one particular axial load value P_n such that the M_{nx}/M_{ny} ratio would have the same value as the M_{ox}/M_{oy} as detailed by Parme and associates. Such simplification leads to

$$\left(\frac{M_{nx}}{M_{ox}}\right)^{\alpha} + \left(\frac{M_{ny}}{M_{oy}}\right)^{\alpha} = 1.0 \tag{9.49}$$

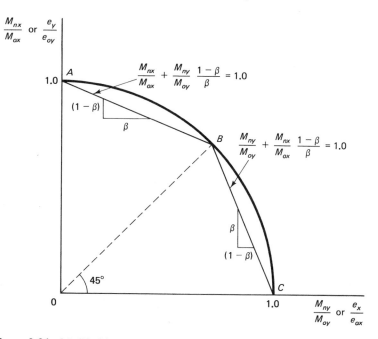

Figure 9.34 Modified interaction contour plot of constant P_n for biaxially loaded column.

where α would have a value of ($\log 0.5/\log \beta$). Figure 9.34 gives a contour plot ABC from Eq. 9.49.

For design purposes, the contour is approximated by two straight lines BA and BC and Eq. 9.49 can be simplified to two conditions:

1. For AB when $M_{ny}/M_{oy} < M_{nx}/M_{ox}$:

$$\frac{M_{nx}}{M_{ox}} + \frac{M_{ny}}{M_{oy}}\left[\frac{1-\beta}{\beta}\right] = 1.0 \qquad (9.50a)$$

2. For BC when $M_{ny}/M_{oy} > M_{nx}/M_{ox}$:

$$\frac{M_{ny}}{M_{oy}} + \frac{M_{nx}}{M_{ox}}\left[\frac{1-\beta}{\beta}\right] = 1.0 \qquad (9.50b)$$

In both Eqs. 9.50(a) and (b), the *actual* controlling equivalent uniaxial moment strength M_{oxn} or M_{oyn} should be at least equivalent to the *required* controlling moment strength M_{ox} or M_{oy} of the chosen column section.

For rectangular sections where the reinforcement is evenly distributed along all the column faces, the ratio $\dfrac{M_{oy}}{M_{ox}}$ can be approximately taken as equal to b/h. Hence

Eqs. 9.50a and 9.50b can be modified as follows:

1. For $\dfrac{M_{ny}}{M_{nx}} > b/h$:

$$M_{ny} + M_{nx}\frac{b}{h}\frac{1-\beta}{\beta} \simeq M_{oy} \qquad (9.51a)$$

2. For $\dfrac{M_{ny}}{M_{nx}} \le b/h$:

$$M_{nx} + M_{ny}\frac{h}{b}\frac{1-\beta}{\beta} \simeq M_{ox} \qquad (9.51b)$$

The controlling required moment strength M_{ox} or M_{oy} for designing the section is the larger of the two values as determined from Eqs. 9.51a or b.

Plots of Fig. 9.35 are used in the selection of β in the analysis and design of such columns. In effect, the modified load-contour method can be summarized in Eq. 9.49 as a method for finding an equivalent required moment strength M_{ox} and M_{oy} for designing the columns as if they were uniaxially loaded.

9.16.3 Step-by-Step Operational Procedure for the Design of Biaxially Loaded Columns

The following steps can be used as a guideline for the design of columns subjected to bending in both the x and y directions. The procedure assumes an equal area of reinforcement on all four faces.

1. Calculate the uniaxial bending moments assuming equal number of bars on each column face. Assume a value of an interaction contour β factor between 0.50 and 0.70. Assume a ratio of h/b. This ratio can be approximated to M_{nx}/M_{ny}. Using Eqs. 9.51a and 9.51b, determine the equivalent required uniaxial moment M_{ox} or M_{oy}. If M_{nx} is larger than M_{ny}, use M_{ox} for the design and vice versa.
2. Assume a cross section for the column and a reinforcement ratio $\rho = \rho' \simeq 0.01$ to 0.02 on each of the two faces parallel to the axis of bending of the larger equivalent moment. Make a preliminary selection of the steel bars.

    Verify the capacity P_n of the assumed column cross section. In the completed design, the same amount of longitudinal steel should be used on all four faces.
3. Calculate the *actual* nominal moment strength M_{oxn} for equivalent uniaxial bending about the x axis when $M_{oy} = 0$. Its value has to be at least equivalent to the *required* moment strength M_{ox}.
4. Calculate the actual nominal moment strength M_{oyn} for the equivalent uniaxial bending moment about the y axis when $M_{ox} = 0$.
5. Find M_{ny} by entering M_{nx}/M_{oxn} and the trial β value into the β factor contour plots of Fig. 9.35.

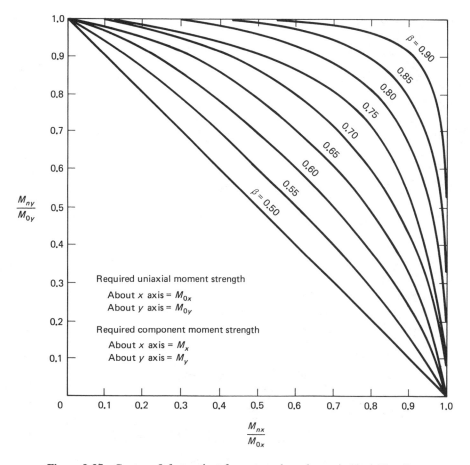

Figure 9.35   Contour β-factor chart for rectangular columns in biaxial bending.

6. Make a second trial and adjustment, increasing the assumed β value if the M_{ny} value obtained from entering the chart is less than the required M_{ny}. Repeat this step until the two values of M_{ny} converge either through changing β or changing the section.
7. Design the lateral ties and detail the section.

9.16.4 Example 9.16: Design of a Biaxially Loaded Column

A corner column is subjected to a factored compressive axial load $P_u = 210{,}000$ lb (945 kN), a factored bending moment $M_{ux} = 1{,}680{,}000$ in.-lb (189.8 kN-m) about the x axis, and a factored bending moment $M_{uy} = 980{,}000$ in.-lb (110.7 kN-m) about the y axis, as shown in Fig. 9.36. Given:

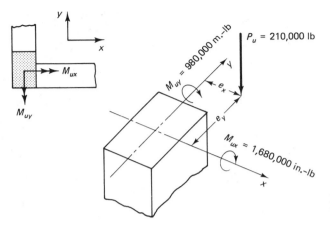

Figure 9.36 Compression and biaxial bending on corner column in Ex. 9.16.

f'_c = 4000 psi (27.6 MPa), normalweight concrete

f_y = 60,000 psi (414 MPa)

Design a rectangular tied column section to resist the biaxial bending moments resulting from the given eccentric compressive load.

Solution

Calculate the equivalent uniaxial bending moments
assuming equal numbers of bars on all faces (Step 1)

Assume that ϕ = 0.70 for tied columns.

$$\text{required nominal } P_n = \frac{210,000}{0.70} = 300,000 \text{ lb (1350 kN)}$$

$$\text{required nominal } M_{nx} = \frac{1,680,000}{0.70} = 2,400,000 \text{ in.-lb (271.2 kN-m)}$$

$$\text{required nominal } M_{ny} = \frac{980,000}{0.70} = 1,400,000 \text{ in.-lb (158.2 kN-m)}$$

Analyze for equivalent moment and equivalent eccentricity about the x axis since the larger of the two biaxial moments is M_{nx} = 2,400,000 in.-lb about the x axis.

$$\frac{M_{nx}}{M_{ny}} = \frac{2,400,000}{1,400,000} = 1.71$$

Since the column dimensions are proportional to the applied moments, assume that $h/b \simeq 1.71$ or b = 12 in. and h = 20 in. to give h/b = 1.67. Assume that the interaction contour factor β = 0.61.

$$\text{equivalent } M_{ox} = M_{nx} + M_{ny}\frac{h}{b}\frac{1-\beta}{\beta}$$

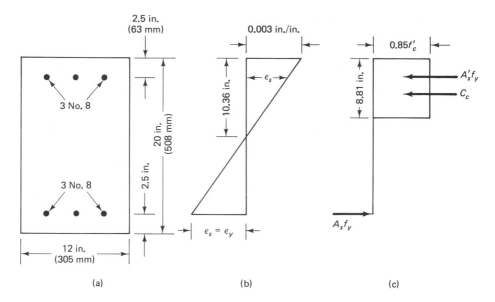

Figure 9.37 Equivalent column geometry—strain and stress diagrams (balanced failure): (a) cross section; (b) strains; (c) stresses.

or

$$M_{ox} = 2,400,000 + 1,400,000(1.67)\left(\frac{1 - 0.61}{0.61}\right) = 3,894,787 \text{ in.-lb (440.1 kN-m)}$$

($M_{ox} = 3,891,803$ in.-lb from Sec. 13.12.5.)

Verify capacity P_n of the assumed column section (Step 2)

From step 1, $b = 12$ in. (304.8 mm) and $h = 20$ in. (508.0 mm). Assume that the steel ratio $\rho = \rho' \simeq 0.012$ and $d' = 2.5$ in. (63.5 mm). $d = 20.0 - 2.5 = 17.5$ in. (445.5 mm).

$$A_s = A'_s = 0.012 \times 12(20.0 - 2.5) = 2.52 \text{ in.}^2$$

Try $A_s = A'_s = 2.37$ in.^{2} (1528.7 mm^2) on each of the two 12-in. faces parallel to the *x* axis of bending in Fig. 9.37.

Analyze the balanced condition:

$$c_b = \frac{87,000}{87,000 + f_y}\, d = \left(\frac{87,000}{87,000 + 60,000}\right)17.5 = 10.36 \text{ in.}$$

$$a_b = 0.85 \times 10.36 = 8.81 \text{ in. (223.7 mm)}$$

$$f'_s = \epsilon_c E_s \frac{c - d'}{c} = 0.003 \times 29 \times 10^6\left(\frac{10.36 - 2.5}{10.36}\right)$$

$$= 66,006 \text{ psi} > f_y = 60,000 \text{ psi}$$

Hence $f'_s = 60,000$ psi.

$$P_{nb} = 0.85 f'_c b a_b + A'_s f_y - A_s f_y$$

$$= 0.85 \times 4000 \times 12.0 \times 8.81 + 2.37 \times 60,000 - 2.37 \times 60,000$$

$$= 359,448 \text{ lb} > P_n = 300,000 \text{ lb}$$

Therefore, tension controls with failure by initial yielding of the tension steel.

Evaluate the axial load capacity P_n:

$$\rho = \frac{A_s}{bd} = \frac{2.37}{12 \times 17.5} = 0.0113$$

$$e = \frac{M_{ox}}{P_n} = \frac{3,894,787}{300,000} = 12.98 \text{ in. (329.8 mm)}$$

$$e' = e + \frac{d - d'}{2} = 12.98 + \frac{17.5 - 2.5}{2} = 20.48 \text{ in.}$$

$$1 - \frac{e'}{d} \quad \text{or} \quad \frac{h - 2e}{2d} = 1 - \frac{20.48}{17.5} = -0.17$$

$$1 - \frac{d'}{d} = 1 - \frac{2.5}{17.5} = 0.857$$

$$m = \frac{f_y}{0.85 f'_c} = \frac{60,000}{0.85 \times 4000} = 17.65$$

The following equation applies for initial tension failure in columns if the compression steel A'_s is assumed to have yielded.

$$P_n = 0.85 f'_c b d \left[\left(1 - \frac{e'}{d} \right) + \sqrt{ \left(1 - \frac{e'}{d} \right)^2 + 2m\rho \left(1 - \frac{d'}{d} \right) } \right]$$

neglecting the area of concrete replaced by the steel; or

$$P_n = 0.85 \times 4000 \times 12 \times 17.5[-0.17$$

$$+ \sqrt{(-0.17)^2 + 2 \times 17.65 \times 0.0113 \times 0.857}]$$

$$= 714,000(0.4389) = 313,368 \text{ lb} > 300,000$$

Accept this for the first trial.

Check assumed value of $\phi = 0.70$:

$$\phi = 0.90 - \frac{2.0 P_u}{f'_c A_g} = 0.90 - \frac{2.0 \times 0.7 \times 313,368}{4000 \times 12 \times 20} = 0.44 < 0.70$$

Assume that $\phi = 0.70$—O.K.

Strain-compatibility check: $P_n = 0.85 f'_c b a$ for $A'_s f_y = A_s f_y$, or

$$a = \frac{P_n}{0.85 f'_c b} = \frac{313,368}{0.85 \times 4,000 \times 12} = 7.68 \text{ in.}$$

$$c = \frac{a}{\beta_1} = \frac{7.35}{0.85} = 9.04 \text{ in.}$$

$$f'_s = 87,000\left(\frac{9.04 - 2.5}{9.04}\right) = 62,940 \text{ psi} > f_y = 60,000$$

Hence A'_s yielded satisfying compatibility of strain requirement for the expression used to evaluate P_n.

Calculate the actual nominal resisting moment M_{oxn} for equivalent uniaxial bending about the x axis when $M_{oy} = 0$ (Step 3)

Required $P_n = 300,000$ lb. Assuming that the compression steel has yielded (to be verified later), $f'_s = 60,000$ and $A_s f'_s - A_s f_y = 0$. Hence $P_n = 0.85 f'_c ba$, or

$$a = \frac{P_n}{0.85 f'_c b} = \frac{300,000}{0.85 \times 4000 \times 12} = 7.35 \text{ in.}$$

$$c = \frac{a}{\beta_1} = \frac{7.35}{0.85} = 8.65 \text{ in.}$$

$$f'_s = 87,000\left(\frac{8.65 - 2.5}{8.65}\right) = 61,855 \text{ psi} > 60,000 \qquad \text{O.K.}$$

A_s at the tension side had to yield since $a = 7.35$ in. $< a_b = 8.81$ in. with the neutral axis shallower than the balanced condition.

$$M_{oxn} = P_n e = 0.85 f'_c ba\left(\bar{y} - \frac{a}{2}\right) + A'_s f_y(\bar{y} - d') + A_s f_y(d - \bar{y})$$

or

$$M_{oxn} = 0.85 \times 4000 \times 12 \times 7.35\left(\frac{20}{2} - \frac{7.35}{2}\right) + 2.37 \times 60,000\left(\frac{20}{2} - 2.5\right)$$

$$+ 2.37 \times 60,000\left(17.5 - \frac{20}{2}\right)$$

$$= 4,029,741 \text{ in.-lb (455.4 kN-m)} > M_{ox}(3,894,787 \text{ in.-lb}) \qquad \text{O.K.}$$

If this calculation showed that $M_{oxn} < M_{ox}$ obtained in step 1, revise the assumed cross section by increasing the steel area or enlarging the section, or both.

Calculate the actual nominal resisting moment M_{oyn} for the equivalent uniaxial bending moment about the y axis when $M_{ox} = 0$ (Step 4)

In this condition, $b = 20$ in., $h = 12$ in., $d = 9.5$ in., and $A_s = A'_s = 2.37$ in.^{2}. By trial and adjustment, choose compressive block depth a such that the calculated P_n approximates the required P_n.

At the third trial, $a = 4.8$ in. and $c = 4.8/0.85 = 5.65$ in.

$$f_s' = 87,000\left(\frac{5.65 - 2.5}{5.65}\right) = 48,504 \text{ psi}$$

$$f_s = 87,000\left(\frac{d - c}{c}\right) = 87,000\left(\frac{9.5 - 5.65}{5.65}\right) = 59,283 \text{ psi}$$

$$P_n = 0.85 \times 4000 \times 20 \times 4.8 + 2.37 \times 48,504 - 2.37 \times 59,283$$

$$= 300,854 \simeq \text{ required } P_n = 300,000 \quad \text{O.K.}$$

Hence use $a = 4.8$ for calculating M_{oyn}.

$$M_{oyn} = 0.85 \times 4000 \times 20 \times 4.8\left(\frac{12}{2} - \frac{4.8}{2}\right) + 2.37 \times 48,504\left(\frac{12}{2} - 2.5\right)$$

$$+ 2.37 \times 59,283\left(9.5 - \frac{12}{2}\right) = 2,069,133 \text{ in.-lb}$$

Find M_{ny} by entering M_{nx}/M_{oxn} and trial β value into the β-factor contour plots in Fig. 9.35 (Step 5)

First trial $\beta = 0.61$. From step 3, $M_{oxn} = 4,029,741$ in.-lb

$$\frac{M_{nx}}{M_{oxn}} = \frac{2,400,000}{4,029,741} = 0.596$$

Enter into Fig. 9.35 the values of $\beta = 0.61$ and $M_{nx}/M_{oxn} = 0.596$, to get

$$\frac{M_{ny}}{M_{oyn}} = 0.62$$

But M_{oyn} from step 4 = 2,069,133 in.-lb, or

$$\frac{M_{ny}}{2,069,133} = 0.62$$

Hence

$$M_{ny} = 0.62 \times 2,069,133 = 1,282,862 \text{ in.-lb}$$

$$< \text{ required } M_{ny} = 1,400,000 \text{ in.-lb}$$

Revise the solution assuming a higher β value. If adjusting β does not give the actual M_{ny} at least = required M_{ny}, increase the reinforcement area or enlarge the section.

Second trial and adjustment (Step 6)

Assume the same section but assume that $\beta = 0.64$.

$$M_{ox} = 2,400,000 + 1,400,000 \times 1.67 \times \frac{1 - 0.64}{0.64} = 3,715,125 \text{ in.-lb}$$

$$\text{equivalent } e = \frac{3,715,125}{300,000} = 12.38 \text{ in.}$$

$$e' = 12.38 + 7.5 = 19.88$$

$$1 - \frac{e'}{d} = 1 - \frac{19.88}{17.5} = -0.136$$

$$1 - \frac{d'}{d} = 0.857 \text{ from step 2}$$

$$P_n = 714,000[-0.136 + \sqrt{(-0.136)^2 + 2 \times 17.65 \times 0.0113 \times 0.857}]$$

$$= 714,000(0.4642) = 331,500 \text{ lb} > 300,000 \qquad \text{O.K.}$$

Strain-compatibility check:

$$a = \frac{P_n}{0.85 f_c' b} = \frac{331,472}{0.85 \times 4000 \times 12} = 8.12 \text{ in.} > \text{first trial } (a = 7.68 \text{ in.}).$$

Hence the neutral axis would be lower, giving a larger value of strain ϵ_s' or $f_s' = f_y$ = yield strength assumed in calculating P_n. From step 3, $M_{oxn} = 4,029,741$ in.-lb since this value does not change as long as the section and its reinforcement remain the same. $M_{nx}/M_{oxn} = 0.596$ from before. $\beta = 0.64$. From contour plots in Fig. 9.35, $M_{ny}/M_{oyn} = 0.68$. From step 4, $M_{oyn} = 2,069,133$.

$$M_{ny} = 0.68 \times 2,069,133 = 1,407,010 \text{ in.-lb} > \text{required } M_{ny} = 1,400,000$$

Adopt the design.

It should be stated that use of handheld computers, minicomputers, or charts in steps 2 and 6 reduces the calculation effort for biaxially loaded columns almost to that for the design of uniaxially loaded columns.

Select the longitudinal and lateral reinforcement (Step 7)

Longitudinal bars: Provide three No. 8 bars (25.4 mm diameter) on each of the two 12-in.-wide faces. Provide one No. 8 bar at the center of the 20-in.-wide face so that each face of this column would have an *equal* number of reinforcing bars.

Lateral ties: Try No. 3 bar lateral ties. The spacing s should be the minimum of

$$16 d_b \times \text{longitudinal bar diameter} = 16 \times \frac{8}{8} = 16 \text{ in.}$$

$$48 d_b \times \text{lateral tie diameter} = 48 \times \frac{3}{8} = 18 \text{ in.}$$

The minimum lateral dimension = 12 in. Therefore, provide No. 3 (9.53 mm diameter) lateral ties at 12 in. (304.8 mm) center to center. Reinforcing details are shown in Fig. 9.38.

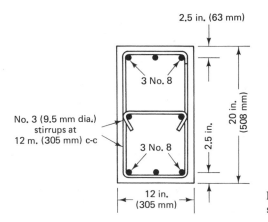

Figure 9.38 Biaxially loaded column section.

SELECTED REFERENCES

9.1 ACI Committee 318, "Building Code Requirements for Reinforced Concrete, ACI Standard 318-83," American Concrete Institute, Detroit, 1983, 111 pp., and the "Commentary on Building Code Requirements for Reinforced Concrete," 1983, 155 pp.

9.2 Hognestad, E., "A Study of Combined Bending and Axial Load in Reinforced Concrete Members," *Bulletin No. 399,* University of Illinois, Urbana, Ill., November 1951.

9.3 ACI Committee 105, "Reinforced Concrete Column Investigation," *Journal of the American Concrete Institute,* Vol. 26, April 1930, pp. 601–612. Also see Vol. 27, pp. 675–676; Vol. 28, pp. 157–158; Vol. 29, pp. 53–56 and 274–284; Vol. 30, pp. 78–90 and 153–156.

9.4 Richart, F. E., Draffin, J. O., Olson, T. A., and Heitman, R. H., "The Effect of Eccentric Loading, Protective Shells, Slenderness Ratios, and Other Variables in Reinforced Concrete Columns," *Bulletin No. 368,* Engineering Experiment Station, University of Illinois, Urbana, Ill., 1947, 130 pp.

9.5 Whitney, C. S., "Plastic Theory of Reinforced Concrete Design," *Transactions of the ASCE,* Vol. 107, 1942, pp. 251–326.

9.6 American Institute of Steel Construction, *"Specifications for the Design, Fabrication and Erection of Structural Steel for Buildings"* and "Commentary" on their Specifications, AISC, New York, 1969, and November 1978, 166 pp.

9.7 Johnston, B. G. (Ed.), *Structural Stability Research Council Guide to Stability Design Criteria for Metal Structures,* Wiley, New York, New York, 1966, 217 pp.

9.8   American Concrete Institute, *Design Handbook in Accordance with the Strength Design Method,* Vol. 2; *Columns,* Publications S-17A (78), ACI, Detroit, 1978, 191 pp.

9.9   Broms, B., and Viest, I. M., "Long Reinforced Concrete Columns," *Symposium Proceedings,* ASCE Transactions Paper No. 3155, January 1958, pp. 309–400.

9.10   Timoshenko, S. P., and Gere, J. M., *Theory of Elastic Stability,* 2nd ed., McGraw-Hill, New York, 1961, 541 pp.

9.11   Bresler, B., "Design Criteria for Reinforced Concrete Columns under Axial Load and Biaxial Bending," Proceedings, *Journal of the American Concrete Institute,* Proc. Vol. 57, November 1960, pp. 481–490.

9.12   Parme, A. L., Nieves, J. M., and Gouens, A., "Capacity of Reinforced Rectangular Columns Subjected to Biaxial Bending," *Journal of the American Concrete Institute,* Proc. Vol. 63, No. 9, September 1966, pp. 911–921.

PROBLEMS FOR SOLUTION

9.1   Calculate the axial load strength P_n for columns having the cross sections shown in Fig. 9.39. Assume zero eccentricity for all cases. Cases (a), (b), (c), and (d) are tied columns. Case (e) is spirally reinforced.

9.2   Calculate P_u and e in Fig. 9.39a and c of Problem 9.1. Assume that the stresses in the tension steel is zero.

9.3   Determine P_{ub} and e_b for the rectangular cross sections of Fig. 9.39 in Problem 9.1.

9.4   For the cross section shown in Fig. 9.39a of Problem 9.1, determine the safe eccentricity e if $P_u = 20,000$ lb and the safe P_u if $e = 15$ in.

9.5   For the cross section shown in Fig. 9.39c of Problem 9.1, determine the safe eccentricity e if $P_u = 1,500,000$ lb. Use the trial-and-adjustment method satisfying the compatibility of strains.

9.6   Repeat Problem 9.5 using Whitney's approximate procedure. Compare the results.

9.7   Construct the load–moment interaction diagram for the cross sections shown in Fig. 9.39a and c of Problem 9.1.

9.8   Obtain the equivalent rectangular column for the cross sections of Fig. 9.39d and e of Problem 9.1. Also calculate the balanced load P_{ub} and the balanced eccentricity e_b using the equivalent cross section.

9.9   For the cross section shown in Fig. 9.39d of Problem 9.1, calculate the design load P_u if $e = 6$ in. Repeat the calculation for $e = 20$ in.

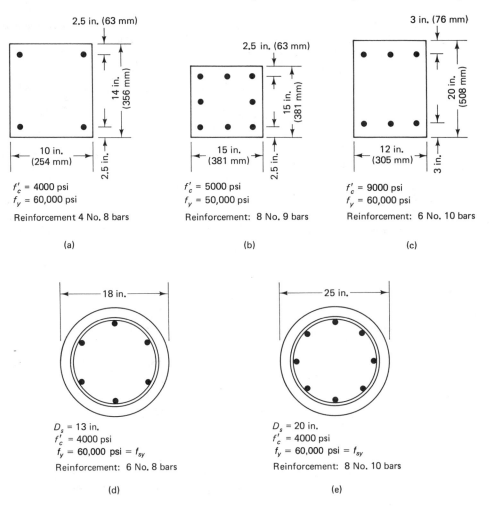

Figure 9.39 Column sections.

9.10 Design the reinforcement for a nonslender 15 in. × 20 in. column to carry the following loading. The factored ultimate axial force P_u = 300,000 lb. The eccentricity e to plastic centroid = 6 in. Given:

$$f'_c = 4,000 \text{ psi}$$
$$f_y = 60,000 \text{ psi}$$

9.11 Design a nonslender column to support the following service loads and moments. P_L = 100 kips, P_D = 50 kips, M_L = 2500 in.-kips, and M_D = 1000 in.-kips. Given:

$$f'_c = 5000 \text{ psi}$$
$$f_y = 60,000 \text{ psi}$$

9.12 Design a nonslender circular column to support a factored ultimate load $P_u = 250,000$ lb and a factored moment $M_u = 5 \times 10^6$ in.-lb. Given:

$f'_c = 6000$ psi, normalweight concrete
$f_y = 60,000$ psi
$d' = 2.50$ in.

9.13 Design the reinforcement for a 14 in. × 20 in. braced rectangular reinforced concrete column which can support a factored axial load $P_u = 500,000$ lb and a factored moment $M_u = 3,500,000$ in.-lb. The unsupported length, l_u, of the column is 10 ft. Assume that the end moments M_1 and M_2 are equal. Given:

$f'_c = 4000$ psi, sand-lightweight concrete
$f_y = 60,000$ psi
$d' = 2.5$ in.

9.14 A rectangular braced column of a multi-story frame building has a floor height $l_u = 25$ ft. It is subjected to service dead load moments $M_2 = 3,500,000$ on top and $M_1 = 2,500,000$ in.-lb at the bottom. The service live load moments are 80 percent of the dead load moments. The column carries a service axial dead load $P_D = 200,000$ lb and a service live load $P_L = 350,000$. Design the cross-section size and reinforcement for this column. Given:

$f'_c = 7000$ psi
$f_y = 60,000$ psi
$\psi_A = 1.3 \quad \psi_B = 0.9$
$d' = 2.5$ in.

9.15 A rectangular unbraced exterior column of a multi-bay, multi-floor frame system is subjected to $P_u = 500,000$ lb, factored end moments $M_1 = 2,500,000$ in.-lb and $M_2 = 3,500,000$ in.-lb. The unbraced length, l_u, of the column = 18 ft. Design this column if

(a) it is subjected to gravity loads with side sway considered as negligible.
(b) it is subjected to wind load resulting in a side sway factored moment $M_u = 2,100,000$ in.-lb. Assume the total loading of all interior and ex-

terior columns in a single floor is $\Sigma P_u = 20 \times 10^6$ lb and $\Sigma P_c = 44 \times 10^6$ lb. Use a section having a width $b = 14$ in. Given:

$f_c' = 6500$ psi, normalweight concrete
$f_y = 60,000$ psi
$\psi_A = 2, \quad \psi_B = 1.2$
$d' = 2.5$ in.

9.16 Design the column in Problem 9.15 using a circular cross section.

9.17 The columns of the first floor in a 9-story 7×3 bays office building have a clear height of 18 ft. (5.49 m). They are not braced against side sway, and the clear height above or below the first floor is 11 ft (3.35 m).
Design a typical interior column in that floor. Given:

EI/l for the connecting beams $= 450 \times 10^6$ in.-lb (50,850 kN-m)
Service loads for *interior columns* (1b) are:
    $D = 360,000, \quad L = 130,000, \quad W = 0$
Service loads for *exterior columns* (1b) are:
    $D = 80,000, \quad L = 65,000, \quad$ and $W = 5,000$
Service moments for the *interior columns* (in.-lb) are:
    Top, $D = 200,000, \quad L = 160,000, \quad W = 600,000$
    Bottom, $D = 500,000, \quad L = 400,000, \quad W = 600,000$
Service moments for *exterior columns* (in.-lb) are:
    Top, $D = 400,000, \quad L = 240,000, \quad W = 300,000$
    Bottom, $D = 700,000, \quad L = 360,000, \quad W = 300,000$
    $f_c' = 5000$ psi
    $f_y = 60,000$ psi
    $d' = 2.5$ in.

9.18 Design a typical exterior column for the structural framing system in Problem 9.15.

9.19 A nonslender square corner column is subjected to biaxial bending about its x and y axes. It supports a factored load $P_u = 200,000$ lb acting at eccentricities $e_x = e_y = 7$ in. Design the column size and reinforcement needed to resist the applied stresses. Given:

$f_c' = 5000$ psi
$f_y = 60,000$ psi
gross reinforcement percentage $\rho_g = 0.03$
$d' = 2.5$ in.

9.20 Design a nonslender rectangular end column to support a factored load $P_u = 200{,}000$ lb acting at eccentricities $e_x = 9.0$ in. and $e_y = 6.0$ in. Try a 12-in.-wide section with a total gross reinforcement percentage not to exceed $\rho_g = 0.025$. Given:

$$f'_c = 5000 \text{ psi, normalweight concrete}$$
$$f_y = 60{,}000 \text{ psi}$$
$$d' = 2.5 \text{ in.}$$

10

Bond Development
of Reinforcing Bars

Photo 54 Ladd Canyon overpass, Oregon. (Courtesy Portland Cement Association.)

10.1 INTRODUCTION

Reinforcement for concrete to develop the strength of a section in tension depends on the compatibility of the two materials to act "together" in resisting the external load. The reinforcing element, such as a reinforcing bar, has to undergo the same strain or deformation as the surrounding concrete in order to prevent the discontinuity or separation of the two materials under load. The modulus of elasticity, the ductility, and the yield or rupture strength of the reinforcement must also be considerably higher than those of the concrete in order to raise the capacity of the reinforced concrete section to a meaningful level. Consequently, materials such as brass, aluminum, rubber, or bamboo are not suitable for developing the bond or adhesion necessary between the reinforcement and the concrete. Steel and fiberglass do possess the principal factors necessary: yield strength, ductility, and bond value.

Bond strength results from a combination of several parameters, such as the mutual adhesion between the concrete and steel interfaces and the pressure of the hardened concrete against the steel bar or wire due to the drying shrinkage of the concrete. Additionally, friction interlock between the bar surface deformations or projections and the concrete caused by the micro movements of the tensioned bar

Photo 55 Coronado High School, Scottsdale, Arizona. (Courtesy Portland Cement Association.)

results in increased resistance to slippage. The total effect of this is known as *bond*. In summary, bond strength is controlled by the following major factors:

1. Adhesion between the concrete and the reinforcing elements
2. Gripping effect resulting from the drying shrinkage of the surrounding concrete and the shear interlock between the bar deformations and the surrounding concrete
3. Frictional resistance to sliding and interlock as the reinforcing element is subjected to tensile stress
4. Effect of concrete quality and strength in tension and compression
5. Mechanical anchorage effect of the ends of bars through development length, splicing, hooks, and cross bars
6. Diameter, shape, and spacing of reinforcement as they affect crack development

The individual contributions of these factors are difficult to separate or quantify. Shear interlock, shrinking confining effect, and the quality of the concrete can be considered as major factors.

10.2 BOND STRESS DEVELOPMENT

Bond stress is primarily the result of the shear interlock between the reinforcing element and the enveloping concrete caused by the various factors previously enumerated. It can be described as a local shearing stress per unit area of the bar surface. This direct stress is transferred from the concrete to the bar interface so as to change the tensile stress in the reinforcing bar along its length.

Three types of tests can determine the bond quality of the reinforcing element: the pull-out test, the embedded-rod test, and the beam test. Figure 10.1 shows the first two types of test. The pull-out test can give a good comparison of the bond efficiency of the various types of bar surfaces and the corresponding embedment lengths. It is, however, not truly representative of the bond stress development in a structural beam. The concrete is subjected to compression and the reinforcing bar acts in tension in this test, whereas both the bar and the surrounding concrete in the beam are subjected to the same stress.

In the embedded-rod test, Fig. 10.1b, the number of cracks, their widths, and their spacing at the various loading levels are a measure of the bond stress development and bond strength. The process resembles more closely the behavior in beams as the progressive increase in crack widths ultimately leads to bar slippage and beam failure.

The progressive slippage of the reinforcing bar in a beam and the redistribution of stresses is represented schematically in Fig. 10.2. As the resistance to slippage over length l_1 becomes larger than the tensile strength of concrete, a new crack forms in that area and a new stress distribution develops around the newly formed crack. The bond stress peak in Fig. 10.2a continues to progress to the right from position A to position

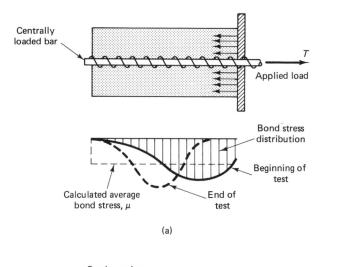

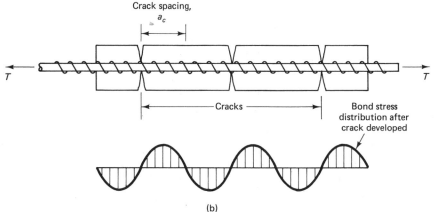

Figure 10.1 Bond stress development: (a) pull-out bond test; (b) embedded rod test.

B, passing the center line between the two potential cracks until a second crack forms at a distance a_c from crack 1.

Consequently, it is important to choose the appropriate length of the reinforcing bars that can minimize cracking and bond slippage. As a result, the reinforcement can attain its full strength in tension, that is, its yield strength within the structural element without bond failure.

10.2.1 Anchorage Bond

Assume l_d in Fig. 10.3a to be the length of the bar embedded in the concrete subjected to a net pulling force dT. If d_b is the diameter of the bar, μ is the average bond stress, and f_s is the stress in the reinforcing bar due to direct pull or bending stresses in a beam,

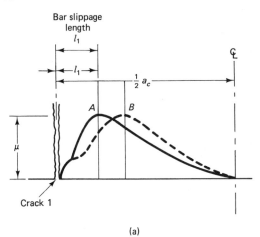

Bar slippage length l_1

$\frac{1}{2} a_c$

$\mathcal{C}$

A B

μ

Crack 1

(a)

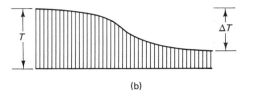

T

ΔT

(b)

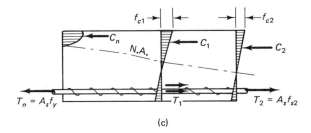

f_{c1} f_{c2}

C_n C_1 C_2

N.A.

$T_n = A_s f_y$ T_1 $T_2 = A_s f_{s2}$

(c)

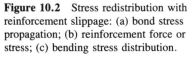

Figure 10.2 Stress redistribution with reinforcement slippage: (a) bond stress propagation; (b) reinforcement force or stress; (c) bending stress distribution.

the anchorage pulling force dT would be $\mu \pi d_b l_d$ and equal to the tensile force dT on the bar cross section, that is,

$$dT = \frac{\pi d_b^2}{4} f_s$$

Hence

$$\mu \pi d_b l_d = \pi \frac{d_b^2}{4} f_s$$

from which the average bond stress

$$\mu = \frac{f_s d_b}{4 l_d} \qquad (10.1a)$$

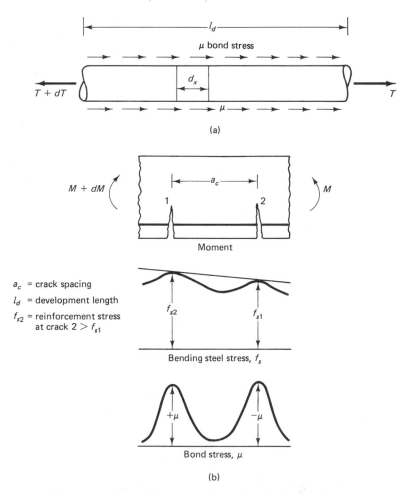

Figure 10.3   Bond stress across a reinforcing bar: (a) pull-out anchorage bond in a bar; (b) flexural bond.

and the development length

$$l_d = \frac{f_s}{4\mu} d_b \tag{10.1b}$$

10.2.2 Flexural Bond

The change in stress along the length of a bar in a beam due to the variation of moment along the span is shown schematically in Fig. 10.3b. If jd is the lever arm of the couple T due to moment M, then $T = M/jd$. In terms of the moment difference between cracked sections 1 and 2,

$$dT = \frac{dM}{jd} \qquad (10.2a)$$

Also,

$$dT = \mu d \times \Sigma o \qquad (10.2b)$$

where Σo is the total circumference of all the reinforcement subjected to the bond stress pull, to get $dM/dx = \mu \Sigma ojd$; since $dM/dx = $ shear V,

$$\mu = \frac{V}{\Sigma ojd} \qquad (10.2c)$$

Equation 10.2c is primarily of academic importance since it is indirectly accounted for in the development length approach given in Eq. 10.1b and the expressions to follow.

10.3 BASIC DEVELOPMENT LENGTH

From the discussion in the preceding section it can be concluded that the development length l_d as a function of the size and yield strength of the reinforcement determines the resistance of the bars to slippage, hence the magnitude of the beam's failure capacity. It has been verified by tests that the bond strength μ is a function of the compressive strength of concrete such that

$$\mu = k\sqrt{f_c'} \qquad (10.3a)$$

where k is a constant. If the bond strength equals or exceeds the yield strength of a bar of cross-sectional area $A_b = \pi d_b^2/4$, then

$$\pi d_b l_d \mu \geq A_b f_y \qquad (10.3b)$$

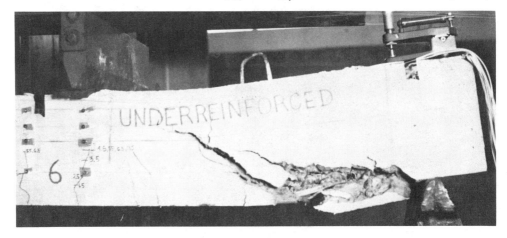

Photo 56   Bond failure at support of simply supported beam. (Tests by Nawy et al.)

Photo 57 Bond failure and destruction of concrete cover at rupture load. (Tests by Nawy et al.)

From Eqs. 10.1b, 10.3a, and 10.3b,

$$l_{db} = k_1 \frac{A_b f_y}{\sqrt{f_c'}} \qquad (10.4)$$

where k_1 is a function of the geometrical property of the reinforcing element and the relationship between bond strength and compressive strength of concrete.

Equation 10.4 is consequently the basic model for defining the minimum development length of bars in structural elements. The ACI code specifies k_1 values for various bar sizes and bond stresses in tension as well as in compression. These values are the result of extensive experimental verifications, particularly those reported by Ferguson et al. in Refs. 10.3 and 10.4.

10.3.1 Development of Deformed Bars in Tension

A reinforcing bar should have a sufficient embedment length l_d to prevent bond failure. The factor k_1 in Eq. 10.4 has different values for different bar sizes. The ACI code includes modifying multipliers for top bars since such bars have a lesser confining cover effect, hence lesser bond strength than bottom bars. Modifying multipliers are also provided for lightweight concrete and for bars of yield strength higher than 60,000 psi. The basic development length should be as follows but not less than 12 in. (304.8 mm).

No. 11 bars and smaller:

$$l_{db} = 0.04 \, \frac{A_b f_y}{\sqrt{f_c'}} \qquad (10.5a)$$

and

$$l_{db} \geq 0.0004 \, d_b f_y \qquad (10.5b)$$

where d_b is the bar diameter.

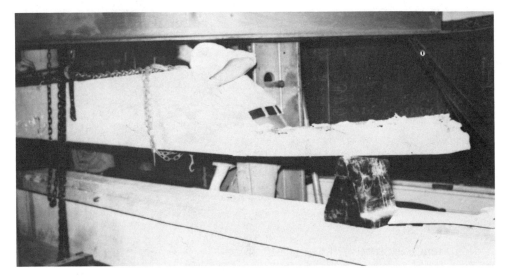

Photo 58   Bond failure in an over-reinforced concrete beam. (Tests by Nawy et al.)

No. 14 bar:

$$l_{db} = 0.085 \, \frac{f_y}{\sqrt{f'_c}}$$ (10.5c)

No. 18 bar:

$$l_{db} = 0.110 \, \frac{f_y}{\sqrt{f'_c}}$$ (10.5d)

Deformed wire (but not including wire fabric):

$$l_{db} = 0.03 \, \frac{d_b f_y}{\sqrt{f'_c}}$$ (10.5e)

10.3.2 Modifying Multipliers of Development Length l_d for Bars in Tension

The basic development length l_{db} must be multiplied by the applicable λ_d factor for the particular conditions to achieve the necessary embedment length, so that development length $l_d = l_{db}\lambda_d$.

1. *Top reinforcement:*

$$\lambda_d = 1.4$$

2. *Reinforcement with $f_y > 60,000$ psi:*

$$\lambda_d = 2 - \frac{60,000}{f_y}$$

3. *Lightweight aggregate concrete:* When the tensile splitting strength f_{ct} is specified and the mix designed in accordance with ACI:

$$\lambda_d = \frac{6.7\sqrt{f_c'}}{f_{ct}} \geq 1.0$$

If f_{ct} is not specified:
(a) All lightweight concrete: $\lambda_d = 1.33$
(b) Sand-lightweight concrete: $\lambda_d = 1.18$

The basic development length l_d can be reduced by using a reduction multiplier for certain conditions as follows.

1. *Reinforcement spaced laterally:* At least 6 in. on center and 3 in. minimum from face of member to edge bar: $\lambda_d = 0.8$.
2. *Required A_s < provided A_s:*

$$\lambda_d = \frac{\text{required } A_s}{\text{provided } A_s}$$

3. *Confined reinforcement:* confinement within spiral reinforcement not less than $\frac{1}{4}$ in. diameter and not more than 4 in. of pitch: $\lambda_d = 0.75$.

10.3.3 Development of Deformed Bars in Compression and the Modifying Multipliers

Bars in compression require shorter development length than bars in tension. This is due to the absence of the weakening effect of the tensile cracks. Hence the expression for the basic development length is

$$l_{db} = 0.02\frac{d_b f_y}{\sqrt{f_c'}} \qquad (10.6a)$$

and

$$l_{db} \geq 0.0003 d_b f_y \qquad (10.6b)$$

with the modifying multiplier for

(1) Excess reinforcement: $\lambda_d = $ required $A_s/$provided A_s

(2) Spirally enclosed reinforcement: $\lambda_d = 0.75$

If bundled bars are used in tension or compression, l_d has to be increased by 20% for three-bar bundles and 33% for four-bar bundles.

10.3.4 Example 10.1: Embedment Length of Deformed Bars

Calculate the required embedment length of the deformed bars of the following three cases:

(a) No. 7 bar (22.2 mm diameter) top reinforcement:
$f_y = 60,000$ psi (413.7 MPa)
$f_c' = 4500$ psi (31.03 MPa)

The section was overdesigned by providing a steel area $A_s = 7.2$ in.^{2} (4644 mm^2) instead of the required $A_s = 6.9$ in.^{2}.

(b) 1/8-in.-diameter deformed wire in tension:
$f_y = 80,000$ psi (551.6 MPa)
$f_c' = 5000$ psi (34.47 MPa)

(c) No. 8 bar (28.6 mm diameter) in compression:
$f_y = 60,000$ psi (413.7 MPa)
$f_c' = 4000$ psi (27.6 MPa)

Solution

$$\text{(a)} \quad l_{db} = \frac{0.04 A_b f_y}{\sqrt{f_c'}} = \frac{0.04 \times 0.6 \times 60,000}{\sqrt{4500}} = 21.47 \text{ in.}$$

or

$$l_{db} = 0.0004 d_b f_y = 0.0004 \times 0.875 \times 60,000 = 21.0 \text{ in.}$$

Therefore,

controlling basic development length $= 21.47$ in.

modifying multiplier for top reinforcement $\lambda_d = 1.4$

modifier for overdesign $\lambda_d = \dfrac{6.9}{7.2} = 0.958$

Therefore,

minimum embedment or development length $l_{db} = 1.4 \times 0.958 \times 21.47$

$= 28.8$ in. (732 mm)   say 30 in. $> 12.$ in.   O.K.

$$\text{(b)} \quad l_{db} = \frac{0.03 d_b f_y}{\sqrt{f_c'}} = \frac{0.03 \times 0.125 \times 80,000}{\sqrt{5000}} = 4.24 \text{ in.}$$

For $f_y > 60,000$ psi,

$$\lambda_d = 2 - \frac{60,000}{80,000} = 1.25$$

minimum development length $l_d = 1.25 \times 4.24 = 5.30$ in. < 12.0 in.

Use $l_d = 12$ in. (305 mm).

$$\text{(c)} \quad l_{db} = \frac{0.02 d_b f_y}{\sqrt{f_c'}} = \frac{0.02 \times 1.0 \times 60,000}{\sqrt{4500}} = 17.89 \text{ in.}$$

or

$$l_{db} = 0.0003 d_b f_y = 0.0003 \times 1.0 \times 60,000 = 18.0 \text{ in.}$$

Therefore,

$$\text{minimum development length } l_d = 18.0 \text{ in. (457 mm)}$$

10.3.5 Mechanical Anchorage and Hooks

Hooks are used when space limitation in a concrete section does not permit the necessary straight embedment length. Hooks in structural members are placed relatively close to the free surface of a concrete element, where splitting forces proportional to the total bar force may determine the hook capacity. The standard hook does *not* develop the tension yield strength of the bar. If l_{hb} is the basic development length for the standard hook in tension, additional embedment length has to be incorporated to give a total length l_{dh} not less than $8d_b$ or 6 in., whichever is greater. l_{dh} length is shown in Fig. 10.4. The length l_{hb} varies with the bar size, reinforcement yield strength, and the compressive strength of concrete. For $f_y = 60,000$-psi steel,

$$l_{hb} = 1200 \, \frac{d_b}{\sqrt{f_c'}} \tag{10.7}$$

where d_b is the diameter of the hook bar.

Modifying Multipliers for Hooks in Tension

1. *Yield strength f_y effect:* For a yield strength different than 60,000, $\lambda_d = f_y/60,000$.
2. *Concrete cover effect:* For No. 11 bars and smaller, side cover normal to the plane of hook not less than $2\frac{1}{2}$ in. and for 90° hook with cover on bar extension beyond the hook not less than 2 in., $\lambda_d = 0.7$ (see Fig. 10.4c).
3. *Ties of stirrups:* For No. 11 bars and smaller, hook enclosed vertically or horizontally within or stirrup ties not greater than $3d_b$, where d_b is diameter of hook bar, $\lambda_d = 0.8$.
4. *Excess reinforcement:* Where anchorage or development for f_y is not specifically required but the reinforcement area A_s used is in excess of A_s required for analysis,

$$\lambda_d = \frac{\text{required } A_s}{\text{provided } A_s}$$

5. *Bars developed by standard hooks at discontinuous ends:* If the concrete cover is less than $2\frac{1}{2}$ in., bars *should* be *enclosed* within ties or stirrups along the full

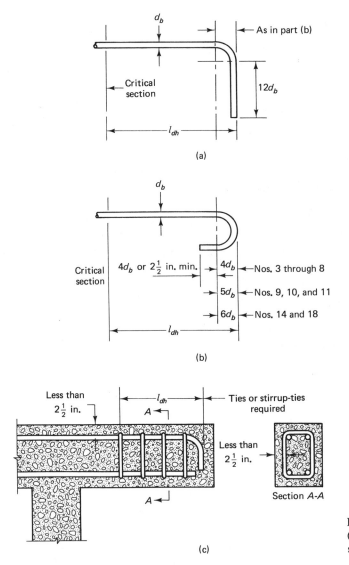

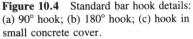

Figure 10.4 Standard bar hook details: (a) 90° hook; (b) 180° hook; (c) hook in small concrete cover.

development length l_{dh} spaced at no greater than $3d_b$, for this case $\lambda_d = 0.8$ from item 3 above; modifying multiplier shall not apply.

6. *Lightweight concrete:* $\lambda_d = 1.3$. It should be noted that hooks cannot be considered effective in developing bars in compression. The total development or embedment length

$$l_{dh} = l_{hb} \times \lambda_d \tag{10.8}$$

Figure 10.4a and b shows details of standard 90° and 180° hooks used in axial tension or bending tension and Fig. 10.4c gives details of bar hooks susceptible to

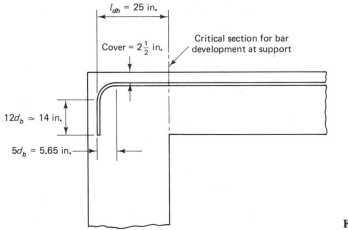

Figure 10.5 Hook embedment detail.

concrete splitting when the cover is small, namely less than $2\frac{1}{2}$ in. Confinement is enhanced through the use of closed ties or stirrups. No distinction is made between a top bar and a bottom bar if hooks are used.

10.3.6 Example 10.2: Embedment Length for a Standard 90° Hook

Compute the development length required for the top bars of a lightweight concrete beam extending into the column support shown in Fig. 10.5 assuming No. 9 reinforcing bars (28.6 mm diameter) hooked at the end. The concrete cover is 2 in (508 mm). Given:

$$f_c' = 5000 \text{ psi (34.47 MPa)}$$
$$f_y = 60,000 \text{ psi (413.7 MPa)}$$

Solution

Top bars for hooks behave similarly to bottom bars; hence no modifier is needed. For No. 9 bars, $d_b = 1.128$ in. (28.65 mm).

$$\text{basic development length } l_{hb} = 1200 \frac{d_b}{\sqrt{f_c'}}$$

or

$$l_{hb} = \frac{1200 \times 1.128}{\sqrt{5000}} = 19.14 \text{ in.}$$

For lightweight concrete, $\lambda_d = 1.3$.

$$l_{dh} = 1.3 \times 19.14 = 24.88 \text{ in.} > 8d_b \text{ or 6 in.} \qquad \text{O.K.}$$

Use a 90° hook with embedment length $l_{dh} = 25$ in. (635 mm) beyond the critical section (face of support). Figure 10.5 shows the geometrical details of the hook.

10.4 DEVELOPMENT OF FLEXURAL REINFORCEMENT IN CONTINUOUS BEAMS

As discussed earlier, reinforcing bars should be adequately embedded in order to prevent serious bar slippage resulting in bond pull-out failure. The critical locations for bar discontinuance are points along the structural member where there is a rapid drop in the bending moment or stress, such as the inflection points in a bending moment diagram of a continuous beam.

Tension reinforcement can be developed by bending the lower tension bars at a 45° inclination across the web of the beam, and be anchored or made continuous with the reinforcing bars on the top of the member. To ensure full development, reinforcement has to be extended beyond the point at which it is no longer required to resist flexure for a distance equal to the effective depth d or 12 d_b, whichever is greater, except for supports of simple span beams or at the free end of a cantilever. Figure 10.6 shows details of flexural reinforcement development in typical continuous beams for both the positive and the negative steel reinforcement.

The following are general guidelines for full development of the reinforcement and for ensuring continuity in the case of continuous beams.

1. At least *one-third* of the positive moment reinforcement in simple beams and *one-fourth* of the positive moment reinforcement in continuous beams should be extended at least 6 in. into the support without being bent.
2. At simple supports as in Fig. 10.7a and at points of inflection as in Fig. 10.7b, the positive moment reinforcement should be limited to such a diameter that the development length

$$l_d \leq \frac{M_n}{V_u} + l_a \qquad (10.9)$$

where M_n = nominal moment strength where all the reinforcement is stressed to f_y

V_u = factored shear force at the section under consideration

inflection point l_a = effective depth d or 12 d_b, where d_b is the bar diameter, whichever is greater

Equation 10.9 imposes a design limitation on the flexural bond stress in areas of large shear and small moment in order to prevent splitting. Such a condition exists in short-span heavily loaded simple beams. Thus the bar diameter for positive moment is so chosen that even if length AC to the critical section in Fig. 10.7a is larger than length AB, the bar size must be limited such that $l_d \leq$ 1.3 ($M_n/V_u + l_a$). For confining reactions such as at simple supports the value M_n/V_u in Eq. 10.9 is increased by 30%.

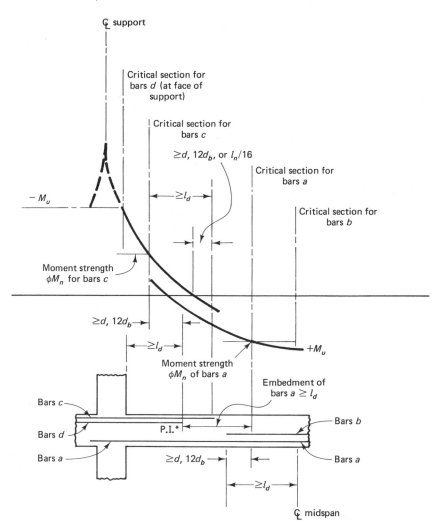

Figure 10.6 Development of reinforcement in continuous beams.

3. At least *one-third* of the total tension reinforcement provided for negative bending moment at the support should be extended beyond the inflection point not less than the effective depth d of the member, $12d_b$, or $\frac{1}{16}$ of the clear span, whichever has the largest value.

4. Web stirrups have to be carried as close to the compression and tension surfaces of the member as the minimum concrete cover requirements allow. The ends of

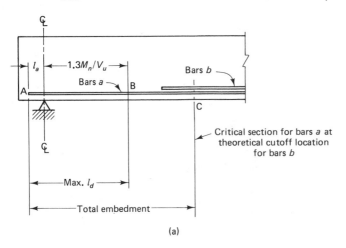

(a)

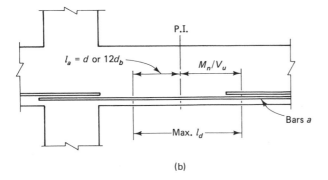

(b)

Figure 10.7 Cutoff points for reinforcement: (a) simply supported beams; (b) continuous beams.

the stirrups without hooks should have an embedment of at least $d/2$ above or below the compression side of the member for full development length l_d but not less than 12 in. or 24 d_b. For stirrups with hook ends, the total embedment length should equal $0.5l_d$ plus the standard hook.

A typical detail of cutoff points for continuous one-way beam and joist construction from Ref. 10.6 is given in Fig. 10.8. Typical cutoff points for one-way slabs are shown in Fig. 10.9 and for beams with diagonal tension stirrups are given in Fig. 10.10.

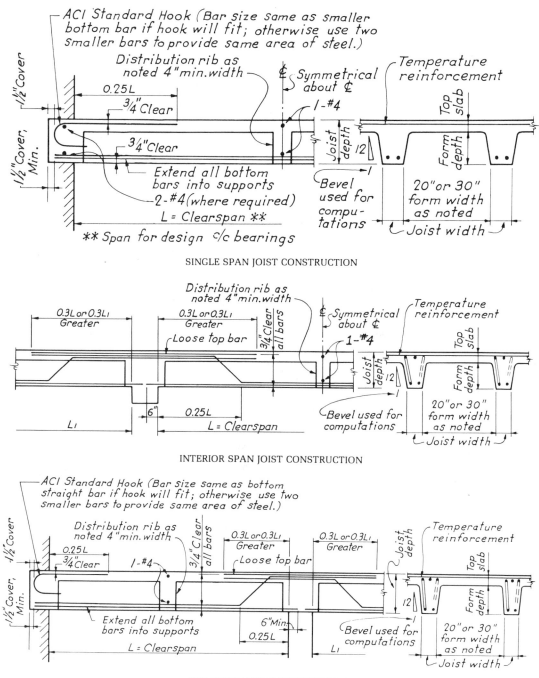

Figure 10.8 Cutoff point for one-way joist construction. (From Ref. 10.6.)

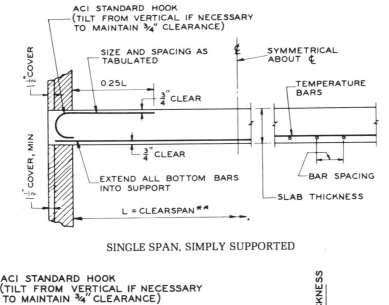

SINGLE SPAN, SIMPLY SUPPORTED

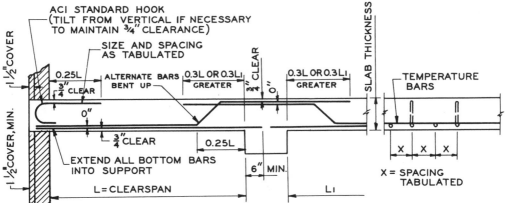

END SPAN, SIMPLY SUPPORTED

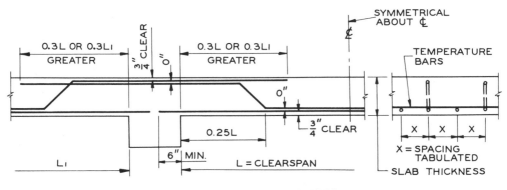

INTERIOR SPAN, CONTINUOUS

Figure 10.9 Cutoff points for one-way slabs. (From Ref. 10.6.)

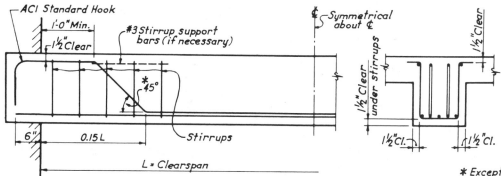

SINGLE SPAN BEAM, SIMPLY SUPPORTED

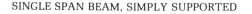

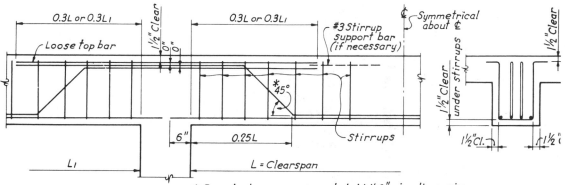

INTERIOR SPAN OF CONTINUOUS BEAM

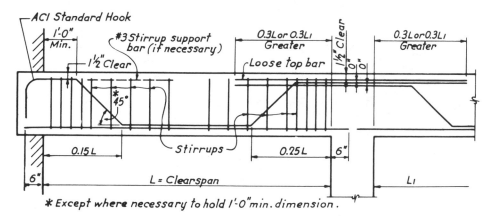

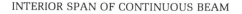

END SPAN OF SIMPLY SUPPORTED BEAM

Figure 10.10 Reinforcing details for continuous beams with diagonal tension steel. (From Ref. 10.6.)

10.5 SPLICING OF REINFORCEMENT

Steel reinforcing bars are produced in standard lengths controlled by transportability and weight considerations. In general, 60-ft lengths are normally produced. But it is impractical in beams and slabs spanning over several supports to interweave bars of such lengths on site over several spans. Consequently, bars are cut to shorter lengths and lapped at the least critical bending moment locations for bar sizes No. 11 or smaller. A general rule of thumb for maximum bar length is about 40 ft for shipping purposes. The most effective means of continuity in reinforcement is to weld the cut pieces without reducing the mechanical or strength properties of the welded bar at the weld. However, cost considerations require alternatives. There are basically three types of splicing:

1. *Lap splicing:* This depends on full bond development of the two lapping bars at the lap for bars of size not larger than No. 11.
2. *Welding by fusion of the two bars at the connection:* This process can be economically justifiable for bar sizes larger than No. 11 bars.
3. *Mechanical connecting:* This can be achieved by mechanical sleeves threaded on the ends of the bars to be interconnected. Such connectors should have a yield strength at least 1.25 times the yield strength of the bars they interconnect. They are also more commonly used for large-diameter bars.

10.5.1 Lap Splicing

Figure 10.11a shows a bar lap splice and the force and stress distribution along the splice length l_s. Failure of the concrete at the splice region develops by a typical splitting mechanism as shown in Fig. 10.11b. At failure, one bar slips relative to the other. The *idealized* tensile stress distribution in the bars along the splice length l_s has a maximum value f_y at the splice end and $\frac{1}{2}f_y$ at $l_s/2$. At failure, the expected magnitude of slip is approximately $(0.5 f_y/E_s)$ × half splice length l_s in Fig. 10.11a.

10.5.2 Splices of Deformed Bars and Deformed Wires in Tension

Three classes of lap splices are specified by the ACI code. The minimum length l_s but not less than 12 in. is

Class A: $1.0l_{db}$
Class B: $1.3l_{db}$
Class C: $1.7l_{db}$

Table 10.1 gives the maximum percentage of tensile steel area A_s that can be spliced.

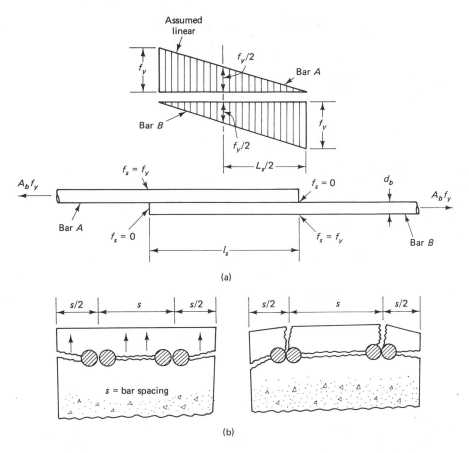

Figure 10.11   Reinforcing bars splicing: (a) lap splice idealized stress distribution; (b) splice splitting failure.

Splicing should be avoided at maximum tensile stress if at all possible; splicing may be by simple lapping of bars either in contact or separated by concrete. However, every effort should be made to stagger the splice rather than having all the bars spliced within the required lap length.

TABLE 10.1   MAXIMUM TENSION STEEL AREA TO BE SPLICED

$\dfrac{\text{Provided } A_s}{\text{Required } A_s}$	Maximum percentage of A_s spliced within the required lap length		
	60	75	100
≥ 2.0	Class A	Class A	Class B
< 2.0	Class B	Class C	Class C

10.5.3 Splices of Deformed Bars in Compression

The lap length l_s should be equal to at least the development length in compression as given in Section 10.3.3 and Eqs. 10.6a and 10.6b and the modifiers. l_s should also satisfy the following, but not less than 12 in.:

$$f_y \leq 60{,}000 \text{ psi} \qquad l_s \geq 0.0005f_yd_b \qquad\qquad (10.10a)$$
$$f_y > 60{,}000 \text{ psi} \qquad l_s \geq (0.0009f_y - 24)d_b \qquad (10.10b)$$

If the compressive strength f_c' of the concrete is less than 3000 psi, such as might occur in foundations, the splice length l_s has to be increased by one-third.

Modifying multipliers with values less than 1.0 are used in heavily reinforced tied compression members (0.83) and in spirally reinforced columns (0.75), but the lap length should not be less than 12 in.

10.5.4 Splices of Deformed Welded Wire Fabric

The minimum lap length l_s measured between the ends of the two lapped fabric sheets of welded deformed wire has to be $\geq 1.7l_d$ or 8 in. (204 mm), whichever is greater. Additionally, the overlap measured between the outermost cross wires of each fabric sheet should not be less than 2 in. (50.8 mm).

10.6 EXAMPLES OF EMBEDMENT LENGTH AND SPLICE DESIGN FOR BEAM REINFORCEMENT

10.6.1 Example 10.3: Embedment Length at Support of a Simply Supported Beam

Calculate the maximum development length that can be used for bars a at the support of the simply supported superstructure beam in Fig. 10.12 if the distance AC from the theoretical cutoff point of bars b is 42 in. (340.7 mm). The beam is not integral with the support (nonconfining). Assume that the reinforcing bars used for moment strength are (a) No. 8 deformed (25.4 mm), and (b) No. 14 deformed (43.0 mm), if the beam was a raft foundation beam (a maximum No. 11 bar size is usually used in superstructure normal-size beams). Given:

$$V_u = 100{,}000 \text{ lb (444.8 kN)}$$
$$M_n = 2{,}353{,}000 \text{ in.-lb (25.6 kNm)}$$
$$f_c' = 4000 \text{ psi (27.58 MPa), normalweight concrete}$$
$$f_y = 60{,}000 \text{ psi (413.7 MPa)}$$
$$l_a = 12 \text{ in. (304.8 mm)}$$

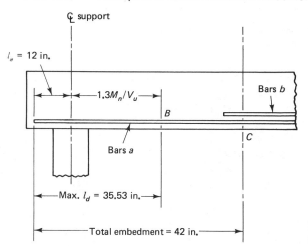

Figure 10.12 Embedment at end of simply supported beam.

Solution

(a) No. 8 bars: $A_b = 0.79$ in.^{2} per bar. From Eqs. 10.5a and 10.5b,

$$l_{db} = \frac{0.04A_b f_y}{\sqrt{f'_c}} = \frac{0.04 \times 0.79 \times 60{,}000}{\sqrt{4000}} = 29.98 \text{ in.}$$

and

$$l_{db} = 0.0004d_b f_y = 0.0004 \times 1.0 \times 60{,}000 = 24.0 \text{ in.}$$

Controlling $l_d = 29.98$, say 30 in. (762 mm). From Eq. 10.9,

$$\frac{M_n}{V_u} = \frac{2{,}353{,}000}{100{,}000} = 23.53 \text{ in.}$$

$$l_a = \text{embedment length beyond support center } = 12 \text{ in.}$$

The maximum allowable $l_d = 23.53 + 12 = 35.53$ in., which is within the distance AC, hence O.K. for bar cutoff. The actual needed $l_d = 30$ in. < 35.53 in. Therefore, the size No. 8 bar chosen for this design is safe for bond development without slippage using $l_d = 30$ in.

(b) No. 11 bars:

$$A_b = 2.25 \text{ in.}^2$$

$$l_d = \frac{0.085f_y}{\sqrt{f'_c}} = \frac{0.085 \times 60{,}000}{\sqrt{4000}} = 80.6 \text{ in.}$$

$$> \left(\frac{M_n}{V_u} + l_a = 35.53 \text{ in.} \right)$$

Consequently, No. 14 bars are too large to satisfy bond development without slippage failure and a smaller size has to be used.

10.6.2 Example 10.4: Embedment Length at Support of a Continuous Beam

A continuous reinforced concrete beam has clear spans $l_{nr} = 36$ ft (10.97 m) and $l_{nl} = 22$ ft (6.7 m) and the bending moment diagram segment at an interior support as shown in Fig. 10.13. Calculate the cutoff lengths of the negative moment top reinforcement bars to satisfy the development length requirements at the cutoff points. The beam is singly reinforced and has the dimensions $h = 27$ in. (685.8 mm), $d = 23.5$ in. (596.9 mm), and $b = 15$ in. (381.0 mm). It is subjected to a factored negative bending at the center of the intermediate support.

$$-M_u = 6,127,000 \text{ in.-lb (692.4 kNm)}$$

Given:

$f'_c = 4000$ psi (27.58 MPa), normalweight concrete
$f_y = 60,000$ psi (413.7 MPa)

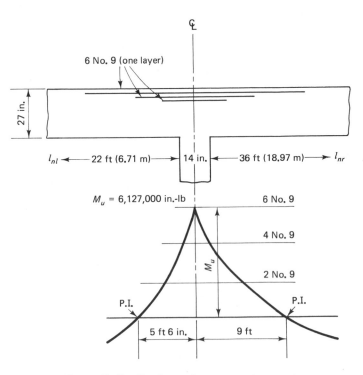

Figure 10.13 Continuous beam support moment.

required $A_s = 5.62$ in.^{2}

provided $A_s = 6.00$ in.^{2} (six No. 9 bars)

Solution

$A_b = 1.0$ in.^{2} per bar. The basic development length from Eqs. 10.5a and 10.5b,

$$l_{db} = \frac{0.04 A_b f_y}{\sqrt{f_c'}} = \frac{0.04 \times 1.0 \times 60,000}{\sqrt{4000}} = 37.94 \text{ in.}$$

or

$$l_{db} = 0.0004 d_b f_y = 0.0004 \times 1.128 \times 60,000 = 27.07 \text{ in.}$$

The basic $l_d \simeq 38$ in. controls. The development length for top bars is $l_d = 38 \times 1.4 = 53$ in. The reduction of l_d due to steel area adjustment

$$l_d = \frac{\text{required } A_s}{\text{provided } A_s} \times 53 = \frac{5.62}{6.00} \times 53 = 49.6 \text{ in.} \qquad \text{say 50 in.}$$

Use $l_d = 50$ in. (1270 mm) for the six No. 9 bars.

Cutoff points

At least one-third of the bars have to extend beyond the point of inflection by the largest of $\frac{1}{16}$ (span l_n), d, or $12 d_b$.

$$\tfrac{1}{3} A_s = \text{two No. 9 bars} \qquad 12 d_b = 12 \times 1\tfrac{1}{8} = 13.5 \text{ in.}$$

1. Right span $l_{nr} = 36$ ft:

$$\tfrac{1}{16} l_{nr} = \tfrac{36}{16} \times 12 = 27.0 \text{ in. controls}$$

2. Left span $l_{nl} = 22$ ft:

$$\tfrac{1}{16} l_{nl} = \tfrac{22}{16} \times 12 = 16.5 \text{ in.} \qquad d = 23.5 \text{ in. controls} \qquad \text{say 24 in.}$$

As given in Fig. 10.6, details of the development length dimensions at all cutoff points for this continuous beam are shown in Fig. 10.14.

10.6.3 Example 10.5: Splice Design for Tension Reinforcement

Calculate the lap splice length for No. 7 tension bars (22.2 mm diameter). The ratio of the provided A_s to the required A_s is (a) >2.0, (b) <2.0, and the maximum percentage of A_s spliced within the section is 75%. Given:

$f_c' = 5000$ psi (34.47 MPa)

$f_y = 60,000$ psi (413.7 MPa)

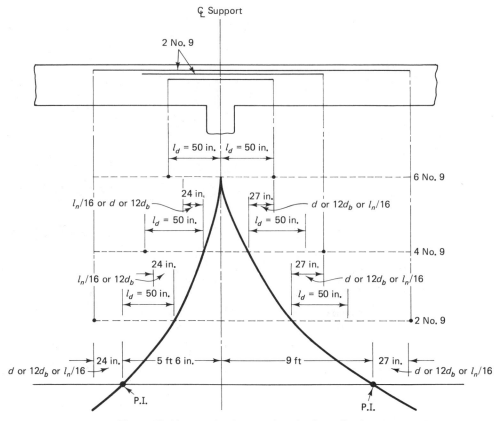

Figure 10.14 Bar development length of cutoff points.

Solution

From Eqs. 10.5a and 10.5b for tension bars,

$$l_{db} = \frac{0.04 A_b f_y}{\sqrt{f_c'}} = \frac{0.04 \times 0.60 \times 60{,}000}{\sqrt{5000}} = 20.4 \text{ in.}$$

or

$$l_{db} = 0.0004 d_b f_y = 0.0004 \times 0.875 \times 60{,}000 = 21.0 \text{ in.}$$

$$l_d = 21 \text{ in.} \qquad \text{controls}$$

(a) For provided A_s/required $A_s > 2.0$, class A splice:$l_s = 1 \times l_d = 21$ in.
(b) For provided A_s/required $A_s < 2.0$, class C splice:

lap splice length $l_s = 1.7 l_d = 1.7 \times 21 = 35.7 \approx 36$ in. (914.4 mm)

10.6.4 Example 10.6: Splice Design for Compression Reinforcement

Calculate the lap splice length for a No. 9 compression deformed bar (28.7 mm diameter). Given:

$$f_c' = 7000 \text{ psi (48.26 MPa)}$$
$$f_y = 80,000 \text{ psi (551.6 MPa)}$$

Solution

d_b = 1.128 in. From Eqs. 10.6a and 10.6b for compression bars, the basic development length is

$$l_{db} = \frac{0.02 d_b f_y}{\sqrt{f_c'}} = \frac{0.02 \times 1.128 \times 80,000}{\sqrt{7000}} = 21.57 \text{ in.}$$

or

$$l_{db} = 0.0003 d_b f_y = 0.0003 \times 1.128 \times 80,000 = 27.07 \text{ in.}$$

Controlling $l_d = 27.07 \simeq 28$ in. From Eqs. 10.10b, the minimum lap splice length for $f_y = 70,000$ steel is

$$l_s = (0.0009 f_y - 24) d_b = (0.0009 \times 80,000 - 24)1.128 = 54.14 \text{ in.}$$
$$> 28 \text{ in.}$$

Use lap splice length $l_s = 55$ in. (1397 mm).

10.7 TYPICAL DETAILING OF REINFORCEMENT AND BAR SCHEDULING

The design examples for bond development length, lap splicing, and spacing reinforcement are applied in Figs. 10.15 to 10.19 from Ref. 10.6. Additional examples from the author's parking-garage working drawing details are given in Figs. 10.20 to 10.24. These representative examples can serve as a good guideline for producing correct engineering working drawings. It should be recognized that successful execution of a designed system is directly dependent on the availability of clear and correct detailing and the avoidance of any congestion of the reinforcement. Such congestion can only lead to honeycombing in the concrete, resulting in possible cracking, capacity reduction, and even failure. Consequently, equal attention has to be given to detailing as to design if a constructed system is to perform the structural functions for which it is intended.

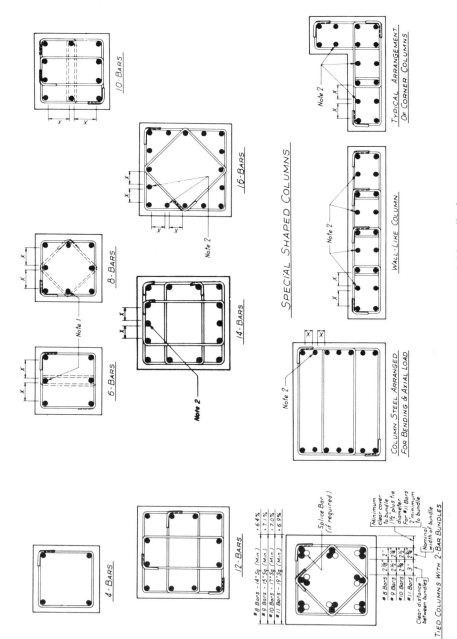

Figure 10.15 Column ties for preassembled lap-spliced cages.

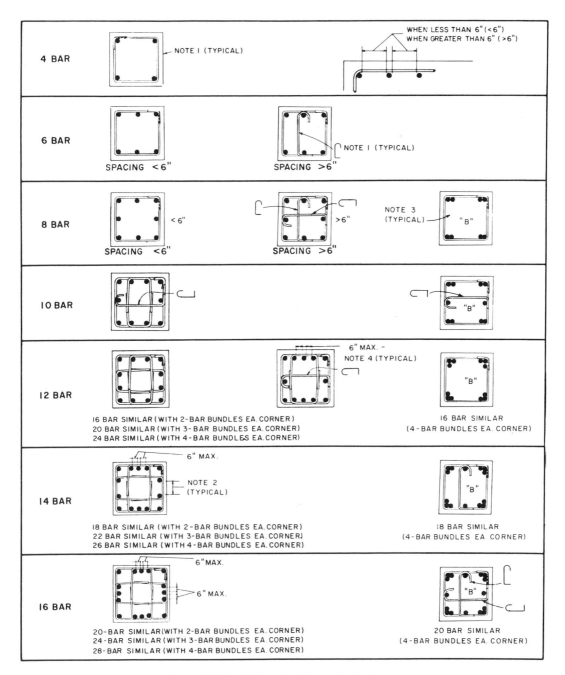

Figure 10.16 Column ties for standard columns.

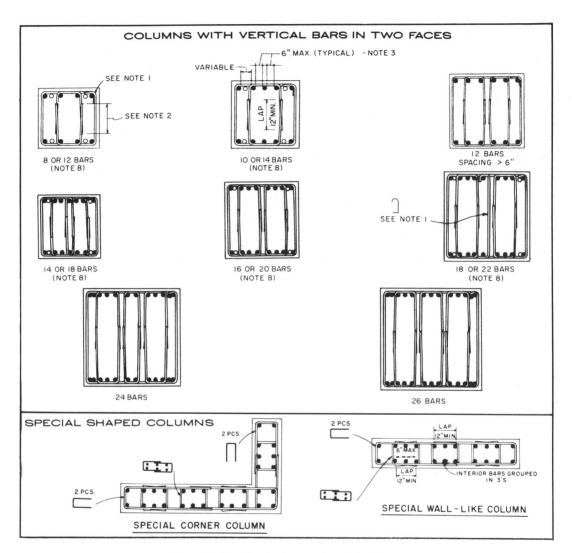

Figure 10.17 Ties for large and special columns.

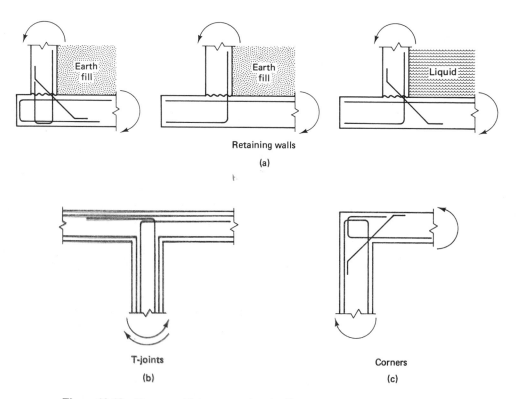

Figure 10.18 Corner and joint connection details: (a) retaining walls; (b) T joints; (c) corners.

Earth
fill

Earth
fill

Liquid

Retaining walls

(a)

T-joints

(b)

Corners

(c)

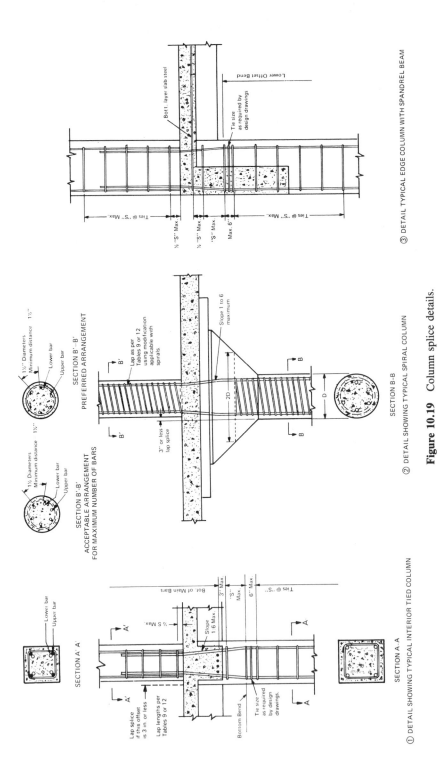

Figure 10.19 Column splice details.

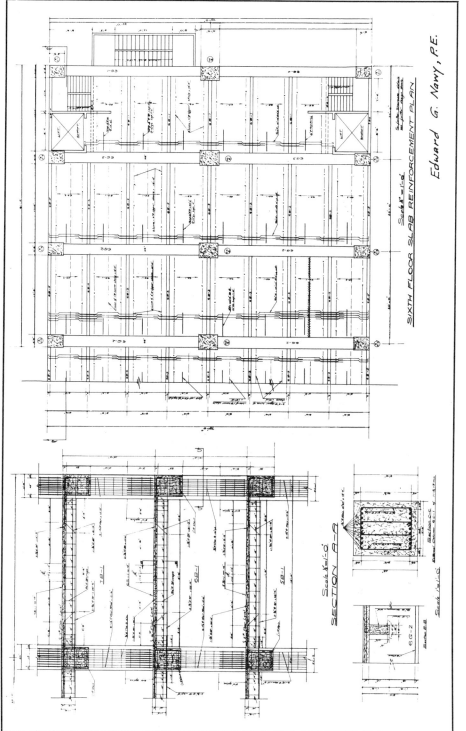

Figure 10.20 Typical beam and slab reinforcing working drawing. (Design by E. G. Nawy.)

413

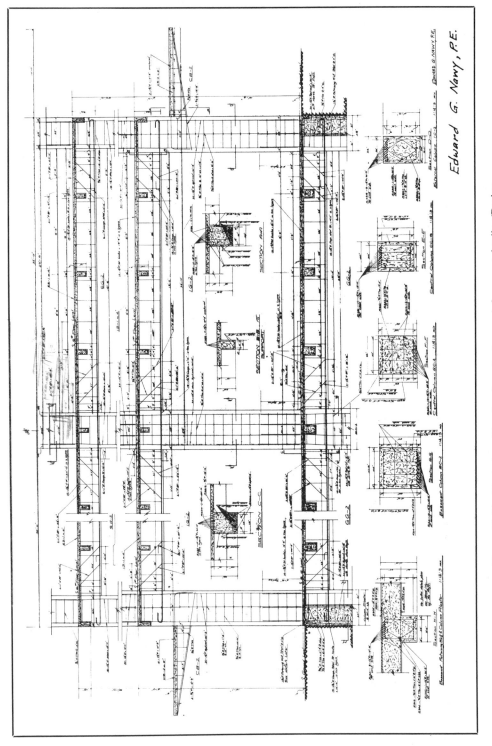

Figure 10.21 Typical working drawing of column reinforcement details. (Design by E. G. Nawy.)

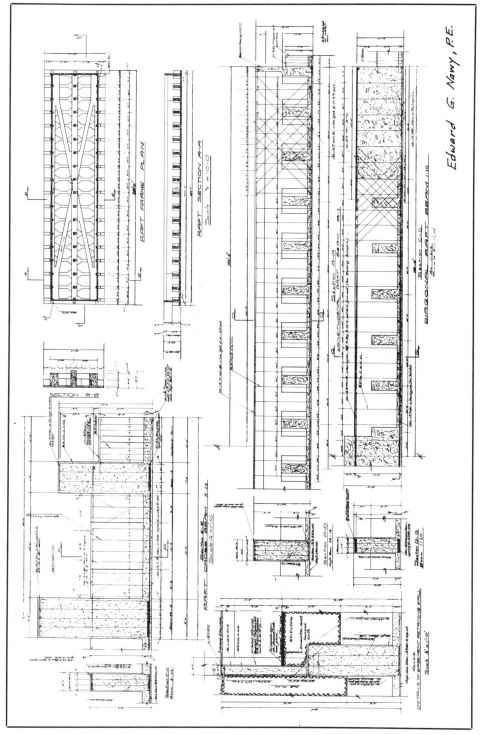

Figure 10.22 Raft foundation details. (Design by E. G. Nawy.)

Edward G. Nawy, P.E.

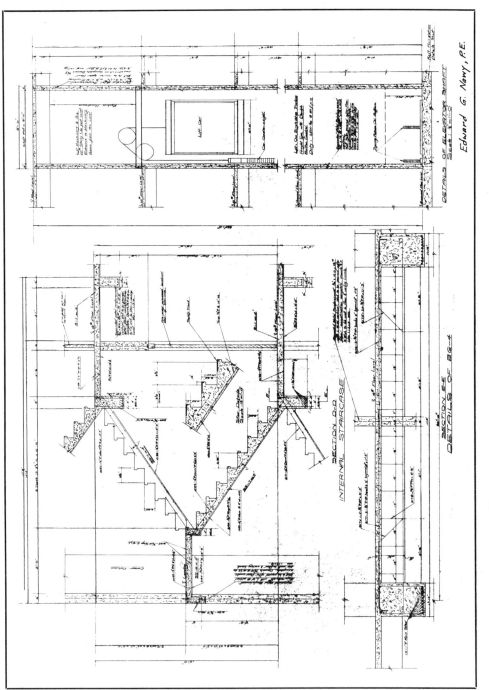

Figure 10.23 Elevator and stairwell details. (Design by E. G. Nawy.)

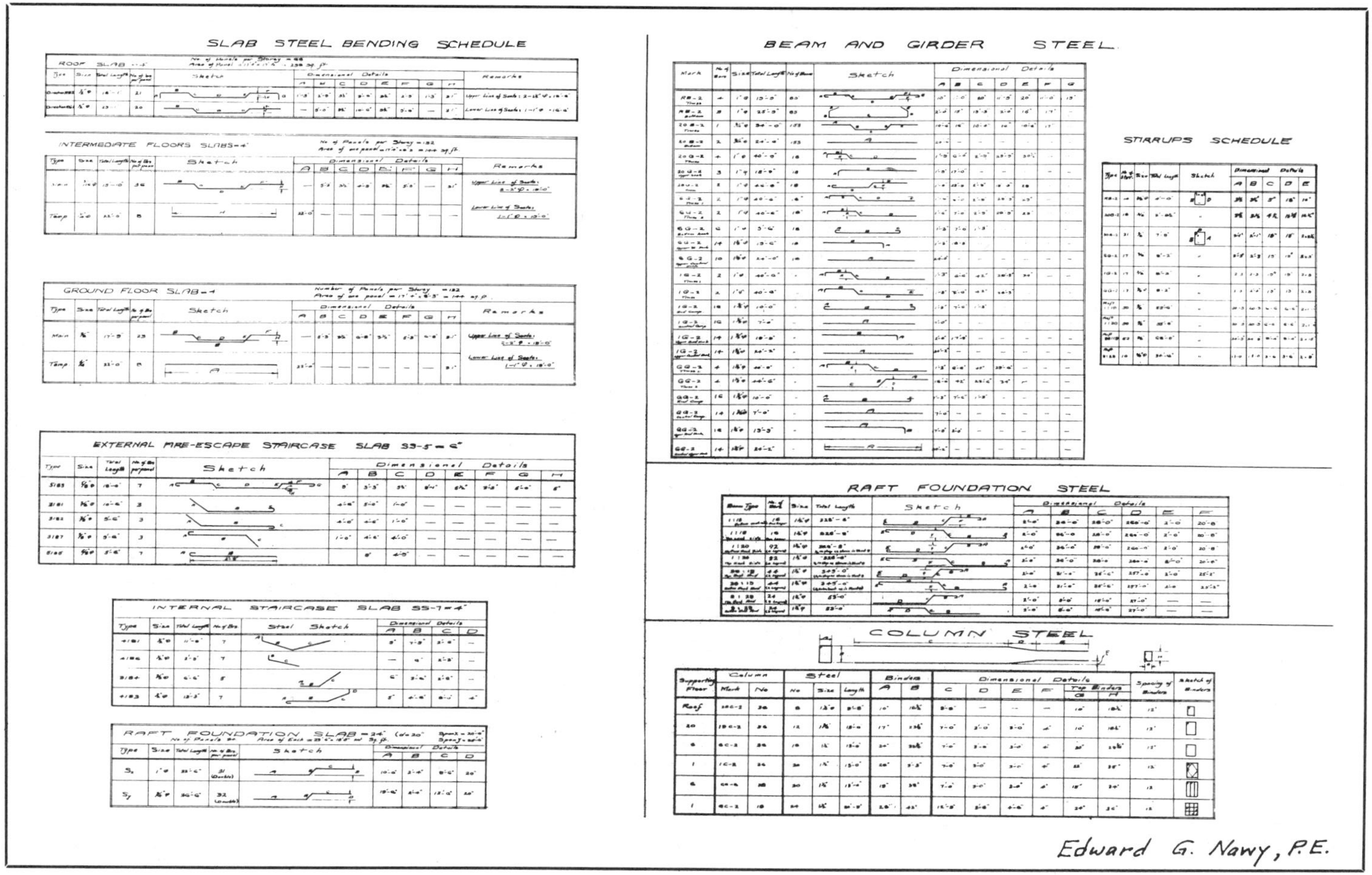

Figure 10.24 Typical reinforcement bar bending schedule. (Design by E. G. Nawy.)

SELECTED REFERENCES

10.1 Mathey, R. G., and Watstein, D., "Investigation of Bond in Beam Pull-Out Specimens with High Yield Strength Deformed Bars," *Journal of the American Concrete Institute,* Proc. Vol. 57, No. 9, March 1961, pp. 1071–1090.

10.2 Jirsa, J. O., Lutz, L. A., and Gergely, P., "Rationale for Suggested Development, Splice and Standard Hook Provisions for Deformed Bars in Tension," *Concrete International: Design and Construction* (American Concrete Institute), Vol. 1, No. 7, July 1979, pp. 47–61.

10.3 Ferguson, P. M., Breen, J. E., and Thompson, J. N., "Pullout Tests on High Strength Reinforcing Bars," Part 1, *Journal of the American Concrete Institute,* Proc. Vol. 62, No. 8, August 1965, pp. 933–950.

10.4 Ferguson, P. M., and Breen, J. E., "Lapped Splices for High Strength Reinforcing Bars," *Journal of the American Concrete Institute,* Proc. Vol. 62, No. 9, September 1965, pp. 1063–1078.

10.5 ACI Committee 408, "Suggested Development, Splice, and Hook Provisions for Deformed Bars in Tension," ACI 408-1 R-79, American Concrete Institute, Detroit, 1979, 3 pp.

10.6 ACI Committee 315, *ACI Detailing Manual—1980,* Special Publication SP-66, American Concrete Institute, Detroit, 1980, 206 pp.

10.7 Wire Reinforcement Institute, *Reinforcement Anchorages and Splices,* 3rd ed., WRI Publication, Mclean, Va., 1979, 32 pp.

PROBLEMS FOR SOLUTION

10.1 Calculate the basic development lengths in tension for the following deformed bars embedded in normalweight concrete.
(a) No. 5, No. 8. Given:

$$f_c' = 5000 \text{ psi } (34.47 \text{ MPa})$$
$$f_y = 60,000 \text{ psi } (413.7 \text{ MPa})$$

(b) No. 14, No. 18. Given:

$$f_c' = 4000 \text{ psi}$$
$$f_y = 60,000 \text{ psi}$$
$$f_y = 80,000 \text{ psi}$$

10.2 Calculate the total embedment length for the bars in Problem 10.1 if they are used as compression reinforcement and the concrete is sand lightweight.

10.3 Design the cutoff length for the continuous beam in Ex. 10.4 if eight No. 8 bars are used instead of six No. 9 bars.

10.4 Design the compression lap splice for a column section 16 in. × 16 in. (406 mm × 406 mm) reinforced with eight No. 9 bars (eight bars diameter 28.7 mm) equally spaced around all faces.

(a) $f_c' = 5000$ psi (34.47 MPa)

$\quad f_y = 60{,}000$ psi (413.7 MPa)

(b) $f_c' = 7000$ psi (48.26 MPa)

$\quad f_y = 80{,}000$ psi (551.6 MPa)

10.5 An 18-ft (5.49-m) normalweight concrete cantilever beam is subjected to a factored $M_u = 3{,}500{,}000$ in.-lb (396 kN-m) and a factored shear $V_u = 32{,}400$ lb (144 kN) at the face of the support. Design the top reinforcement and the appropriate embedment and 90° hook into the concrete wall to sustain the external shear and moment. Given:

$$f_c' = 4500 \text{ psi}$$
$$f_y = 60{,}000 \text{ psi}$$

10.6 Design the beam reinforcement in Problem 10.5 if it was simply supported having a span $l_n = 36$ ft (10.97 m) and subjected to the same factored M_u value at midspan and the shear V_u at the face of the support. Evaluate the required embedment length at the support to ensure that no bond failure due to slippage can develop. Assume (a) confining beam reaction, and (b) beam not monolithic with its support.

11

Design of
Two-Way Slabs & Plates

Photo 59 Sydney Opera House, Sydney, Australia. (Courtesy of Australian Information Service.)

11.1 INTRODUCTION: REVIEW OF METHODS

Supported floor systems are usually constructed of reinforced concrete cast in place. Two-way slabs and plates are those panels in which the dimensional ratio of length to width is less than 2. The analysis and design of framed floor slab systems represented in Fig. 11.1 encompasses more than one aspect. The present state of knowledge permits reasonable evaluation of (1) the moment capacity, (2) the slab–column shear capacity, and (3) serviceability behavior as determined by deflection control and crack control. It is to be noted that flat plates are slabs supported directly on columns without beams, as shown in Fig. 11.1a compared to Fig. 11.1b for slabs on beams, or Fig. 11.1c for waffle slab floors.

The evolution of the state of knowledge in slab design in the last 40 years will be briefly reviewed. The analysis of slab behavior in flexure up to the 1940s and early 1950s followed the classical theory of elasticity, particularly in the United States. The small deflections theory of plates, assuming the material to be homogeneous and isotropic, formed the basis of ACI code recommendations with moment coefficient tables. The work, principally by Westergaard, which empirically allowed limited moment redistribution, guided the thinking of the code writers. Hence the elastic solutions, complicated even for simple shapes and boundary conditions when no computers were available, made it mandatory to idealize and sometimes empiricize conditions beyond economic bounds.

In 1943, Johansen presented his yield-line theory for evaluating the collapse capacity of slabs. Since that time, extensive research into the ultimate behavior of reinforced concrete slabs has been undertaken. Studies by many investigators, such as those of Ockleston, Mansfield, Rzhanitsyn, Powell, Wood, Sawczuk, Gamble-Sozen-Siess, and Park, contributed immensely to further understanding of the limit-state behavior of slabs and plates at failure as well as serviceable load levels.

The various methods that are used for the analysis (design) of two-way action slabs and plates can be summarized in the following.

11.1.1 The Semielastic ACI Code Approach

The ACI approach gives two alternatives for the analysis and design of a framed two-way action slab or plate system: the direct design method and the equivalent frame method. Both methods are discussed in more detail in Section 11.3.

11.1.2 The Yield-Line Theory

Whereas the semielastic code approach applies to standard cases and shapes and has an inherent, excessively large safety factor with respect to capacity, the yield-line theory is a plastic theory easy to apply to irregular shapes and boundary conditions.

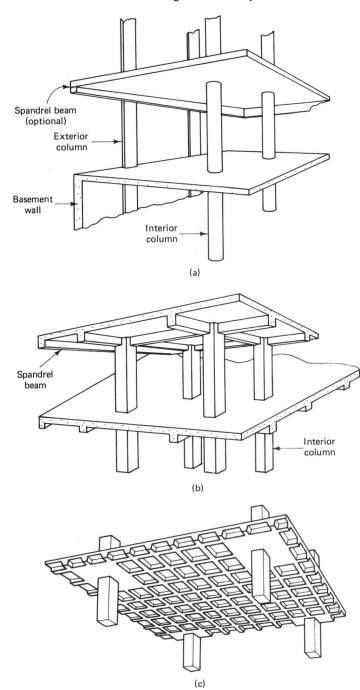

Figure 11.1 Two-way-action floor systems: (a) two-way flat-plate floor; (b) two-way slab floor on beams; (c) waffle slab floor.

Provided that serviceability constraints are applied, Johansen's yield-line theory is the simplest approach that the designer can use, representing the true behavior of reinforced concrete slabs and plates. It permits evaluation of the bending moments from an assumed collapse mechanism which is a function of the type of external load and the shape of the floor panel. This topic will be discussed in more detail in Section 11.8.

11.1.3 The Limit Theory of Plates

The interest in developing a limit solution became necessary due to the possibility of finding a variation in the collapse field which can give a lower failure load. Hence an upper-bound solution requiring a valid mechanism when supplying the work equation was sought, as well as a lower-bound solution requiring that the stress field satisfies everywhere the differential equation of equilibrium, that is,

$$\frac{\partial^2 M_x}{\partial x^2} - 2 \frac{\partial^2 M_{xy}}{\partial x\, \partial y} + \frac{\partial^2 M_y}{\partial y^2} = -w \qquad (11.1)$$

where M_x, M_y, and M_{xy} are the bending moments and w is the unit intensity of load. Variable reinforcement permits the lower-bound solution still to be valid. Wood, Park, and other researchers have given more accurate semiexact predictions of the collapse load.

For limit-state solutions, the slab is assumed to be completely rigid until collapse. Further work at Rutgers by the author incorporated the deflection effect at high load levels as well as the compressive membrane force effects in predicting the collapse load.

11.1.4 The Strip Method

This method was proposed by Hillerborg, attempting to fit the reinforcement to the strip fields. Since practical considerations require the reinforcement to be placed in orthogonal directions, Hillerborg set twisting moments equal to zero and transformed the slab into intersecting beam strips, hence the name "strip method."

Except for Johansen's yield-line theory, most of the other solutions are lower bound. Johansen's upper-bound solution can give the highest collapse load as long as a valid failure mechanism is used in predicting the collapse load.

11.1.5 Summary

The direct design method will be primarily discussed. This is necessitated by the scope of the textbook and the need for the student in a first course in reinforced concrete to acquire rapidly the minimum tools needed to use effectively the ACI building code for the design of two-way slabs and plate systems. A limited discussion of the equivalent frame method will, however, be presented in conjunction with some of the limitations of the direct design method.

11.2 FLEXURAL BEHAVIOR OF TWO-WAY SLABS AND PLATES

11.2.1 Two-Way Action

A single rectangular panel supported on all four sides by unyielding supports such as shear walls or stiff beams is first considered. The purpose is to visualize the physical behavior of the panel under gravity load. The panel will deflect in a dishlike form under the external load and its corners will lift if it is not monolithically cast with the supports. The contours shown in Fig. 11.2a indicate that the curvatures and consequently the moments at the central area C are more severe in the shorter direction y with its steep contours than in the longer direction x.

Evaluation of the division of moments in the x and y directions is extremely complex as the behavior is highly statically indeterminate. The discussion of the

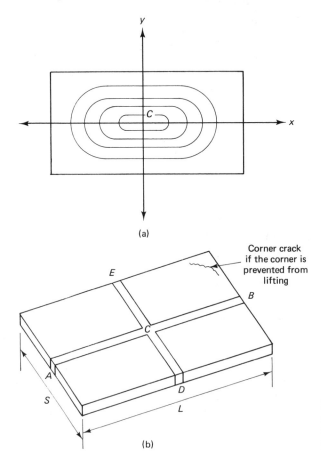

Figure 11.2 Deflection of panels and strips: (a) curvature and deflection contours in a floor panel; (b) central strips in a two-way slab panel.

simple case of the panel in Fig. 11.2a is expanded further by taking strips AB and DE at midspan as in Fig. 11.2b such that the deflection of both strips at central point C is the same.

The deflection of a simply supported uniformly loaded beam is $5wl^4/384EI$, namely, $\Delta = kwl^4$, where k is a constant. If the thickness of the two strips is the same, the deflection of strip AB would be $kw_{AB}L^4$ and the deflection of strip DE would be $kw_{DE}S^4$ where w_{AB} and w_{DE} are the portions of the total load intensity w transferred to strips AB and DE, respectively, namely $w = w_{AB} + w_{DE}$. Equating the deflections of the two strips at the central point C, we get

$$w_{AB} = \frac{wS^4}{L^4 + S^4} \tag{11.2a}$$

and

$$w_{DE} = \frac{wL^4}{L^4 + S^4} \tag{11.2b}$$

It is seen from the two relationships w_{AB} and w_{DE} in Eqs. 11.2a and 11.2b that the shorter span S of strip DE carries the heavier portion of the load. Hence the shorter span of such a slab panel on unyielding supports is subjected to the larger moment, supporting the foregoing discussion of the steepness of the curvature contours in Fig. 11.2a.

11.2.2 Relative Stiffness Effects

Alternatively, one has to consider a slab panel supported by flexible supports such as beams and columns, or flat plates supported by a grid of columns. The distribution of moments in the short and long directions are considerably more complex. The complexity arises from the fact that the degree of stiffness of the yielding supports determines the intensity of steepness of the curvature contours in Fig. 11.2a in both the x and y directions and the redistribution of moments.

The ratio of the stiffness of the beam supports to the slab stiffness can result in curvatures and moments in the long direction larger than those in the short direction, as the total floor behaves as an orthotropic *plate* supported on a grid of columns without beams. The moment values in the long and short directions in Ex. 11.1 and 11.2 arithmetically illustrate this discussion. If the long span L in such floor systems of slab panels without beams is considerably larger than the short span S, the maximum moment at the center of a plate panel would approximate the moment at the middle of a uniformly loaded strip of span L and clamped at both ends.

In summary, as slabs are flexible and highly under-reinforced, redistribution of moments in both the long and short directions depends on the relative stiffnesses of the supports and the supported panels. Overstress in one region is reduced by such redistribution of moments to the lesser stressed regions.

11.3 THE DIRECT DESIGN METHOD

The following discussion of the direct design method of analysis for two-way systems summarizes the ACI code approach for evaluation and distribution of the total moments in a two-way slab panel. The various moment coefficients are taken directly from the ACI code provisions. An assumption is made that vertical planes cut through an entire rectangular in plan multistory building along lines *AB* and *CD* in Fig. 11.3 midway between columns. A rigid frame results in the *x* direction. Similarly, vertical planes *EF* and *HG* result in a rigid frame in the *y* direction. A solution of such an idealized frame consisting of horizontal beams or equivalent slabs and supporting columns enables the design of the slab as the beam part of the frame. Approximate determinations of the moments and shears using simplified coefficients is presented throughout the direct design method. The equivalent frame method treats the idealized frame in a manner similar to an actual frame, hence more exact, and has fewer limitations than the direct design method. It basically involves a full moment distribution of many cycles as compared to the direct design method, which involves a one-cycle-moment distribution approximation.

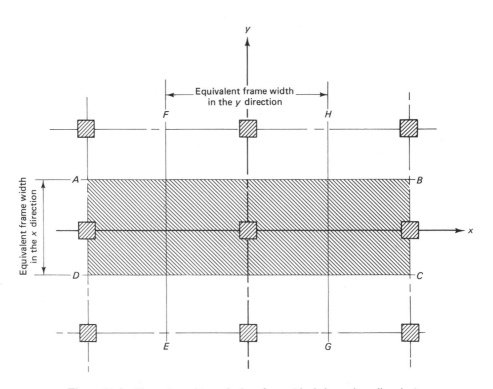

Figure 11.3   Floor plan with equivalent frame (shaded area in *x* direction).

11.3.1 Limitations of the Direct Design Method

The following are the limitations of this method:

1. A minimum of three continuous spans in each direction.
2. The ratio of the longer to the shorter span within a panel should not exceed 2.0.
3. Successive span lengths in each direction should not differ by more than one-third of the longer span.
4. Columns may be offset a maximum of 10% of the span in the direction of the offset from either axis between center lines of successive columns.
5. All loads shall be due to gravity only and uniformly distributed over the entire panel. The live load shall not exceed three times the dead load.
6. If the panel is supported by beams on all sides, the relative stiffness of the beams in two perpendicular directions shall not be less than 0.2 nor greater than 5.0.

It should be noted that the majority of normal floor systems satisfy these conditions.

11.3.2 Determination of the Factored Total Statical Moment M_o

There are basically four major steps in the design of the floor panels.

1. Determination of the total factored statical moment in each of the two perpendicular directions.
2. Distribution of the total factored design moment to the design of sections for negative and positive moment.
3. Distribution of the negative and positive design moments to the column and middle strips and to the panel beams, if any. A column strip is a width = 25% of the equivalent frame width on each side of the column center line and the middle strip is the balance of the equivalent frame width.
4. Proportioning of the size and distribution of the reinforcement in the two perpendicular directions.

Hence correct determination of the values of the distributed moments becomes a principal objective. Consider typical interior panels having center line dimensions l_1 in the direction of the moments being considered and dimensions l_2 in the direction perpendicular to l_1 as shown in Fig. 11.4. The clear span l_n extends from face to face of columns, capitals, or walls. Its value should not be less than $0.65\,l_1$ and circular supports shall be treated as square supports having the same cross-sectional area. The total statical moment of a uniformly loaded simply supported beam as a one-dimensional member is $M_0 = wl^2/8$. In a two-way slab panel as a two-dimensional member, the idealization of the structure through conversion to an equivalent frame

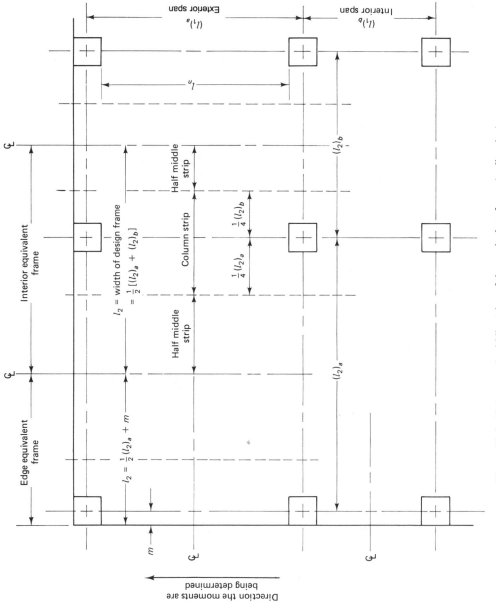

Figure 11.4 Column and middle strips of the equivalent frame (y direction).

Photo 60 Sydney Opera House during construction.

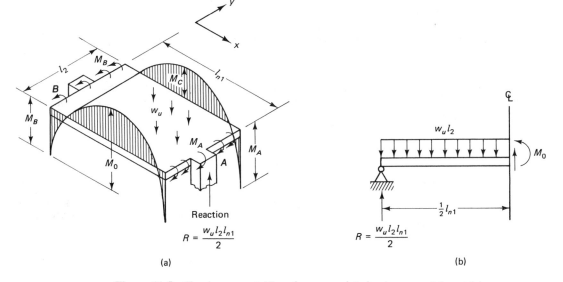

Figure 11.5 Simple moment M_0 acting on an interior two-way slab panel in x direction: (a) moment on panel; (b) free-body diagram.

makes it possible to calculate M_0 once in the x direction and again in the orthogonal y direction. If one takes as a free-body diagram the typical interior panel shown in Fig. 11.5a, symmetry reduces the shears and twisting moments to zero along the edges of the cut segment. If no restraint existed at ends A and B, the panel would be considered simply supported in span l_n direction. If one cuts at midspan as in Fig. 11.5b and considers half the panel as a free-body diagram, the moment M_0 at midspan would be

$$M_0 = \frac{w_u l_2 l_{n1}}{2} \frac{l_{n1}}{2} - \frac{w_u l_2 l_{n1}}{2} \frac{l_{n1}}{4}$$

or

$$M_0 = \frac{w_u l_2 (l_{n1})^2}{8} \tag{11.3}$$

Due to the actual existence of restraint at the supports, M_0 in the x direction would be distributed to the supports and midspan such that

$$M_0 = M_C + \tfrac{1}{2}(M_A + M_B)$$

The distribution would depend on the degree of stiffness of the support. In a similar manner, M_0 in the y direction would be the sum of the moments at midspan and the average of the moments at the supports in that direction.

The distribution of the statical factored moment M_0 to the column strip of the equivalent frame leads to the proportioning of the reinforcement in those strips.

11.4 DISTRIBUTED FACTORED MOMENTS AND SLAB REINFORCEMENT

11.4.1 Negative and Positive Factored Design Moments

From Fig. 11.6a, the negative factored moment factor in interior spans is 0.65 and the positive factor is 0.35 of the total statical moment M_0. For end spans of flat-plate floor panels, the M_0 factors are given in Table 11.1.

11.4.2 Factored Moments in Column Strips

A column strip is a design strip with a width on each side of the column equal to $0.25 l_2$ or $0.25 l_1$, whichever is *less* as shown in Figs. 11.4 and 11.6. The strip includes beams, if any. The middle strip is a design strip bound by the two column strips of the panel being analyzed.

Interior Panels

For interior negative moments, column strips have to be proportioned to resist the following portions in percent of the *interior* negative factored moments, with linear interpolation made for intermediate values.

l_2/l_1	0.5	1.0	2.0
$\alpha_1 (l_2/l_1) = 0$	75	75	75
$\alpha_1 (l_2/l_1) \geq 1.0$	90	75	45

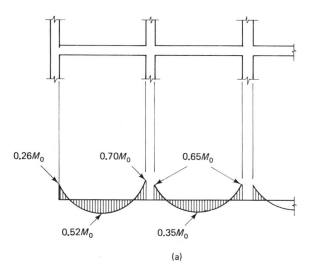

$0.26M_0$ $0.70M_0$ $0.65M_0$

$0.52M_0$ $0.35M_0$

(a)

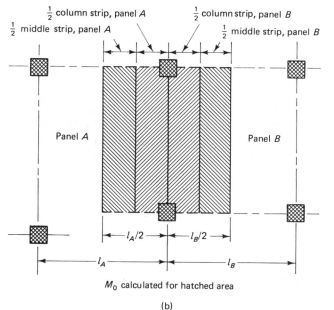

$\frac{1}{2}$ column strip, panel A $\frac{1}{2}$ column strip, panel B

$\frac{1}{2}$ middle strip, panel A $\frac{1}{2}$ middle strip, panel B

Panel A Panel B

$l_A/2$ $l_B/2$

l_A l_B

M_0 calculated for hatched area

(b)

Figure 11.6 Distribution of the statical factored moments M_0 into negative and positive moments: (a) moment coefficients for multispans; (b) slab areas for which M_0 is calculated.

α_1 in these tables is α in the direction of span l_1 for cases of two-way slabs on beams and is equal to the ratio of flexural stiffness of the beam section to the flexural stiffness of a width of slab bound laterally by centerlines of adjacent panels, if any, on each side of the beam $\alpha_1 = E_{cb}I_b/E_{cs}I_s$, where E_{cb} and E_{cs} are the modulus values of concrete and I_b and I_s are the moments of inertia of the beam and the slab, respectively. The factored moments in beams between supports have to be proportioned to resist 85% of the column strip moment when $\alpha_1(l_2/l_1) \geq 1.0$. Linear interpolation between 85 and 0% needs to be made for cases where $\alpha_1(l_2/l_1)$ varies between 1.0 and 0.

TABLE 11.1 MOMENT FACTORS FOR M_0 DISTRIBUTION IN EXTERIOR SPANS

	Exterior edge unre-strained	Slab with beams between all supports	Slab without beams between interior supports		Exterior edge fully restrained
			Without edge beam	With edge beam	
Interior negative factored moment	0.75	0.70	0.70	0.70	0.65
Positive factored moment	0.63	0.57	0.52	0.50	0.35
Exterior negative factored moment	0	0.16	0.26	0.30	0.65

Exterior Panels

For exterior negative moments, the column strips should be proportioned to resist the following portions in percent of the *exterior* negative factored moments with linear interpolation made for the intermediate values, where β_t is the torsional stiffness ratio. β_t = ratio of torsional stiffness of the edge beam section to the flexural stiffness of a width of a slab equal to span length of beam center to center of supports.

l_2/l_1		0.5	1.0	2.0
$\alpha_1 (l_2/l_1) = 0$	$\beta_t = 0$	100	100	100
	$\beta_t \geq 2.5$	75	75	75
$\alpha_1 (l_2/l_1) \geq 1.0$	$\beta_t = 0$	100	100	100
	$\beta_t \geq 2.5$	90	75	45

Positive Moments

For positive moments, the column strips have to be proportioned to resist the following portions in percent of the positive factored moments with linear interpolation being made for intermediate values.

l_2/l_1	0.5	1.0	2.0
$\alpha_1 (l_2/l_1) = 0$	60	60	60
$\alpha_1 (l_2/l_1) \geq 1.0$	90	75	45

11.4.3 Factored Moments in Middle Strips

That portion of the negative and positive factored moments not resisted by the column strips would have to be proportionately assigned to the corresponding half of the middle strips. Adjacent spans do not necessarily have to be equal such that the two halves of the column strip flanking a row of columns need not be equal in width. Hence each middle strip has to be proportioned to resist the sum of the moments assigned to its two half middle strips. A middle strip adjacent to and parallel with an edge supported by a wall must be proportioned to resist twice the moment assigned to the half middle strip corresponding to the first row of interior columns.

11.4.4 Effects of Pattern Loading on Increase of Positive Moments

The direct design method is sensitive to the increase in positive moments at midspans of a multi-panel floor system when not all the spans are simultaneously loaded. When alternate spans are loaded, the change in the negative moments at the supports is small, whereas the positive moments at midspans can increase considerably. If the ratio of the live load to dead load is high, such an increase in the positive moment can be as high as 50%, compared to the case where all spans are uniformly loaded. This increase in moment can result in excessive deflection and cracking of interior panels. It can be minimized only through stiffening of the columns, as seen in Fig. 11.7b, compared to Fig. 11.7a.

The ACI code permits up to a 33% increase in the positive moment value due to the expected redistribution of moments in a multipanel slab system from regions of high negative moments at the supports to regions of lower positive moments at midspan. However, the code stipulates that if the ratio of the unfactored live load to dead loads exceeds a value of 0.5, the stiffness ratio α_c has to be equal to or greater than the minimum stiffness ratio α_{min} given in Table 11.2.

If α_c is less than α_{min}, the positive factored moments in the spans of the panels supported by such columns have to be multiplied by a multiplier δ_s larger than 1.0, where

$$\delta_s = 1 + \frac{2 - \beta_a}{4 + \beta_a}\left(1 - \frac{\alpha_c}{\alpha_{min}}\right) \tag{11.4}$$

11.4.5 Shear-Moment Transfer to Columns Supporting Flat Plates

11.4.5.1 Shear Strength

The shear behavior of two-way slabs and plates is a three-dimensional stress problem. The critical shear failure plane follows the perimeter of the loaded area and is located at a distance that gives a minimum shear perimeter b_0. Based on extensive

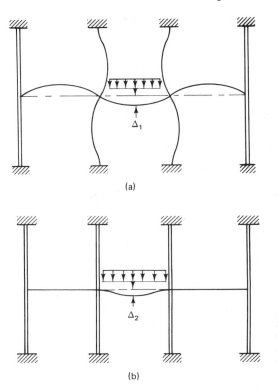

(a)

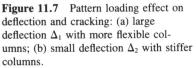

(b)

Figure 11.7 Pattern loading effect on deflection and cracking: (a) large deflection Δ_1 with more flexible columns; (b) small deflection Δ_2 with stiffer columns.

analytical and experimental verification, the shear plane should not be closer than a distance $d/2$ from the concentrated load or reaction area.

If no special shear reinforcement is provided, the nominal shear strength V_c of the section as required by the ACI is

$$V_c = \left(2 + \frac{4}{\beta_c}\right)\sqrt{f_c'}\,b_0 d \le 4\sqrt{f_c'}\,b_0 d \qquad (11.5)$$

where β_c is the ratio of the longer side to the shorter side of the loaded area and b_0 is the perimeter of the critical section. It is clear from Eq. 11.5 that the shear strength provided by the plain concrete cannot exceed $4\sqrt{f_c'}$, which is almost double the shear strength allowed in one-way members such as beams and one-way slabs.

If special shear reinforcement is provided, the maximum nominal shear strength V_n cannot exceed $6\sqrt{f_c'}\,b_0 d$, provided that the value used for V_c does not exceed $2\sqrt{f_c'}\,b_0 d$.

11.4.5.2 Shear-Moment Transfer

The unbalanced moment at the column face support of a slab without beams is one of the more critical design considerations in proportioning a flat plate or a flat slab. To ensure adequate shear strength requires moment transfer to the column by flexure

TABLE 11.2 VALUES OF α_{min} [a]

β_a	Aspect ratio, l_2/l_1	Relative Beam Stiffness, α				
		0	0.5	1.0	2.0	4.0
2.0	0.5–2.0	0	0	0	0	0
1.0	0.5	0.6	0	0	0	0
	0.8	0.7	0	0	0	0
	1.0	0.7	0.1	0	0	0
	1.25	0.8	0.4	0	0	0
	2.0	1.2	0.5	0.2	0	0
0.5	0.5	1.3	0.3	0	0	0
	0.8	1.5	0.5	0.2	0	0
	1.0	1.6	0.6	0.2	0	0
	1.25	1.9	1.0	0.5	0	0
	2.0	4.9	1.6	0.8	0.3	0
0.33	0.5	1.8	0.5	0.1	0	0
	0.8	2.0	0.9	0.3	0	0
	1.0	2.3	0.9	0.4	0	0
	1.25	2.8	1.5	0.8	0.2	0
	2.0	13.0	2.6	1.2	0.5	0.3

$$[a]\ \beta_a = \frac{\text{unfactored dead load per unit area}}{\text{unfactored live load per unit area}}$$

$$\alpha_c = \frac{\Sigma K_c}{\Sigma(K_b + K_s)}$$

$$= \frac{\text{sum of stiffness of columns above and below slab}}{\text{sum of stiffness of beams and slabs framing into the joint}}$$
in the direction of the span for which the moments are
being determined

across the perimeter of the column and by eccentric shearing stress such that approximately 60% is transferred by flexure and 40% by shear.

The fraction γ_v of the moment transferred by eccentricity of the shear stress decreases as the width of the face of the critical section resisting the moment increases such that

$$\gamma_v = 1 - \frac{1}{1 + \frac{2}{3}\sqrt{\dfrac{c_1 + d}{c_2 + d}}} \tag{11.6a}$$

where $(c_2 + d)$ is the width of the face of the critical section resisting the moment and $(c_1 + d)$ is the width of the face at right angles to $(c_2 + d)$.

The remaining portion γ_f of the unbalanced moment transferred by flexure is given by

$$\gamma_f = \frac{1}{1 + \frac{2}{3}\sqrt{\dfrac{c_1 + d}{c_2 + d}}} \quad \text{or } \gamma_f = 1 - \gamma_v \tag{11.6b}$$

and acting on an effective slab width between lines that are $1\frac{1}{2}$ times the total slab thickness h on both sides of the column support.

The distribution of shear stresses around the column edges is as shown in Fig. 11.8. It is considered to vary linearly about the centroid of the critical section. The factored shear force V_u and the unbalanced factored moment M_u, both assumed acting at the column face, have to be transferred to the centroidal axis c–c of the critical section. Thus, the axis position has to be located, thereby obtaining the shear force arm g (distance from the column face to the centroidal axis plane) of the critical section c–c for the shear moment transfer.

For calculating the maximum shear stress sustained by the plate in the edge column region, the ACI Code requires using the full nominal moment strength M_n provided by the column strip in Eqs. 11.7a, b, and c as the unbalanced moment, multiplied by the transfer fraction factor γ_v. This unbalanced moment $M_n \geq \dfrac{M_{ue}}{\phi}$ is composed of two parts: the negative end panel moment $M_{ne} = M_e/\phi$ at the face of the column and the moment $(V_u/\phi)g$, due to the eccentric factored perimetric shear force V_u. The limiting value of the shear stress intensity is expressed as

$$\frac{v_{u(AB)}}{\phi} = \frac{V_u}{\phi A_c} + \frac{\gamma_v M_{ue} c_{AB}}{\phi J_c} \tag{11.7a}$$

$$\frac{v_{u(CD)}}{\phi} = \frac{V_u}{\phi A_c} - \frac{\gamma_v M_{ue} c_{CD}}{\phi J_c} \tag{11.7b}$$

where the nominal shear strength intensity is

$$v_n = v_u/\phi \tag{11.7c}$$

and where A_c = Area of concrete of assumed critical section

$$= 2d\,(c_1 + c_2 + 2d)\ \text{for an interior column}$$

and J_c = Property of assumed critical section analogous to polar moment of inertia.

The value of J_c for an interior column is:

$$J_c = \frac{d(c_1 + d)^3}{6} + \frac{d^3(c_1 + d)}{6} + \frac{d(c_2 + d)(c_1 + d)^2}{2}$$

The value of J_c for an edge column with bending parallel to the edge is:

$$J_c = \frac{(c_1 + d/2)(d)^3}{6} + \frac{2(d)}{3}(c^3{}_{AB} + c^3{}_{CD}) + (c_2 + d)(d)(c_{AB})^2$$

It can be recognized from basic principles of mechanics of materials that the shearing stress

$$v_u = \frac{V_u}{A_c} + \gamma_v \frac{Mc}{J}$$

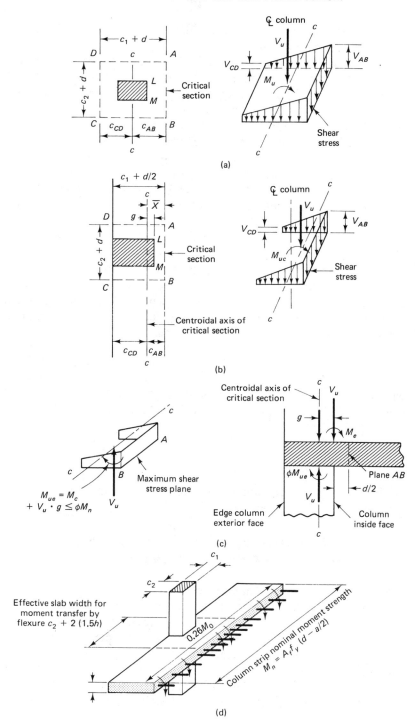

Figure 11.8   Shear stress distribution around column edges: (a) interior column; (b) end column; (c) critical surface; (d) transfer nominal moment strength M_n.

where the second part of the term is the shearing stress resulting from the torsional moment at the column face.

If the nominal moment strength M_n of the shear moment transfer zone after the design of the reinforcement results in a larger value than M_{ue}/ϕ, the M_n value should be used in Eqs. 11.7a and b in lieu of M_{ue}/ϕ. In such a case, namely where the moment strength value $M_n = M_{ne} + (V_u/\phi)g$ is increased because of the use of flexural reinforcement in excess of what is needed to resist M_{ue}/ϕ, the slab stiffness is raised, thereby increasing the transferred shear stress v_u calculated from Eqs. 11.7a and b for development of full moment transfer. Consequently, it is advisable to maintain a design moment M_{ue} with a value close to the factored moment value M_{ue} if an increase in the shear stress due to additional moment transfer needs to be avoided and a possible resulting need for additional increase in the plate design thickness prevented.

Numerical Ex. 11.1 illustrates the procedure for calculating the limit perimeter shear stress in the plate at the edge column region.

A higher perimetric shearing stress v_u can occur than evaluated by Eqs. 11.7a or b when adjoining spans are unequal or unequally loaded in the case of an interior column. The ACI Code stipulates in the slab section pertaining to factored moments in columns and walls that the supporting element such as a column or a wall has to resist an unbalanced moment M', such that

$$M' = 0.07[(w_{nd} + 0.5w_{nl})l_2 l_{n2} - w'_{nd}l'_2 (l'_n)^2]$$

where w'_{nd}, l'_2, l'_n refer to the shorter span.
Hence, an additional term is added to Eqs. 11.7a or b in such cases so that

$$v_u = \frac{V_u}{A_c} + \frac{\gamma_v M_u c_{AB}}{J_c} + \frac{\gamma_v M' c}{J'_c}$$

where J'_c in the polar moment of inertia with moment areas taken in a direction perpendicular to that used for J_c.

11.4.6 Deflection Requirements for Minimum Thickness—An Indirect Approach

Serviceability of a floor system can be maintained through deflection control and crack control. Since deflection is a function of the stiffness of the slab as a measure of its thickness, a minimum thickness has to be provided irrespective of the flexural thickness requirement. Table 11.3 gives the maximum permissible computed deflections to safeguard against plaster cracking and to maintain an aesthetic appearance. Deflection computations for two-way action slabs can be made using the analytical procedures described in Section 11.6 in order to determine whether the analysis gives long-term deflections within the limitations of Table 11.3.

TABLE 11.3 MINIMUM PERMISSIBLE RATIOS OF SPAN (l) TO DEFLECTION (a)

(l = longer span)

Type of member	Deflection, a, to be considered	$(l/a)_{min}$
Flat roofs not supporting and not attached to nonstructural elements likely to be damaged by large deflections	Immediate deflection due to live load L	180[a]
Floors not supporting and not attached to nonstructural elements likely to be damaged by large deflections	Immediate deflection due to live load L	360
Roof or floor construction supporting or attached to nonstructural elements likely to be damaged by large deflections	That part of total deflection occurring after attachment of nonstructural elements; sum of long-time deflection due to all sustained loads (dead load plus any sustained portion of live load) and immediate deflection due to any additional live load[b]	480[c]
Roof or floor construction supporting or attached to nonstructural elements not likely to be damaged by large deflections		240[c]

[a]Limit not intended to safeguard against ponding. Ponding should be checked by suitable calculations of deflection, including added deflections due to ponded water, and considering long-term effects of all sustained loads, camber, construction tolerances, and reliability of provisions for drainage.

[b]Long-term deflection has to be determined but may be reduced by the amount of deflection calculated to occur before attachment of nonstructural elements. This reduction is made on the basis of accepted engineering data relating to time–deflection characteristics of members similar to those being considered.

[c]Ratio limit may be lower if adequate measures are taken to prevent damage to supported or attached elements, but should not be lower than tolerance of nonstructural elements.

Approximate empirical limitation on deflection through determination of the minimum thickness of the slab or plate can be obtained from the following ACI expressions for the total thickness h:

$$h = \frac{l_n(800 + 0.005f_y)}{36,000 + 5000\beta[\alpha_m - 0.5(1 - \beta_s)(1 + 1/\beta)]} \tag{11.8}$$

but not less than

$$h = \frac{l_n(800 + 0.005f_y)}{36,000 + 5000\beta(1 + \beta_s)} \tag{11.9}$$

and need not be more than

$$h = \frac{l_n(800 + 0.005f_y)}{36,000} \tag{11.10}$$

The total thickness h (in.) has to be the larger of the two values from Eqs. 11.8 and 11.9 but need not exceed the value obtained from Eq. 11.10. Thickness h, however, cannot be less than the following values:

Slabs without beams or drop panels	5 in.
Slabs without beams, but with drop panels	4 in.
Slabs with beams on all four edges with a value of α_m at least equal to 2.0	3.5 in.

h also has to be increased by at least 10% for flat-plate floors if the end panels have no edge beams.

In addition, in the equations above,

α = ratio of flexural stiffness of beam section to flexural stiffness of a width of slab bounded laterally by centerline of adjacent panel (if any) on each side of beam

α_m = average value of α for all beams on edges of a panel

β = ratio of clear spans in long to short direction of two-way slabs

β_s = ratio of length of continuous edges to total perimeter of a slab panel

It has to be emphasized that a deflection check is critical for the construction loading condition. Shoring and reshoring patterns can result in dead load deflection in excess of the normal service load state at a time when the concrete has only a seven-day strength or less and not the normal design 28-day strength. The stiffness EI in such a state is less than the design value. Flexural cracking lowers further the stiffness values of the two-way slab or plate with a possible increase in long-term deflection several times the anticipated design deflection. Consequently, reinforced concrete two-way slabs and plates have to be constructed with a camber of one-eighth inch in 10 ft. span or more and crack control exercised as in Section 11.7 in order to counter the effects of excessive deflection at the construction loading stage.

11.5 DESIGN AND ANALYSIS PROCEDURE

11.5.1 Operational Steps

Figure 11.9 gives a logic flowchart for the following operational steps.

1. Determine whether the slab geometry and loading allow the use of the direct design method as listed in Section 11.3.1.
2. Select slab thickness to satisfy deflection and shear requirements. Such calculations require a knowledge of the supporting beam or column dimensions. A

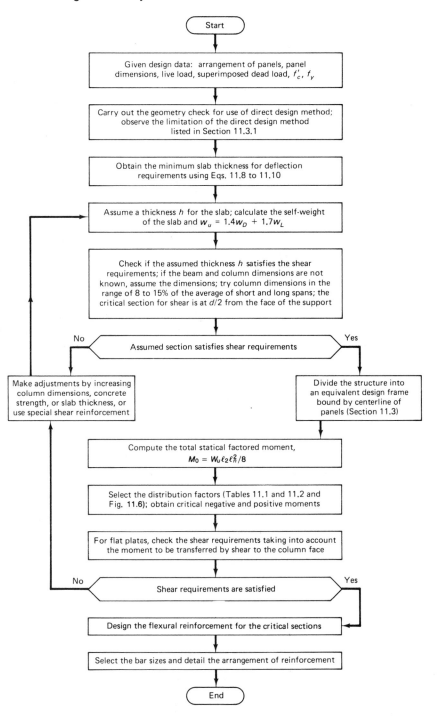

Figure 11.9   Flowchart for design sequence in two-way slabs and plates.

reasonable value of such a dimension of columns or beams would be 8 to 15% of the average of the long and short span dimensions, namely $\frac{1}{2}(l_1 + l_2)$. For shear check, the critical section is at a distance $d/2$ from the face of the support. If the thickness shown for deflection is not adequate to carry the shear, use one or more of the following:

(a) Increase the column dimension.
(b) Increase concrete strength.
(c) Increase slab thickness.
(d) Use special shear reinforcement.
(e) Use drop panels or column capitals to improve shear strength.

3. Divide the structure into equivalent design frames bound by centerlines of panels on each side of a line of columns.
4. Compute the total statical factored moment $M_0 = (w_u l_2 l_{n1}^2)/8$.
5. Select the distribution factors of the negative and positive moments to the exterior and interior columns and spans as in Fig. 11.5 and Table 11.1 and calculate the respective factored moments.
6. Distribute the factored equivalent frame moments from step 4 to the column and middle strips.
7. Determine whether the trial slab thickness chosen is adequate for moment-shear transfer in the case of flat plates at the interior column junction computing that portion of the moment transferred by shear and the properties of the critical shear section at distance $d/2$ from column face.
8. Design the flexural reinforcement to resist the factored moments in step 6.
9. Select the size and spacing of the reinforcement to fulfill the requirements for crack control, bar development lengths, and shrinkage and temperature stresses.

11.5.2 Example 11.1: Design of Flat Plate without Beams

A three-story building is four panels by four panels in plan. The clear height between the floors is 12 ft and the floor system is a reinforced concrete flat-plate construction with no edge beams. The dimensions of the end panels as well as the size of the supporting columns are shown in Fig. 11.10. Given:

> Live load = 50 psf (2.39 kPa)
> f'_c = 4000 psi (27.6 MPa), normalweight concrete
> f_y = 60,000 psi (414 MPa)

The building is not subject to earthquake; consider gravity loads only. Design the end panel and the size and spacing of the reinforcement needed. Consider flooring weight to be 10 psf in addition to the floor self-weight.

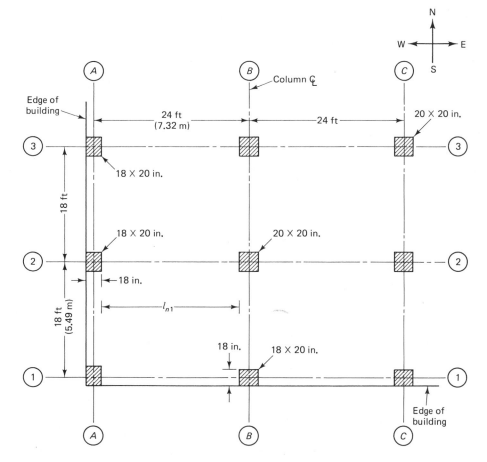

Figure 11.10   Floor plan of end panels in a three-story building.

Solution

Geometry check for use of direct design method (Step 1)

(a) Ratio $\dfrac{\text{longer span}}{\text{shorter span}} = \dfrac{24}{18} = 1.33 < 2.0$, hence two-way action

(b) More than three spans in each direction and successive spans in each direction the same and columns are not offset.

(c) Assume a thickness of 9 in. and flooring of 10 psf.

$$w_d = 10 + \frac{9}{12} \times 150 = 122.5 \text{ psf} \qquad 3w_d = 367.5 \text{ psf}$$

$$w_l = 50 \text{ psf} < 3w_d \qquad \text{O.K.}$$

Hence the direct design method is applicable.

Minimum slab thickness for deflection requirement (Step 2)

$$\text{E–W direction } l_{n1} = 24 \times 12 - \frac{18}{2} - \frac{20}{2} = 269 \text{ in. (6.83 m)}$$

$$\text{N–S direction } l_{n2} = 18 \times 12 - \frac{20}{2} - \frac{20}{2} = 196 \text{ in. (4.98 m)}$$

$$\text{Ratio of longer to shorter clear span } \beta = \frac{269}{196} = 1.37$$

$$\beta_s = \frac{24 + 18 + 24}{2(24 + 18)} = 0.79$$

$$\alpha_m = 0 \quad \text{since there are no edge beams}$$

The larger span should be used in the calculation of the plate thickness as required for deflection control, using Eqs. 11.8, 11.9, and 11.10. Equation 11.8 contains in its denominator the stiffness expression

$$\alpha_m - 0.5(1 - \beta_s)\left(1 + \frac{1}{\beta}\right)$$

Since α_m in this case is zero, the denominator in Eq. 11.8 becomes smaller as the bracketed stiffness expression becomes negative. Consequently, Eq. 11.8 will not control since the slab thickness h cannot exceed the value determined by Eq. 11.10. It is noted that Eq. 11.10 always governs for slabs without beams. Hence

$$h = \frac{l_n(800 + 0.005f_y)}{36,000} = 8.22 \text{ in.} \qquad (11.10)$$

Since there is no edge beam, h has to be increased by at least 10%, which equals $8.22 \times 1.10 \simeq 9.1$ in. Try a $9\frac{1}{2}$-in. (241.3-mm)-thick plate. This thickness is larger than the absolute minimum thickness of 5 in. required in the code for flat plates; hence O.K. Assume $d \simeq h - 1'' = 8.5$ in.

$$\text{new } w_d = 10 + \frac{9.5}{12} \times 150 = 128.75 \text{ psf}$$

Therefore,

$$3w_d = 386.25 \text{ psf}$$

$$w_l = 50 \text{ psf} < 3w_d \qquad \text{O.K.}$$

Shear thickness requirement (Step 2)

$$w_u = 1.7L + 1.4D = 1.7 \times 50 + 1.4 \times 128.75$$

$$= 265.25 \text{ psf} \quad \text{say 266 psf (12.74 kPa)}$$

Interior Column: The controlling critical plane of maximum perimetric shear stress is at a distance $d/2$ from the column faces; hence, the net factored perimetric shear force is

$$V_u = [(l_1 \times l_2 - (c_1 + d)(c_2 + d)]w_u$$

$$= \left(18 \times 24 - \frac{20 + 8.5}{12} \times \frac{20 + 8.5}{12}\right)266 = 113{,}412 \text{ lb}$$

From Fig. 11.11, perimeter of the critical shear failure surface is

$$b_0 = 2(c_1 + d + c_2 + d) = 2\,(c_1 + c_2 + 2d)$$

Perimetric shear surface $A_c = b_0 d = 2d(c_1 + c_2 + 2d) = 2 \times 8.5(20 + 20 + 17)$

$$= 969 \text{ in}^2 \ (625{,}005 \text{ mm}^2)$$

Since moments are not known at this stage, only a preliminary check for shear can be made. If

$$\beta_c = \text{Ratio of longer to shorter side of columns}$$

$$= \frac{20}{20} = 1.0$$

For

$$\beta_c \le 2.0 \qquad V_c \le 4\sqrt{f_c'}\,A_c$$

$$\beta_c > 2.0 \qquad V_c \le \left(2 + \frac{4}{\beta_c}\right)\sqrt{f_c'}\,A_c$$

$$V_n = \frac{V_u}{\phi} = \frac{113{,}412}{0.85} = 133{,}425 \text{ lb } (593.5 \text{ kN})$$

$$V_c = 4\sqrt{4000} \times 969 = 245{,}140 \quad \text{lb} > 133{,}425 \quad \text{lb;} \quad \text{o.k.} \quad \text{for}$$
preliminary calculations.

Exterior Column: Include weight of exterior wall, assuming its service weight to be 270 plf. Net factored perimetric shear force is

$$V_u = \left[18 \times \left(\frac{24}{2} + \frac{18}{2 \times 12}\right) - \frac{(18 + 4.25)(20 + 8.5)}{144}\right]266$$

$$+ \left(18 - \frac{20}{12}\right) \times 270 \times 1.4 = 66{,}050 \text{ lb.} \qquad V_n = \frac{66{,}050}{0.85} = 77{,}706 \text{ lb.}$$

Consider the line of action of V_u to be at the column face *LM* in Fig. 11.12 for shear moment transfer to the centroidal plane c–c. This approximation is adequate since V_u acts *perimetrically* around the column faces and not along line *AB* only.
From Fig. 11.12,

$$A_c = d(2c_1 + c_2 + 2d) = 8.5(2 \times 18 + 20 + 17)$$

$$= 621 \text{ in}^2 \ (400{,}322 \text{ mm}^2)$$

$$V_c = 4\sqrt{4000} \times 621 = 157{,}102 \text{ lb} > 77{,}706 \text{ lb} \qquad \text{O.K. for}$$
preliminary check.

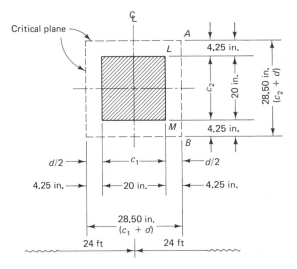

Figure 11.11 Critical plane for shear moment transfer in Ex. 11.1 interior column (line B-B, Fig. 11.10).

Statical moment computation (Steps 3 to 5)

$$\text{E–W: } l_{n1} = 269 \text{ in.} = 22.42 \text{ ft}$$

$$\text{N–S: } l_{n2} = 196 \text{ in.} = 16.33 \text{ ft}$$

$$0.65\, l_1 = 0.65 \times 24 = 15.6 \text{ ft} \qquad \text{Use } l_{n1} = 22.4 \text{ ft}$$

$$0.65\, l_2 = 0.65 \times 18 = 11.7 \text{ ft} \qquad \text{Use } l_{n2} = 16.33 \text{ ft}$$

(a) *E–W direction:*

$$M_0 = \frac{w_u l_2 l_{n1}^2}{8} = \frac{266 \times 18(22.42)^2}{8} = 300{,}840 \text{ ft-lb (408 kN-m)}$$

For end panel of a flat plate without end beams, the moment distribution factors as in Table 11.1 are:

$$-M_u \text{ at first interior support} = 0.70 M_0$$

$$+M_u \text{ at midspan of panel} = 0.52 M_0$$

$$-M_u \text{ at exterior face} = 0.26 M_0$$

$$\text{negative design moment } -M_u = 0.70 \times 300{,}840$$

$$= 210{,}588 \text{ ft-lb (286 kN-m)}$$

$$\text{positive design moment } +M_u = 0.52 \times 300{,}840$$

$$= 156{,}437 \text{ ft-lb (212 kN-m)}$$

$$\text{negative moment at exterior } -M_u = 0.26 \times 300{,}840$$

$$= 78{,}218 \text{ ft-lb (106 kN-m)}$$

(b) *N–S direction:*

$$M_0 = \frac{w_u l_1 l_{n2}^2}{8} = \frac{266 \times 24(16.33)^2}{8} = 212{,}802 \text{ ft-lb (289 kN-m)}$$

$$\text{negative design moment } -M_u = 0.70 \times 212{,}802$$
$$= 148{,}961 \text{ ft-lb (202 kN-m)}$$

$$\text{positive design moment } +M_u = 0.52 \times 212{,}802$$
$$= 110{,}657 \text{ ft-lb (150 kN-m)}$$

$$\text{negative design moment at exterior face } -M_{u1} = 0.26 \times 212{,}802$$
$$= 55{,}329 \text{ ft-lb (75 kN-m)}$$

Note that the smaller moment factor 0.35 could be used for the positive factored moment in the N–S direction in this example if the exterior edge is fully restrained.

Moment distribution in the column and middle strips (Steps 6 and 7)

At column E there is no torsional edge beam, hence torsional stiffness ratio β_t of an edge beam to the columns is zero. Hence $\alpha_1 = 0$. From the *exterior* factored moments tables for the column strip in Section 11.4.2, the distribution factor for the negative moment at the exterior support is 100%, the positive midspan moment 60% and the interior negative moment is 75%.

Check the shear moment transfer capacity at the exterior column supports

$$-M_c \text{ at interior column 2-}B = 210{,}588 \text{ ft-lb}$$

$$-M_e \text{ at exterior column 2-}A = 78{,}218 \text{ ft-lb}$$

$$V_u = 66{,}050 \text{ lb acting at the face of the column}$$

TABLE 11.4 MOMENT DISTRIBUTION OPERATIONS TABLE

	E–W direction l_2/l_1: 18/24 = 0.75 $\alpha_1(l_2/l_1)$: 0			N–S direction 24/18 = 1.33 0		
Column strip	Interior negative moment	Positive midspan moment	Exterior negative moment	Interior negative moment	Positive midspan moment	Exterior negative moment
M_u ft-lb	210,588	156,437	78,218	148,961	110,657	55,329
Distribution factor (%)	75	60	100	75	60	100
Column strip design moments (ft-lb)	0.75 × 210,588 / 157,941	0.60 × 156,437 / 93,862	1.0 × 78,218 / 78,218	.75 × 148,961 / 111,721	0.60 × 110,657 / 66,394	1.0 × 55,329 / 55,329
Middle strip design moments (ft-lb)	210,588 − 157,941 / 52,647	156,437 − 93,862 / 62,575	78,218 − 78,218 / 0	148,961 − 111,721 / 37,240	110,657 − 66,394 / 44,263	55,329 − 55,329 / 0

The ACI Code stipulates that the nominal moment strength be used in evaluating the unbalanced transfer moment at the edge column, namely, using M_n based on $-M_e = 78,218$ ft-lb.

Factored shear force at the edge column adjusted for the interior moment is

$$V_u = 66,050 - \frac{210,588 - 78,218}{24 - \frac{9 + 10}{12}} = 60,145 \text{ lb}$$

$V_n = 60,145/0.85 = 70,759$ lb, assuming that the design M_u has the same value as the factored M_u.

$$A_c \text{ from before } = 621 \text{ in}^2$$

From Figs. 11.8c and 11.12, taking the moment of area of the critical plane about axis AB,

$$d(2c_1 + c_2 + 2d)\bar{x} = d\left(c_1 + \frac{d}{2}\right)^2$$

where $\bar{x}$ is the distance to the centroid of the critical section or

$$(2 \times 18 + 20 + 17)\bar{x} = \left(18 + \frac{8.5}{2}\right)^2$$

$$\bar{x} = \frac{495.06}{73} = 6.78 \text{ in. (172.2 mm)}$$

$$g = 6.78 - \frac{8.5}{2} = 2.53 \text{ in. where } g \text{ is the distance from}$$

the column face to the centroidal axis of the section.

To transfer the shear V_u from the face of column to the centroid of the critical section adds an additional moment to the value of $M_e = 78,218$ ft-lb. Therefore, the total external factored moment $M_{ue} = 78,218 + 60,145(2.53/12) = 90,899$ ft-lb. Total required minimum unbalanced moment strength:

$$M_n = \frac{M_{ue}}{\phi} = \frac{90,899}{0.90} = 100,999 \text{ ft-lb}$$

The fraction of nominal moment strength M_n to be transferred by shear is

$$\gamma_v = 1 - \frac{1}{1 + \frac{2}{3}\sqrt{\frac{c_1 + d}{c_2 + d}}} = 1 - \frac{1}{1 + 0.59} = 0.37$$

It should be noted that the dimension $(c_1 + d)$ for the end column in the above expression becomes $(c_1 + d/2)$. Hence $M_{nv} = 0.37 M_n$. Moment of inertia of sides parallel to the moment direction about y–y axis:

$$I_1 = \left(\frac{bh^3}{12} + Ad^2 + \frac{hb^3}{12}\right)2 \text{ for both faces}$$

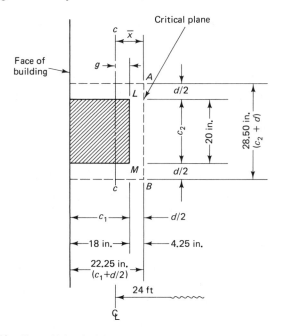

Figure 11.12 Centroidal axis for shear moment transfer in Ex. 11.1 end column (line A-A or 1-1, Fig. 11.10).

$$I_1 = \left[\frac{8.5(22.25)^3}{12} + (8.5 \times 22.25)\left(\frac{22.25}{2} - 6.78\right)^2 + \frac{22.25(8.5)^3}{12} \right]2$$

$$= (7,802 + 3,571 + 1,139)2 = 25,024 \text{ in}^4$$

Moment of inertia of sides perpendicular to the moment direction about y–y axis:

$$I_2 = Ad^2 = [(20 + 8.5)8.5](6.78)^2 = 11,136 \text{ in}^4$$

Therefore,

$$\text{torsional moment of inertia } J_c = 25,024 + 11,136$$

$$= 36,160 \text{ in}^4$$

Shearing stress due to perimeter shear, effect of M_n and weight of wall is

$$v_n = \frac{V_u}{\phi A_c} + \frac{\gamma_v c_{AB} M_n}{J_c} \qquad \text{where } M_{nv} = \gamma_v \times M_n$$

$$= \frac{60,145}{0.85 \times 621} + \frac{0.37 \times 6.78 \times 100,999 \times 12}{36,160}$$

$$= 113.94 + 84.08 = 198.02 \text{ psi}$$

$$\text{Max allowable } v_c = 4\sqrt{f_c'} = 4\sqrt{4,000} = 253.0 \text{ psi}$$

$$v_n < v_c$$

Therefore, accept plate thickness. For the corner panel column, special shear-head provision or an enlarged column or capital might be needed to resist the high shear stresses at that location.

Design of reinforcement in the slab area at column face for the unbalanced moment transferred to the column by flexure

From Eq. 11.6b

$$\gamma_f = 1 - \gamma_v = 1 - 0.37 = 0.63$$

$$M_{nf} = \gamma_f M_n = 0.63 \times 100,999 \times 12 = 763,552 \text{ in.-lb}$$

This moment has to be transferred within $1.5h$ on each side of the column as in Fig. 11.8d.

$$\text{Transfer width} = (1.5 \times 9.5) 2 + 20 = 48.5 \text{ in.}$$

$$M_{nf} = \phi A_s f_y \left(d - \frac{a}{2} \right); \qquad \text{Assume } \left(d - \frac{a}{2} \right) \approx 0.9d$$

or $763,552 = A_s \times 60,000(8.5 \times 0.9)$ gives

$$A_s = 1.66 \text{ in}^2 \text{ over a strip width} = 48.5 \text{ in.}$$

Verifying A_s:

$$a = \frac{1.66 \times 60,000}{0.85 \times 4,000 \times 48.5} = 0.60 \text{ in}^2$$

Therefore,

$$763,552 = A_s \times 60,000 \left(8.5 - \frac{0.60}{2} \right)$$

$$A_s = 1.55 \text{ in.}^2 \cong 5 \text{ #5 bars @ } 4'' \text{ center to center}$$

are to be used in the 20 in. column width strip and anchored into the column as required for bond length development.

This additional steel will have to be used to effect the moment transfer. Reinforcement to carry the total edge column strip moment $M_e = 78,218$ ft-lb, namely $M_{ne} = 86,909$ ft-lb, is proportioned in the next section.

Checks have to be made in a similar manner for the shear moment transfer at the face of the interior column C. As also described in Sec. 11.4.5.2, checks are sometimes necessary for pattern loading conditions and for cases where adjoining spans are not equal or not equally loaded.

Proportioning of the plate reinforcement (Steps 8 and 9)

(a) *E–W direction (long span)*
1. *Summary of moments in column strip (ft-lb):*

$$\text{interior column negative } M_n = \frac{157{,}941}{\phi = 0.9} = 175{,}490$$

$$\text{midspan positive } M_n = \frac{93{,}862}{0.9} = 104{,}291$$

$$\text{exterior column negative } M_{ne} = \frac{78{,}218}{0.9} = 86{,}909$$

2. *Summary of moments in middle strip (ft-lb):*

$$\text{interior column negative } M_n = \frac{52{,}647}{0.9} = 58{,}497$$

$$\text{midspan positive } M_n = \frac{62{,}575}{0.9} = 69{,}528$$

$$\text{exterior column negative } M_n = 0$$

3. *Design of reinforcement for column strip:*

$-M_n = 175{,}490$ ft-lb acts on a strip width of $2(0.25 \times 18) = 9.0$ ft

$$\text{unit } -M_n \text{ per 12 in. width strip} = \frac{175{,}490 \times 12}{9.0} = 233{,}987 \text{ in.-lb}$$

$$\text{unit } +M_n = \frac{104{,}291 \times 12}{9.0} = 139{,}055 \text{ in.-lb/ 12 in. width strip}$$

Negative steel:

$$M_n = A_s f_y \left(d - \frac{a}{2} \right) \qquad \text{or} \qquad 233{,}987 = A_s \times 60{,}000 \left(8.5 - \frac{a}{2} \right)$$

Assume that moment arm $(d - a/2) \simeq 0.9d$ for first trial and $d = h - 3/4$ in. $- \frac{1}{2}$ diameter of bar $\simeq 8.5$ in. for all practical purposes. Therefore,

$$A_s = \frac{233{,}987}{60{,}000 \times 0.9 \times 8.5} = 0.51 \text{ in.}^2$$

$$a = \frac{A_s f_y}{0.85 f_c' b} = \frac{0.51 \times 60{,}000}{0.85 \times 4000 \times 12} = 0.75 \text{ in.}$$

For the second trial-and-adjustment cycle,

$$233{,}987 = A_s \times 60{,}000 \left(8.5 - \frac{0.75}{2} \right)$$

Therefore, required $-A_s$ per 12-in.-wide strip $= 0.48$ in.^{2}. Try No. 5 bars (area per bar $= 0.305$ in.^{2}).

$$\text{spacing } s = \frac{\text{area of one bar}}{\text{required } A_s \text{ per 12-in. strip}}$$

Therefore,

$$s \text{ for negative moment} = \frac{0.305}{0.48/12} = 7.63 \text{ in. c-c (194 mm)}$$
(No. 5 bars)

$$s \text{ for positive moment} = 7.63 \times \frac{233,987}{139,055} = 12.84 \text{ in. } c\text{-}c \text{ (326 mm)}$$

The maximum allowable spacing $= 2h = 2 \times 9.5 = 19$ in. (483 mm). Try No. 4 bars for positive movement ($A_s = 0.20$ in.^{2}).

$$A_s = \frac{139,055}{233,987} \times 0.48 = 0.29 \text{ in.}^2 \text{ per 12-in. strip}$$

$$s = \frac{0.20}{0.29/12} = 8.28 \text{ in. c-c (210 mm)}$$

For an external negative moment, use No. 4 bars.

$$s = 8.28 \times \frac{104,291}{86,909} = 9.94 \text{ in. c-c}$$

Use 14 No. 5 bars at 7 1/2 in. center to center for negative moment at interior column side; 12 No. 4 bars at 8 in. center to center for positive moment; and 10 No. 4 bars at 9 1/2 in. center to center for the exterior negative moment M_e with 8 of these bars to be placed outside the shear moment transfer band of width 48.5 in. as seen in Fig. 11.13(b)

4. *Design of reinforcement for middle strip:*

$$\text{unit} -M_n = \frac{52,647}{0.9} = 58,497 \text{ acting on a strip width of } 18.0 - 9.0 = 9.0 \text{ ft}$$

$$\text{unit} -M \text{ per 12-in.-width strip} = \frac{58,497 \times 12}{9} = 77,996 \text{ lb-in.}$$

$$77,996 = A_s \times 60,000(8.5 \times 0.9)$$

$$A_s = 0.17 \text{ in.}^2 \qquad a = \frac{0.17 \times 60,000}{0.85 \times 4000 \times 12} = 0.25 \text{ in.}$$

Second cycle:

$$77,996 = A_s \times 60,000\left(8.5 - \frac{0.25}{2}\right)$$

$$A_s = 0.16 \text{ in.}^2/12\text{-in. strip}$$

Trying No. 3 bars ($A_s = 0.11$ in.^{2} per bar).

$$s \text{ for negative moment} = \frac{0.11}{0.16/12} = 8.25 \text{ in. c-c}$$

$$s \text{ for positive moment} = 8.25 \times \frac{58,497}{69,528} = 6.94 \text{ in. c-c}$$

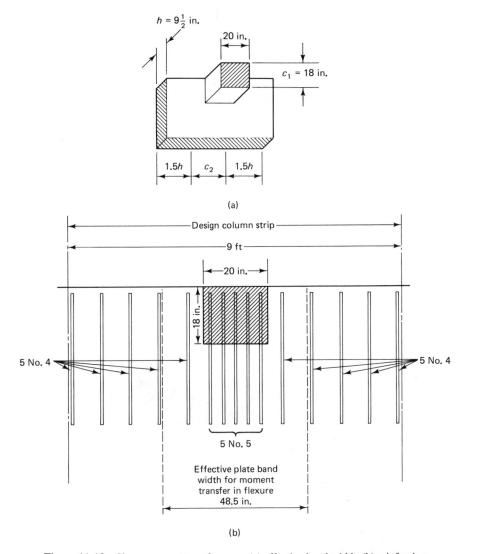

Figure 11.13   Shear moment transfer zone: (a) effective band width; (b) reinforcing details.

Use No. 3 bars at 8 in. center to center for negative moment and No. 3 bars at $6\frac{1}{2}$ in. center to center for positive moment.

(b) *N–S direction (short span):* The same procedure has to be followed as for the E–W direction. The width of the column strip on one side of the column = $0.25l_1$ = 0.25×24 = 6 ft, which is greater than $0.25l_2$ = 4.5 ft; hence a width of 4.5 ft controls. The total width of the column strip in the N–S direction = 2×4.5 = 9.0 ft. The width of the middle strip $= 24.0 - 9.0 = 15.0$ ft. Also, the effective depth d_2 would be smaller; $d_2 = (h - 3/4$ in. cover $- 0.5$ in. $- 0.5/2) = 8.0$ in. The following are the moment values and the bar size and distribution for the panel in the N–S direction as well as the E–W directions. It is recommended for crack-control purposes that a minimum of No. 3 bars at 12 in. center to center be used and that bar spacing not exceed 12 in. center to center using smaller bars at closer spacing if necessary.

Strip	Moment type	E–W			N–S		
		Moment (lb-in./12 in.)	A_s req'd	Bar size and spacing	Moment (lb-in./12 in.)	A_s req'd	Bar size and spacing
Column	Interior negative	175,490	0.48	No. 5 at $7\frac{1}{2}$	124,134	0.36	No. 4 at 6
	Exterior negative	86,909	0.24	No. 4 at $9\frac{1}{2}$	61,477	0.18	No. 3 at 7
	Midspan positive	104,291	0.29	No. 4 at 8	73,771	0.21	No. 3 at 6
Middle	Interior negative	58,497	0.16	No. 3 at 8	41,378	0.07	No. 3 at 12
	Exterior negative	0	0	No. 3 at 12	0	0	No. 3 at 12
	Midspan positive	64,528	0.19	No. 3 at $6\frac{1}{2}$	49,181	0.08	No. 3 at 12

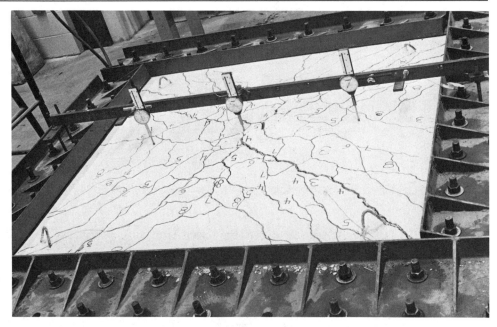

Photo 61 Flexural cracking in restrained one-panel reinforced concrete slab. (Tests by Nawy et al.)

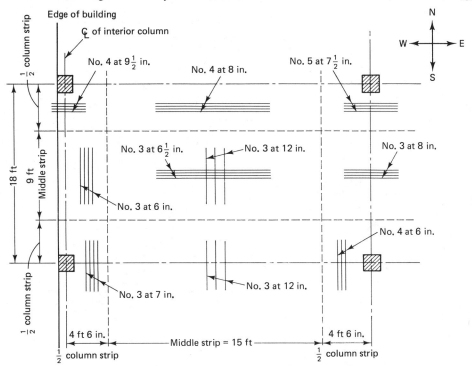

Figure 11.14   Schematic reinforcement distribution.

It should be noted that the choice of size and spacing of the reinforcement is a matter of engineering judgment. As an example, the designer could have chosen for the positive moment in the middle strip No. 4 bars at 12 in. center to center instead of No. 3 bars at $6\frac{1}{2}$ in. center to center as long as the maximum permissible spacing is not exceeded and practicable bar sizes are used for the middle strip.

The placing of the reinforcement is schematically shown in Fig. 11.14. The minimum bent point location of the reinforcing bars to ensure the required development of bond for reinforcement in flat-plate floors is given in Fig. 11.15.

11.5.3 Example 11.2: Design of Two-Way Slab on Beams

A two-story factory building is three panels by three panels in plan, monolithically supported on beams. Each panel is 18 ft (5.49 m) center to center in the N–S direction and 24 ft (7.32 m) center to center in the E–W direction as shown in Fig. 11.16. The clear height between the floors is 16 ft. The dimensions of the supporting beams and columns are also shown in Fig. 11.16, and the building is subject to gravity loads only. Given:

Live load = 115 psf (5.45 kPa)

f'_c = 4000 psi (27.6 MPa), normal-weight concrete

f_y = 60,000 psi (414 MPa)

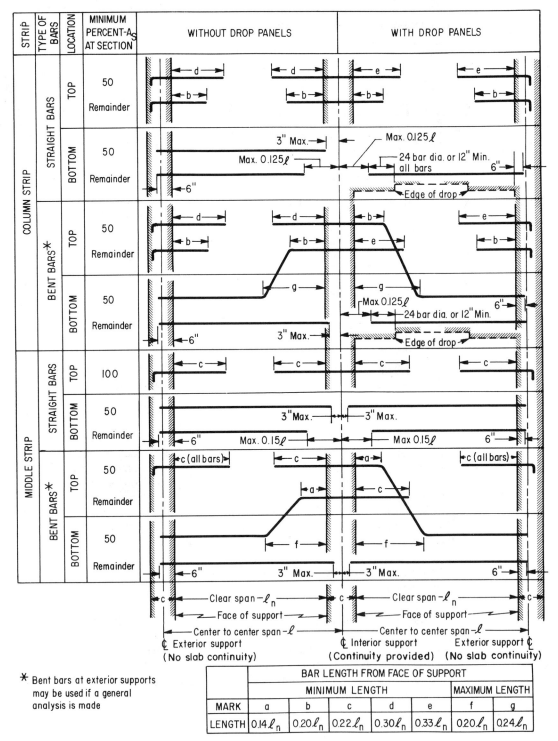

Figure 11.15 Reinforcement cutoffs for flat plates.

	BAR LENGTH FROM FACE OF SUPPORT						
	MINIMUM LENGTH					MAXIMUM LENGTH	
MARK	a	b	c	d	e	f	g
LENGTH	$0.14\ell_n$	$0.20\ell_n$	$0.22\ell_n$	$0.30\ell_n$	$0.33\ell_n$	$0.20\ell_n$	$0.24\ell_n$

* Bent bars at exterior supports
   may be used if a general
   analysis is made

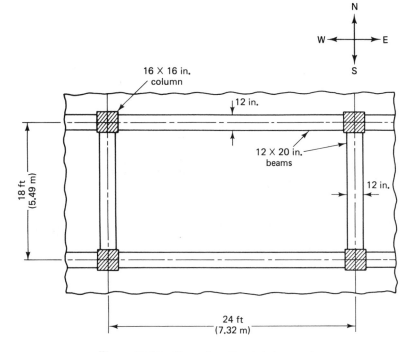

Figure 11.16   Floor plan of an interior panel.

Design the interior panel and the size and spacing of reinforcement needed. Consider flooring weight to be 14 psf in addition to the slab self-weight.

Solution

Geometry check for use of direct design method (Step 1)

(a) Ratio $\dfrac{\text{longer span}}{\text{shorter span}} = \dfrac{24}{18} = 1.33 < 2.0$; hence two-way action.

(b) More than three panels in each direction.

(c) Assume a thickness of 7 in.

$$w_d = 14 + \frac{7}{12} \times 150 = 101.5 \text{ psf}$$

$$3w_d = 304.5 \text{ psf}$$

$$w_l = 115 \text{ psf} < 3w_d$$

Hence the direct design method is applicable.

Minimum slab thickness for deflection requirement (Step 2)

$$l_n(\text{E–W}) = 24 \times 12 - 2 \times 6 = 276 \text{ in.}$$

$$l_n(\text{N–S}) = 18 \times 12 - 2 \times 6 = 204 \text{ in.}$$

$$\beta = \frac{276}{204} = 1.35 \qquad \beta_s = 1 \text{ since all the edges are continuous}$$

From Eq. 11.9,

$$h = \frac{l_n(800 + 0.005 f_y)}{36,000 + 5000\beta(1 + \beta_s)}$$

$$= \frac{276(800 + 5 \times 60)}{36,000 + 5000 \times 1.35(1 + 1)}$$

$$= 6.13 \text{ in. } (156 \text{ mm})$$

To check h from Eq. 11.8, the stiffness ratio is needed. Since this is an interior panel, the end and corner adjacent panel would necessitate larger thickness, Try $h = 7$ in.
    To locate the beam centroid for the section in Fig. 11.17,

$$(38 \times 7)(\bar{y} + 3.5) + \frac{12(\bar{y}^2)}{2} = \frac{12(13 - \bar{y})^2}{2}$$

$$\bar{y} = 0.20 \text{ in.}$$

$$I_b = \tfrac{1}{3} \times 12(0.20)^3 + \tfrac{1}{12} \times 38(7)^3 + 38 \times 7(0.20 + 3.5)^2$$

$$+ \tfrac{1}{3} \times 12(13 - 0.20)^3 = 13{,}116.4 \text{ in.}^4$$

$I_s = h^3/12 \times$ width of slab bound laterally by the centerline of the adjacent panel on each side of the beam section shown in Fig. 11.16.

$$I_{s1}(\text{N–S}) = \frac{(7)^3}{12} \times 24 \times 12 = 8232 \text{ in.}^4$$

$$I_{s2}(\text{E–W}) = \frac{(7)^3}{12} \times 18 \times 12 = 6174 \text{ in.}^4$$

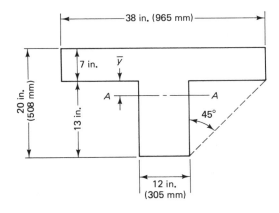

Figure 11.17 Effective flanged beam section.

Therefore,

$$\alpha_1 = \frac{13,116.4}{8232} = 1.59 \qquad \alpha_2 = \frac{13,116.4}{6174} = 2.12$$

$$\alpha_m = \frac{1.59 \times 2 + 2.12 \times 2}{4} = 1.86$$

From Eq. 11.8,

$$h = \frac{l_n(800 + 0.005 f_y)}{36,000 + 5000\beta[\alpha_m - 0.5(1 - \beta_s)(1 + 1/\beta)]}$$

$$= \frac{276(800 + 0.005 \times 60,000)}{36,000 + 5000 \times 1.35(1.86 - 0)} = 6.25 \text{ in.}$$

Choose h from Eq. 11.9 as h should not exceed the thickness obtained from Eq. 11.10 which gives

$$h = \frac{l_n(800 + 0.005 f_y)}{36,000} = \frac{276(800 + 0.005 \times 60,000)}{36,000} = 8.43 \text{ in.}$$

Therefore, for deflection, use $h = 7$ in. assumed at the beginning (178 mm).

Statical moment computation (Steps 3 to 5)

Given the flooring weighs 14 psf.

$$w_u = 1.4D + 1.7L$$

$$= 1.4\left(\frac{7}{12} \times 150 + 14\right) + 1.7 \times 115 = 338 \text{ psf}$$

E–W $l_{n_1} = 276$ in. $= 23.0$ ft

N–S $l_{n_2} = 204$ in. $= 17.0$ ft

$0.65l_1 = 15.6$ ft   use $l_1 = 23.0$ ft

$0.65l_2 = 11.7$ ft   use $l_2 = 17.0$ ft

(a) *E–W direction:*

$$M_0 = \frac{w_u l_2 l_{n_1}^2}{8} = \frac{338 \times 18.0(23)^2}{8} = 402,305 \text{ lb-ft}$$

Moment distribution factors for interior panels from Fig. 11.6:

$$-M_u = 0.65M_0 = 0.65 \times 402,305 = 261,498 \text{ lb-ft}$$

$$+M_u = 0.35M_0 = 0.35 \times 402,305 = 140,807 \text{ lb-ft}$$

(b) *N–S direction:*

$$M_0 = \frac{w_u l_1 l_{n_2}^2}{8} = \frac{338 \times 24.0(17)^2}{8} = 293,046 \text{ lb-ft (398 kN-m)}$$

Moment distribution factors for interior panel from Fig. 11.6 or Table 11.1:

$$-M_u = 0.65M_0 = 0.65 \times 293{,}046 = 190{,}480 \text{ lb-ft (258 kN-m)}$$

$$+M_u = 0.35M_0 = 0.35 \times 293{,}046 = 102{,}566 \text{ lb-ft (139 kN-m)}$$

Moment distribution in the column and middle strips (Steps 5 to 7)

(a) *E–W stiffness ratio (long span):*

$$\alpha = \frac{E_{cb}I_{b2}}{E_{cs}I_{s2}} = \frac{13{,}116.4}{6174} = 2.12$$

$$\frac{l_2}{l_1} = \frac{18}{24} = 0.75 \qquad \alpha\frac{l_2}{l_1} = 1.59 > 1.0$$

Moment factors for the column strip for this panel from the factored moment coefficients for the column strip of an interior panel (Section 11.4.2, interior panels and positive moments) are linearly interpolated to give the following:

$$-M: \quad 0.75 + \frac{0.90 - 0.75}{2} = 0.83$$

$$+M: \quad 0.75 + \frac{0.90 - 0.75}{2} = 0.83$$

(b) *N–S Stiffness ratio α:*

$$\alpha = \frac{E_{cb}I_{b1}}{E_{cs}I_{s1}} = \frac{13{,}116.4}{8232} = 1.59$$

$$\frac{l_2}{l_1} = \frac{24}{18} = 1.33 \qquad \alpha\frac{l_2}{l_1} = 2.12 > 1.0$$

Hence moment factors are in this case using the same tables by linear interpolation.

$$-M: \quad 0.75 - (0.75 - 0.45)\tfrac{1}{3} = 0.65$$

$$+M: \quad 0.75 - (0.75 - 0.45)\tfrac{1}{3} = 0.65$$

The distributed moments are then evaluated using the interpolated factors above to produce a moment distribution operations table (Table 11.5).

It should be noted from this table that the stiffness ratio of the slab to the supporting beams for the span ratio in this example has resulted in middle strip moments in the N–S direction larger than the moments in the E–W direction.

Check slab thickness for shear capacity

$$\alpha_1\frac{l_2}{l_1} = 1.59 > 1.0$$

Hence shear will be transferred to the beams surrounding the slab according to a tributary

TABLE 11.5   MOMENT DISTRIBUTION OPERATIONS TABLE

	E–W direction l_2/l_1: $18/24 = 0.75$ $\alpha_1(l_2/l_1)$: $2.12 \times 0.75 = 1.59$		N–S direction $24/18 = 1.33$ $1.59 \times 1.33 = 2.12$	
Column strip	Negative moment	Positive moment	Negative moment	Positive moment
M_u(ft-lb)	261,498	140,807	190,480	102,566
Distribution factor (%)	83	83	65	65
Total column strip design moment (ft-lb)	217,043	116,870	123,812	66,668
Beam moment 85%	184,487	99,340	105,240	56,668
Slab moment (ft-lb)	32,556	17,530	18,572	10,000
Total middle strip design moment (ft-lb)	261,498 $\times$ 0.17 44,455	140,807 $\times$ 0.17 23,937	190,480 $\times$ 0.35 66,668	102,566 $\times$ 0.35 35,898

area bound by 45° lines drawn from the corners of the panel and the centerline of the panel parallel to the long side.

The largest part of the load has to be carried in the short direction with the largest value at the face of the first interior support. The factored shear on a 12-in.-wide strip spanning in the short direction can be approximated as

$$V_u = 1.15 \frac{w_u l_{n2}}{2} = \frac{1.15 \times 338.0(17 \times 12)}{2 \times 12} = 3304 \text{ lb/ft width}$$

where the value 1.15 is the continuity factor.

slab effective $d = 7 - 0.75 - 0.25 = 6.0$ in. (152.4 mm)

$$\phi V_c = \phi(2\sqrt{f_c'}\, bd)$$

$$= 0.85 \times 2\sqrt{4000} \times 12 \times 6 = 7741 \text{ lb}$$

$$V_u < \phi V_c \qquad \text{hence safe}$$

Proportioning of the slab reinforcement (Steps 7 and 8)

As was done in Ex. 11.1, the moments per 12-in.-width strip have to be evaluated.
(a) *E–W direction:*

Column strip:

$$-M_n = \frac{32,556}{\phi = 0.9} = 36,173 \text{ ft-lb}$$

$$0.25l_2 = 0.25 \times 18 \text{ ft} = 4.5 \text{ ft} < 0.25 \times 24 \text{ ft}$$

Hence, the half column strip = 4.5 ft controls. The net width of the slab in the column strip on which moments act = $2 \times 4.5 - 38 \text{ in.}/12 = 5.83$ ft.

$$\text{required unit} - M \text{ per 12-in. strip} = \frac{36,173 \times 12}{5.83} = 74,456 \text{ in.-lb}$$

$$\text{required unit} + M \text{ per 12-in. strip} = \frac{17,530 \times 12}{0.9 \times 5.83} = 40,091 \text{ in.-lb}$$

Middle strip:

width of strip = $18 - 9.0 = 9.0$ ft

$$\text{required unit} - M \text{ per 12-in. strip} = \frac{44,455 \times 12}{0.9 \times 9.0} = 65,859 \text{ in.-lb}$$

$$\text{required unit} + M \text{ per 12-in. strip} = \frac{23,937 \times 12}{0.9 \times 9.0} = 35,462 \text{ in.-lb}$$

(b) *N–S direction (short span):* From before, the maximum allowable width of the half column strip = 4.5 ft.

Column strip:

net width of slab in column strip on which moments act

$$= 2 \times 4.5 - \frac{38}{12} = 5.83 \text{ ft.}$$

$$\text{required unit} - M \text{ per 12-in. strip} = \frac{18,572 \times 12}{0.9 \times 5.83} = 42,474 \text{ in.-lb}$$

$$\text{required unit} + M \text{ per 12-in. strip} = \frac{10,000 \times 12}{0.9 \times 5.83} = 22,870 \text{ in.-lb}$$

Middle strip:

width of strip = $24 - 9.0 = 15.0$ ft

$$\text{required unit} - M \text{ per 12-in. strip} = \frac{66,668 \times 12}{0.9 \times 15.0} = 59,260 \text{ in.-lb}$$

$$\text{required unit} + M \text{ per 12 in. strip} = \frac{35,898 \times 12}{0.9 \times 15.0} = 31,909 \text{ in.-lb}$$

Selection of size and spacing of reinforcement (Step 9)

The maximum unit moment in the negative moment region of the column strip in the E–W direction $= 74{,}456$ lb-in. per 12-in-wide strip.

$$M_n = A_s f_y \left(d - \frac{a}{2} \right)$$

Hence

$$74{,}456 = A_s \times 60{,}000 (\approx 0.9d)$$

$$A_s = \frac{74{,}456}{60{,}000 \times 0.9 \times 6.0} = 0.23 \text{ in.}^2$$

Adjustment trial:

$$a = \frac{A_s f_y}{0.85 f'_c b} = \frac{0.23 \times 60{,}000}{0.85 \times 4000 \times 12} = 0.34 \text{ in.}$$

Hence

$$74{,}456 = A_s \times 60{,}000 \left(6.0 - \frac{0.34}{2} \right)$$

Therefore, required $A_s = 0.21$ in.² per 12-in. strip. Try No. 4 bars (0.20 in.²) (12.7 mm diameter)

$$s = \frac{\text{area of one bar}}{\text{required area per 12-in. strip}} = \frac{0.20}{0.21/12} = 11.43 \text{ in. c-c}$$

Therefore, use No. 4 bars at 11 in. center to center (12.7 mm diameter at 280 mm center to center).

In the same manner, calculate the area of steel needed in each direction for both the column and beam strips. Note that the effective depth d in the N–S direction would be $= 7.0 - (0.75 + 0.5 + 0.25) = 5.5$ in. since it is assumed in this design that the E–W grid of reinforcement is closest to the concrete surface.

	Column strip				Middle strip			
	Support		Midspan		Support		Midspan	
Direction	A_s 12 in.	Bar size and spacing (in. c-c)	A_s 12 in.	Bar size and spacing (in. c-c)	A_s 12 in.	Bar size and spacing (in. c-c)	A_s 12 in.	Bar size and spacing (in. c-c)
E–W ($d = 6.0$)	0.21	No. 4 at 11	0.11	No. 3 at 12	0.19	No. 4 at 12	0.10	No. 3 at 12
N–S ($d = 5.5$)	0.14	No. 3 at 9	0.06	No. 3 at 12	0.20	No. 4 at 12	0.11	No. 3 at 12

Maximum spacing of bars should not exceed 12 in. center to center for crack control.

Compare the reinforcement areas obtained in this example with those of Ex. 11.1 in conjunction with the discussion in Section 11.2.1 on two-way action and moment redistribution as a function of stiffness ratios. It should be noted that when the slab or plate panel is either supported on flexible supports or on columns only, the moments are not necessarily more severe in the shorter direction.

Carry the reinforcement at the same spacing for each respective strip up to the webs of the supporting beams. Also, as the next step, design (analyze) the supporting beams in the usual manner as discussed in Chapter 5.

More refinements could have been obtained using the equivalent frame method for moment calculations. Also, for cases where limitations exist, such as horizontal loads and others discussed in Section 11.3.1, the equivalent frame method would have to be used for moment calculations.

The scope of this book prevents adequate coverage of the equivalent frame method for moment distributions. However, the reader, in becoming familiar with this chapter would have acquired the tools necessary for an easier and faster understanding of the equivalent frame moment analysis documented in several other texts and works of reference.

11.6 DIRECT METHOD OF DEFLECTION EVALUATION

11.6.1 The Equivalent Frame Approach

As in the direct design method discussed in detail in the preceding sections, the structure is divided into continuous frames centered on the column lines in each of the two perpendicular directions. Each frame would be composed of a row of columns and

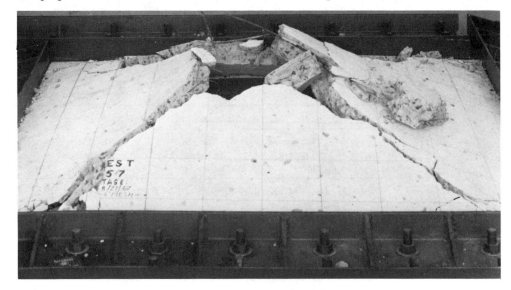

Photo 62   Rectangular concrete slab at rupture. (Tests by Nawy et al.)

a *broad* band of slab together with column line beams, if any, between panel center-lines.

By the requirement of statics, the applied load must be accounted for in each of the two perpendicular (orthogonal) directions. In order to account for the torsional deformations of the support beams, an *equivalent* column is used whose flexibility is the *sum* of the flexibilities of the actual column and the torsional flexibility of the transverse beam or slab strips (stiffness is the inverse of flexibility). In other words,

$$\frac{1}{K_{ec}} = \frac{1}{\Sigma\, K_c} + \frac{1}{K_t} \qquad (11.11)$$

where K_{ec} = flexural stiffness of the equivalent column; bending moment per unit rotation

$\Sigma\, K_c$ = sum of flexural stiffnesses of upper and lower columns; bending moment per unit rotation

K_t = torsional stiffness of the transverse beam or slab strip; torsional moment per unit rotation

The value of K_{ec} would thus have to be known in order to calculate the deflection by this procedure.

The slab–beam strips are considered supported *not* on the columns but on *transverse* slab–beam strips on the column centerlines. Figure 11.18a illustrates this point. Deformation of a typical panel is considered in *one direction at a time*. There-after, the contribution in each of the two directions, x and y, is added to obtain the total deflection at any point in the slab or plate.

First, the deflection due to bending in the x direction is computed (Fig. 11.18b). Then the deflection due to bending in the y direction is found. The midpanel deflection can now be obtained as the sum of the center-span deflections of the column strip in one direction and that of the middle strip in the orthogonal direction (Fig. 11.18c).

The deflection of each panel can be considered as the sum of three components:

1. Basic midspan deflection of the panel, assumed fixed at both ends, given by

$$\delta' = \frac{wl^4}{384 E_c I_{\text{frame}}}$$

This has to be proportioned to separate deflection δ_c of the column strip and δ_s of the middle strip, such that

$$\delta_c = \delta' \frac{M_{\text{col strip}}}{M_{\text{frame}}} \frac{E_c I_{cs}}{E_c I_c}$$

$$\delta_s = \delta' \frac{M_{\text{slab strip}}}{M_{\text{frame}}} \frac{E_c I_{cs}}{E_c I_s}$$

where I_{cs} is the moment of inertia of the total frame, I_c the moment of inertia of the column strip, and I_s the moment of inertia of the middle slab strip.

2. Center deflection, $\delta''_{\theta L} = \frac{1}{8}\, \theta L$, due to rotation at the left end while the right end

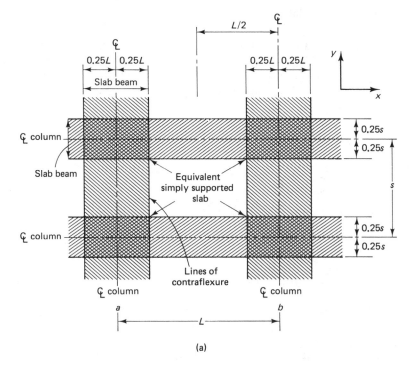

(a)

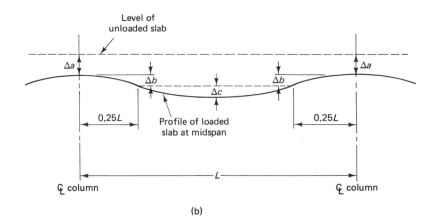

(b)

Figure 11.18 Equivalent frame method for deflection analysis: (a) plate panel transferred into equivalent frames; (b) profile of deflected shape at centerline; (c) deflected shape of panel.

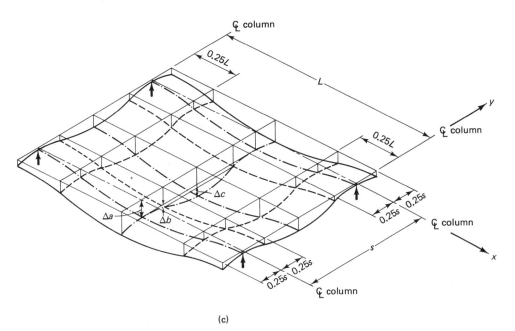

(c)

Figure 11.8 (*cont.*)

is considered fixed, where θ_L = left M_{net}/K_{ec} and K_{ec} is the flexural stiffness of equivalent column (moment per unit rotation).

3. Center deflection, $\delta''_{\theta R} = \frac{1}{8}\theta L$ due to rotation at the right end while the left end is considered fixed, where θ_L = right M_{net}/K_{ec}. Hence

$$\delta_{cx} \text{ or } \delta_{cy} = \delta_c + \delta''_{\theta L} + \delta''_{\theta R} \tag{11.12a}$$

$$\delta_{sx} \text{ or } \delta_{sy} = \delta_s + \delta''_{\theta L} + \delta''_{\theta R} \tag{11.12b}$$

(Use in Eqs. 11.12a and 11.12b the value of δ_c, $\delta''_{\theta L}$, and $\delta''_{\theta R}$ which correspond to the applicable span directions. From Figs. 11.18b and c, the total deflection is

$$\Delta = \delta_{sx} + \delta_{cy} = \delta_{sy} + \delta_{cx} \tag{11.13}$$

11.6.2 Example 11.3: Central Deflection Calculations of a Slab Panel on Beams

A 7-in. (177.8 mm) slab of a five panel by five panel floor system spanning 25 ft in the E–W direction (7.62 m) and 20 ft in the N–S direction (6.10 m) is shown in Fig. 11.19a. The panel is monolithically supported by beams 15 in. × 27 in. in the E–W direction (381 mm × 686 mm) and 15 in. × 24 in. in the N–S direction (381 mm × 610 mm). The floor is subjected to a time-dependent deflection due to an equivalent uniform working load intensity w = 450 psf (21.5 kPa). Material properties of the floor are:

$f'_c = 4000$ psi (27.6 MPa)

$f_y = 60,000$ psi (414 MPa)

$E_c = 3.6 \times 10^6$ psi (24.8 $\times$ 10 kPa)

Assume

1. Net moment M_w from adjacent spans (ft-lb)

	E–W	N–S
Support 1:	20×10^3	Support 1: 40×10^3
Support 2:	5×10^3	Support 4: 20×10^3

2. Equivalent column stiffness $K_{ec} \simeq 400 \ E_c$ lb-in. per radian in both directions. Find the maximum central deflection of the panel due to the long-term loading and determine if its magnitude is acceptable if the floor supports sensitive equipment which can be damaged by large deflections.
3. Cracked moment of inertia:

$$\text{E–W: } I_{cr} = 45,500 \text{ in.}^4$$

$$\text{N–S: } I_{cr} = 32,500 \text{ in.}^4$$

Solution

Calculate the gross moments of inertia (in.^{4}) of the sections in Fig. 11.19, namely, the total equivalent frame I_{cs} in part (b), the column strip beam I_c in part (c), and the middle strip slab I_s in part (d). These values are:

	I_{cs}	I_c	I_s
E–W	63,600	53,700	3430
N–S	47,000	40,000	4288

Next, calculate factors $\alpha_1 l_2/l_1$ and $\alpha_2 l_1/l_2$ as in Ex. 11.2. In both cases they are greater than 1.0. Hence the factored moments coefficients (percent) obtained from the tables in Section 11.4.2 are as follows:

	Column strip (+ and −)	Middle strip (+ and −)
E–W	81.0	19.0
N–S	67.5	32.5

E–W direction deflections (span = 25 ft)

Long-term $w_w = 450$ psf

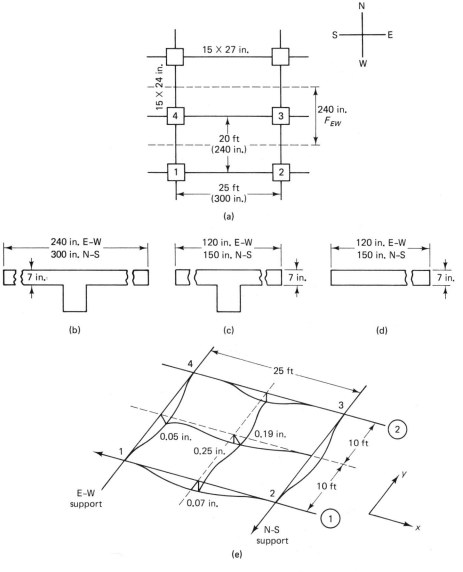

Figure 11.19 Example on equivalent frame deflection evaluation.

$$\delta'_{25} = \frac{450 \times 20(25)^4 \times 1728}{384 \times 3.6 \times 10^6 \times 63,600} = 0.0691 \text{ in.}$$

$$\delta_c = 0.0691 \times 0.81 \times \frac{63,600}{53,700} = 0.0663 \text{ in.}$$

$$\delta_s = 0.0691 \times 0.19 \times \frac{63,600}{3,430} = 0.243 \text{ in.}$$

Rotation at end 1 is

$$\theta_1 = \frac{M_1}{K_{ec}} = \frac{20 \times 10^3 \times 12}{400 \times 3.6 \times 10^6} = 1.67 \times 10^{-4} \text{ rad}$$

and rotation at end 2 is

$$\theta_2 = \frac{M_2}{K_{ec}} = \frac{5 \times 10^3 \times 12}{400 \times 3.6 \times 10^6} = 0.42 \times 10^{-4} \text{ rad}$$

where θ is the rotation at one end if the other end is fixed.

δ'' = deflection adjustment due to rotation at supports

$$1 \text{ and } 2 = \frac{\theta l}{8}$$

$$\delta'' = \frac{(1.67 + 0.42) \times 10^{-4} \times 300}{8} = 0.0078 \text{ in.}$$

Therefore,

$$\text{net } \delta_{cx} = 0.0663 + 0.0078 = 0.0741 \qquad \text{say } 0.07 \text{ in.}$$

$$\text{net } \delta_{sx} = 0.243 + 0.0078 = 0.2508 \qquad \text{say } 0.25 \text{ in.}$$

N–S direction deflections (span = 20 ft)

$$\delta'_{20} = \frac{450 \times 25(20)^4 \times 1728}{384 \times 3.6 \times 10^6 \times 47,000} = 0.0479 \text{ in.}$$

$$\delta_c = 0.0479 \times 0.675 \times \frac{47,000}{40,000} = 0.038 \text{ in.}$$

$$\delta_s = 0.0479 \times 0.325 \times \frac{47,000}{4288} = 0.171 \text{ in.}$$

$$\text{rotation } \theta_1 = \frac{M_1}{K_{ec}} = \frac{40 \times 10^3 \times 12}{400 \times 3.6 \times 10^6} = 3.3 \times 10^{-4} \text{ rad}$$

$$\text{rotation } \theta_4 = \frac{M_4}{K_{ec}} = \frac{20 \times 10^3 \times 12}{400 \times 3.6 \times 10^6} = 1.67 \times 10^{-4} \text{ rad}$$

$$\delta'' = \frac{\theta l_2}{8} = \frac{(3.3 + 1.67)10^{-4} \times 240}{8} = 0.0149 \text{ in.}$$

Therefore,

$$\text{net } \delta_{cy} = 0.038 + 0.0149 = 0.0529 \qquad \text{say } 0.05 \text{ in.}$$

$$\delta_{sy} = 0.171 + 0.0149 = 0.1859 \qquad \text{say } 0.19 \text{ in.}$$

$$\text{total central deflection } \Delta = \delta_{sx} + \delta_{cy} = \delta_{sy} + \delta_{cx}$$

$$\Delta_{E-W} = \delta_{sx} + \delta_{cy} = 0.25 + 0.05 = 0.30 \text{ in.}$$

$$\Delta_{N-S} = \delta_{sy} + \delta_{cx} = 0.19 + 0.07 = 0.26 \text{ in.}$$

Hence the average deflection at the center of the interior panel

$$\tfrac{1}{2}\,(\Delta_{E-W} + \Delta_{N-S}) = 0.28 \text{ in. (7.1 mm)}$$

Adjustment for cracked section: Use Branson's effective moment of inertia equation,

$$I_e = \left(\frac{M_{cr}}{M_a}\right)^3 I_g + \left[1 - \left(\frac{M_{cr}}{M_a}\right)^3\right] I_{cr}$$

as discussed in Chapter 8. Calculation of ratio M_{cr}/M_a:

$$M_{cr} = \frac{f_r I_g}{y_t}$$

where f_r = modulus of rupture of concrete

y_t = distance of center of gravity of section from outer tension fibers

E–W (240-in. flange width): $y_t = 21.54$ in.

N–S (300-in. flange width): $y_t = 19.20$ in.

$$f_r = 7.5\sqrt{f_c'} = 7.5\sqrt{4000} = 474 \text{ psi}$$

Hence,

$$M_{cr}(\text{E–W}) = \frac{474 \times 63{,}600}{21.54} \times \frac{1}{12} = 1.17 \times 10^5 \text{ ft-lb}$$

$$M_{cr}(\text{N–S}) = \frac{474 \times 47{,}000}{19.20} \times \frac{1}{12} = 0.97 \times 10^5 \text{ ft-lb}$$

$$\text{interior panel } M_a = \frac{w_w l^2}{16} = \frac{20 \times 450(25)^2}{16} \qquad \text{for E–W}$$

$$= 3.52 \times 10^5 \text{ ft-lb}$$

$$= \frac{25 \times 450(20)^2}{16} \qquad \text{for N–S}$$

$$= 2.81 \times 10^5 \text{ ft-lb}$$

Note that the moment factor $\tfrac{1}{16}$ is used to be on the safe side, although the actual moment coefficients for two-way action would have been smaller.

E–W effective moment of inertia I_e:

$$\frac{M_{cr}}{M_a} = \frac{1.17 \times 10^5}{3.52 \times 10^5} = 0.332$$

$$\left(\frac{M_{cr}}{M_a}\right)^3 = 0.037$$

$$I_e = 0.037 \times 63{,}600 + (1 - 0.037)45{,}500 = 46{,}170 \text{ in.}^4$$

N–S effective moment of inertia I_e:

$$\frac{M_{cr}}{M_a} = \frac{0.97 \times 10^5}{2.81 \times 10^5} = 0.345$$

$$\left(\frac{M_{cr}}{M_a}\right)^3 = 0.041$$

$$I_e = 0.041 \times 47,000 + (1 - 0.041)32,500 = 33,095 \text{ in.}^4$$

$$\text{Average } \frac{I_g}{I_e} = \frac{1}{2}\left(\frac{63,600}{46,170} + \frac{47,000}{33,095}\right) = 1.40$$

Adjusted central deflection for cracked section effect

$$= 1.40 \times 0.28 = 0.39 \text{ in. (9.9 mm)}$$

$$\frac{l}{\Delta} = \frac{25 \text{ ft} \times 12}{0.39} = 769 > 480 \qquad \text{allowed in Table 11.3}$$

Hence the long-term central deflection is acceptable.

11.7 CRACKING BEHAVIOR AND CRACK CONTROL IN TWO-WAY-ACTION SLABS AND PLATES

11.7.1 Flexural Cracking Mechanism and Fracture Hypothesis

Flexural cracking behavior in concrete structural floors under two-way action is significantly different from that in one-way members. Crack-control equations for beams underestimate the crack widths developed in two-way slabs and plates, and do not tell the designer how to space the reinforcement. Cracking in two-way slabs and plates is controlled primarily by the steel stress level and the spacing of the reinforcement in the two perpendicular directions. In addition, the clear concrete cover in two-way slabs and plates is nearly constant [$\frac{3}{4}$ in. (19 mm) for interior exposure], whereas it is a major variable in the crack-control equations for beams. The results from extensive tests on slabs and plates by Nawy et al. demonstrate this difference in behavior in a fracture hypothesis on crack development and propagation in two-way plate action. As seen in Fig. 11.20, stress concentration develops initially at the points of intersection of the reinforcement in the reinforcing bars and at the welded joints of the wire mesh, that is, at grid nodal points A_1B_1, A_1A_2, A_2B_2, and B_2B_1. The resulting fracture pattern is a total repetitive cracking grid, provided that the spacing of the nodal points A_1, B_1, A_2, and B_2 is close enough to generate this preferred initial fracture mechanism of orthogonal cracks narrow in width, as a preferred fracture mechanism.

If the spacing of the reinforcing grid intersections is too large, the magnitude of the stress concentration and the energy absorbed per unit grid is too low to generate cracks along the reinforcing wires or bars. As a result, the principal cracks follow

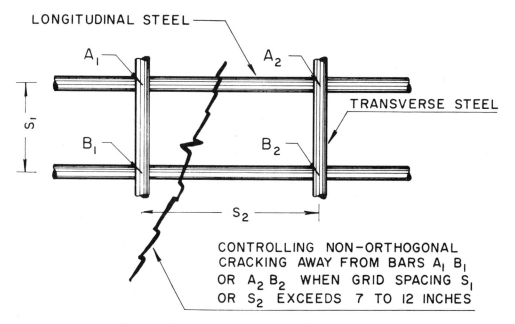

Figure 11.20   Grid unit in two-way-action reinforcement.

diagonal yield-line cracking in the plain concrete field away from the reinforcing bars early in the loading history. These cracks are wide and few.

This hypothesis also leads to the conclusion that surface deformations of the individual reinforcing elements have little effect in arresting the generation of the cracks or controlling their type or width in a two-way-action slab or plate. In a similar manner, one may conclude that the scale effect on two-way-action cracking behavior is insignificant, since the cracking grid would be a reflection of the reinforcement grid if the preferred orthogonal narrow cracking widths develop. Therefore, to control cracking in two-way-action floors, the major parameter to be considered is the reinforcement spacing in two perpendicular directions. Concrete cover has only a minor effect, since it is usually a small, constant value of 0.75 in. (20 mm).

For a constant area of steel determined for bending in one direction, that is, for energy absorption per unit slab area, the smaller the spacing of the transverse bars or wires, the smaller should be the diameter of the longitudinal bars. The reason is that less energy has to be absorbed by the individual longitudinal bars. If one considers that the magnitude of fracture is determined by the energy imposed per specific volume of reinforcement acting on a finite element of the slab, a proper choice of the reinforcement grid size and bar size can control cracking into preferred orthogonal grids.

It must be emphasized that this hypothesis is important for serviceability and reasonable overload conditions. In relating orthogonal cracks to yield-line cracks, the failure of a slab ultimately follows the generally accepted rigid-plastic yield-line criteria.

11.7.2 Crack Control Equation

The basic equation (Section 8.11) for relating crack width to strain in the reinforcement is

$$w = \alpha a_c^{\beta} \epsilon_s^{\gamma} \tag{11.14}$$

The effect of the tensile strain in the concrete between the cracks is neglected as insignificant. a_c is the crack spacing, ϵ_s the unit strain in the reinforcement, and α, β, and γ are constants. As a result of this fracture hypothesis, the mathematical model in Eq. 11.14 and the statistical analysis of the data of 90 slabs tested to failure the following crack-control equation emerged:

$$w = K\beta f_s \sqrt{\frac{d_{b_1} s_2}{Q_{i_1}}} \tag{11.15}$$

where the quantity under the radical, $G_1 = d_{b_1} s_2 / Q_{i_1}$, is termed the grid index, and can be transformed into

$$G_1 = \frac{s_1 s_2 d_c}{d_{b1}} \frac{8}{\pi}$$

where K = fracture coefficient, having a value of $K = 2.8 \times 10^{-5}$ for uniformly loaded restrained two-way-action square slabs and plates. For concentrated loads or reactions, or when the ratio of short to long span is less than 0.75 but larger than 0.5, a value of $K = 2.1 \times 10^{-5}$ is applicable. For a span aspect ratio of 0.5, $K = 1.6 \times 10^{-5}$. Units of coefficient K are in square inch per lb.

β = ratio of the distance from the neutral axis to the tensile face of the slab to the distance from the neutral axis to the centroid of the reinforcement grid (to simplify the calculations use $\beta = 1.25$, although it varies between 1.20 and 1.35)

f_s = actual average service load stress level, or 40% of the design yield strength, f_y (ksi)

d_{b1} = diameter of the reinforcement in direction 1 closest to the concrete outer fibers (in.)

s_1 = spacing of the reinforcement in direction 1 in.

s_2 = spacing of the reinforcement in perpendicular direction 2 in.

1 = direction of the reinforcement closest to the outer concrete fibers; this is the direction for which crack control check is to be made

Q_{i_1} = active steel ratio

= $\dfrac{\text{area of steel } A_s \text{ per ft width}}{12(d_{b1} + 2c_1)}$

where c_1 is clear concrete cover measured from the tensile face of the concrete to the nearest edge of the reinforcing bar in direction 1

w = crack width at face of concrete caused by flexural load (in.)

Subscripts 1 and 2 pertain to the directions of reinforcement. Detailed values of the fraction coefficients for various boundary conditions are given in Table 11.6.

TABLE 11.6 FRACTURE COEFFICIENTS FOR SLABS AND PLATES

Loading type[a]	Slab shape	Boundary condition[b]	Span ratio,[c] S/L	Fracture coefficient, 10^{-5} K
A	Square	4 edges r	1.0	2.1
A	Square	4 edges s	1.0	2.1
B	Rectangular	4 edges r	0.5	1.6
B	Rectangular	4 edges r	0.7	2.2
B	Rectangular	3 edges r, 1 edge h	0.7	2.3
B	Rectangular	2 edges r, 2 edges h	0.7	2.7
B	Square	4 edges r	1.0	2.8
B	Square	3 edges r, 1 edge h	1.0	2.9
B	Square	2 edges r, 2 edges h	1.0	4.2

[a]Loading type: A, concentrated; B, uniformly distributed.

[b]Boundary condition: r, restrained; s, simply supported; h, hinged.

[c]Span ratio: S, clear short span; L, clear long span.

A graphical solution of Eq. 11.15 is given in Fig. 11.21 for

$$f_y = 60,000 \text{ psi } (414 \text{ MPa})$$

$$f_s = 40\% \, f_y$$

$$f_s = 24,000 \text{ psi } (165.5 \text{ MPa})$$

for rapid determination of the reinforcement size and spacing method for crack control.

The grid index, G_1, specifies the size and spacing of the bars in the two perpendicular directions of any concrete floor system, and w_{max} is the maximum allowable crack width.

The crack control equation and guidelines presented are important not only for the control of corrosion in the reinforcement but also for deflection control. The reduction of the stiffness EI of the two-way slab or plate due to orthogonal cracking when the limits of permissible crack widths in Table 8.4 are exceeded, can lead to excessive deflection both short-term and long-term. Deflection values several times those anticipated in the design, including deflection due to construction loading, can be reasonably controlled through camber and control of the flexural crack width in the slab or plate. Proper selection of the *reinforcement spacing* s_1 and s_2 in both perpendicular directions as discussed in this section, and not exceeding twelve inches center to center, can maintain good serviceability performance of a slab system under normal

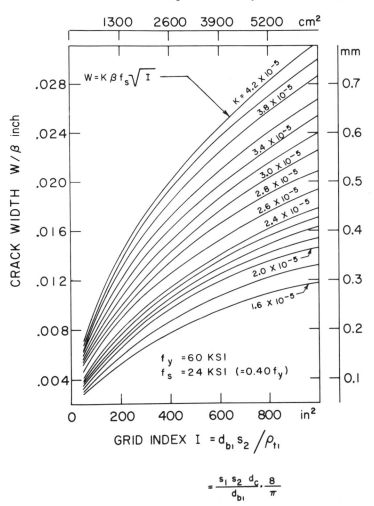

Figure 11.21 Crack control reinforcement distribution in two-way-action slabs and plates for all exposure conditions: f_y = 60,000 psi, f_s = 24,000 psi (= $0.40f_y$).

and reasonable overload conditions. The 1985 Australian Code and other codes generally follow the principles presented on the selection of reinforcement size and spacing in slabs and plates and good engineering practice mandates such caution.

11.7.3 Example 11.4: Crack-Control Evaluation for Serviceability in an Interior Two-Way Panel

Check the bar size and spacing used in an interior panel to determine if it satisfies serviceability through crack control. As shown in Fig. 11.22, the panel has a width/length ratio l_s/l_l = 1.0 and its thickness is 5 in. (125 mm). The floor is subjected

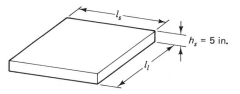

Figure 11.22 Square panel.

to normal weather conditions. Reinforcement used for flexure is No. 4 bars at 9 in. center to center in each direction. Given:

$$\beta = 1.25$$
$$w_{max} = 0.016 \text{ in.}$$
$$f_y = 60 \text{ ksi } (60{,}000 \text{ psi})$$
$$K = 2.8 \times 10^{-5} \text{ in.}^2/\text{lb}$$

Solution

$$f_s = 0.40 f_y = 0.40(60) = 24 \text{ ksi}$$

The fracture coefficient K for this panel aspect ratio is $K = 2.8 \times 10^{-5}$. The maximum permissible crack width for normal interior conditions is $w_{max} = 0.016$ in. (0.4 mm) (Table 8.4).

$$w_{max} = k\beta f_s \sqrt{G_I}$$

$$\text{grid index } G_I = \frac{s_1 s_2 d_c}{d_{b1}} \times \frac{8}{\pi}$$

$$0.016 = 2.8 \times 10^{-5} \times 1.25 \times 24\sqrt{G_I}$$

to give

$$G_I = 363 \text{ in.}^2 = \frac{s_2 s_1 d_c}{d_{b_1}} \times \frac{8}{\pi}$$

If $s_1 = s_2$ for this square panel, cover $d_c = 0.75 + 0.25 = 1.0$ in. to the center of the first reinforcement layer, $d_{b1} = 0.5$ in. = diameter of the No. 4 bar.

$$363 = \frac{s^2 \times 1.0}{0.5} \times \frac{8}{\pi} \qquad \text{giving } s = 8.4 \text{ in. or } 8.5 \text{ in. maximum}$$

Hence the 9 in. center-to-center spacing specified for flexure is not satisfactory. Reduce the spacing of reinforcement to No. 4 bars at $8\frac{1}{2}$ in. (216 mm) center to center for crack control (12.7 mm diameter at 216 mm center to center).

11.7.4 Example 11.5: Crack-Control Evaluation for Serviceability in a Rectangular Panel Subjected to Severe Exposure Conditions

Select the bar size and spacing necessary for crack control at the column reaction region of a 7-in.-thick slab shown in Fig. 11.23 that is uniformly loaded. Select the bar size for two conditions:

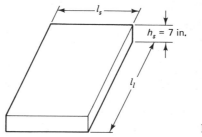

Figure 11.23 Rectangular panel.

Condition A: Floor is subjected to severe exposure of humidity and moist air.

Condition B: Floor sustains an aggressive chemical environment where the design working stress level in the reinforcement is limited to 15 ksi (15,000 psi).

Given:

$$\beta = 1.20$$
$$l_s/l_l = 0.8$$
$$f_y = 60 \text{ ksi (414 MPa)}$$

Solution

Condition A: Humidity and moist air

Permissible $w_{max} = 0.012$ in. (0.3 mm) (Table 8.4). Try No. 4 bars $d_b = 0.5$, $d_c = 0.75 + 0.25 = 1.0$ in. Assume that $s_1 = s_2 = s$ for the given panel. The aspect ratio $l_s/l_l = 0.8$. $K = 2.1 \times 10^{-5}$ for concentrated reaction at the column support (Table 11.6).

$$0.012 = 2.1 \times 10^{-5} \times 1.20 \times 0.4 \times 60\sqrt{G_I}$$

to give $G_I = 394$ in.^{2}. Therefore,

$$394 = \frac{s^2 d_c}{d_{b_1}} \times \frac{8}{\pi} = \frac{s^2 \times 1.0}{0.5} \times \frac{8}{\pi}$$

$$s = 8.8 \text{ in.}$$

Hence use No. 4 bars at $8\frac{1}{2}$ in. center to center each way for crack control.

Condition B: Aggressive chemical environment

Permissible $w_{max} = 0.007$ in (0.18 mm) (Table 8.2) $f_s = 15$ ksi to be used as a low stress level for sanitary or water-retaining structures instead of $0.4 f_y$. Try No. 5 bars ($d_{b1} = 0.625$ in.)

$$0.007 = 2.1 \times 10^{-5} \times 1.20 \times 15.0\sqrt{G_I}$$

to give a grid index $G_I = 343$ in.^{2}.

$$d_{c1} = 0.75 + 0.312 = 1.06 \text{ in.}$$

$$G_I = 343 = \frac{s^2 \times 1.06}{0.625} \times \frac{8}{\pi} \qquad \text{to get } s = 8.9 \text{ in.}$$

Use No. 5 bars at 9 in. (229 mm) center-to-center spacing each way for crack control.

Reinforcement Summary

> *Condition A*: No. 4 bars at 8 1/2 in. c-c (12.7 mm diameter at 222 mm c-c)
> *Condition B*: No. 5 bars at 9 in. c-c (15.9 mm diameter at 222 mm c-c)

11.8 YIELD-LINE THEORY FOR TWO-WAY ACTION PLATES

A study of the hinge-field mechanism in a slab or plate at loads close to failure aids the engineering student in developing a feel for the two-way-action behavior of plates. Hinge fields are succession of hinge bands which are idealized by lines; hence the name *yield-line theory* by K. W. Johansen.

To do justice to this subject, an extensive discussion over several chapters or a whole textbook is necessary. The intention of this chapter is only to introduce the reader to the fundamentals of the yield-line theory and its application.

The yield-line theory is an upper-bound solution to the plate problem. This means that the predicted moment capacity of the slab has the highest expected value in comparison with test results. Additionally, the theory assumes a totally rigid-plastic

Photo 63   Testing setup of four-panel prestressed concrete floor. (Tests by Nawy et al.)

behavior, namely, that the plate stays planer at collapse, producing rigid planer failure systems. Consequently, deflection is not accounted for, nor are the compressive membrane forces that will act in the plane of the slab or plate considered. The plates are assumed to be considerably under-reinforced such that the maximum reinforcement percentage ρ does not exceed $1/2\%$ of the section bd.

Since the solutions are upper bound, the slab thickness obtained by this process is in many instances thinner than what is obtained by the other lower-bound solutions, such as the direct design method. Consequently, it is important to apply rigorously the serviceability requirements for deflection control and for crack control in conjunction with the use of the yield-line theory as given in Sections 11.6 and 11.7.

One distinct advantage in this theory is that solutions are possible for any shape of a plate, whereas most other approaches are applicable only to the rectangular shapes with rigorous computations for boundary effects. The engineer can, with ease, find the moment capacity for a triangular, trapezoidal, rectangular, circular, and any other conceivable shape provided that the failure mechanism is known or predictable. Since most failure patterns are presently identifiable, solutions can be readily obtained, as seen in Section 11.8.2.

11.8.1 Fundamental Concepts of Hinge-Field Failure Mechanisms in Flexure

Under action of a two-dimensional system of bending moments, yielding of a rigid-plastic plate occurs when the principal moments satisfy Johansen's square yield criterion as shown in Fig. 11.24.

In this criterion, yielding is considered to have occurred when the numerically greater of the principal moments reaches the value of $\pm M$ at the yield-line cracks. The directions of the principal curvature rates is considered to coincide with the curvatures of the principal moments. The idealized moment–curvature relationship is shown as the solid line in Fig. 11.25.

Line OA is considered almost vertical at point 0 and strain hardening is neglected.

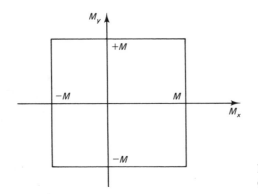

Figure 11.24   Johansen's square yield criterion.

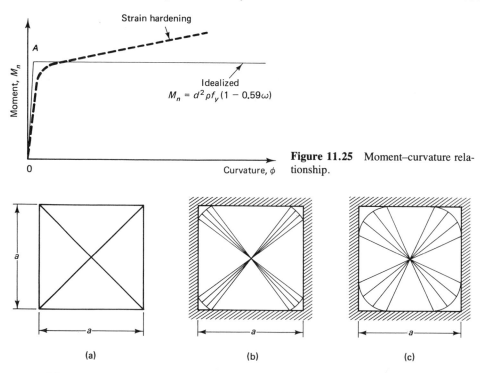

Figure 11.25 Moment–curvature relationship.

Figure 11.26 Failure mechanism of a square slab: (a) $i = 0$; (b) $i = 0.5$; (c) $i = 1.0$.

If one considers the simplest case of a square slab with supports, degree of fixity i varying from $i = 0$ for simply supported to $i = 1.0$ for fully restrained on all four sides, the failure mechanism would be as shown in Fig. 11.26, when a uniformly distributed load is applied.

Take the simply supported case (a). The yield-line moments along the yield lines are the principal moments. Hence the twisting moments are zero in the yield lines and in most cases the shearing forces are also zero. Consequently, only moment m per unit length of the yield line acts about the lines AD and BE in Fig. 11.27. The total moments can be represented by a vector in the direction of the yield line whose value is $M \times$ length of the yield line, that is, $M(a/2 \cos \theta)$ in Fig. 11.27c. The virtual work of the yield moments of the shaded triangular segment ABO is the scalar product of the two moment vectors $Ma/2 \cos \theta$ on fracture lines AO and BO and a rotation θ. In other words, the internal work

$$E_I = \sum \overline{M} \overline{\theta}$$

If the displacement of the shaded segment at its center of gravity c is δ, the external work

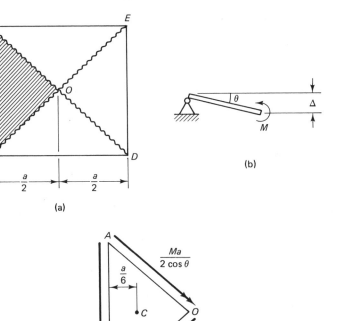

(a)

(b)

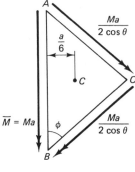

(c)

Figure 11.27 Vector moments on slab segment at failure.

$$E_E = \text{force} \times \text{displacement}$$

$$= \sum \iint w_u \, dx \, dy \, \delta$$

where w_u is the intensity of external load per unit area. But $E_I = E_E$, hence

$$\sum \overline{M}\overline{\theta} = \sum \iint w_u \, dx \, dy \, \delta \tag{11.16}$$

Applying Eq. 11.16 to the particular case under discussion gives us

$$\overline{M}\overline{\theta} = Ma\frac{\Delta}{a/2}$$

since angle θ in Fig. 11.27b is small, where $\theta = \Delta/(a/2)$.
Work per one triangular segment:

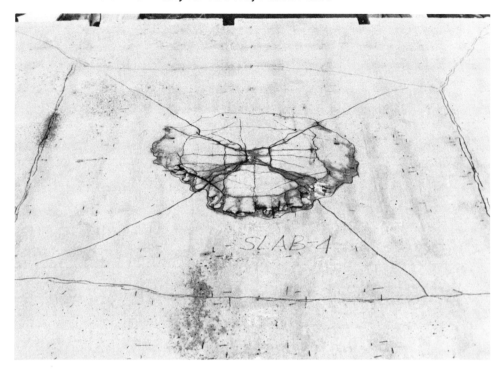

Photo 64   Yield-line pattern at failure at column reaction and panel boundaries of a two-way multipanel floor. (Tests by Nawy, Chakrabarti, et al.)

$$E_I = \overline{M}\overline{\theta} = 2M\Delta$$

$$E_E = \frac{w_u a^2}{4} \times \frac{\Delta}{3}$$

where deflection at center of gravity of the triangle $= \Delta/3$. Therefore,

$$4(2M\Delta) = 4\left(\frac{w_u a^2}{12}\Delta\right)$$

or

$$\text{unit } M = \frac{w_u a^2}{24} \tag{11.17}$$

If the square slab was fully fixed on all four sides $E_I = 4(4M\Delta)$ since fracture lines develop around not only the diagonals but also the four edges, as shown in Fig. 11.26c. Hence,

$$\text{unit } M = \frac{w_u a^2}{48} \tag{11.18}$$

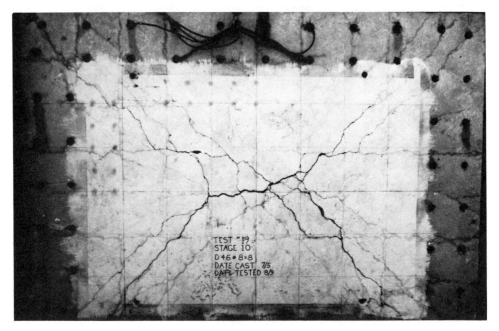

Photo 65 Yield-line patterns at failure at tension face of rectangular restrained panel. (Tests by Nawy et al.)

It is to be noted that a lower-bound solution as proposed by Mansfield's failure pattern in Fig. 11.26c gives a value $M = w_u a^2 / 42.88$. Hence, for a uniformly loaded square slab with load intensity w_u per unit area and degree of support fixity i on all sides,

$$w_u a^2 = M[24(1 + i)] \tag{11.19}$$

The general equation for the yield-line moment capacity of a rectangular isotropic slab on beams and having dimensions $(a \times b)$ as shown in Fig. 11.28, with side a being the shorter dimension is

$$\text{unit } M \frac{\text{ft-lb}}{\text{ft}} = \frac{w_u a_r^2}{24} \left[\sqrt{3 + \left(\frac{a_r}{b_r}\right)^2} - \frac{a_r}{b_r} \right]^2 \tag{11.20}$$

where $a_r = \dfrac{2a}{\sqrt{1 + i_2} + \sqrt{1 + i_4}}$

$b_r = \dfrac{2b}{\sqrt{1 + i_1} + \sqrt{1 + i_3}}$

i = degree of restraint depending on stiffness ratios as discussed in Section 11.2

Note that Eq. 11.20 reduces to the simplified form of Eq. 11.18 or 11.19 for the case of a square slab restrained on all four sides ($i = 1.0$).

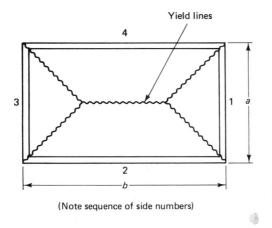
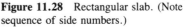

(Note sequence of side numbers)

Figure 11.28   Rectangular slab. (Note sequence of side numbers.)

Affine Slabs

Slabs that are reinforced differently in the two perpendicular directions are called *orthotropic slabs* or *plates*. The moment in the x direction equals M and in the y direction equals μM, where μ is a measure of the degree of orthotropy or the ratio

$$\frac{M_y}{M_x} = \frac{(A_s)_y}{(A_s)_x}$$

To simplify the analysis, the slab should be converted to an affine (isotropic) slab where the strength and reinforcement area in both the x and y directions are the same. Such conversion can be made as follows:

1. *Divide* the linear dimension in the M direction by $\sqrt{\mu}$ of the positive moment for a slab to be reinforced for a moment M in both directions using the same unit load intensity w_u per unit area.
2. In the case of concentrated loads or total loads, also divide such loads by $\sqrt{\mu}$.
3. In the case of line loads, the line load has to be divided by $\sqrt{\mu \cos^2 \theta + \sin^2 \theta}$, where θ is the angle between the line load and the M direction.

If the slab is to be analyzed as an affine slab with the moment μM in both directions, the dimension in the μM direction would have to be *multiplied* by $\sqrt{\mu}$. In either case, the result would of course have to be the same (see Ex. 11.6).

11.8.2 Failure Mechanisms and Moment Capacities of Slabs of Various Shapes Subjected to Distributed or Concentrated Loads

The preceding concise introduction to the virtual-work method of yield-line moments evaluation should facilitate good understanding of the mathematical procedures of most standard rectangular shapes subjected to uniform loading. More complicated slab shapes and other types of symmetrical and nonsymmetrical loading require additional

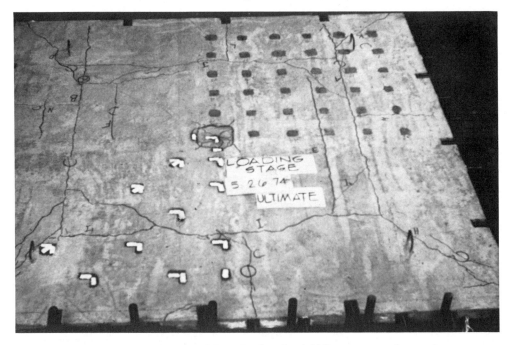

Photo 66   Four-panel slab at failure showing the yield-line patterns at the negative compression face of the supports. (Tests by Nawy and Chakrabarti.)

and more advanced knowledge of the subject as discussed in the introduction. Also, the assumed failure shape and minimization energy principles can give values for particular cases that can differ slightly from one author to another depending on the mathematical assumptions made with respect to the failure shape.

The following summary of failure patterns and the respective moment capacities in terms of load, many of them due to Mansfield (Ref. 11.9), should give the reader adequate coverage in a capsule of solutions to most cases expected in today's and tomorrow's structures.

1. Point load to corner of rectangular cantilever plates:

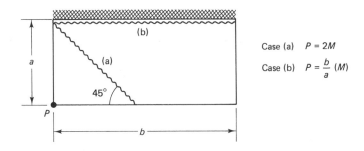

Case (a) $P = 2M$

Case (b) $P = \dfrac{b}{a}\,(M)$

2. Square plate centrally loaded having boundaries simply supported against both downward and upward movements:

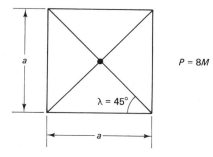

$P = 8M$

3. Regular n-sided plate with simply supported edges and centrally loaded ($n > 4$):

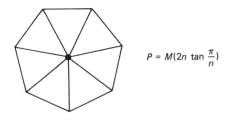

$$P = M(2n \tan \frac{\pi}{n})$$

4. Square plates centrally loaded, having boundaries simply supported against downward movement but free for upward movement:

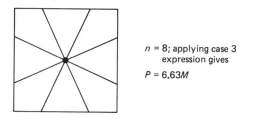

$n = 8$; applying case 3
expression gives

$P = 6.63M$

5. Circular centrally loaded plate simply supported along the edges:

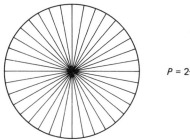

$P = 2\pi M$

6. Circular plate with fully restrained edges and centrally loaded by point load P:

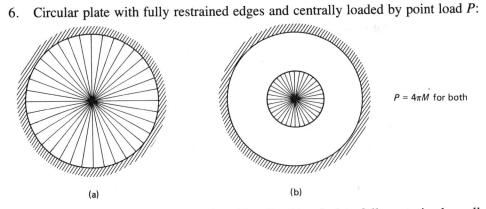

(a)   (b)

$P = 4\pi M$ for both

7. Point load P applied anywhere in arbitrarily shaped plate fully restrained on all boundries:

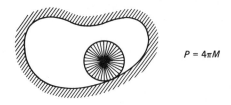

$P = 4\pi M$

8. Equilateral triangular plate with simply supported edges and centrally loaded by point load P:

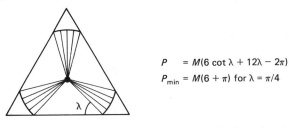

$$P = M(6 \cot \lambda + 12\lambda - 2\pi)$$
$$P_{min} = M(6 + \pi) \text{ for } \lambda = \pi/4$$

9. Acute-angled triangular plate on simply supported edges loaded with point load P at the center of the inscribed circle:

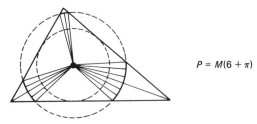

$P = M(6 + \pi)$

10. Obtuse-angled triangular plate with simply supported edges and load P at the center of the inscribed circle:

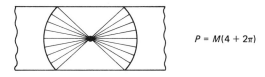

$P = M(4 + 2\lambda + 2 \cot 1/2\lambda)$,
where λ is in radians

As λ approaches π, the plate
degenerates into case 11

11. A long strip simply supported along edges and load with point P midway between edges;

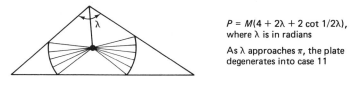

$P = M(4 + 2\pi)$

12. Simply supported strip with equal loads P between edges:

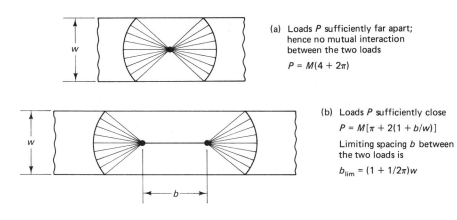

(a) Loads P sufficiently far apart;
hence no mutual interaction
between the two loads

$P = M(4 + 2\pi)$

(b) Loads P sufficiently close

$P = M[\pi + 2(1 + b/w)]$

Limiting spacing b between
the two loads is

$b_{lim} = (1 + 1/2\pi)w$

13. Strip simply supported with unequal loads P and kP midway between the edges where $k < 1.0$ and the loads sufficiently apart:

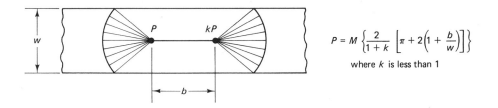

$P = M \left\{ \dfrac{2}{1 + k} \left[\pi + 2\left(1 + \dfrac{b}{w}\right) \right] \right\}$

where k is less than 1

14. Uniformly loaded square slab with degree of fixity i varying between zero and 1.0:

$i = 0$ and no upward movement
$w_u a^2 = 24M$

(a)

$i = 0$ and free upward movement
$w_u a^2 = 22.20M$
[for $\lambda_{min} = 1/2 \tan^{-1}(3)$]

(b)

$i = 0.5$ (partial restraint)
$w_u a^2 = 34.72M$

(c)

$i = 1.0$ (full restraint)
$w_u a^2 = 48M$ (upper bound)

(d)

15. Equilateral triangular plate ($\lambda = 60°$) uniformly loaded:

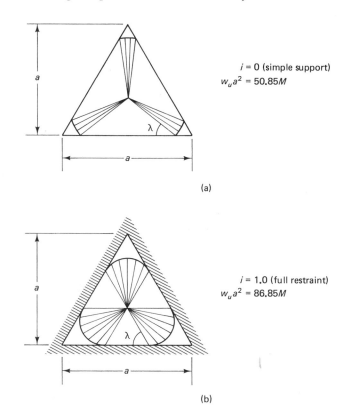

$i = 0$ (simple support)
$w_u a^2 = 50.85 M$

(a)

$i = 1.0$ (full restraint)
$w_u a^2 = 86.85 M$

(b)

16. Rectangular slab uniformly loaded with unit load of intensity w_u supported on all four sides with degree of restraint i varying from zero to 1.0 (note sequence of numbers assigned to panel sides):

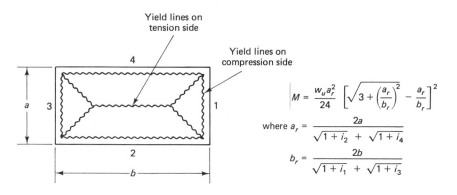

Yield lines on tension side

Yield lines on compression side

$$\left| M = \frac{w_u a_r^2}{24} \left[\sqrt{3 + \left(\frac{a_r}{b_r}\right)^2} - \frac{a_r}{b_r} \right]^2 \right.$$

where $a_r = \dfrac{2a}{\sqrt{1+i_2} + \sqrt{1+i_4}}$

$b_r = \dfrac{2b}{\sqrt{1+i_1} + \sqrt{1+i_3}}$

General note: Load P is assumed in the foregoing expression acting at a point. To adjust for the fact that P acts on a finite area, assume that it acts over a circular area of radius ρ. For a slab fully restrained on all boundaries, the hinge field would be bound by a circle touching the slab boundary (circle radius $= r$). In such a case,

$$M + M' = \frac{P}{2\pi}\left(1 - \frac{2\rho}{3r}\right) \tag{11.21}$$

where M = positive unit moment
   M' = negative unit moment

Reaction of columns supporting flat plates can be similarly considered for analyzing the flexural local capacity of the plate in the column area. For rectangular supports, an approximation to equivalent circular support can be made in the use of Eq. 11.21.

11.8.3 Example 11.6: Rectangular Slab Yield-Line Design

A reinforced concrete slab shown in Fig. 11.29 is 14 ft 6 in. $\times$ 24 ft in plan (4.42 m $\times$ 7.32 m). It carries an external factored ultimate uniform load $w_u = 220$ psf (10.5 kPa), including its self-weight. It is simply supported on one long edge and the adjacent short edge and built in on the opposite edges. Let the reinforcement spanning the short direction be twice the reinforcement spanning the long direction. Also assume the reinforcement on the built-in edges to be equal to the strong reinforcement. Design the slab structure for flexure, including the reinforcement needed and its spacing, using the yield-line theory. Given

$f_c' = 4000$ psi (27.6 MPa), normal-weight concrete
$f_y = 60,000$ psi (414 MPa)

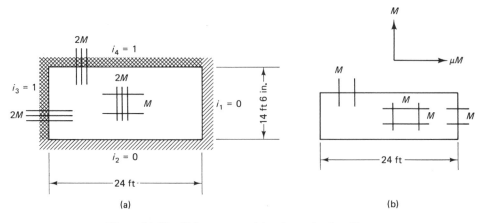

Figure 11.29   Slab geometry: (a) orthotropic; (b) affine.

Solution

μ = ratio of reinforcement in the strong direction to the weak direction = 2. From Eq. 11.20, the expression for the unit moment in an affine rectangular slab is

$$M_u = \frac{w_u a_r^2}{24} \left[\sqrt{3 + \left(\frac{a_r}{b_r}\right)^2} - \frac{a_r}{b_r} \right]^2$$

Change to affine slab converting the span dimension

$$a = 14.5 \times \frac{1}{\sqrt{u}} = 14.5 \times \frac{1}{\sqrt{2}} = 10.25 \text{ ft}$$

$$a_r = \frac{2a}{\sqrt{i_2 + 1} + \sqrt{i_4 + 1}} = \frac{2 \times 10.25}{\sqrt{0 + 1} + \sqrt{1 + 1}}$$

$$= \frac{20.50}{2.414} = 8.492$$

$$b_r = \frac{2b}{\sqrt{1 + i_1} + \sqrt{1 + i_3}} = \frac{2 \times 24.0}{\sqrt{1 + 0} + \sqrt{1 + 1}}$$

$$= \frac{48.0}{2.414} = 19.884$$

$$\frac{a_r}{b_r} = \frac{8.492}{19.884} = 0.427$$

$$w_n = \frac{w_u}{\phi = 0.9} = 244 \text{ psf}$$

$$M_n = \frac{w_n (8.492)^2}{24} [\sqrt{3 + (0.427)^2} - 0.427]^2$$

$$= \frac{w_n \times 72.11}{24} (1.841) = 5.532 w_n = 1350 \text{ lb}$$

$$\mu M_n = 2 \times 1350 = 2700 \text{ lb or ft-lb/ft}$$

Assume that d = 4 in. (h = 5 in.). $M_n = d^2 \rho f_y (1 - 0.59\omega)$, where $\omega = \rho(f_y/f_c')$ (see chapter 5); or

$$2700 = (4)^2 \rho \times 60,000 \left(1 - 0.59 \rho \frac{60,000}{4000} \right)$$

to get ρ = 0.00289.

reinforcement A_s on 12-in. (305-mm) strip = 0.00289 $\times$ 4 $\times$ 12 = 0.139 in.^{2}/12 in.

This steel area is less than the balanced ratio $0.75\overline{\rho}_b$, hence O.K. No. 3 bars at $9\frac{1}{2}$ in. center to center $\simeq$ 0.139 in^2. Hence use No. 3 bars at $9\frac{1}{2}$ in. center to center in the short direction and on the tension top face of the fixed supports (9.53-mm bars at 241 mm center to center).

Maximum allowable spacing $s = 2h = 2 \times 5 = 10$ in. (254 mm)

Use No. 3 bars at 10 in. center to center for the other direction.

Alternate affine slab in the perpendicular direction

Multiply the M direction by $\sqrt{\mu}$ to get μM in both directions. Affine $b = 24.0\sqrt{2} = 33.9$ ft.

$$a_r = \frac{2 \times 14.5}{2.414} = 12.0 \text{ ft} \qquad b_r = \frac{2 \times 33.9}{2.414} = 28.1 \text{ ft}$$

$$\frac{a_r}{b_r} = \frac{12.0}{28.1} = 0.427 \qquad \text{(same in the preceding solution)}$$

Hence

$$M_n = \frac{w_n(12.0)^2}{24}[\sqrt{3 + (0.427)^2} - 0.427]^2$$

$$= 6w_n(1.839) = 2700 \text{ lb} \quad \text{(as before)}$$

Check of optimum reinforcement distribution

A reasonably optimum reinforcement percentage can be specified using the following expression:

$$\rho_0 = (1 + \mu)\left[\sqrt{3.0 + \left(\frac{a}{b}\right)^2 \mu} - \frac{a}{b}\sqrt{\mu}\right]^2 \tag{11.22}$$

where ρ_0 is the optimum reinforcement percentage. From before, a/b for this affine slab $= 0.427$.

$$\rho_0 = (1 + 2)[\sqrt{3 + (0.427)^2 \times 2} - 0.427\sqrt{2}]^2$$

$$= 4.54\% \text{ over 12-in. strip} = 0.378\%$$

$$\text{actual } \rho = \frac{A_s}{b_d} = \frac{0.139}{12 \times 4} = 0.289\%$$

Hence the reinforcement can be considered close to optimal from an economical viewpoint. To achieve $\rho \simeq \rho_0$ would have necessitated using a thinner section, which might not have satisfied deflection requirements. A check of the thickness for minimum deflection and crack-control requirements would have to be made before the design is complete.

11.8.4 Example 11.7: Moment Capacity and Yield-Line Design of a Triangular Balcony Slab

A balcony floor in Fig. 11.30 is triangular in shape and is supported at the two perpendicular sides and carries a factored uniform line load of intensity $p = 400$ lb per linear foot (5.84 kN/m) acting on the triangle hypotenuse. The reinforcement in the short direction is three times the reinforcement in the long direction ($\mu = 3.0$). Analyze this

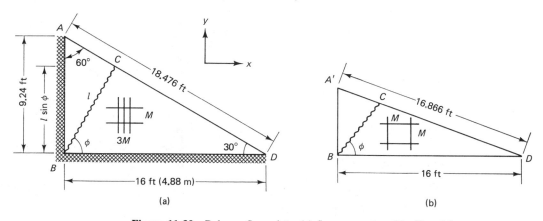

Figure 11.30   Balcony floor plate: (a) floor geometry; (b) affine slab.

floor moment capacity by the yield-line theory and design the thickness and reinforcement spacing in both directions if the longer of the two perpendicular supports is 16 ft (4.88 m) and it subtends an angle of 30° with the hypotenuse. The shorter side is 9.24 ft (2.82 m). Given:

$$f_c' = 4000 \text{ psi (27.58 kPa), normalweight concrete}$$
$$f_y = 60,000 \text{ psi (413.7 kPa)}$$

Assume that the self-weight of the plate can be neglected in the solution on the assumption that it is small compared to the line load.

Solution

$$AB = 16 \tan 30° = 9.24 \text{ ft}$$
$$AD = \sqrt{16^2 + 9.24^2} = 18.48 \text{ ft}$$

The yield line at failure is expected to be line BC subtending an angle ϕ with side BD. Its components in the x and y directions are $l \cos \phi$ and $l \sin \phi\check{}$, respectively.

Assume that point C deforms by a magnitude Δ, causing the center of gravity of the load on segment DC at its centerline to deflect $\frac{1}{2}\Delta$. Summing vectorially the work in the x and y directions in Fig. 11.30, gives

$$E_I = 3Ml \cos \phi \, \frac{1}{l \sin \phi} + Ml \sin \phi \, \frac{1}{l \cos \phi}$$

$$E_E = 18.48P \times \tfrac{1}{2} = 9.24p$$

Since $E_I = E_E$,

$$\frac{p}{M} = 0.324 \cot \phi + 0.108 \tan \phi$$

$$\frac{d(p/M)}{d\phi} = -0.324 \csc^2 \phi + 0.108 \sec^2 \phi = 0$$

$$\frac{\sec^2 \phi}{\csc^2 \phi} = 3.0 \quad \text{or} \quad \tan \phi = 1.732 \qquad \phi_{\min} = 60°$$

Hence

$$p = 0.324 \times \frac{1}{1.732} + 0.108 \times 1.732 = 0.375M$$

Affine slab solution

A'B in Fig. 11.30b becomes $9.24/\sqrt{3} = 5.335$ ft. Free edge A'D becomes 16.866 ft. If

$$p' = \text{affine linear load} = \frac{p}{\sqrt{\mu \cos^2 \theta + \sin^2 \theta}}$$

where θ = angle between the line load and M direction, or

$$p' = \frac{p}{\sqrt{3 \cos^2 30 + \sin^2 30}} = 0.632p$$

$$E_I = M \cot \theta + M \tan \theta$$

$$E_E = P' \times 16.866 \times \tfrac{1}{2} = 8.433p'$$

But $E_I = E_E$; therefore,

$$\frac{p'}{M} = 0.1186 \cot \phi + 0.1186 \tan \phi$$

$$\phi_{\min} = 45° \qquad \tan \phi = 1.0$$

Therefore,

$$p' = 2 \times 0.1186M = 0.237M$$

or

$$p = \frac{0.237}{0.632}M = 0.375M \quad \text{(as before)}$$

Design of reinforcement

$400 = 0.375M$ to give $M = 1067$ lb or $\mu M = 3 \times 1067 = 3200$ lb or ft-lb/ft. Assume that $h = 5$ in. ($d = 4$ in. $= 100$ mm)

$$3200 = d^2 \rho f_y (1 - 0.59\omega)$$

or

$$3200 = 4^2 \rho \times 60{,}000 \left(1 - 0.59\rho \times \frac{60{,}000}{4000} \right)$$

to give $\rho = 0.0034$. The required $A_s = 0.0034 \times 4 \times 12 = 0.163$ in.^{2}. Use No. 3 bars at 8 in. center to center in the short direction $\simeq 0.165$ in.^{2} (9.52 mm diameter at 165 mm center to center) and No. 3 bars at $2h = 9$ in. center to center in the long direction to satisfy moment requirements.

SELECTED REFERENCES

11.1   ACI Committee 318, "Building Code Requirements for Reinforced Concrete, ACI Standard 318–83," American Concrete Institute, Detroit, 1983, pp. 111, and "Commentary on Building Code Requirements for Reinforced Concrete," American Concrete Institute, 1983, 155 pp.

11.2   CEB-FIP,"Concrete Design—U.S. and European Practices," Joint ACI-CEB Symposium, *Bulletin d'Information No. 113,* Comité Euro-International du Béton, Paris, February 1979, 345 pp.

11.3   Gamble, W. L., Sozen, M. A., and Siess, C. P., "An Experimental Study of Reinforced Concrete Two-Way Floor Slab," *Civil Engineering Studies, Structural Research Series No. 211,* University of Illinois, 1961, 304 pp.

11.4   Corley, W. G., and Jirsa, J. D., "Equivalent Frame Analysis for Slab Design," *Journal of the American Concrete Institute,* Proc. Vol. 67, No. 11, November 1970, pp. 875–884.

11.5   Wang, C. K., and Salmon, C. G., *Reinforced Concrete Design,* 3rd ed., Harper & Row, New York, 918 pp.

11.6   Branson, D. E., *Deformation of Concrete Structures,* McGraw-Hill, New York, 1977, 546 pp.

11.7   Nilson, A. H., and Walters, D. B., "Deflection of Two-Way Floor Systems by the Equivalent Frame Method," *Journal of the American Concrete Institute,* Proc. Vol. 72, No. 5, May 1975, pp. 210–218.

11.8   Nawy, E. G., and Chakrabarti, P., Deflection of Prestressed Concrete Flat Plates," *Journal of the Prestressed Concrete Institute,* Vol. 21, No. 2, March–April 1976, pp. 86–102.

11.9   Mansfield, E. H., "Studies in Collapse Analysis of Rigid-Plastic Plates with a Square Yield Diagram," *Proceedings of the Royal Society,* Vol. 241, August 1957, pp. 311–338.

11.10 Wood, R. H., *Plastic and Elastic Design of Slabs and Plates,* Thames and Hudson, London, 1961, pp. 225–261.

11.11 Johansen, K. W., *Yield-Line Theory,* Cement and Concrete Association, London, 1962, 181 pp.

11.12 Hognestad, E., "Yield-Line Theory for the Ultimate Flexural Strength of Reinforced Concrete Slabs," *Journal of the American Concrete Institute,* Proc. Vol. 49, No. 7, March 1953, pp. 637–655.

11.13 Hung, T. Y., and Nawy, E. G., "Limit Strength and Serviceability Factors in Uniformly Loaded, Isotropically Reinforced Two-Way Slabs," *Symposium on Cracking, Deflection, and Ultimate Load of Concrete Slab Systems,* Special Publication SP-30, American Concrete Institute, Detroit, 1972, pp. 301–324.

11.14 Nawy, E. G., and Blair, K. W., "Further Studies on Flexural Crack Control in Structural Slab Systems," *Symposium on Cracking, Deflection, and Ultimate Load of Concrete Slab Systems,* Special Publication SP-30, American Concrete Institute, Detroit, 1972, pp. 1–41.

11.15 Nawy, E. G., "Crack Control through Reinforcement Distribution in Two-Way Acting Slabs and Plates," *Journal of the American Concrete Institute,* Proc. Vol. 69, No. 4, April 1972, pp. 217–219.

11.16 ACI Committee 340, *Design Hand Book: Beams, Slabs, Etc.,* Special Publication SP-17 (81), American Concrete Institute, Detroit, Vol. 1, 1982, 508 pp.

11.17 Hawkins, N., and Corley, W. G., "Transfer of Unbalanced Moment and Shear from Plates to Columns," *Symposium on Cracking Deflection and Ultimate Load of Concrete Slab Systems,* Special Publication SP-30, American Concrete Institute, Detroit, 1972, pp. 147–176.

11.18 Park, R., and Gamble, W. L., *Reinforced Concrete Slabs,* Wiley, 1980, 618 pp.

11.19 Warner, R. F., Rangan, B. V., and Hall, H. S., "Reinforced Concrete," Pitman Australia, 1982, 471 pp.

11.20 Standards Association of Australia, "SAA Concrete Structures Code, 1985," Standards Assoc. of Australia, Sydney, 1985, pp. 1–158.

11.21 Nawy, E. G., "Strength, Serviceability and Ductility," Chapter 12 in *Handbook of Structural Concrete,* Pitman Books, London/McGraw-Hill, 1983, 1968 pp.

PROBLEMS FOR SOLUTION

11.1 An end panel of a floor system supported by beams on all sides carries a uniform service live load $w_L = 75$ psf and an external dead load $w_D = 20$ psf in addition to its self-weight. The centerline dimensions of the panel are 18 ft $\times$ 20 ft (the dimension of the discontinuous side is 18 ft). Design the panel and the size and spacing of the reinforcement using the ACI code method. Given:

$f'_c = 4000$ psi, normalweight concrete

$f_y = 60,000$ psi

Column sizes at each corner 12 in. $\times$ 12 in.

Width of the supporting beam webs = 12 in.

Assume reinforcement ratio $\rho \simeq 0.4\, \rho_b$ for the supporting beams.

11.2 An interior flat-plate panel is supported on columns spaced 18 ft $\times$ 20 ft. The panel dimensions, loading, and material properties are the same as those in Problem 11.1. Design the panel and size and spacing of the reinforcement by the ACI code method.

11.3 Calculate the time-dependent deflection at the center of the panel in (a) Problem 11.1, and (b) Problem 11.2. Check also if the panels satisfy the serviceability requirements for deflection control and crack control for aggressive environment. Assume $K_{ec}/E_c = 350$ in.^{3} per rad. for part (a) and $= 225$ in.^{3} per rad. for part (b).

11.4 Use the yield-line theory to evaluate the slab thickness needed in the column zone of the flat plate in Problem 11.2 for flexure, assuming that the hinge field would have a radius of 24 in.

11.5 An isotropically reinforced long strip is simply supported on the edges. A concentrated load P acts on the minor axis of the slab midway between the long edges. Prove that the magnitude of the collapse load is $P = M(4 + 2\pi)$.

11.6 A slab 21 ft × 13 ft 6 in. carries an external factored ultimate load of 200 lb per square foot, including its self-weight. It is simply supported on one long edge and the adjacent short edge and built in on the opposite edges. Let the reinforcement spacing the short way be three times the reinforcement spanning the long way. Also assume the reinforcement on the built-in edges to be equal to the strong reinforcement. Design the slab structures and the reinforcement needed and its spacing using the yield-line theory. Given:

$f'_c = 4000$ psi, normalweight concrete
$f_y = 60,000$ psi

11.7 Calculate the maximum crack width in a two-way interior panel of a reinforced concrete floor system. The slab thickness is 8 inches (203.2 mm) and the panel size is 20 ft. × 28 ft. (6.10 m. × 8.53 m.). Also design the size and spacing of the reinforcement necessary for crack control assuming (a) the floor is exposed to normal environment, and (b) the floor is part of a parking garage. Given: $f_y = 60.0$ ksi(414 MPa)

12

Footings

12.1 INTRODUCTION

Cumulative floor loads of a superstructure are supported by foundation substructures in direct contact with the soil. The function of the foundation is to transmit safely the high concentrated column and/or wall reactions or lateral loads from earth-retaining walls to the ground without causing unsafe differential settlement of the supported structural system or soil failure.

If the supporting foundations are not adequately proportioned, one part of a structure can settle more than an adjacent part. Various members of such a system become overstressed at the column–beam joints due to *uneven* settlement of the supports leading to large deformations. The additional bending and torsional moments in excess of the resisting capacity of the members can lead to excessive cracking due to yielding of the reinforcement and ultimately to failure.

If the total structure undergoes *even* settlement, little or no overstress occurs. Such behavior is observed when the foundation is excessively rigid and the supporting soil highly yielding such that a structure behaves similar to a floating body that can sink or tilt without breakage. Numerous examples of such structures can be found in such locations as Mexico City with buildings on mat foundations or rigid supports which sank several feet over the years due to the high consolidation of the supporting soil. Examples of other famous cases of very slow and relatively uneven consolidation process can be cited. Gradual loss of stability of a structure undergoing tilting with time, as the leaning Tower of Pisa, is an example of foundation problems resulting from uneven bearing support.

Layouts of structural supports vary widely and soil conditions differ from site to site and within a site. As a result, the type of foundation to be selected has to be governed by these factors and by optimal cost considerations. In summary, the structural engineer has to acquire the maximum economically feasible soil data on the site before embarking on a study of the various possible alternatives for site layout.

Basic knowledge of soil mechanics and foundation engineering is assumed in presenting the topic of design of footings in this chapter. Background knowledge of the methodology of determining the resistance of cohesive and noncohesive soils is necessary to select the appropriate bearing capacity value for the particular site and the particular foundation system under consideration.

The bearing capacity of soils is usually determined by borings, test pits, or other soil investigations. If these are not available for the preliminary design, representative values at the footing level can normally be used from Table 12.1.

12.2 TYPES OF FOUNDATIONS

There are basically six types of foundation substructures, as shown in Fig. 12.1. The foundation area must be adequate to carry the column loads, the footing weight, and any overburden weight within the permissible soil pressure.

TABLE 12.1   PRESUMPTIVE BEARING CAPACITY (TONS/FT²)

Type of soil	Bearing capacity
Massive crystalline bedrock, such as granite, diorite, gneiss, and trap rock	100
Foliated rocks, such as schist or slate	40
Sedimentary rocks, such as hard shales, sandstones, limestones, and siltstones	15
Gravel and gravel–sand mixtures (GW and GP soils)	
Densely compacted	5
Medium compacted	4
Loose, not compacted	3
Sands and gravely sands, well graded (SW soil)	
Densely compacted	$3\frac{3}{4}$
Medium compacted	3
Loose, not compacted	$2\frac{1}{4}$
Sands and gravely sands, poorly graded (SP soil)	
Densely compacted	3
Medium compacted	$2\frac{1}{2}$
Loose, not compacted	$1\frac{3}{4}$
Silty gravels and gravel–sand–silt mixtures (GM soil)	
Densely compacted	$2\frac{1}{2}$
Medium compacted	2
Loose, not compacted	$1\frac{1}{2}$
Silty sand and silt–sand mixtures (SM soil)	2
Clayey gravels, gravel–sand–clay mixtures, clayey sands, sand–clay mixtures (GC and SC soils)	2
Inorganic silts, and fine sands; silty or clayey fine sands and clayey silts, with slight plasticity; inorganic clays of low to medium plasticity; gravely clays; sandy clays; silty clays; lean clays (ML and CL soils)	1
Inorganic clays of high plasticity, fat clays; micaceous or diatomaceous fine sandy or silty soils, elastic silts (CH and MH soils)	1

1. *Wall footings.* Such footings comprise a continuous slab strip along the length of the wall having a width larger than the wall thickness. The projection of the slab footing is treated as a cantilever loaded up by the distributed soil pressure. The length of the projection is determined by the soil bearing pressure, with the critical section for bending being at the face of the wall. The main reinforcement is placed perpendicular to the wall direction.

2. *Independent isolated column footings.* They consist of rectangular or square slabs of either constant thickness or sloping toward the cantilever tip. They are reinforced in both directions and are economical for relatively small loads or for footings on rock.

3. *Combined footings.* Such footings support two or more column loads. They are necessary when a wall column has to be placed on a building line and the footing slab cannot project outside the building line. In such a case, an independent footing would be eccentrically loaded, causing apparent tension on the foundation soil.

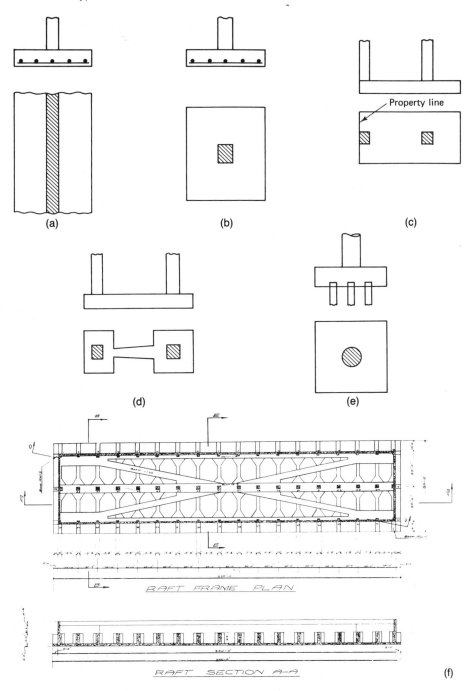

Figure 12.1 Types of foundations: (a) wall footing; (b) isolated footing; (c) combined footing; (d) strap footing; (e) pile foundation; (f) raft foundation.

In order to achieve a relatively uniform stress distribution, the footing for the exterior wall column can be combined with the footing of the adjoining interior column. Additionally, combined footings are also used when the distance between adjoining columns is relatively small, such as in the case of corridor columns, when it becomes more economical to build a combined footing for the closely spaced columns.

4. *Cantilever or strap footings.*   These are similar to the combined footings except that the footings for the exterior and interior columns are built independently. They are joined by a strap beam to transmit the effect of the bending moment produced by the eccentric wall column load to the interior column footing area.

5. *Pile foundations.*   This type of foundation is essential when the supporting ground consists of structurally unsound layers of material to large depths. The piles may be driven either to solid bearing on rocks or hardpan or deep enough into the soil to develop the allowable capacity of the pile through skin frictional resistance or a combination of both. The piles could be either precast and hence driven into the soil, or cast in place by drilling a caisson and subsequently filling it with concrete. The precast piles could be reinforced or prestressed concrete. Other types of piles are made of steel or treated wood. In all types, the piles have to be provided with appropriately designed concrete caps reinforced in both directions.

6. *Raft or floating foundations.*   Such foundation systems are necessary when the allowable bearing capacity of soil is very low to great depths, making pile foundations uneconomical. In this case it becomes necessary to have a deep enough excavation with sufficient depth of soil removed that the net bearing pressure of the soil on the foundation is almost equivalent to the structure load. It becomes necessary to spread the foundation substructure over the entire area of the building such that the superstructure is considered to be theoretically floating on a raft. Continuously consolidating soils require such a substructure, which is basically an inverted floor system. Otherwise, friction piles or piles driven to rock become mandatory.

12.3 SHEAR AND FLEXURAL BEHAVIOR OF FOOTINGS

To simplify foundation design, footings are assumed to be rigid and the supporting soil layers elastic. Consequently, uniform or uniformly varying soil distribution can be assumed. The net soil pressure is used in the calculation of bending moments and shears by subtracting the footing weight intensity and the surcharge from the total soil pressure. If a column footing is considered as an *inverted* floor segment where the intensity of net soil pressure is considered to be acting as a column-supported cantilever slab, the slab would be subjected to both bending and shear in a similar manner to a floor slab subjected to gravity loads.

When heavy concentrated loads are involved, it has been found that shear rather than flexure controls most foundation designs. The mechanism of shear failure in

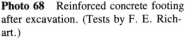

Photo 68   Reinforced concrete footing after excavation. (Tests by F. E. Richart.)

footing slabs is similar to that in supported floor slabs. However, the shear capacity is considerably higher than that of beams, as will be discussed in the next section. Since the footing in most cases bends in double curvature, shear and bending about both principal axes of the footing plan have to be considered.

The state of stress at any element in the footing is due primarily to the combined effects of shear, flexure, and axial compression. Consequently, a basic understanding of the fundamental behavior of the footing slab and the cracking mechanism involved is essential. It enables developing a background feeling for the underlying hypothesis used in the analysis and design requirements of footings both in shear and in flexure.

12.3.1 Failure Mechanism

The inclined shear cracks develop in essentially the same manner as in beams, stabilizing at approximately 65% of the ultimate load and extending rapidly toward the neutral axis. Thereafter, the cracks propagate slowly toward the compression zone such that a very shallow depth in compression remains at failure.

The inclined cracks always form close to the concentrated load or column reaction in two-way slabs or footings, as seen in Fig. 12.2a. This is due partly to the heavy concentration of bending moments in the region close to the column face, forming a truncated pyramid at the foot of the column region. The column can perimetrically punch through the slab in this failure form if the slab is not adequately designed to resist shear failure (also called *diagonal tension* or *punching shear*). The action of the confining surrounding punched slab on the column base interface punching through the slab in Fig. 12.2b can be represented by the resulting shear forces V_1 and V_2, the compressive forces C_1 and C_2, and the tensile forces T_1 and T_2, in addition to the internal dowel and membrane action of the slab.

Figure 12.2c shows an infinitesimal element taken from the compression zone above the inclined crack. The element is subjected to the following four stress components: (1) vertical shear stress v_0, (2) direct compressive stress f_c, (3) vertical compressive stress f_3, and (4) lateral compressive stress f_2.

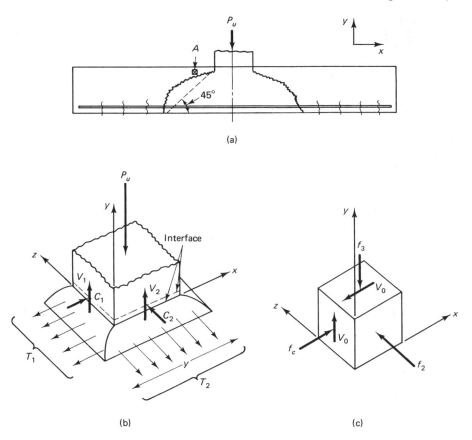

Figure 12.2 Two-way-action failure mechanism in slabs and footings: (a) footing elevation; (b) failure pyramid; (c) element A in the compression zone.

The vertical shearing stress v_0 is the result of the total shear that has to be entirely transmitted by the compression zone above the inclined crack. The direct compressive stress f_c, which varies along the length of the critical section, results from the bending moments. The vertical compressive force f_3 is due to the heavy concentrated column load. It has a major influence on increasing the shear capacity of the slab, as demonstrated in Ref. 12.2 for pressure in an infinite semielastic solid loaded at the surface. The lateral compressive stress f_2 is the result of the bending moment about an axis perpendicular to the critical section. It contributes further to the increase in the compressive strength of the concrete as a result of the triaxial state of stress. Consequently, the existence of the multiaxial forces and stresses in Fig. 12.2c explains why the shear capacity of a slab subjected to concentrated loads in considerably higher than that of a beam.

In addition, the inclined crack generating *close* to the critical section in two-way slabs and footings due to the high moment concentration justifies considering the critical section to be at a distance of $d/2$ from the face of the column in slabs and footings, while in beams and one-way slabs and footings, the ACI code specifies the critical section at a distance d from the face of the column support. The nominal shearing stress at failure varies between $6\sqrt{f_c'}$ and $9\sqrt{f_c'}$ for the slabs, whereas it does not exceed $2\sqrt{f_c'}$ to $4\sqrt{f_c'}$ in beams. The code, however, allows a maximum nominal resisting shear strength of plain concrete not to exceed $v_c = 4\sqrt{f_c'}$ for the supported two-way slab or the footing, and $v_c = 2\sqrt{f_c'}$ for beams and one-way-action footings. For plain concrete footings cast against soil, the effective thickness used in computing stresses is taken as the overall thickness minus 3 in. The overall thickness should not be less than 8 in.

12.3.2 Loads and Reactions

Based on the foregoing discussion, it is essential to make the correct assumptions for evaluating all the combined forces acting on the foundation. The footing slab has to be proportioned to sustain all the applied factored loads and induced reactions which include axial loads, shears, and moments to be resisted at the base of the footing.

After the permissible soil pressure is determined from the available site data and the principles of soil mechanics and the local codes, the footing area size is computed on the basis of the *unfactored* (service) loads, such as dead, live, wind, or earthquake loads in whatever combination governs the design.

The minimum eccentricity requirement for column slenderness considerations is neglected in the design of footings or pile caps and only the computed end moments that exist at the base of a column are considered to have been transferred to the footing. In cases where eccentric loads or moments exist due to any loading combinations, the extreme soil pressure resulting from such loading conditions has to be within such permissible bearing values as those in Table 12.1 or as determined by actual soil tests.

Once the size of a footing or pile cap for a single pile or a group of piles is determined, the design of the footing geometry becomes possible using the principles and methodologies presented in the preceding chapters for shear and flexure design. The external *service* loads and moments used to determine the size of the foundation area are converted to their ultimate *factored* values using the appropriate load factors and strength reduction factors ϕ for determining the nominal resisting values to be used in the analysis and proportioning the size and reinforcement distribution in the footing.

12.4 SOIL BEARING PRESSURE AT BASE OF FOOTINGS

The distribution of soil bearing pressure on the footing depends on the manner in which the column or wall loads are transmitted to the footing slab and the degree of rigidity of the footing. The soil under the footing is assumed to be a homogeneous elastic material and the footing is assumed to be rigid as a most common type of foundation. Consequently, the soil bearing pressure can be considered uniformly

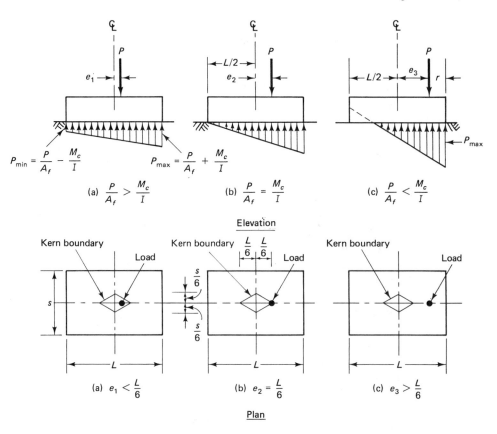

Figure 12.3 Eccentrically loaded footings.

distributed if the reaction load acts through the axis of the footing slab area. If the load is not axial or symmetrically applied, the soil pressure distribution becomes trapezoidal due to the combined effects of axial load and bending.

12.4.1 Eccentric Load Effect on Footings

As indicated in Section 12.2, exterior column footings and combined footings can be subjected to eccentric loading. When the eccentric moment is very large, tensile stress on one side of the footing can result since the bending stress distribution is dependent on the magnitude of load eccentricity. It is always advisable to proportion the area of these footings such that the load falls within the middle kern, as shown in Figs. 12.3 and 12.4. In such a case, the location of the load is in the middle third of the footing dimension in each direction, thereby avoiding tension in the soil that can theoretically occur prior to stress redistribution.

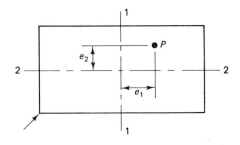

Figure 12.4 Biaxial loading of footing.

1. *Eccentricity case $e < L/6$* (Fig. 12.3a).   In this case, the direct stress P/A_f is larger than the bending stress M_c/I. The stress

$$p_{max} = \frac{P}{A_f} + \frac{Pe_1 c}{I} \tag{12.1a}$$

$$p_{min} = \frac{P}{A_f} - \frac{Pe_1 c}{I} \tag{12.1b}$$

2. *Eccentricity case $e_2 = L/6$* (Fig. 12.3b):

$$\text{direct stress} = \frac{P}{A_f} = \frac{P}{sL} \tag{12.2a}$$

$$\text{bending stress} = \frac{M_c}{I} = Pe_2 \times \frac{c}{I} \tag{12.2b}$$

$$\frac{c}{I} = \frac{\dfrac{L}{2}}{s(L^3/12)} = \frac{1}{s(L^2/6)} = \frac{6}{sL^2} \tag{12.2c}$$

s and L are the width and the length of the footing, respectively. In order to find the limiting case where *no* tension exists on the footing, the direct stress P/A_f has to be equivalent to the bending stress so that

$$\frac{P}{A_f} - Pe_2 \frac{c}{I} = 0 \tag{12.2d}$$

Substituting for P/A_f and C/I from Eqs. 12.2a and 12.2c into Eq. 12.2d,

$$\frac{P}{sL} - Pe_2 \times \frac{6}{sL^2} = 0 \quad \text{or} \quad e_2 = \frac{L}{6}$$

Consequently, the eccentric load has to act within the middle third of the footing dimension to avoid tension on the soil.

3. *Eccentricity case $e_3 > L/6$* (Figure 12.3c).   As the load acts outside the middle third, tensile stress results at the left side of the footing, as shown in Fig. 12.3c. If the maximum bearing pressure p_{max} due to load P does not exceed the allowable

bearing capacity of the soil, no uplift is expected at the left end of the footing and the center of gravity of the triangular bearing stress distribution *coincides* with the point of action of load p in Fig. 12.3c.

The distance from the load P to the tip of footing is $r = (L/2) - e_3 =$ distance of the centroid of the stress triangle from the base of the triangle. Therefore, the width of the triangle is $3r = 3[(L/2) - e_3]$. Hence the maximum compressive bearing stress is

$$p_{max} = \frac{P}{\dfrac{3r \times s}{2}} = \frac{2P}{3s\left(\dfrac{L}{2} - e_3\right)} \qquad (12.3a)$$

4. *Eccentricity about two axes, biaxial loading* (Fig. 12.4).   In the case where a concentrated load has an eccentricity in two directions (both within their respective kern points), the stresses are

$$p_{max} = \frac{P}{A_f} \pm \frac{Pe_1 c_1}{I_1} \pm \frac{Pe_2 c_2}{I_2} \qquad (12.3b)$$

12.4.2 Example 12.1: Concentrically Loaded Footings

A column support transmits axially a total service load of 400,000 lb (1779 kN) to a square footing at the frost line (3 ft below grade), as shown in Fig. 12.5. The frost line is the subgrade soil level below which the groundwater does not freeze throughout the year. Test borings indicate a densely compacted gravel–sand soil. Determine the required area of the footing and the net soil pressure intensity p_n to which it is subjected. Given:

  Unit weight of soil $\gamma = 135$ lb/ft^3 (21.1 kN/m^3)
  Footing slab thickness $= 2$ ft (0.61 m)

Solution

Since the footing is concentrically loaded, the soil bearing pressure is considered uniformly distributed assuming the footing is rigid. From the soil test borings and Table 12.1, the presumptive bearing capacity of the soil is 5 tons/ft^2 at the level of the footing, that is, 10,000 lb/ft^2 (478.8 kPa). Assume that the average weight of the soil and concrete above the footing is 135 pcf. Since the top of the footing has to be below the frost line (minimum 3 ft below grade), the net allowable pressure

  $p_n = 10,000 - (5 \times 135 + 100$ psf for surcharge paving$) = 9225$ psf

  minimum area of footing $A_f = \dfrac{400,000}{9,225} = 43.36$ ft^2

Use square footing 6 ft 8 in. $\times$ 6 ft 8 in. (2.03 m $\times$ 2.03 m),

$$A_f = 44.44 \text{ ft}^2 (4.13 \text{ m}^2) > 43.36 \text{ ft}^2$$

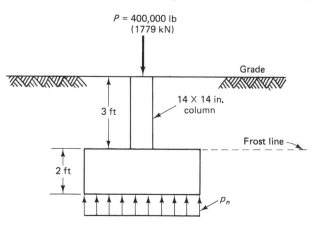

Figure 12.5 Concentrically loaded footing.

12.4.3 Example 12.2: Eccentrically Loaded Footings

A reinforced concrete footing supports a 14 in. × 14 in. column reaction $P = 400,000$ lb (1779 kN) at the frost line (3 ft below grade). The load acts at an eccentricity $e_1 = 0.4$ ft, $e_2 = 1.3$ ft, and $e_3 = 2.2$ ft. Select the necessary area of footing assuming that it is rigid and has a thickness $h = 2\frac{1}{2}$ ft. Soil test borings have indicated that the bearing area is composed of layers of shale and clay to a considerable depth below the foundation. Use a unit weight $\gamma = 140$ lb/ft³.

Solution

From Table 12.1, assume an allowable bearing capacity $p_g = 6.5$ tons/ft² (13,000 lb/ft²) at the footing base level.

Eccentricity $e_1 = 0.4$ ft

By trial and adjustment, assume a footing 5 ft × 9 ft (1.52 m × 2.74 m), $A_f = 45$ ft². Assume that the footing base is 6 ft below grade and that a slab on grade surcharge weighs 120 psf. Assume that the average weight of the soil and footing is ≈ 140 pcf.

$$\text{net allowable bearing pressure } p_n = 13,000 - (6 \times 140 + 120)$$

$$= 12,040 \text{ lb/ft}^2 \text{ (576.5 KPa)}$$

Stress due to the service eccentric column load is

$$p = \frac{P}{A_f} \pm \frac{P_e}{I/c} = \frac{400,000}{45} \pm \frac{400,000 \times 0.4 \times 6}{5(9)^2}$$

$$= 8889 \pm 2370 = 11,259 \text{ lb/ft}^2 \text{ (C) and } 6,519 \text{ lb/ft}^2 \text{ (C)} < 12,040 \text{ lb/ft}^2$$

The distribution of the bearing pressure is as shown in Fig. 12.6a, therefore, O.K.

Eccentricity $e_2 = 1.3$ ft

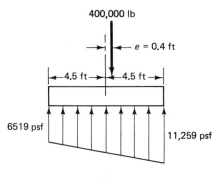

400,000 lb

$e = 0.4$ ft

4.5 ft   4.5 ft

6519 psf

11,259 psf

(a)

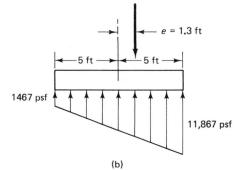

400,000 lb

$e = 1.3$ ft

5 ft   5 ft

1467 psf

11,867 psf

(b)

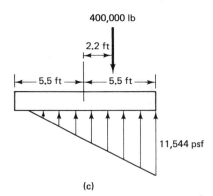

400,000 lb

2.2 ft

5.5 ft   5.5 ft

11,544 psf

(c)

Figure 12.6 Bearing area and bearing stress distribution in Ex. 12.2.

By trial and adjustment, assume a footing 6 ft $\times$ 10 ft (1.83 m $\times$ 3.05 m), $A_f = 60$ ft^2 (5.57 m^2). The actual service load-bearing pressure is

$$p = \frac{400,000}{60.0} \pm \frac{400,000 \times 1.3 \times 6}{6(10)^2}$$

$$= 6667 \pm 5200 = 11,867 \text{ lb/ft}^2 \text{ (C) and } 1467 \text{ lb/ft}^2 \text{ (C)}$$

$$< 12,040 \text{ lb/ft}^2 \qquad \text{therefore, O.K.}$$

Notice in comparing the two cases that as the moment increases leading to larger eccentricities, the minimum bearing pressure decreases, as seen from Fig. 12.6a and b.

Eccentricity e_3 = 2.2 ft

By trial and adjustment, try a footing 7 ft $\times$ 11 ft (2.13 m $\times$ 3.35 m), A_f = 77 ft^2 (7.15 m^2).

$$p = \frac{400,000}{77.0} \pm \frac{400,000 \times 2.2 \times 6}{7(11)^2}$$

$$= 5195 \pm 6234 = 11,429 \text{ lb/ft}^2 \text{ (C) and } -1039 \text{ lb/ft}^2 \text{ (T)}$$

Check by Eq. 12.3 for $e > L/6 > 11.0/6 = 1.83$ ft:

$$p = \frac{2P}{3S[(L/2) - e_3]} = \frac{2 \times 400,000}{3 \times 7[(11/2) - 2.2]} = 11,544 \text{ lb/ft } 2 < 12,040 \text{ lb/ft}^2$$

O.K.

Figure 12.6c shows that as the load acts outside the middle third of the base, only part of the footing is subjected to compressive bearing stress.

12.5 DESIGN CONSIDERATIONS IN FLEXURE

The maximum external moment on any section of a footing is determined on the basis of computing the factored moment of the forces acting on the entire area of footing on *one side* of a vertical plane assumed to pass through the footing. This plane is taken at the following locations:

1. At the face of column, pedestal, or wall for an isolated footing as in Fig. 12.7a
2. Halfway between the middle and edge of wall for footing supporting masonry wall as in Fig. 12.7b
3. Halfway between face of column and edge of steel base for footings supporting a column with steel base plates

12.5.1 Reinforcement Distribution

In one-way footings and in two-way square footings, the flexural reinforcement should be uniformly distributed across the entire width of the footing. This recommendation is conservative, particularly if the soil bearing pressure is not uniform. However, no meaningful saving can be accomplished if refinement is made in the bending moment assumptions.

In two-way rectangular footings supporting one column, the bending moment in the short direction is taken as equivalent to the bending moment in the long direction. The distribution of reinforcement differs in the long and short directions. The effective depth is assumed without meaningful loss of accuracy to be equal in both the short and

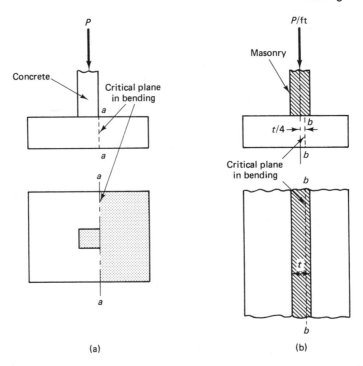

Figure 12.7 Critical planes in flexure: (a) concrete column; (b) masonry wall.

long directions, although it differs slightly because of the two-layer reinforcing mats. The following is the recommended reinforcement distribution:

1. Reinforcement in the long direction is to be uniformly distributed across the entire width of the footing.

2. For reinforcement in the short direction, a central band of width equal to the width of footing in the short direction shall contain a major portion of the reinforcement total area as in Eq. 12.4 uniformly distributed along the band width:

$$\frac{\text{reinforcement in band width}}{\text{total reinforcement in short direction, } A_s} = \frac{2}{\beta + 1} \qquad (12.4)$$

where β is the ratio of long to short side of footing. The remainder of the reinforcement required in the short direction is uniformly distributed outside the center band of the footing.

  In all cases, the depth of the footing above the reinforcement has to be at least 6 in. (152 mm) for footings on soil and at least 12 in. (305 mm) for footings on piles (footings on piles must always be reinforced). A practical depth for column footings should not be less than 9 in. (229 mm).

12.6 DESIGN CONSIDERATIONS IN SHEAR

As discussed in Section 12.3.1, the behavior of footings in shear is not different from that of beams and supported slabs. Consequently, the same principles and expressions as those used in Chapter 6 on shear and diagonal tension are applicable to the shear design of foundations.

The shear strength of slabs and footings in the vicinity of column reactions is governed by the more severe of the following two conditions.

12.6.1 Beam Action

The critical section for shear in slabs and footings is assumed to extend in a plane across the entire width and located at a distance d from the face of the concentrated load or reaction area. In this case, if only shear and flexure act, the nominal shear strength of the section is

$$V_c = 2\sqrt{f_c'}\, b_w d \tag{12.5}$$

where b_w is the footing width.

V_c must always be larger than the nominal shear force $V_n = V_u/\phi$ unless shear reinforcement is provided.

12.6.2 Two-Way Action

The plane of the critical section perpendicular to the plane of the slab is assumed to be so located that it has a minimum perimeter b_0. This critical section need not be closer than $d/2$ to the perimeter of the concentrated load or reaction area. The fundamental shear failure mechanism in two-way action as presented in Section 12.3.1 demonstrates that the critical section occurs at a distance $d/2$ from the face of the support and not at d as in beam action.

The shear strength of the section in this case is

$$V_c = \left(2 + \frac{4}{\beta_c}\right)\sqrt{f_c'}\, b_0 d \le 4\sqrt{f_c'}\, b_0 d \tag{12.6}$$

where $\beta_c = \dfrac{\text{long side } c_l}{\text{short side } c_s}$ of the concentrated load or reaction area

b_0 = perimeter of the critical section, that is, the length of the idealized failure plane

Figure 12.8 gives the relationship of the column side ratio β_c to the shear strength V_c of the footing. V_c must always be larger than the nominal shear force $V_n = V_u/\phi$ unless shear reinforcement is provided.

In cases of both one-way and two-way action, if shear reinforcement consisting of bars or wires are used,

$$V_n = V_c + V_s \le 6\sqrt{f_c'}\, b_0 d \tag{12.7}$$

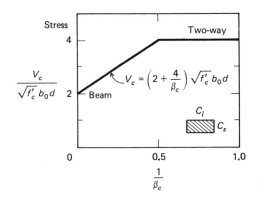

Figure 12.8 Shear strength in footings.

where $V_c = 2\sqrt{f'_c}\, b_o d$ and V_s is based on the shear reinforcement size and spacing as described in Chapter 6 unless shear heads made from steel I or channel shapes are used.

It is worthwhile to keep in mind that in most footing slabs, as in most supported superstructure slabs or plates, the use of shear reinforcement is not popular, due to practical considerations and the difficulty of holding the shear reinforcement in position.

12.6.3 Force and Moment Transfer at Column Base

The forces and moments at the base of a column or wall are transferred to the footing by bearing on the concrete and by reinforcement, dowels, and mechanical connectors. Such reinforcement can transmit the compressive forces which exceed the concrete bearing strength of the footing or the supported column as well as any tensile force across the interface.

The permissible bearing stress on the actual loaded area of the column base or footing top area of contact is

$$f_b = \phi(0.85f'_c) \qquad \text{where } \phi = 0.70 \tag{12.8a}$$

or

$$f_b = 0.60f'_c \tag{12.8b}$$

Hence the permissible bearing stress on the column can normally be considered $0.60f'_c$ for the column concrete. The compressive force which *exceeds* that developed by the permissible bearing stress at the base of the column or at the top of the footing has to be carried by dowels or extended longitudinal bars.

If the footing supporting surface is wider on all sides than the loaded area, the code allows the design bearing strength on the loaded area to be multiplied by $\sqrt{A_2/A_1}$, but the value of $\sqrt{A_2/A_1}$ cannot exceed 2.0. A_1 is the loaded area and A_2 is the maximum area of the supporting surface that is geometrically similar and concentric with the loaded area.

A minimum area of reinforcement of $0.005A_g$ (but not less than four bars) has to be provided across the interface of the column and the footing even when the concrete bearing strength is not exceeded, A_g (in.^{2}) being the gross area of the column cross section.

Lateral forces due to horizontal normal loads, wind, or earthquake can be resisted by shear-friction reinforcement, as described in Section 6.10.

12.7 OPERATIONAL PROCEDURE FOR THE DESIGN OF FOOTINGS

The following sequence of steps can be used for the selection and geometrical proportioning of the size and reinforcement spacing in footings.

1. Determine the allowable bearing capacity of the soil based on site boring test data and soil investigations.
2. Determine the service loads and bending moments acting at the base of the columns supporting the superstructure. Select the controlling service load and moment combinations.
3. Calculate the required area of the footing by dividing the total controlling service load by the selected allowable bearing capacity of the soil if the load is concentric or by also taking into account the controlling bending stress if combined load and bending moments exist.
4. Calculate the factored loads and moments for the controlling loading condition and find the required nominal resisting values by dividing the factored loads and moments by the applicable strength reduction factors ϕ.
5. By trial and adjustment, determine the required effective depth d of the section which has adequate punching shear capacity at a distance d from the support face for one-way action and at a distance $d/2$ for two-way action such that $V_c = 2\sqrt{f_c'}\, b_w d$ for one-way action and $V_c = (2 + 4/\beta_c)\, \sqrt{f_c'}\, b_o d \leq 4\sqrt{f_c'}\, b_o d$ for two-way action, where b_w is the footing width for one-way action and b_o is the perimeter of the failure planes in two-way action. Use an average value of d, since there are two reinforcing mats in the footing. If the footing is rectangular, check the beam shear capacity in each direction on planes at a distance d from the face of the column support.
6. Calculate the factored moment of resistance M_u on a plane at the face of the column support due to the controlling factored loads from that plane to the extremity of the footing. Find $M_n = M_u/(\phi = 0.9)$. Select a total reinforcement area A_s based on M_n and the applicable effective depth.
7. Determine the size and spacing of the flexural reinforcement in the long and short directions:
    (a) Distribute the steel uniformly across the width of the footing in the long direction.

(b)   Determine the portion A_{s1} of the total steel area A_s determined in step 6 for the short direction to be uniformly distributed over the central band:

$$A_{s1} = \frac{2}{\beta + 1} A_s$$

Distribute uniformly the remainder of the reinforcement $(A_s - A_{s1})$ outside the center band of the footing. Verify that the area of steel in each principal direction of the footing plan exceeds the minimum value required for temperature and shrinkage: $A_s = 0.0018b_w d$ for sections reinforced with grade 60 steel and $0.0020b_w d$ with grade 40 steel.

8.   Check the development length and anchorage available to verify that bond requirements are satisfied (see Chapter 10).

9.   Check the bearing stresses on the column and the footing at their area of contact such that the bearing strength P_{nb} for both is larger than the nominal value of column reaction $P_n = P_u/(\phi = 0.70)$. For footing bearing $P_{nb} = \sqrt{A_2/A_1}\,(0.85f'_c A_1)$, $\sqrt{A_2/A_1}$ not to exceed 2.0.

10.   Determine the number and size of the dowel bars that transfer the column load to the footing slab.

Figure 12.9 presents a flowchart for the sequence of calculation operations.

12.8 EXAMPLES OF FOOTING DESIGN

12.8.1 Example 12.3: Design of Two-Way Isolated Footing

Design the footing thickness and reinforcement distribution for the isolated square footing in Ex. 12.1 if the total service load $P = 400,000$ comprises 230,000 lb (1023 kN) dead load and 170,000 lb (756 kN) live load. Given:

> $f'_c = 3000$ psi (20.68 MPa), normal-weight concrete (footing)
> $f'_c = 5,500$ psi (37.91 MPa) in column
> $f_y = 60,000$ psi (413.7 MPa)

Solution

Factored load intensity (Step 4)

Data from Ex. 12.1:

column size = 14 in. × 14 in. (355.6 mm × 355.6 mm)
footing area = 6 ft 8 in. × 6 ft 8 in. (2.03 m × 2.03 m), $A_f = 44.49$ ft²
assumed footing slab thickness $h = 2$ ft
factored load $U = 1.4 \times 230,000 + 1.7 \times 170,000 = 611,000$ lb

factored load intensity $= q_s = \dfrac{U}{A_f} = \dfrac{611,000}{44.49} = 13,733$ lb/ft² (657.6 kPa)

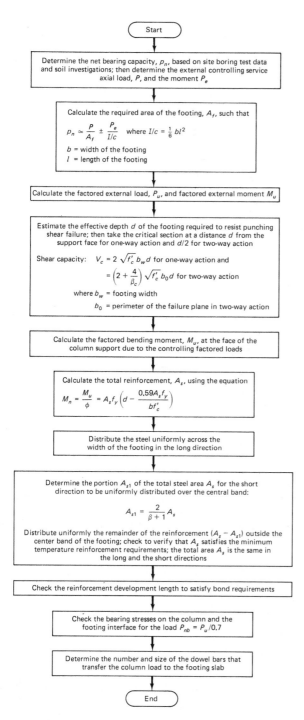

Figure 12.9 Flowchart for footing design.

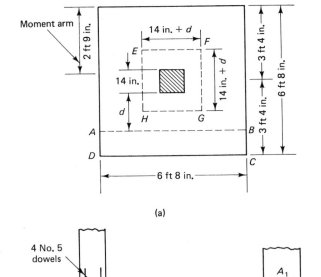

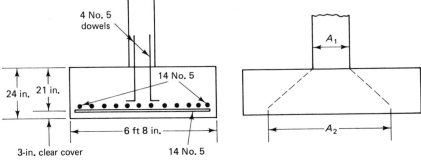

Figure 12.10 Details of footing in Ex. 12.3.

Shear capacity (Step 5)

Assume that the thickness of the footing slab ≈ 2 ft. The average depth $d = h - 3$ in. minimum cover $\approx 0.75 \approx 20$ in.

Beam action (at d from support face): The area to be considered for factored shear V_u is shown as *ABCD* in Fig. 12.10.

$$\text{factored } V_u = 13{,}733\left(\frac{6 \text{ ft 8 in.}}{2} - \frac{14}{2 \times 12} - \frac{20}{12}\right)(6 \text{ ft 8 in.}) = 99{,}340 \text{ lb}$$

$$\text{required } V_n = \frac{V_u}{\phi} = \frac{99{,}340}{0.85} = 116{,}871 \text{ lb}$$

$$b_w = 6 \text{ ft 8 in.} = 80 \text{ in. } (7.82 \text{ m})$$

$$\text{available } V_c = 2\sqrt{f'_c}\, b_w d = 2\sqrt{3000} \times 80 \times 20 = 175{,}271 \text{ lb}$$

Two-way action (at d/2 from support face): The area to be considered for factored shear V_u is equal to the total area of footing less area *EFGH* of the failure zone.

$$\text{factored } V_u = 13{,}733\left[44.49 - \left(\frac{14+20}{12}\right)^2\right] = 500{,}736 \text{ lb}$$

$$\text{required } V_n = \frac{V_u}{\phi} = 589{,}100 \text{ lb (2620 kN)}$$

$$b_o = \text{perimeter of failure zone } EFGH = (14+20)4 = 136 \text{ in.}$$

$$\beta_c = \frac{14}{14} = 1.0$$

The available nominal shear strength from Eq. 12.6:

$$V_c = \left(2 + \frac{4}{\beta_c}\right)\sqrt{f_c'}\, b_o d \le 4\sqrt{f_c'}\, b_o d$$

Since $\beta_c = 1.0$, $V_c = 4\sqrt{f_c'}\, b_o d$ has the lower value, hence controls.

$$V_c = 4\sqrt{3000} \times 136 \times 20 = 595{,}922 \text{ lb (2650.7 kN)} > 589{,}100 \qquad \text{O.K.}$$

Therefore, $d = 20$ in. is adequate for shear.

Bending moment capacity (Steps 6 and 7)

The critical section is at the face of the column.

$$\text{moment arm} = \frac{6 \text{ ft } 8 \text{ in.}}{2} - \frac{14}{2 \times 12} = 2 \text{ ft } 9 \text{ in.}$$

$$\text{factored moment } M_u = 13{,}733 \times 6.67\left[\frac{(2 \text{ ft } 9 \text{ in.})^2}{2}\right]$$

$$= 346{,}359.1 \text{ ft-lb} = 4{,}156{,}310 \text{ in.-lb}$$

$$M_n = \frac{M_u}{\phi} = \frac{4{,}156{,}310}{0.90} = 4{,}618{,}122 \text{ in.-lb (521.8 kN-m)}$$

$$M_n = A_s f_y\left(d - \frac{a}{2}\right)$$

Assume that $(d - a/2) \approx 0.9d$. Use average $d = 20$ in.

$$4{,}618{,}122 = A_s \times 60{,}000 \times 0.9 \times 20$$

or

$$A_s = \frac{4{,}618{,}122}{60{,}000 \times 0.9 \times 20} = 4.28 \text{ in.}^2/80\text{-in. band}$$

$$a = \frac{A_s f_y}{0.85 f_c' b} = \frac{4.28 \times 60{,}000}{0.85 \times 3000 \times 80} = 1.26 \text{ in.}$$

$$4{,}618{,}122 = A_s \times 60{,}000\left(20.0 - \frac{1.26}{2}\right)$$

$$A_s = 3.98 \text{ in.}^2 \qquad \rho = \frac{A_s}{bd} = \frac{3.98}{80 \times 20} = 0.0025$$

Minimum allowable temperature and shrinkage steel:

$$\rho_{min} = 0.0018 < \rho \qquad \text{O.K.}$$

Use 14 No. 5 bars ($A_s = 4.27$ in.^{2}) each way spaced at $\simeq 5\frac{1}{2}$ in. (139.7 mm) center to center.

Development of reinforcement (Step 8)

The critical section for development-length determination is the same as the critical section in flexure, namely, at the face of the column. From Eq. 10.5a, for No. 5 bars,

$$l_d = \frac{0.04 A_b f_y}{\sqrt{f_c'}}$$

$$\leq 0.0004 d_b f_y$$

Note that a reduction multiplier $\lambda_d = 0.8$ cannot be used since the bar spacing is less than 6 in. Or

$$l_d = \frac{0.04 \times 0.305 \times 60,000}{\sqrt{3000}} = 13.36 \text{ in.}$$

$$l_d = 0.0004 \times 0.625 \times 60,000 = 15.0 \text{ in.} \qquad \text{controls}$$

The projection length of each bar beyond the column face

$$\frac{1}{2}(6 \text{ ft } 8 \text{ in.} - 14 \text{ in.}) - 3 \text{ in. cover} = 30 \text{ in.} > 15 \text{ in.} \qquad \text{O.K.}$$

Force transfer at interface of column and footing (Step 9)

Column $f_c' = 5500$ psi. Factored $P_u = 611,000$ lb.
(a) Bearing strength on column using Eq. 12.8b:

$$\phi P_{nb} = 0.70 \times 0.85 f_c' A_1 = 0.60 f_c' A_1$$

or

$$\phi P_{nb} = 0.60 f_c' A_1 = 0.60 \times 5500 \times 14 \times 14$$

$$= 646,800 \text{ lb} > 611,000 \qquad \text{O.K.}$$

From step 9 of the design operational procedure on bearing strength on footing concrete:

$$\sqrt{\frac{A_2}{A_1}} = \sqrt{\frac{(6 \text{ ft } 8 \text{ in.}) \times (6 \text{ ft } 8 \text{ in.})}{(14 \times 14)/144}} = 5.714 > 2.0 \qquad \text{use } 2.0$$

$$\phi P_{nb} = 2.0(0.60 f_c' A_1) = 2.0 \times 0.60 \times 3000 \times 14 \times 14 = 705,600 \text{ lb}$$

$$> 611,000 \qquad \text{O.K.}$$

Dowel bars between column and footing (Step 10)

Even though the bearing strength at the interface between the column and the footing slab is adequate to transfer the factored P_u, a minimum area of reinforcement is necessary across the interface. The minimum $A_s = 0.005 (14 \times 14) = 0.98$ in.^{2} but not less than four bars. Use four No. 5 bars as dowels ($A_s = 1.22$ in.^{2}).

Development of dowel reinforcement in compression: From Eqs. 10.6a and 10.6b for No. 5 bars

$$l_{db} = \frac{0.02 d_b f_y}{\sqrt{f_c'}}$$

and $l_{db} \geq 0.0003 d_b f_y$, where d_b is the dowel bar diameter. Within column:

$$l_d = \frac{0.02 \times 0.625 \times 60,000}{\sqrt{5500}} = 10.11 \text{ in.}$$

$$0.0003 \times 0.625 \times 60,000 = 11.25 \text{ in.} \qquad \text{controls}$$

Within footing:

$$l_d = \frac{0.02 \times 0.625 \times 60,000}{\sqrt{3000}} = 13.69 \text{ in.}$$

Available length for development above the footing reinforcement assuming column bars size to be the same as the dowel bars size:

$$l = 24 - 3 \text{ (cover)} - 2 \times 0.625 \text{ (footing bars)} - 0.625 \text{ (dowels)}$$

$$= 19.13 > 13.69 \text{ in.} \qquad \text{O.K.}$$

12.8.2 Example 12.4: Design of Two-Way Rectangular Isolated Footing

Determine the size and distribution of the bending reinforcement of an isolated rectangular footing subjected to a concentrated concentric factored column load $P_u = 770,000$ lb (3425 kN) and having an area 10 ft × 15 ft (3.05 m × 4.57 m). Given:

$f_c' = 3000$ psi (20.68 MPa), footing
$f_y = 60,000$ psi (413.7 MPa)
Column size = 14 in. × 18 in.

Solution

$$\text{factored load intensity } q_s = \frac{770,000}{10 \times 15} = 5134 \text{ lb/ft}^2$$

Shear capacity (Step 5)

Through trial and adjustment, assume that the footing slab is 2 ft 4 in. thick.

Beam action (at distance d from column face): Average effective depth $\simeq$ 2 ft 4 in. − 3 in. (cover) − $\frac{3}{4}$ in. (diameter of bars in first layer) $\simeq$ 24 in.

From Fig. 12.11, length CD subjected to bearing intensity q_s in one-way beam action

$$\frac{15 \text{ ft}}{2} - \frac{18 \text{ in.}}{2 \times 12} - \frac{24 \text{ in.}}{12} = 4 \text{ ft } 9 \text{ in.} = 57 \text{ in.}$$

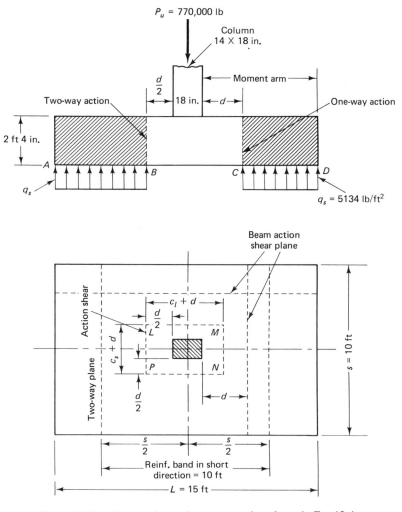

Figure 12.11 Beam action and two-way action planes in Ex. 12-4.

factored V_u = 5134 × 10 ft × 4 ft 9 in. = 243,865 lb

required $V_n = \dfrac{V_u}{\phi} = \dfrac{243{,}865}{0.85}$ = 286,900 lb

available $V_n = 2\sqrt{f_c'}\,b_w d = 2\sqrt{3000} \times 120 \times 24$

$= 315{,}488 \text{ lb} > 286{,}900$ O.K.

Notice that the shorter side length was used for b_w to give the lower available V_n value.

Two-way action (at distance d/2 from column face):

loaded area outside the failure zone *LMNP* in Fig. 12.11

$$= 15 \times 10 - (c_l + d)(c_s + d)$$

$$= 150 - \frac{(18 + 24)(14 + 24)}{144}$$

$$= 138.92 \text{ ft}^2$$

$$\text{factored } V_u = 5134 \times 138.92 = 713,215 \text{ lb}$$

$$\text{required } V_n = \frac{713,215}{0.85} = 839,077 \text{ lb (3732 kN)}$$

$$\text{perimeter of shear failure plane } b_o = 2[(c_l + d) + (c_s + d)]$$

$$= 2[(18 + 24) + (14 + 24)] = 160 \text{ in.}$$

From Eq. 12.6,

$$V_c = \left(2 + \frac{4}{\beta_c}\right)\sqrt{f_c'}\, b_o d \le 4\sqrt{f_c'}\, b_o d$$

$$\beta_c = \frac{18}{14} = 1.286 \qquad 2 + \frac{4}{1.286} > 4$$

Hence $V_c = 4\sqrt{f_c'}\, b_o d$ controls.

$$\text{available } V_c = 4\sqrt{f_c'}\, b_o d = 4\sqrt{3000} \times 160 \times 24$$

$$= 841,302 \text{ lb} > 839,077 \qquad \text{O.K.}$$

Design of two-way reinforcement

The critical section for bending is at the face of the column. The controlling moment arm is in the long direction:

$$\frac{15 \text{ ft}}{2} - \frac{18 \text{ in.}}{2 \times 12} = 6.75 \text{ ft (2.06 m)}$$

$$\text{factored moment } M_u = 5134 \times \frac{10(6.75)^2}{2}$$

$$= 1,169,589 \text{ ft-lb} = 14,035,073 \text{ in.-lb (1585.96 kN-m)}$$

$$M_n = \frac{14,035,073}{0.9} = 15,594,526 \text{ in.-lb (1762.18 kN-m)}$$

Assume that $(d - a/2) \simeq 0.9d$.

$$M_n = A_s f_y\left(d - \frac{a}{2}\right) \qquad \text{or} \qquad 15,594,526 = A_s \times 60,000 \times 0.9 \times 24$$

$$A_s = \frac{15,594,526}{60,000 \times 0.9 \times 24} = 12.03 \text{ in.}^2/\text{10-ft-wide strip}$$

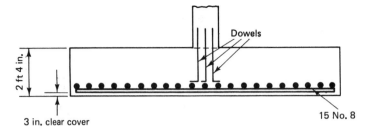

Section

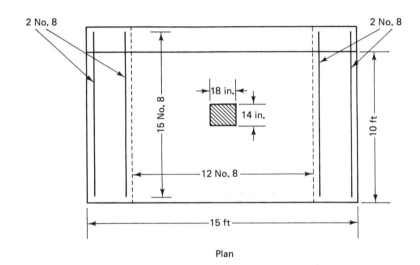

Plan

Figure 12.12 Footing reinforcement details of Ex. 12.4.

Check:

$$a = \frac{A_s f_y}{0.85 f_c' b} = \frac{12.03 \times 60,000}{0.85 \times 3000 \times 120} = 2.36 \text{ in.}$$

$$15,594,526 = A_s \times 60,000 \left(24 - \frac{2.36}{2} \right)$$

$$A_s = 11.39 \text{ in.}^2 = \frac{11.39}{10 \text{ ft}} = 1.14 \text{ in.}^2/\text{ft width}$$

Try No. 8 bars, $A_s = 0.79$ in.^{2} per bar.

$$\text{number of bars in the short direction} = \frac{11.39}{0.79} = 14.42$$

Use 15 bars.

Reinforcement in the short direction

The band width $= s = 10$ ft (Fig. 12.11). From Eq. 12.4,

$$\beta = \frac{15}{10} = 1.5$$

$$\frac{A_{s1}}{A_s} = \frac{2}{\beta + 1} \quad \text{or} \quad \frac{A_{s1}}{11.39} = \frac{2}{1 + 1.5}$$

Therefore,

$$A_{s1} = \frac{2 \times 11.39}{2.5} = 9.11 \text{ in.}^2$$

to be placed in the central 10-ft-wide band and the balance ($11.39 - 9.11 = 2.28$ in.^{2}) to be placed in the remainder of the footing. Use 12 No. 8 bars in the central band $= 9.48$ in.^{2} and two No. 8 bars at each side of the band, as in Fig. 12.12. To complete the design, a check of the development length, bearing stress at the column–footing interface, and dowel action has to be made, as in Ex. 12.3.

12.8.3 Example 12.5: Proportioning of a Combined Footing

A combined footing has the layout shown in Fig. 12.13. Column L at the property line is subjected to a total service axial load $P_L = 200,000$ lb (889.6 kN) and the internal column R is subjected to a total service load $P_R = 350,000$ lb (1556.8 kN). The live load is 35% of the total load. The bearing capacity of the soil at the level of the footing base is 4000 lb/ft^2 (191.5 kPa) and the average value of the soil and footing unit weight $\gamma = 120$ pcf (1922 kg/m^3). A surcharge of 100 lb/ft^2 results from the slab on grade. Proportion the footing size and select the necessary size and distribution of the footing slab reinforcement. Given:

$f'_c = 3000$ psi (20.68 MPa)
$f_y = 60,000$ psi (413.7 MPa)
Base of footing at 7 ft below grade

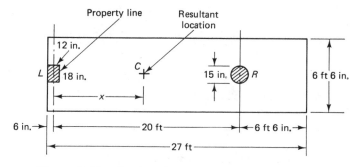

Figure 12.13 Combined footing plan geometry in Ex. 12.5.

Solution

$$\text{total columns load} = 200{,}000 + 350{,}000 = 550{,}000 \text{ lb } (2446.4 \text{ kN})$$

net allowable soil capacity $p_n = p_g - 120(7 \text{ ft height to base of footing}) - 100$

or

$$p_n = 4000 - 120 \times 7 - 100 = 3060 \text{ lb/ft}^2$$

$$\text{minimum footing area } A_f = \frac{P}{p_n} = \frac{550{,}000}{3060} = 179.8 \text{ ft}^2$$

Center of gravity of column loads from the property line:

$$\bar{x} = \frac{200{,}000 \times 0.5 + 350{,}000 \times 20.5}{550{,}000} = 13.23 \text{ ft}$$

$$\text{length of footing } L = 2 \times 13.23 = 26.46 \text{ ft}$$

Use $L = 27$ ft.

$$\text{width of footing } S = \frac{179.8}{27.0} = 6.66 \text{ ft}$$

Use $S = 6$ ft 6 in. as shown in Fig. 12.13.

Factored shears and moments

Column L:

$$P_D = 0.65 \times 200{,}000 = 130{,}000 \text{ lb}$$

$$P_L = 200{,}000 - 130{,}000 = 70{,}000 \text{ lb}$$

$$P_U = 1.4 \times 130{,}000 + 1.7 \times 70{,}000 = 301{,}000 \text{ lb}$$

Column R:

$$P_D = 227{,}500 \text{ lb}$$

$$P_L = 122{,}500 \text{ lb}$$

$$P_U = 1.4 \times 227{,}500 + 1.7 \times 122{,}500 = 526{,}750 \text{ lb}$$

The net factored soil bearing pressure for footing structural design is

$$q_s = \frac{P_u}{A_f} = \frac{301{,}000 + 526{,}750}{6.5 \times 27.0} = 4716.5 \text{ lb/ft}^2$$

Assume that the column loads are acting through their axes.

Factored bearing pressure per foot width $= q_s \times S = 4{,}716.5 \times 6.5 = 30{,}658$ lb/ft

$$V_u \text{ at centerline of column } L = 301{,}000 - 30{,}658 \times \frac{6}{12} = 285{,}671 \text{ lb}$$

$$V_u \text{ at centerline of column } R = 526{,}750 - 30{,}658 \times 6.5 = 327{,}473 \text{ lb}$$

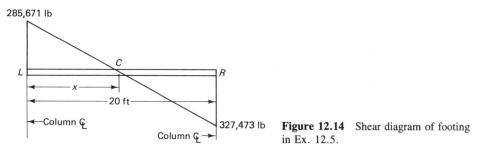

Figure 12.14   Shear diagram of footing in Ex. 12.5.

The maximum moment is at the point C of zero shear in Fig. 12.14 x (ft) from the center of the left column L.

$$x = \frac{285{,}671 \text{ lb}}{30{,}658 \text{ plf}} = 9.32 \text{ ft}$$

Taking a free-body diagram to the left of a section through C, the factored moment at point C is

$$M_{uc} = \frac{w_u l^2}{2} - P_{ul} x$$

$$M_u \text{ from left side} = 30{,}658 \frac{(9.32 + 0.50)^2}{2} - 301{,}000 \times 9.32$$

$$= -1{,}327{,}108 \text{ ft-lb} = -15{,}925{,}293 \text{ in.-lb} \qquad \text{(Fig. 12.15)}$$

$$M_u \text{ from right side} = 30{,}658 \frac{(27.0 - 9.82)^2}{2} - 526{,}750(20.0 - 9.32)$$

$$= 1{,}101{,}299 \text{ ft-lb} = -13{,}215{,}588 \text{ in.-lb}$$

Hence M_u from the left side controls. Note that M_u from the right side differs from M_u from the left side because the footing length of 27 ft is used instead of the computed length of 26.46 ft and because x is rounded off. Therefore, the load is not exactly uniform due to the small eccentricity.

Design of the footing in the longitudinal direction

(a) *Shear:* The combined footing is considered as a beam in the shear computations. Hence the critical section is at a distance d from the face of the support. Controlling V_n at the column centerline

$$\frac{V_u}{\phi} = \frac{327{,}473}{0.85} = 385{,}262 \text{ lb}$$

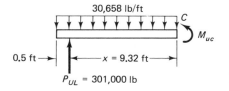

Figure 12.15   Free-body diagram.

Assume that the total footing thickness = 3 ft (0.92 m). The effective footing depth d = 3 ft 3 in. minimum cover ≃ 1 in. for steel = 32 in. For the controlling interior column R, the equivalent rectangular column size $\sqrt{\pi(15)^2/4}$ = 13.29 in.

$$\text{required } V_n \text{ at } d \text{ section} = 385{,}262 - \frac{(13.29/2 + d)}{12} \times \frac{30{,}658}{\phi}$$

$$= 385{,}262 - \frac{38.65 \times 30{,}658}{12 \times 0.85} = 269{,}107 \text{ lb (1196.9 kN)}$$

$$V_c = 2\sqrt{f'_c}\, b_w d = 2\sqrt{3000} \times 6.5 \times 12 \times 32$$

$$= 273{,}423 \text{ lb (1216.2 kN)} > 269{,}107 \qquad \text{O.K.}$$

(b) *Moment and reinforcement in the longitudinal direction (Step 4):* The distribution of shear and moment in the longitudinal direction is shown in Fig. 12.16. The critical section for moment is taken at the face of the columns.

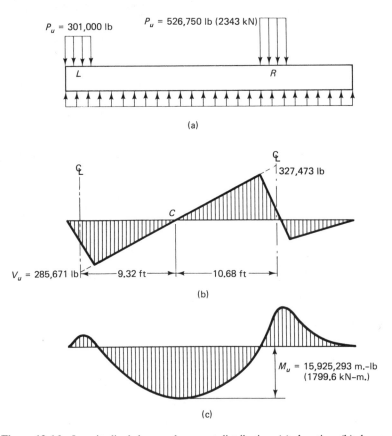

Figure 12.16 Longitudinal shear and moment distribution: (a) elevation; (b) shear; (c) moment.

controlling moment $M_n = \dfrac{M_u}{\phi} = \dfrac{15{,}925{,}293}{0.9} = 17{,}694{,}770$ in.-lb (1999.5 kNm)

$$M_n = A_s f_y \left(d - \frac{a}{2} \right)$$

Assume that $(d - a/2) \simeq 0.9d$.

$$17{,}694{,}770 = A_s \times 60{,}000(0.9 \times 32)$$

or

$$A_s = \frac{17{,}694{,}770}{60{,}000 \times 0.9 \times 32} = 10.24 \text{ in.}^2$$

$$a = \frac{A_s f_y}{0.85 f_c' b} = \frac{10.24 \times 60{,}000}{0.85 \times 3{,}000 \times 6.5 \times 12} = 3.09 \text{ in.}$$

$$17{,}694{,}770 = A_s \times 60{,}000 \left(32 - \frac{3.09}{2} \right)$$

$$A_s = 9.68 \text{ in.}^2 \ (6245 \text{ mm}^2)$$

Use 22 No. 6 bars at the top for the middle span.

$$A_s = 9.68 \text{ in.}^2 \ (22 \text{ bars } 19.1 \text{ mm diameter})$$

Design of footing in the transverse direction

Both columns are treated as isolated columns. The width of the band should not be larger than the width of the column plus half the effective depth d on *each* side of the column. This assumption is on the safe side since the actual bending stress distribution is highly indeterminate. It is, however, possible to assume that the flexural reinforcement in the transverse direction can raise the shear punching capacity within the $d/2$ zone from the face of the rectangular left column L and the *equivalent* rectangular right column R. Figure 12.17 shows the transverse band widths for both columns L and R determined on the basis of this discussion.

$$\text{band width } b_L = 12 + \frac{32}{2} = 28 \text{ in.} = 2.33 \text{ ft}$$

The rectangular column size equivalent to the circular interior 15-in.-diameter column = 13.29 in.

$$\text{band width } b_R = 13.29 + 2 \left(\frac{32}{2} \right) = 45.3 \text{ in.} = 3.77 \text{ ft}$$

Column L transverse band reinforcement:

$$\text{moment arm} = \frac{6 \text{ ft } 6 \text{ in.}}{2} - \frac{18}{2 \times 12} = 2.50 \text{ ft} = 30.0 \text{ in.}$$

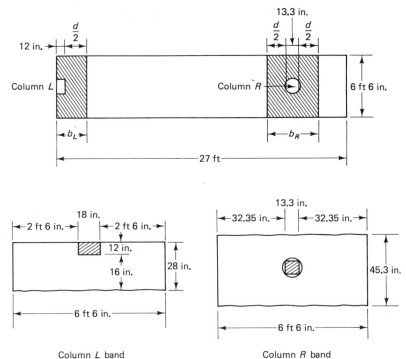

Figure 12.17   Footing transverse band widths.

The net factored bearing pressure in the transverse direction,

$$q_s = \frac{301,000}{6.5} = 46,308 \text{ lb/ft}$$

$$M_u = q_s \frac{l^2}{2} = 46,308 \frac{(2.50)^2}{2} = 144,713 \text{ ft-lb} = 1,736,550 \text{ in.-lb}$$

$$M_n = \frac{M_u}{\phi} = \frac{1,736,550}{0.90} = 1,929,500 \text{ in.-lb (218.0 kN-m)}$$

$$M_n = A_s f_y \left(d - \frac{a}{2} \right)$$

or

$$1,929,500 = A_s \times 60,000 \times 0.9 \times 32$$

$$A_s = 1.12 \text{ in.}^2$$

$$a = \frac{A_s f_y}{0.85 f_c' b} = \frac{1.12 \times 60,000}{0.85 \times 3,000 \times 28} = 0.94 \text{ in.}$$

$$1,929,500 = A_s \times 60,000 \left(32 - \frac{0.94}{2} \right)$$

$$A_s = 1.02 \text{ in.}^2 \ (658 \text{ mm}^2)$$

$$\text{min. } A_s = 0.0018 b_w d = 0.0018 \times 28 \times 32 = 1.62 \text{ in.}^2$$

$$\rho = \frac{1.02}{28 \times 32} = 0.00114$$

Use six No. 5 bars, $A_s = 1.86 \text{ in.}^2$ (six bars 15.9 mm diameter) equally spaced in the band which is to be centered under the column.

Column R transverse band reinforcement: Equivalent square column size = 13.3 in. $\times$ 13.3 in.

$$\text{moment arm} = \frac{6 \text{ ft 6 in.}}{2} - \frac{13.3 \text{ in.}}{2 \times 12} = 2.69 \text{ ft} = 32.35 \text{ in.}$$

Net factored bearing pressure in the transverse direction,

$$q_s = \frac{526,750}{6.50} = 81,038 \text{ lb/ft}$$

$$M_u = q_s \frac{l^2}{2} = 81,038 \frac{(2.69)^2}{2} = 293,200 \text{ ft-lb} = 3,518,400 \text{ in.-lb}$$

$$M_n = \frac{M_u}{\phi} = \frac{3,518,400}{0.90} = 3,909,333 \text{ in.-lb} \ (441.8 \text{ kN-m})$$

$$M_n = A_s f_y \left(d - \frac{a}{2} \right) \qquad \text{assume that } d - \frac{a}{2} \approx 0.90d$$

or

$$3,909,333 = A_s \times 60,000 \times 0.9 \times 32$$

$$A_s = 2.26 \text{ in.}^2 \qquad a = \frac{A_s f_y}{0.85 f_c' b} = \frac{2.26 \times 60,000}{0.85 \times 3,000 \times 45.3} = 1.17 \text{ in.}$$

$$3,909,333 = A_s \times 60,000 \left(32 - \frac{1.17}{2} \right)$$

$$A_s = 2.07 \text{ in.}^2 \ (3347 \text{ mm}^2)$$

$$\rho = \frac{2.07}{45.3 \times 32} = 0.0014 < \rho_{min}$$

where $\rho_{min} = 0.0018$ (shrinkage temperature reinforcement)

minimum $A_s = 0.0018 \times 45.3 \times 32 = 2.61 \text{ in.}^2$

Use nine No. 5 bars, $A_s = 2.79 \text{ in.}^2$ (nine bars 15.9 mm diameter) equally spaced.

Development length check for bars in tension

(a) *Longitudinal top steel:* From Eq. 10.5,

$$\text{min. } l_{db} = \frac{0.04 A_b f_y (1.4)}{\sqrt{f_c'}}$$

or

$$l_{db} = 0.04 \times 0.44 \times 60,000 \times \frac{1.4}{\sqrt{3000}} = 26.99 \text{ in.} = 2.25 \text{ ft} \qquad \text{(governs)}$$

or

$$l_{db} = 0.0004d_b f_y = 0.0004 \times 0.750 \times 60,000 = 18.0 \text{ in.} = 1.50 \text{ ft}$$

Modifying multiplier L_d for top reinforcement = 1.4; hence the minimum development length l_{db} = 1.4 × 26.99 = 37.79 in. = 3.15 ft. The distance from point C at the maximum moment in Fig. 12.14 to the center of the left column = 9.32 + 0.50 = 9.82 ft > 3.15—O.K.

(b) *Transverse bottom steel:*

$$l_{db} = 0.04 \times 0.305 \times \frac{60,000}{\sqrt{3000}} = 13.36 \text{ in.}$$

$$l_{db} = 0.004 \times 0.625 \times 60,000 = 15.00 \text{ in.}$$

available development length = (32.35 − 3.0) in. > 15.00 in.   O.K.

$$\text{Required modifier for column } L = \frac{1.62}{1.86} = 0.87$$

$$\text{Required modifier for column } R = \frac{2.61}{2.79} = 0.94$$

Minimum development length l_{db} = 0.94 × 15.00 = 14.10 in L available

Therefore, adopt reinforcement as in Fig. 12.18. Check for dowel steel from the columns to the footing slab.

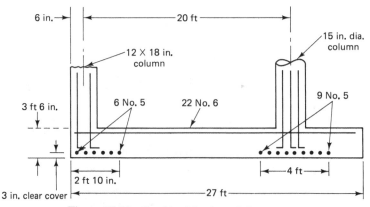

Figure 12.18 Combined footing reinforcement.

12.9 STRUCTURAL DESIGN OF OTHER TYPES OF FOUNDATIONS

From the discussion and the examples given in the foregoing sections, it is clear that the design of foundation substructures follows all the hypotheses and procedures used in proportioning the superstructures once the intensity and distribution of the soil bearing pressure is determined. If a cluster of piles supports a very heavy reaction through a pile cap, the analysis reduces to determining the punching load for each pile and determining the corresponding thickness of the cap. A determination of the center of gravity of the resultant of all pile forces has to be made if the system is subjected to bending in addition to axial load in order to choose the appropriate pile cap layout.

When raft foundations are necessary in poor soil conditions and deep excavations, the design of such a substructure is not too different from the design of any heavily loaded floor system. Once the soil pressure distribution is determined, the design becomes that of an inverted floor supported by deep beams longitudinally and transversely.

Variations are to be expected in the described foundation types particularly in cases of specialized or unique structures. Through an understanding of the basic principles presented, the student and the designer should have no difficulty in utilizing the soil data developed by the geotechnical engineer in selecting and proportioning the appropriate foundation substructure.

SELECTED REFERENCES

12.1   Richart, F. E., "Reinforced Concrete Walls and Column Footings," *Journal of the American Concrete Institute,* Proc. Vol. 45, October and November 1948, pp. 97–127 and 237–245.

12.2   Timoshenko, S., and Woinowsky-Kreiger, *Theory of Plates and Shells,* 2nd ed., 1968, 580 pp.

12.3   Balmer, G. G., Jones, V. and McHenry, D., "Shearing Strength of Concrete Under High Triaxial Stress," *Structural Research Laboratory Report SP-23,* U.S. Dept. of Interior, Bureau of Reclamation, 1949, 26 pp.

12.4   American Insurance Association, *The National Building Code,* 1976 Edition, New York, Dec. 1977, 767 pp.

12.5   Moe, J., "Shearing Strength of Reinforced Concrete Slabs and Footings under Concentrated Load," *Publ. Portland Cement Association,* Bulletin D47, April 1961, 134 pp.

12.6   Furlong, R. W., "Design Aids for Square Footings", *Journal of the American Concrete Institute,* Vol. 62, Proc., March 1965, pp. 363–371.

12.7   Hawkins, N. M., Chairman, ASCE-ACI Committee 426, "The Shear Strength of Reinforced Concrete Members—Slabs," *Journal Structural Division, American Society of Civil Engineers,* Proc. Vol. 100, August 1974, pp. 1543–1591.

12.8   Sowers, G. B. and Sowers, G. F., *Introductory Soil Mechanics and Founda-tions,* 3rd ed., Macmillan, New York, 556 pp.

12.9   Bowles, J. E., *Foundation Analysis and Design,* McGraw–Hill, New York, 1982, 816 pp.

12.10   Winterkorn, H. F. and Fang, H. Y., *Foundation Engineering Handbook,* Van Nostrand Reinhold, 1975, 751 pp.

12.11   Baker, A. L. L., "Raft Foundations—The Soil Line Method of Design," *Concrete Publications LTD,* 1948, 141 pp.

PROBLEMS FOR SOLUTION

12.1   Design a reinforced concrete square isolated footing to support an axial col-umn service live load $P_L = 300,000$ lb (1334 kN) and service dead load $P_D = 625,000$ lb (2780 kN). The size of the column is 30 in. $\times$ 24 in. (0.76 m $\times$ 0.61 m). The soil test borings indicate that it is composed of medium compacted sands and gravely sands, poorly graded. The frost line is assumed to be 3 ft. below grade.

Given: Average weight of soil and concrete above the footing,

$$\gamma = 130 \text{ pcf } (20.41 \text{ kN/m}^3)$$
$$\text{Footing } f'_c = 3,000 \text{ psi } (20.68 \text{ MPa})$$
$$\text{Column } f'_c = 4,000 \text{ psi } (27.58 \text{ MPa})$$
$$f_y = 60,000 \text{ psi } (413.7 \text{ MPa})$$
$$\text{Surcharge} = 120 \text{ psf } (5.7 \text{ kPa})$$

12.2   Design a reinforced concrete wall footing for (a) 10 in (0.25 m) reinforced concrete wall, (b) 12 in. (0.30 m) masonry wall. The intensity of service linear dead load is $W_D = 20,000$ lb/ft (292.0 kN/m) and a service linear live load $W_L = 15,000$ lb/ft (219.0 kN/m) of wall length. Assume an evenly distributed soil bearing pressure and that the average soil bearing pressure at the base of the footing is 3 tons/ft^2 (87.6 kN/m). The frost line is assumed to be 2 ft. below grade.

Given: Avg. wt. of soil and footing above base = 125 pcf (19.6 kN/m^3)
$$\text{Footing } f'_c = 3,000 \text{ psi } (20.68 \text{ MPa})$$
$$\text{Column } f'_c = 5,000 \text{ psi } (34.47 \text{ MPa})$$
$$f_y = 60,000 \text{ psi } (413.7 \text{ MPa})$$

12.3   A combined footing is subjected to an exterior 16 in. $\times$ 16 in. (0.4 m $\times$ 0.4 m) column abutting the property line carrying a total service load

P_w = 300,000 lb (1334.4 kN) and an interior column 20 in. $\times$ 20 in. (0.5 m $\times$ 0.5 m) carrying a total factored load P_w = 400,000 lb (1779.2 kN). The live load is 30% of the total load. The center line distance between the two columns is 22′–0″ (6.71 m). Design the appropriate reinforced concrete footing on a soil weighing 135 pcf (21.2 kN/m^3). The bearing capacity of the soil at the level of the footing base is 6000 lb per sq. ft. The frost line is assumed to be at 3′–6″ (1.07 m) below grade. Assume a surcharge of 125 psf (19.62 kN/m^3) at grade level.

Given: Footing f_c' = 3,500 psi   (24.13 MPa)

Column f_c' = 5,500 psi   (37.42 MPa)

f_y = 60,000 psi (413.7 MPa)

12.4 Redesign the isolated reinforced concrete footing in Problem 12.1 if the load is applied at an eccentricity (a) e = 0.5 ft. (0.15 m); (b) e = 1.8 ft. (0.55 m).

12.5 Redesign the combined reinforced concrete footing in Problem 12.3 if the center line distance between the two columns is 15′–0″ (4.6 m).

13

Handheld & Desktop Computer Programming for Analysis & Design of Reinforced Concrete Sections

13.1 INTRODUCTION

As handheld computers are becoming more accessible in terms of cost to almost all engineers and engineering students, the need has become more apparent to utilize this powerful tool for simplifying engineering design work. The principles for the use of handheld and other computers in the analysis and design of reinforced concrete structural members are presented in this chapter.

Extensive programs are presented for handheld computers, using the Hewlett-Packard HP-41 series. This is thought to be advisable in view of the extreme ease of portability and use of such a computer-calculator under field conditions, particularly for instantaneous checking of designs or choice of sections. Yet the trend toward wide use of personal desktop computers such as the Apple, the Hewlett-Packard, and the IBM, and the transportable and lightweight portables warranted the inclusion of several programs using BASIC language as examples to simulate for developing other programs for personal computers. The fundamental feature of this chapter, however, is the operational flowcharts presented on each topic and based on the latest ACI 318-83 code provisions. The student and the design engineer can write any program using any language once an understanding is acquired of the fundamentals of the topics throughout the book and appropriately follow the logic steps given in the detailed flowcharts of this chapter.

The topics covered in this chapter include programs and solutions for rectangular and flanged beams in flexure, shear, and torsion, and the design of deep beams, corbels, and rectangular and circular columns subjected to combined bending and axial forces. It is hoped that the user, through a good understanding of the material presented, will be able to acquire the logic and skills that are necessary for developing efficient solutions to most structural analysis and design problems.

13.1.1 Handheld Computers versus Mainframe Computers

Although many similarities exist between handheld computers (programmable calculators) and mainframe computers, at least several major differences should be noted.

1. The handheld computers available today are slow compared to mainframe computers. These can handle data transfers to and from memory at high speeds, use many computer languages, develop ultra-high-speed control processing and handle extremely complex programming. Contrasted with this, handheld computers are easier to use, considerably less expensive and troublesome, require little or no education in areas such as languages, internal operation, and control, and perform very efficiently without awkward or expensive peripherals or programming. They are very portable and can be taken almost anywhere.

2. Handheld computers are designed primarily for calculator–human interaction. Until recently, computers were designed primarily for automated nonhuman interaction during data processing. A programmable calculator can be used like

a computer to run a program without human intervention, or it can be used as a simple adding machine to add up a chain of numbers. It would be impractical to have a large computer function as a simple adding machine.

3. Mainframe computers are structured to operate as word-oriented straight binary machines. They transfer data and operations on parallel paths. For example, a 32-bit word machine carries data and addresses, say, on 32 wires or paths.

Handheld computers are character-oriented decimal machines, using binary circuit internal logic, but representing numbers one decimal digit at a time as a binary-coded decimal. The advantage to this is that at the end of every computation in a program the result is in a decimal-digit form, making it easy for the program to be debugged.

Groups of bits, usually 8 for handheld computers, are called bytes. In the standard binary-coded-decimal machines, 8 bits can represent two numerals. Each numeral would be made up of a combination of four ones and/or zeros.

Handheld computers can be programmed to solve a problem with the same sequence of steps or logic as a person would follow. To formulate the most efficient sequence of steps required in the solution of a problem, a flowchart should be drawn. It consists of a diagram connecting boxes that contain statements of operation or decision. The boxes represent the steps in the procedure and the connecting lines represent the flow through the steps. The diagrams are usually drawn so that one starts on the top and moves toward the bottom. Arrows on connecting lines show the direction of logic flow. Every group of program instructions that processes information is represented by boxes. Decision functions, where alternate paths are possible based on a decision, are represented by a diamond ($\diamondsuit$ or $\bigcirc$).

One normally starts with input data (such as f'_c, f_y, A_s, loads, etc.) which is then used in the calculation process. As a result, some information is placed into pre-arranged variables. These variables are then tested against some criterion. If the test is satisfied, the program continues to the next stage or branch and/or an output is provided. If the test is not met, the process returns to the calculation phase and continues or follows an alternate path from that taken if the test was satisfied.

The program of a standard computer uses a code to translate instructions into logic that can be interpreted by the machine. Codes used in the handheld computers or calculators employ one instruction for each line of code. The instruction is usually abbreviated. For example, STO 12 means store the contents of the previous step or input in register (a storage location) number 12 for use at a later time. "X=Y?" instructs the computer to compare the contents of the X register with the contents of the Y register. If they are equal, the program performs the next instruction (which might instruct the handheld computer to go to another branch or portion of the program and proceed). If they are not equal, the instrument skips the next instruction and proceeds with the following one. In this way, the computer can perform such functions as addition or multiplication of any chain of numbers, obtain a result and then compare the result with another result or piece of information and make a decision. Based on that decision, the programmable calculator or computer automatically proceeds to a specified location in the program and continues until the final result or results are obtained.

Once a flowchart is developed for a given problem, the next step is to gather all the necessary equations that will be needed for its solution. At this point, the program can be written. The main points to remember are that every program should be given a name or label as should any subroutines. A stop command should be used at the end. A subroutine is a series of instructions that might be executed several times in several places in a program. Rather than repeat the instructions over and over, an instruction in the program causes the subroutine to be executed upon which a result is obtained and execution returns back to the main program. In between the name and stop commands are placed the series of commands or keystrokes that one uses to solve an engineering problem. After the commands are placed in the handheld computer, the program should be executed and checked for mistakes.

The set of instructions for a handheld computer or programmable calculator is quite versatile and powerful, since the instructions provide for conditional transfers, loops, and many internal functions such as trigonometric and logarithmic functions. The instruction set, or programming language, used for a handheld computer is a powerful computer language. However, it is actually a problem-oriented language, designed for the convenient expression of a given class of problem. The problem-oriented language limits the types of problems that can be solved to those of a mathematical nature.

One cannot easily use a handheld computer at this time to store names and addresses and then list them in alphabetical order. A mainframe or personal desktop computer can accomplish such a task. As can be seen from the preceding discussion, handheld computers can be programmed to solve almost any reinforced concrete problem using a logic similar to the logic used in standard computers. However, both have their distinct differences in terms of cost and capability, which should be recognized. Within their respective domains, their limits are measured by the capability of the programmer.

13.1.2 Programming Basics for the Hewlett-Packard HP41 Series

Since a large number of programs presented in this chapter are written primarily for the HP41C or HP41 CV/41CX, some programming basics are presented to assist the user. The operating manual for the HP41 series of handheld computers should be referred to in order to have a complete understanding of the instrument's operation.

13.1.2.1 Loading a Program

To load a program into the calculator manually:

1. Set the number of data storage registers required for the program to be loaded by pressing [XEQ] [ALPHA] SIZE [ALPHA] followed by a three-digit number (e.g., 032).

2. Switch the computer to the program mode by pressing $\boxed{\text{PRGM}}$.

3. Press $\boxed{\text{▨}}$ $\boxed{\text{GTO}}$ $\boxed{\text{•}}$ $\boxed{\text{•}}$ to set the computer to an unused portion of program memory.

4. Press $\boxed{\text{▨}}$ $\boxed{\text{LBL}}$ followed by the alpha characters comprising the program name (maximum of seven characters: e.g., R̲E̲C̲B̲E̲A̲M̲).

5. Key in the program steps (functions, numbers, or alpha strings) just as in the normal mode. In the program mode, however, these instructions are not executed, but they are remembered by the computer in the order that they are keyed in.

6. The last step keyed in should be $\boxed{\text{XEQ}}$ $\boxed{\text{ALPHA}}$ END $\boxed{\text{ALPHA}}$.

In the program mode, the HP41C/41CV/41 CX display is set to one line of program memory at a time. Each line contains a complete instruction consisting of a function or an alpha string of up to 15 characters or a complete number (up to 10 digits, or up to 10 digits plus a two-digit exponent of 10).

Lines are created automatically as instructions are loaded in the program mode. Each line is assigned a number to indicate its position within the program. Each separate program in the HP41C/41CV has its own set of line numbers.

If a function to be entered into a program line does not appear on any keys, the function must be entered by pressing $\boxed{\text{XEQ}}$ $\boxed{\text{ALPHA}}$, the function name (e.g., X < Y?), and then $\boxed{\text{ALPHA}}$. IF $\boxed{\text{XEQ}}$ is not pressed first, the alpha characters will not be recognized as a function name; instead, they will be treated as alpha data and will be entered into the ALPHA register when the program line is executed. Any time a wrong key is inadvertently pressed, it can be erased by pressing the ← key.

If a program is already stored on magnetic cards, it can be entered via a card reader by simply pressing $\boxed{\text{▨}}$ $\boxed{\text{GTO}}$ $\boxed{\text{•}}$ $\boxed{\text{•}}$ to pack the registers, setting in the user mode, and then feeding the cards. It is recommended that the instrument be placed in the user mode so that if the program was assigned to a key before the cards were programmed, the program will be assigned automatically to that key as it is now being entered via the card reader. Execution of the program can then be accomplished by simply pressing that key while in the user mode.

If not enough space is available in the memory to enter the entire program, a previous program(s) must first be deleted from the memory. Once all the steps are input they should be checked carefully and then they may be stored permanently on magnetic cards or tape. The HP41C/41CV/CX operating manual needs to be referred to for details.

13.1.2.2 Running a Program

To run a program, the computer is first switched out of the program mode, and the required input data are placed in the proper storage registers. A program may then be run in a number of ways:

1. By executing the program using ⌊XEQ⌋ ⌊ALPHA⌋ followed by the label name of the program (maximum of seven characters) and then ⌊ALPHA⌋ .

2. By using ASN to assign the program to a key. Press ▨ ⌊ASN⌋ ⌊ALPHA⌋ followed by the label name of the program and then ⌊ALPHA⌋ and then a key on the keyboard (e.g., ⌊LOG⌋). This will assign that program to the last key pressed and execution of the program can be done by pressing that key while in the user mode.

When a program is run, the instructions in program memory are executed until a ⌊STOP⌋ or ⌊END⌋ instruction is executed or until the program is halted by pressing of the ⌊R/S⌋ key by the operator.

Note: The instrument and any accessory attached to it should be turned off before plugging the accessory into the computer ports in order to prevent possible damage.

13.1.2.3 Writing a Program

As noted previously, a program for a handheld computer is little more than a series of keystrokes that one would press to solve a problem manually—except that when programming, the instrument remembers the keystrokes as one enters them, then it executes all of the specified keystrokes when needed.

As an example, a program is presented to calculate the bending moment at any section for a simply supported uniformly loaded beam, shown in Fig. 13.1. Starting from the left support, the equation for moment at any section, M_x, can be written as

$$M_x = \frac{wl}{2}x - \frac{wx^2}{2} = \frac{w}{2}(lx - x^2)$$

Since this problem is very simple, no flowchart is required to assist in writing the program.

The first line should be the program name or label. For the HP41C/41CV/CX, the label should not exceed seven characters in length, including spaces. A label "MOMENT" will be used for this problem. This is followed by the keystrokes needed to solve the equation manually. The last line should be an END statement to define the end of the program. A STOP statement could precede the END statement but is

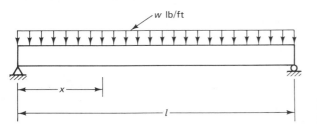

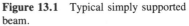

Figure 13.1 Typical simply supported beam.

not required. First, the register locations for the data (specified values of the variables) should be decided. Assume that the values of l (ft), x (ft), and w (lb/ft) will be stored in registers 00, 01, and 02, respectively.

The program steps to solve for the moment at any section for the simply supported uniformly loaded beam are:

01 ▨ [LBL] [ALPHA] MOMENT [ALPHA]		Assigns the name (MOMENT) to and defines the beginning of the program
02 [RCL] 00		Recalls span length l
03 [RCL] 01		Recalls x
04 X		Multiplies l times x
05 [RCL] 01		Recalls x
06 x^2		Squares x
07 −		Subtracts x^2 from $(l \cdot x)$
08 [RCL] 02		Recalls w
09 ×		Multiplies w times $(lx - x^2)$
10 2		Summons 2
11 ÷		Divides $w(lx - x^2)$ by 2
12 [STO] 03		Stores the result $w/2(lx - x^2)$ in register 03
13 [ALPHA] MOMENT =		Labels the result
14 ▨ [RCL] 03 [ALPHA]		Places the contents of register 03 after MOMENT =
15 [XEQ] [ALPHA] PRA [ALPHA]		Tells the computer to print "MOMENT = contents of register 03"
16 [XEQ] [ALPHA] [END] [ALPHA]		Defines end of program space in memory and stops execution of the program

To enter and run the program; the handheld computer has to be in the program mode and the program steps entered. After entering the program steps, the computer is taken out of the program mode. Then the values of the input variables, namely l, x, and w, are stored in registers 00, 01, and 03, respectively. This step is considered the "input of variables." The input variables have to be specified *before* starting execution. Once the variables are specified, the program can be executed by pressing [XEQ] [ALPHA] MOMENT [ALPHA].

The program will calculate and print the bending moment value. If PROMPT is used instead of PRA in line 15, the computer will display the answer regardless of whether or not the printer is attached to the calculator. If PRA statements are used, the computer must be connected to the printer. If PROMPT statements are used with the printer attached, the result will be displayed and printed and program execution will stop at that point. To continue execution, the $\boxed{\text{R/S}}$ key must then be pressed.

13.2 RECTANGULAR BEAMS

RECBEAM is a program that analyzes any singly or doubly reinforced concrete beam. The program computes M_u, the ultimate design moment for any given cross section of a beam. The program also computes β_1; checks the strain in the steel for over-reinforcement (i.e., tension steel has not yielded); calculates the depth of the compression block a, $0.75\rho_b$, ρ, and ρ_{min}; and then checks to see if $\rho_{min} < \rho < 0.75\rho_b$. If any of these parameters are not satisfied, an error message is displayed.

The user can use this program for design as well as analysis. Knowing the required moment strength $M_n = M_u/\phi$, the user can assume a section and its reinforcement, execute the program, and compare the capacity of the beam to the moment to be resisted. Adjustments can be made to the section and/or steel until the most economical section is found (i.e., when the beam capacity is equal to or slightly larger than the required resisting moment).

13.2.1 Design Equations, Flowchart, and Program Steps

(Refer to Figs. 13.2, 13.3, and 13.4.)

$$\beta_1 = 0.85 \qquad\qquad \text{if } f'_c \leq 4000 \text{ psi}$$

$$= 0.65 \qquad\qquad \text{if } f'_c \geq 8000 \text{ psi}$$

$$= 0.85 - 0.05\frac{f'_c - 4000}{1000} \qquad \text{if } 4000 \text{ psi} < f'_c < 8000 \text{ psi}$$

$$a = \frac{A_s f_y - A'_s f'_s}{0.85 f'_c b} \qquad \epsilon'_s = 0.003\frac{a - \beta_1 d'}{a}$$

$$f'_s = \begin{cases} E_s \epsilon'_s & \text{if } \epsilon'_s < \epsilon_y \\ f_y & \text{if } \epsilon'_s > \epsilon_y \end{cases} \qquad \epsilon_s = 0.003\frac{\beta_1 d - a}{a}$$

$$\rho_{min} \leq \rho \leq 0.75\rho_b \qquad \text{for practical designs: } \rho_{min} \leq \rho \leq 0.5\rho_b$$

$$\rho_{min} = \frac{200}{f_y} \qquad \rho_b = \frac{0.85 f'_c \beta_1}{f_y} \times \frac{87,000}{87,000 + f_y} + \frac{\rho' f'_s}{f_y}$$

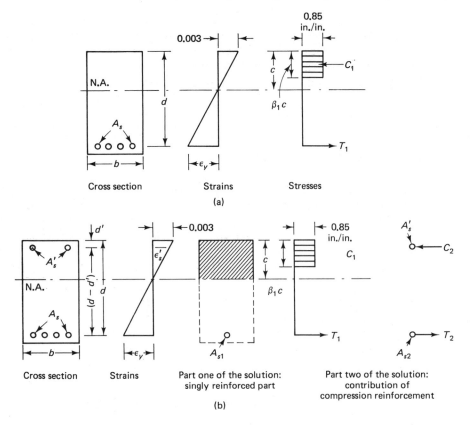

Figure 13.2 Typical rectangular beam cross sections: (a) singly reinforced section; (b) doubly reinforced section.

$$M_u = \phi\left[0.85 f'_c ba\left(d - \frac{a}{2}\right) + A'_s f'_s(d - d')\right]$$

The flowchart and the program steps are presented in Figs. 13.3 and 13.4.

13.2.2 Instructions to Run the Program

Step 1: Set at least 26 data storage registers.

(XEQ ALPHA SIZE ALPHA 026)

Step 2: Load the program from magnetic cards or enter the program steps.
Step 3: Input the following design data.

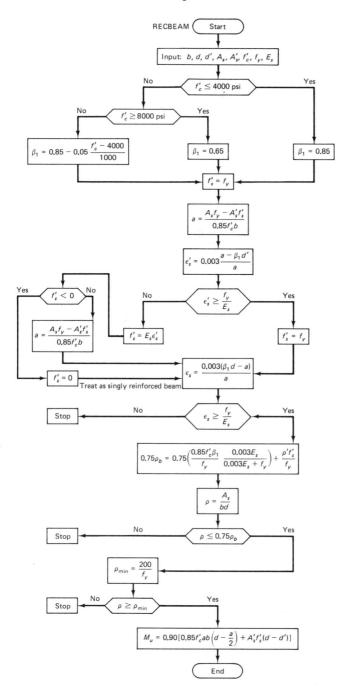

Figure 13.3 Flowchart: flexural analysis of rectangular beams.

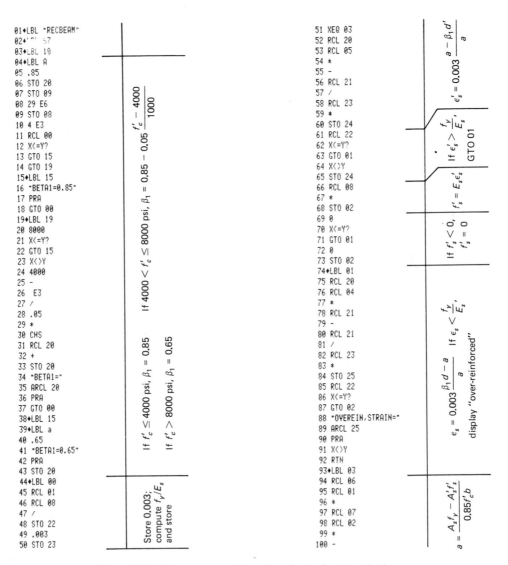

Figure 13.4 Program steps: flexural analysis of rectangular beams.

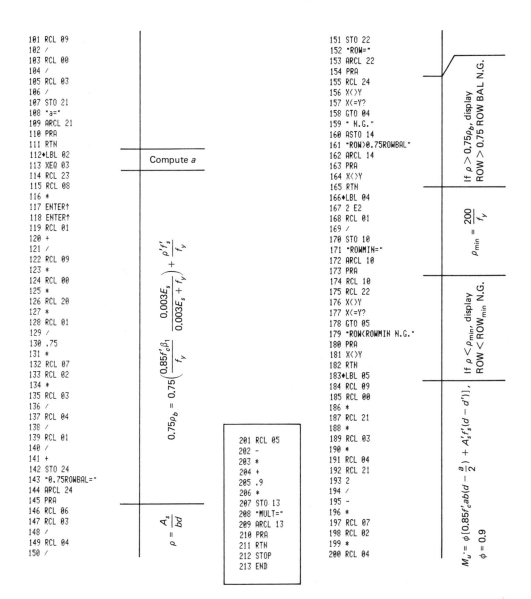

```
101 RCL 09                                                    151 STO 22
102 /                                                         152 "ROW="
103 RCL 00                                                    153 ARCL 22
104 /                                                         154 PRA
105 RCL 03                                                    155 RCL 24
106 /                                                         156 X<>Y
107 STO 21                                                    157 X<=Y?
108 "a="                                                      158 GTO 04
109 ARCL 21                                                   159 " N.G."
110 PRA                                                       160 ASTO 14
111 RTN                                                       161 "ROW>0.75ROWBAL"
112◆LBL 02          Compute a                                 162 ARCL 14
113 XEQ 03                                                    163 PRA
114 RCL 23                                                    164 X<>Y
115 RCL 08                                                    165 RTN
116 *                                                         166◆LBL 04
117 ENTER↑                                                    167 2 E2
118 ENTER↑                                                    168 RCL 01
119 RCL 01                                                    169 /
120 +                                                         170 STO 10
121 /                                                         171 "ROWMIN="
122 RCL 09                                                    172 ARCL 10
123 *                                                         173 PRA
124 RCL 00                                                    174 RCL 10
125 *                                                         175 RCL 22
126 RCL 20                                                    176 X<>Y
127 *                                                         177 X<=Y?
128 RCL 01                                                    178 GTO 05
129 /                                                         179 "ROW<ROWMIN N.G."
130 .75                                                       180 PRA
131 *                                                         181 X<>Y
132 RCL 07                                                    182 RTN
133 RCL 02                                                    183◆LBL 05
134 *                                                         184 RCL 09
135 RCL 03                                                    185 RCL 00
136 /                                                         186 *
137 RCL 04                                                    187 RCL 21
138 /                                                         188 *
139 RCL 01                                                    189 RCL 03
140 /                                                         190 *
141 +                                                         191 RCL 04
142 STO 24                                                    192 RCL 21
143 "0.75ROWBAL="                                             193 2
144 ARCL 24                                                   194 /
145 PRA                                                       195 -
146 RCL 06                                                    196 *
147 RCL 03                                                    197 RCL 07
148 /                                                         198 RCL 02
149 RCL 04                                                    199 *
150 /                                                         200 RCL 04
```

```
201 RCL 05
202 -
203 *
204 +
205 .9
206 *
207 STO 13
208 "MULT="
209 ARCL 13
210 PRA
211 RTN
212 STOP
213 END
```

$$0.75\rho_b = 0.75\left(\frac{0.85f'_c\beta_1}{f_y}\cdot\frac{0.003E_s}{0.003E_s+f_y}\right)+\frac{\rho'f'_s}{f_y}$$

$$\rho = \frac{A_s}{bd}$$

If $\rho > 0.75\rho_b$, display ROW > 0.75 ROW BAL N.G.

$$\rho_{min} = \frac{200}{f_y}$$

If $\rho < \rho_{min}$, display ROW $<$ ROW$_{min}$ N.G.

$$M_u = \phi[0.85f'_c abl(d-\frac{a}{2}) + A'_s f'_s(d-d')],$$
$$\phi = 0.9$$

Figure 13.4 (*cont.*)

Variable	Unit	Store in register
f'_c	psi	00
f_y (tension steel)	psi	01
f_y (compression steel)	psi	02
b	in.	03
d	in.	04
d'	in.	05 (store zero for singly reinforced section)
A_s	in.^{2}	06
A'_s (always less than A_s)	in.^{2}	07 (store zero for singly reinforced section)

Step 4: Execute RECBEAM.

($\boxed{\text{XEQ}}$ $\boxed{\text{ALPHA}}$ RECBEAM $\boxed{\text{ALPHA}}$)

If the design data satisfy the code minimum and maximum reinforcement ratio requirements, the program will calculate and print out the ultimate design moment, M_u in in.-lb. Otherwise, the program will print out the appropriate error message: "ROW < ROWMIN N.G." or "ROW > 0.75 ROW BAL N.G." If the section is over-reinforced, the program will printout "OVEREIN, STRAIN = . . .". The strain represents the tensile steel strain, less than the yield strain. The value of f_y in register 02 must be re-input after each run for doubly reinforced beams since during the execution of the program, that is replaced with f'_s.

The entire output will consist of:

Data	Unit
β_1	—
a	in.
$0.75\rho_b$	—
ρ	—
$\rho_{\min}$	—
M_u	in.-lb

13.2.3 Numerical Examples

13.2.3.1 Example 13.1: Flexural Analysis of a Singly Reinforced Beam

A singly reinforced concrete beam has the cross-sectional properties presented below. Determine if the beam is over-reinforced or under-reinforced and if it satisfies the ACI code requirements for maximum and minimum reinforcement ratios. If the section satisfies the code requirements, calculate the design moment, $M_u = \phi M_n$ when (a) $f_y = 60,000$ psi, and (b) $f_y = 40,000$ psi. Given:

$f'_c = 4000$ psi
$b = 10$ in.
$d = 18$ in.
$A_s = 6.0$ in.^{2}

Solution

(a)   INPUT   (b)   INPUT

              4,000.0000 STO 00   4,000.0000 STO 00
             60,000.0000 STO 01   40,000.0000 STO 01
             60,000.0000 STO 02   40,000.0000 STO 02
                10.0000 STO 03   10.0000 STO 03
                18.0000 STO 04   18.0000 STO 04
                 2.5000 STO 05   2.5000 STO 05
                 6.0000 STO 06   6.0000 STO 06
                 0.0000 STO 07   0.0000 STO 07

              OUTPUT   OUTPUT

                XEQ "RECBEAM"   XEQ "RECBEAM"
       BETA1=0.85   BETA1=0.85
       a=10.5882   a=7.0588
       OVEREIN,STRAIN=0.0013   a=7.0588
                                               0.75ROWBAL=0.0371
                                               ROW=0.0333
                                               ROWMIN=0.0050
                                               MULT=3,125,647.060

13.2.3.2 Example 13.2: Nominal Resisting Moment in a Singly Reinforced Beam

For the following rectangular beam cross section shown, calculate the design moment M_u if f_y is 60,000 psi and f'_c is (a) 3000 psi, (b) 5000 psi, and (c) 9000 psi. Given:

$b = 10$ in.
$d = 18$ in.
$A_s = 4.0$ in.^{2}

Solution

(a) INPUT

```
3,000.0000 STO 00
60,000.0000 STO 01
60,000.0000 STO 02
   10.0000 STO 03
   18.0000 STO 04
    2.5000 STO 05
    4.0000 STO 06
    0.0000 STO 07
```

OUTPUT

```
        XEQ "RECBEAM"
BETA1=0.85
a=9.4118
OVEREIN,STRAIN=0.0019
```

(b) INPUT

```
    5,000.0000 STO 00
```

OUTPUT

```
             XEQ "RECBEAM"
BETA1=0.8000
a=5.6471
a=5.6471
0.75ROWBAL=0.0252
ROW=0.0222
ROWMIN=0.0033
MULT=3,278,117.648
```

(c) INPUT

```
    9,000.0000 STO 00
```

OUTPUT

```
             XEQ "RECBEAM"
BETA1=0.65
a=3.1373
a=3.1373
0.75ROWBAL=0.0368
ROW=0.0222
ROWMIN=0.0033
MULT=3,549,176.471
```

Note: For parts (b) and (c), only f'_c is input since the other variables are already placed in the respective storage registers.

13.2.3.3 Example 13.3: Design of a Singly Reinforced Simply Supported Beam for Flexure

A reinforced concrete simply supported beam has a span of 30 ft and is subjected to a service uniform load intensity $w = 1500$ lb/ft. Design a beam section to resist the factored external load. Given:

$$f'_c = 4000 \text{ psi}$$
$$f_y = 60,000 \text{ psi}$$

Solution

Assume a minimum thickness from the ACI code deflection table:

$$\frac{l}{16} = \frac{30 \times 12}{16} = 22.5 \text{ in.}$$

For the purpose of estimating the preliminary self-weight, assume that the total thickness $h = 24.0$ in., effective depth $d = 20$ in., and width of the beam $b = 10$ in. ($r = b/d = 0.5$).

$$\text{beam self-weight} = \frac{24 \times 10}{144} \times 150 = 250 \text{ lb/ft}$$

factored load $U = 1.4D + 1.7L = 1.4 \times 250 + 1.7 \times 1500 = 2900$ lb/ft

factored moment $M_u = \dfrac{w_u l_n^2}{8} = \dfrac{2900 \times 30^2}{8} \times 12 = 3{,}915{,}000$ in.-lb

Trial 1

Try $b = 10$ in., $d = 20$ in., and $A_s = 3$ in.^{2} (three No. 9 bars).

INPUT	OUTPUT
4,000.0000 STO 00	XEQ "RECBEAM"
60,000.0000 STO 01	BETA1=0.85
60,000.0000 STO 02	a=5.2941
10.0000 STO 03	a=5.2941
20.0000 STO 04	0.75ROWBAL=0.0214
0.0000 STO 05	ROW=0.0150
3.0000 STO 06	ROWMIN=0.0033
0.0000 STO 07	MULT=2,811,176.471

$<$ External factored moment $= 3{,}915{,}000$ in.-lb. Revise the section.

Trial 2

Try $b = 12$ in., $d = 23$ in., $h = 26$ in., and $A_s = 3.81$ in.^{2} (three No. 10 bars).

revised self-weight $= \dfrac{12 \times 26}{144} \times 150 = 325$ lb/ft

factored load $U = 1.4 \times 325 + 1.7 \times 1500 = 3005$ lb/ft

factored moment $M_u = \dfrac{3005(30)^2}{8} \times 12 = 4{,}056{,}750$ in.-lb

INPUT	OUTPUT
12.0000 STO 03	XEQ "RECBEAM"
23.0000 STO 04	BETA1=0.85
3.8100 STO 06	a=5.6029
	a=5.6029
	0.75ROWBAL=0.0214
	ROW=0.0138
	ROWMIN=0.0033
	MULT=4,155,645.443

$>$ External factored moment $= 4{,}056{,}750$ in.-lb Adopt the design.

13.2.3.4 Example 13.4: Design of a One-Way Slab for Flexure

A one-way single-span reinforced concrete slab has a simple clear span of 10 ft and carries a live load of 120 psf and a dead load of 20 psf in addition to its self-weight. Design the slab and the size and spacing of the reinforcement at midspan assuming simple support moment. Given:

f'_c = 4000 psi, normal-weight concrete
f_y = 60,000 psi

Assume also that the minimum thickness for deflection = $l/20$.

Solution

$$\text{minimum depth for deflection, } h = \frac{l}{20} = \frac{10 \times 12}{20} = 6 \text{ in.}$$

Assume for flexure an effective depth $d = 5$ in.

$$\text{self-weight of a 12 in. strip} = \frac{6 \times 12}{144} \times 150 = 75 \text{ lb/ft}^2$$

Therefore,

$$\text{factored external load } U = 1.7 \times 120 + 1.4(20 + 75) = 337 \text{ lb/ft}^2$$

$$\text{factored external moment } M_u = \frac{337 \times 10^2}{8} \times 12 \text{ in.-lb} = 50{,}550 \text{ in.-lb}$$

Trial 1

Try $b = 12$ in., $d = 5$ in., and $A_c = 0.2$ in.^{2}/12-in. strip (No. 4 bars at 12 in. center to center).

INPUT	OUTPUT
4,000.0000 STO 00	XEQ "RECBEAM"
60,000.0000 STO 01	BETA1=0.85
60,000.0000 STO 02	a=0.2941
12.0000 STO 03	a=0.2941
5.0000 STO 04	0.75ROWBAL=0.0214
0.0000 STO 05	ROW=0.0033
0.2000 STO 06	ROWMIN=0.0033
0.0000 STO 07	MULT=52,411.7647

> External factored moment = 50,550 in.-lb. Adopt the design.

13.2.3.5 Example 13.5: Analysis of a Doubly Reinforced Beam for Flexure

Calculate the design moment $M_u = \phi M_n$ of the doubly reinforced section using the following beam properties. Given:

$f_c' = 5000$ psi, normal-weight concrete
$f_y = 60,000$ psi
$b = 14$ in.
$d = 18.5$ in.
$d' = 2.5$ in.
$A_s = 5.08$ in.^{2}
$A_s' = 1.20$ in.^{2}

Solution   INPUT   OUTPUT

    5,000.0000 STO 00   XEQ "RECBEAM"
   60,000.0000 STO 01   BETA1=0.8000
   60,000.0000 STO 02   a=3.9126
       14.0000 STO 03   a=4.2650
       18.5000 STO 04   0.75ROWBAL=0.0284
        2.5000 STO 05   ROW=0.0196
        5.0800 STO 06   ROWMIN=0.0033
        1.2000 STO 07   MULT=4,473,056.019

13.2.3.6 Example 13.6: Design of a Doubly Reinforced Beam for Flexure

A doubly reinforced concrete beam section has a maximum effective depth $d = 25$ in. and is subjected to a total factored moment $M_u = 9.5 \times 10^6$ in.-lb, including its self-weight. Design the section and select the appropriate reinforcement at the tension and the compression faces to carry the required load. Given:

$f_c' = 4000$ psi
$f_y = 60,000$ psi and minimum effective cover
$d' = 2.5$ in.

Solution

Try $b = 14$ in., $d = 25$ in., $d' = 2.5$ in., $A_s = 8.0$ in.^{2} (eight No. 9 bars), and $A_s' = 3$ in.^{2} (three No. 9 bars).

```
        INPUT                          OUTPUT

    4,000.0000 STO 00                    XEC "RECBEAM"
   60,000.0000 STO 01          BETA1=0.85
   60,000.0000 STO 02          a=6.3025
       14.0000 STO 03          a=6.4496
       25.0000 STO 04          0.75ROWBAL=0.0296
        2.5000 STO 05          ROW=0.0229
        8.0000 STO 06          ROWMIN=0.0033
        3.0000 STO 07          MULT=9,519,738.768
```

> Given external factored moment = 9.5×10^6 in.-lb. Adopt the design.

13.3 FLANGED BEAMS

T-BEAM 1 is a program that analyzes flanged beams. T beams and L beams are the normal flanged sections in most cases. Since laterally supported L sections are the same as T sections as far as the analysis is concerned, laterally supported L sections can be analyzed as T sections with this program. The user must check that the given section satisfies the restrictions on flange width with respect to the flange thickness, spacing of webs, and the span.

By inputting f_c', f_y, b, b_w, d, h_f, and A_s for any given cross section of beam, the program computes M_u, the ultimate design moment. The program also computes β_1, $0.75\rho_b$, a, c, $0.75\overline{\rho}_b$, ρ_{min}, and ρ and then checks to see if $\rho_{min} < \rho < 0.75\rho_b$ if $c < h_f$ (i.e., a rectangular section) or $\rho_{min} < \rho < 0.75\rho_b$ if $c > h_f$ (i.e., a flanged section). If any of these parameters are not satisfied, an error message is displayed.

As with the rectangular beam program, the user can use this program for design as well, making adjustments to the section and/or steel until the most economical section is found.

13.3.1 Design Equations, Flowchart, and Program Steps

(Refer to Figs. 13.5, 13.6, and 13.7.)

$$\beta_1 = \begin{cases} 0.85 & \text{if } f_c' \leq 4000 \text{ psi} \\ 0.65 & \text{if } f_c' \geq 8000 \text{ psi} \\ 0.85 - 0.05\dfrac{f_c' - 4000}{1000} & \text{if } 4000 \text{ psi} < f_c' < 8000 \text{ psi} \end{cases}$$

$$\bar{\rho}_b = \frac{0.85 \, f_c' \, \beta_1}{f_y} \frac{87,000}{87,000 + f_y}$$

$$A_{sf} = \frac{0.85 \, f_c' \, h_f (b - b_w)}{f_y} \qquad \rho_f = \frac{A_{sf}}{b_w d} \qquad \rho_w = \frac{A_s}{b_w d}$$

$$\rho_b = \frac{b_w}{b} (\bar{\rho}_b + \rho_f)$$

$\rho \le 0.75\rho_b$ for flanged section $\rho \le 0.75\bar{\rho}_b$ for rectangular section

$$a = \frac{A_s f_y}{0.85 f_c' b} \qquad c = \frac{a}{\beta_1} \qquad \rho_{\min} = \frac{200}{f_y}$$

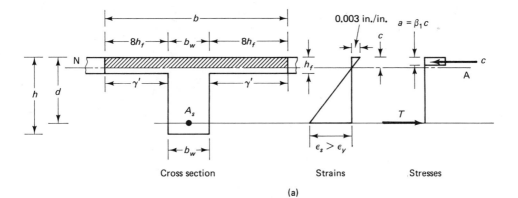

Cross section   Strains   Stresses

(a)

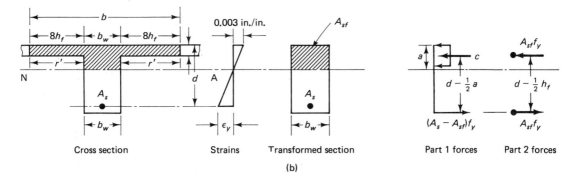

Cross section   Strains   Transformed section   Part 1 forces   Part 2 forces

(b)

Figure 13.5 (a) T-beam section with neutral axis inside the flange ($c < h_f$); (b) stress and strain distribution in flanged sections design (T-beam transfer $c > h_f$).

T-BEAM 1

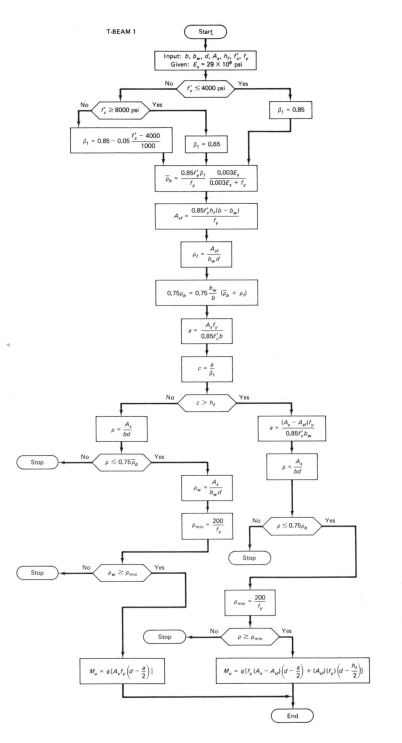

Figure 13.6 Flowchart: flexural analysis of a T beam.

If $c \leq h_f$,
$$M_u = \phi\left[A_s f_y\left(d - \frac{a}{2}\right)\right]$$

If $c \geq h_f$, recompute:
$$a = \frac{(A_s - A_{sf})f_y}{0.85 f_c' b_w}$$

$$M_u = \phi\left[(A_s - A_{sf})f_y\left(d - \frac{a}{2}\right) + A_{sf}f_y\left(d - \frac{h_f}{2}\right)\right]$$

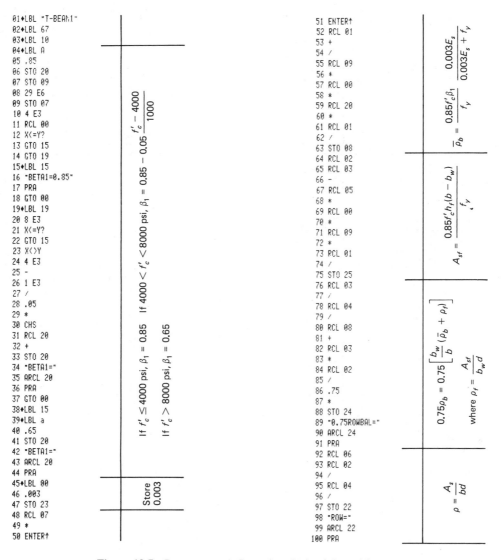

Figure 13.7 Program steps: flexural analysis of flanged beams.

```
101 RCL 02
102 *
103 RCL 03
104 /
105 STO 19
106 "ROW W="
107 ARCL 19
108 PRA
109 RCL 06
110 RCL 01
111 *
112 RCL 09
113 /
114 RCL 00
115 /
116 RCL 02
117 /
118 STO 21
119 "a="
120 ARCL 21
121 PRA
122 RCL 20
123 /
124 STO 13
125 "c="
126 ARCL 13
127 PRA
128 RCL 13
129 RCL 05
130 X<>Y
131 X<=Y?
132 GTO 01
133 "c>FLANGE TH."
134 PRA
135 "TREAT AS T-BEAM"
136 PRA
137 RCL 06
138 RCL 01
139 *
140 RCL 25
141 RCL 01
142 *
143 -
144 RCL 09
145 /
146 RCL 00
147 /
148 RCL 03
149 /
150 STO 21
```

$$\rho_w = \frac{A_s}{b_w d}$$

$$a = \frac{A_s f_y}{0.85 f'_c b}$$

$$c = \frac{a}{\beta_1}$$

If $c \leq h_f$, go to 01 (step 155)

If $c > h_f$ recompute $a = \dfrac{(A_s - A_{sf}) f_y}{0.85 f'_c b_w}$

```
151 "a="
152 ARCL 21
153 PRA
154 GTO 03
155*LBL 01
156 "c<FLANGE TH."
157 PRA
158 "TREAT AS RECBM."
159 PRA
160 RCL 08
161 .75
162 *
163 STO 25
164 "0.75ROWBARBAL="
165 ARCL 25
166 PRA
167 RCL 22
168 X<=Y?
169 GTO 02
170 "BAL NG"
171 ASTO 13
172 "ROW>0.75ROWBAR"
173 ARCL 13
174 PRA
175 X<>Y
176 RTN
177*LBL 02
178 200
179 ENTER↑
180 RCL 01
181 /
182 STO 10
183 "ROWMIN="
184 ARCL 10
185 PRA
186 RCL 10
187 RCL 19
188 X<>Y
189 X<=Y?
190 GTO 05
191 "N N.G."
192 ASTO 18
193 "ROW W<ROWMI"
194 ARCL 18
195 PRA
196 X<>Y
197 RTN
198*LBL 03
199 RCL 22
200 RCL 24
```

Compute $0.75\rho_b$ if rec. beam

If $\rho > 0.75\bar{\rho}_b$, display

$$\rho_{min} = \frac{200}{f_y}$$

If $\rho_w < \rho_{min}$, display

Figure 13.7 (*cont.*)

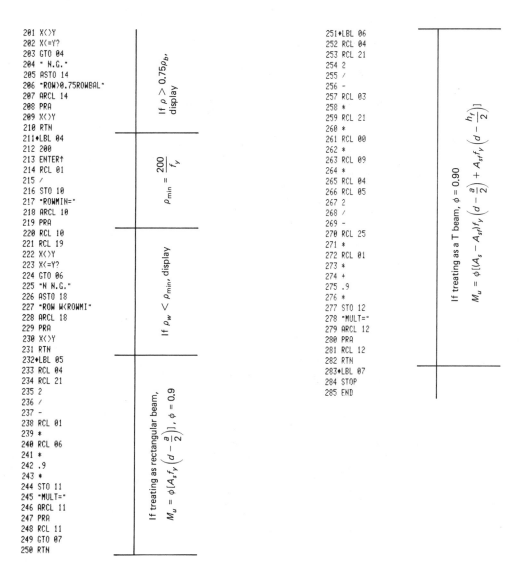

```
201 X<>Y
202 X<=Y?
203 GTO 04
204 " N.G."
205 ASTO 14
206 "ROW>0.75ROWBAL"
207 ARCL 14
208 PRA
209 X<>Y
210 RTN
211+LBL 04
212 200
213 ENTER↑
214 RCL 01
215 /
216 STO 10
217 "ROWMIN="
218 ARCL 10
219 PRA
220 RCL 10
221 RCL 19
222 X<>Y
223 X<=Y?
224 GTO 06
225 "N N.G."
226 ASTO 18
227 "ROW W<ROWMI"
228 ARCL 18
229 PRA
230 X<>Y
231 RTN
232+LBL 05
233 RCL 04
234 RCL 21
235 2
236 /
237 -
238 RCL 01
239 *
240 RCL 06
241 *
242 .9
243 *
244 STO 11
245 "MULT="
246 ARCL 11
247 PRA
248 RCL 11
249 GTO 07
250 RTN
```

If $\rho > 0.75\rho_b$, display

$$\rho_{min} = \frac{200}{f_y}$$

If $\rho_w < \rho_{min}$, display

If treating as rectangular beam,
$$M_u = \phi[A_s f_y \left(d - \frac{a}{2}\right)], \phi = 0.9$$

```
251+LBL 06
252 RCL 04
253 RCL 21
254 2
255 /
256 -
257 RCL 03
258 *
259 RCL 21
260 *
261 RCL 00
262 *
263 RCL 09
264 *
265 RCL 04
266 RCL 05
267 2
268 /
269 -
270 RCL 25
271 *
272 RCL 01
273 *
274 +
275 .9
276 *
277 STO 12
278 "MULT="
279 ARCL 12
280 PRA
281 RCL 12
282 RTN
283+LBL 07
284 STOP
285 END
```

If treating as a T beam, $\phi = 0.90$
$$M_u = \phi[(A_s - A_{sf})f_y \left(d - \frac{a}{2}\right) + A_{sf}f_y \left(d - \frac{h_f}{2}\right)]$$

Figure 13.7 (*cont.*)

In Fig. 13.5, the flange width, b, should not exceed one-fourth the span length and r' should not exceed $8h_f$ or one-half the clear distance to the next web. For beams with a slab on one side only, r' should not exceed one-twelfth the span nor $6h_f$ nor one-half the clear distance to the next web.

The flowchart and the program steps are presented in Figs. 13.6 and 13.7.

13.3.2 Instructions to Run the Program

Step 1: Set at least 26 data storage registers.

(XEQ | ALPHA | SIZE | ALPHA | 026)

Step 2: Load the program from magnetic cards or enter the program steps.
Step 3: Input the following design data.

Variable	Unit	Store in Register
f'_c	psi	00
f_y	psi	01
b	in.	02
b_w	in.	03
d	in.	04
h_f	in.	05
A_s	in.^{2}	06

Step 4: Execute T-BEAM1.

(XEQ | ALPHA | T-BEAM1 | ALPHA |)

If the design data satisfy the code minimum and maximum reinforcement ratio requirements, the program will calculate and print out the ultimate design moment $M_u = \phi M_n$ in.-lb. Otherwise, the program will print out the appropriate error message, "ROW W < ROW MIN N.G." or "ROW > 0.75 ROWBAL N.G." (flanged section) or "ROW > 0.75 ROWBAL N.G." (rectangular section).

The complete output will consist of:

Data	Unit
β_1	—
$0.75\rho_b$	—
a	in.
c	in.
$0.75\bar{\rho}_b$	—
ρ	—
ρ_{min}	—
M_u	in.-lb

The output will also have a statement indicating whether $c > h_f$ or $c < h_f$ and whether the section is treated as a T beam or a rectangular beam.

13.3.3 Numerical Examples

13.3.3.1 Example 13.7: Analysis of a T Beam for Moment Capacity

Calculate the design moment capacity of the precast T beam simply supported over a span of 30 ft and with a distance between webs of 10 ft. Given:

$f_c' = 4000$ psi, normal-weight concrete

$f_y = 60,000$ psi

$b = 40$ in.

$b_w = 10$ in.

$d = 18$ in.

$h_f = 2.5$ in.

Reinforcement area at the tension side:

   (a) $A_s = 4.0$ in.^{2}

   (b) $A_s = 6.0$ in.^{2}

Solution

A flange-width check is not necessary for a precast beam since the precast section can act independently depending on the construction system.

```
INPUT                OUTPUT                    INPUT

4,000.0000 STO 00          XEQ "T-BEAM1"           6.0000 STO 06
 900.0000 STO 01   BETA1=0.85
  40.0000 STO 02   0.75ROWBAL=0.0098               OUTPUT
  10.0000 STO 03   ROW=0.0056
  18.0000 STO 04   ROW W=0.0222                     XEQ "T-BEAM1"
   2.5000 STO 05   a=1.7647                  BETA1=0.85
   4.0000 STO 06   c=2.0761                  0.75ROWBAL=0.0098
                   c<FLANGE TH.               ROW=0.0083
                   TREAT AS RECBM.            ROW W=0.0333
                   0.75ROWBARBAL=0.0214       a=2.6471
                   ROWMIN=0.0033              c=3.1142
                   MULT=3,697,411.766         c>FLANGE TH.
                                              TREAT AS T-BEAM
                                              a=3.0882
                                              ROWMIN=0.0033
                                              MULT=5,399,205.882
```

13.3.3.2 Example 13.8: Design of an End-Span L Beam

A roof-garden floor is composed of a monolithic one-way slab system on beams. The clear span of the beam is 35 ft and all beams are spaced at 7 ft 6 in. center to center. The floor supports 6 ft 4 in. depth of soil in addition to its self-weight. Assume that the slab edges support a 12-in.-wide 7-ft wall weighing 840 lb per linear foot. Design the midspan section of the end spandrel L beam assuming that the moist soil weighs 125 lb/ft³. Given:

$f'_c = 3000$ psi, normal-weight concrete

$f_y = 60,000$ psi

Solution

Slab design

weight of soil = $6.33 \times 125 = 791$ say 800 psf

assume slab thickness $h = 4$ in. = $\dfrac{4}{12} \times 150 = 50$ psf

$d = h - (3/4 \text{ in. cover} + 1/2 \text{ diameter of No. 4 bars}) = 4.0 - 1.0 = 3.0$ in.

factored load $w_u = 1.4(800 + 50) = 1190$ lb/ft²

From the ACI code, the negative moment for the first interior support of a continuous slab is

$$-M_u = \frac{w_u l_n^2}{12} = \frac{1190(7.5)^2}{12} \times 12 = 66,938 \text{ in.-lb}$$

Try $b = 12$ in., $d = 3$ in., and $A_s = 0.496$ in.²/12-in. strip (No. 5 bars at 7.5 in. center to center).

INPUT   OUTPUT

```
3,000.0000 STO 00                    XEQ "RECBEAM"
60,000.0000 STO 01       BETA1=0.85
60,000.0000 STO 02       a=0.9725
   12.0000 STO 03        a=0.9725
    3.0000 STO 04        0.75ROWBAL=0.0160
    0.0000 STO 05        ROW=0.0138
    0.4960 STO 06        ROWMIN=0.0033          External factored moment
    0.0000 STO 07        MULT=67,327.6236       = 66,938 in.-lb.   O.K.
```

Use No. 5 bars at $7\frac{1}{2}$ in. center to center reinforcement.

temperature steel = $0.0018 \, bd = 0.0018 \times 12 \times 3.0 = 0.065$ in.²

maximum allowable spacing = $3h = 3 \times 4 = 12$ in.

Use No. 3 at 12 in. = 0.11 in.^{2} for temperature.

Beam web design

In order to choose a trial web section, assume that

$$d = \frac{l_n}{18} \text{ for deflection} \quad \text{or} \quad d = \frac{35.0 \times 12}{18} = 23.33 \text{ in.}$$

Assume that $h = 26$ in., $d = 22$ in., and $b_w = 14$ in.

$$\text{load area on L beam} = \frac{7.5}{2} + \frac{14}{12} = 4.92 \text{ ft}$$

$$\text{superimposed working } w_w = (4.92 - 1.0) \times 800 = 3136 \text{ lb/ft}$$

$$\text{slab weight} = \frac{4.0}{12} \times 150 \times 4.92 = 246 \text{ lb/ft}$$

$$\text{weight of beam web} = \frac{14(26 - 4)}{144} \times 150 = 321 \text{ lb/ft}$$

7-ft wall weight = 840 lb/ft

total service load = 3136 + 246 + 321 + 840 = 4543 lb/ft

factored load $w_u = 1.4 \times 4543 = 6360$ lb/ft

$$\text{factored external moment } M_u = \frac{w_u l_n^2}{11} = \frac{6360(35.0)^2}{11} \times 12 = 8{,}499{,}273 \text{ in.-lb}$$

$$M_n = \frac{M_u}{\phi} = \frac{8{,}499{,}273}{0.9} = 9{,}443{,}637 \text{ in.-lb}$$

Try $b = 38$ in. $(b_w + 6h_f)$, $b_w = 14$ in., $d = 22.5$ in., $h_f = 4$ in., and $A_s = 8$ in.^{2}
(eight No. 9 bars).

```
   INPUT                    OUTPUT

3,000.0000 STO 00              XEQ "T-BEAM1"
60,000.0000 STO 01    BETA1=0.85
   38.0000 STO 02     0.75ROWBAL=0.0095
   14.0000 STO 03     ROW=0.0094
   22.5000 STO 04     ROW W=0.0254
    4.0000 STO 05     a=4.9536
    8.0000 STO 06     c=5.8277
                      c>FLANGE TH.
                      TREAT AS T-BEAM
                      a=6.5882
                      ROWMIN=0.0033
                      MULT=8,582,061.177
```

> External factored moment
= 8,499,273 in.-lb. Adopt
the design.

13.3.3.3 Example 13.9: Design of an Interior Continuous Floor Beam for Flexure

Design a rectangular interior beam having a clear span 25 ft and carrying a working live load of 8000 lb per linear foot in addition to its self-weight. Assume the beam to have a 4-in. slab cast monolithically with it. Given:

$f'_c = 4000$ psi, normal-weight concrete
$f_y = 60,000$ psi

Solution

For deflection purposes, assume that

$$d = \frac{l}{12} = \frac{25 \times 12}{12} = 25 \text{ in.}$$

Try $d = 25$ in., $b_w = 14$ in., and $h = 28$ in.

$$\text{self-weight} = \frac{14 \times 28}{144} \times 150 = 408 \text{ lb/ft}$$

$$\text{factored load } U = 1.4 \times 408 + 1.7 \times 8000 = 14,171 \text{ lb/ft}$$

Positive factored moment M_u for interior midspan lower fibers (ACI) is

$$+M_u = \frac{w_u l_n^2}{16} = \frac{14,171 \times (25.0)^2}{16} \times 12 = 6,642,656 \text{ in.-lb}$$

The negative factored moment, M_u, at support (tension at top fibers) is

$$-M_u = \frac{14,171 \times (25.0)^2}{11} \times 12 = 9,662,046 \text{ in.-lb}$$

Section at midspan (T Beam)

Assume that $b_w = 14$ in.

$$b \not> 16 \times 4 + 14 = 78 \text{ in.}$$

$$\not> \frac{25 \times 12}{4} = 75 \text{ in.}$$

$\not>$ center-to-center distance of beams not known

Therefore,

$$b = 75 \text{ in. controls}$$

Hence try $b = 75$ in., $d = 25$ in., $b_w = 14$ in., $h_f = 4$ in., and $A_s = 5.08$ in.^{2} (four No. 10 bars).

```
    INPUT                    OUTPUT

  4,000.0000 STO 00          XEQ "T-BEAM1"
 60,000.0000 STO 01     BETA1=0.85
     75.0000 STO 02     0.75ROWBAL=0.0095
     14.0000 STO 03     ROW=0.0027
     25.0000 STO 04     ROW W=0.0145
      4.0000 STO 05     a=1.1953
      5.0800 STO 06     c=1.4062
                        c<FLANGE TH.
                        TREAT AS RECBM.
                        0.75ROWBARBAL=0.0214    > External factored moment
                        ROWMIN=0.0033             = 6,642,656 in.-lb. Adopt
                        MULT=6,694,053.457        the design.
```

*Section at support (doubly reinforced rectangular
section)*

Assume that two No. 10 bars extend from the midspan to the support providing
compression reinforcement. Try $b = 14$ in., $d = 25$ in., $d' = 3$ in., $A_s = 8.89$ in.^{2}
(seven No. 10 bars), and $A_s' = 2.54$ in.^{2} (two No. 10 bars).

```
    INPUT                    OUTPUT

  4,000.0000 STO 00          XEQ "RECBEAM"
 60,000.0000 STO 01     BETA1=0.85
 60,000.0000 STO 02     a=8.0042
     14.0000 STO 03     a=8.0424
     25.0000 STO 04     0.75ROWBAL=0.0286
      3.0000 STO 05     ROW=0.0254
      8.8900 STO 06     ROWMIN=0.0033          $M_u > 9,662,046$ in.-lb.
      2.5400 STO 07     MULT=10,209,469.61     Adopt the section.
```

13.4 SHEAR AND TORSION

SH + TOR1, 2 and 3 is a program that computes the required shear and/or torsion
reinforcement and its spacing for beams. The program is based on the ACI building
code. This subject is extensively presented in Chapter 6.

The program is basically broken into two parts: (1) SH + TOR1 computes
$A_v/s + 2A_t/s$, the total shear reinforcement area required per two legs of shear
reinforcing per inch spacing; (2) SH + TOR2 computes the required stirrup spacing
and longitudinal reinforcement for beams subjected to both shear and torsion or torsion
alone. SH + TOR3 computes the required stirrup spacing for beams subjected to
shear only.

The program can be used to compute the required shear reinforcement for any section by inputting the factored shear and/or torsion at that section. If the beam is subjected to shear only or if the factored torsion is $\leq \phi(0.5\sqrt{f_c'} \Sigma x^2 y)$ and if M_u is input as zero, the program computes V_c, the shear taken by the concrete, using the conservative simplified formula $V_c = 2\sqrt{f_c'} b_w d$. If the actual value of M_u is input for that section, the program uses the more refined formula

$$V_c = \left(1.9\sqrt{f_c'} + 2500\rho_w \frac{V_u d}{M_u}\right) b_w d \leq 3.5\sqrt{f_c'} b_w d$$

For beams subjected to torsion greater than $\phi \times (0.5\sqrt{f_c'} \Sigma x^2 y)$, the program computes V_c and T_c using the appropriate formulas. If the beam is subjected to torsion only, the program sets $V_c = 0$. However, this seldom happens since beams are usually not continuously supported along their entire length and therefore are subjected to at least the shear due to their self-weight.

Once the program computes how much of the shear and/or torsion will be carried by the concrete alone, the remainder is used to compute the amount of torsion and shear web reinforcement needed. The program then computes the spacing of the stirrups based on the shear to be carried by the reinforcing and the size of bars input by the user. The program also ensures that all spacing limitations as specified by ACI code are adhered to.

13.4.1 Design Equations, Flowchart, and Program Steps

(Refer to Figs. 13.8, 13.9 and 13.10.)

$$V_u = \phi V_n \qquad V_n \leq 10\sqrt{F_c'} b_w d$$

$$V_c = 2\sqrt{f_c'} b_w d \quad \text{or} \quad V_c = \left(1.9\sqrt{f_c'} + 2500\rho_w \frac{V_u d}{M_u}\right) b_w d \leq 3.5\sqrt{f_c'} b_w d$$

$$\rho_w = \frac{A_s}{b_w d} \qquad \frac{V_u}{M_u} d \leq 1 \qquad \left|\frac{A_v}{s}\right| = \frac{V_n - V_c}{f_y d}$$

Torsion should be accounted for in design if $T_u \geq \phi(0.5\sqrt{f_c'} \Sigma x^2 y)$:

$$C_t = \frac{b_w d}{\Sigma x^2 y} \qquad T_c = \frac{0.8\sqrt{f_c'} \Sigma x^2 y}{\sqrt{1 + (0.4 V_u / C_t T_u)^2}}$$

$$V_c = \frac{2\sqrt{f_c'} b_w d}{\sqrt{1 + [2.5 C_t (T_u / V_u)]^2}} \qquad \alpha_t = 0.66 + 0.33 \frac{y_1}{x_1} \leq 1.5$$

$$\frac{A_t}{s} = \frac{T_n - T_c}{\alpha_t x_1 y_1 f_y} \qquad \left|\frac{A_t}{s}\right|_{total} = \frac{A_v}{s} + 2\frac{A_t}{s} \leq \frac{50 b_w}{f_y}$$

$$\text{stirrup spacing, } s = \frac{\left|\dfrac{A_t}{s}\right|_{\text{total}}}{A_v} \le \text{ that required for shear force design}$$

$$A_l = 2A_t \frac{x_1 + y_1}{s} \qquad A_l \le \left[\frac{400xs}{f_y} \frac{T_u}{T_u + V_u/3C_t} - 2A_t\right]\frac{x_1 + y_1}{s}$$

$$2A_t \ge \frac{50b_w s}{f_y} \qquad \text{in the second expression for } A_l$$

The flowchart and the program steps are presented in Figs. 13.9 and 13.10, respectively.

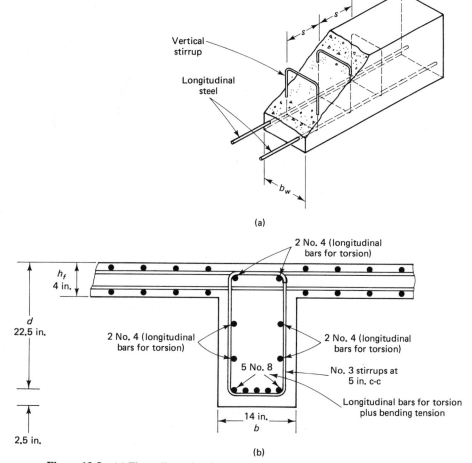

Figure 13.8   (a) Three-dimensional view of vertical stirrups; (b) web reinforcement details for a typical T-beam section subjected to shear and torsion.

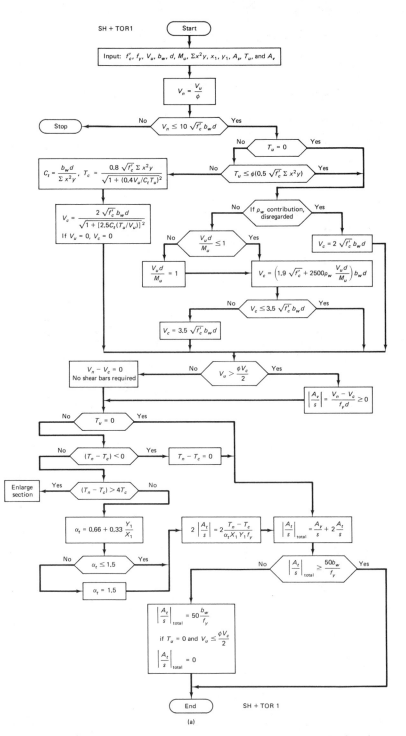

Figure 13.9 (a) Flowchart: analysis for shear and torsion (part I); (b) flowchart: calculation of stirrup spacing (part II).

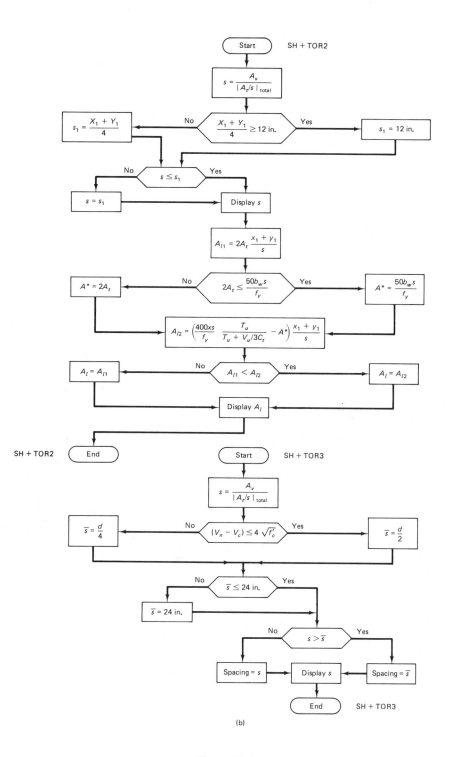

Figure 13.9 (*cont.*)

```
01♦LBL "SH+TOR1"
02 FIX 4
03 RCL 02
04 0.85
05 STO 23
06 /
07 STO 21
08 RCL 00
09 SQRT
10 RCL 03
11 *
12 RCL 04
13 *
14 STO 24
15 10
16 *
17 RCL 21
18 X<=Y?
19 GTO 00
20 "TION"
21 ASTO 26
22 "ENLARGE SEC"
23 ARCL 26
24 PRA
25 X<>Y
26 GTO 11
27♦LBL 00
28 "VN="
29 ARCL 21
30 PRA
31 RCL 09
32 X=0?
33 GTO 02
34 RCL 23
35 0.5
36 *
37 RCL 00
38 SQRT
39 RCL 06
40 *
41 STO 23
42 *
43 RCL 09
44 X<=Y?
45 GTO 01
46 GTO 03
47♦LBL 01
48 "SION"
49 ASTO 26
50 "NEGLECT TOR"
```

$V_n = V_u/\phi$
store 21

If $V_n > 10 \sqrt{f_c'}\, b_w d$, enlarge section

Display V_n

If $T_u = 0$, go to 02 (step 108)

If $T_u < \phi(0.5 \sqrt{f_c'}\, \Sigma\, x^2 y$, neglect torsion and set $T_u = 0$

```
51 ARCL 26
52 PRA
53 0
54 STO 22
55 STO 09
56 GTO 02
57♦LBL 03
58 RCL 23
59 0.8
60 *
61 RCL 02
62 RCL 06
63 *
64 0.4
65 *
66 RCL 09
67 RCL 03
68 *
69 RCL 04
70 *
71 /
72 X↑2
73 1
74 +
75 SQRT
76 /
77 STO 22
78 "Tc="
79 ARCL 22
80 PRA
81 RCL 02
82 X=0?
83 GTO 01
84 RCL 24
85 2
86 *
87 RCL 03
88 RCL 04
89 *
90 RCL 09
91 *
92 2.5
93 *
94 RCL 06
95 RCL 02
96 *
97 /
98 X↑2
99 1
100 +
```

$C_t = \dfrac{b_w d}{\Sigma\, x^2 y}$

$T_c = \dfrac{0.8 \sqrt{f_c'}\, \Sigma\, x^2 y}{\sqrt{1 + (0.4 V_u / C_t T_u)^2}}$

If $V_u = 0$, go to 01 (step 105)

$V_c = \dfrac{2 \sqrt{f_c'}\, b_w d}{\sqrt{1 + (2.5 C_t T_u / V_u)^2}}$

Figure 13.10   Program steps: shear and torsion.

101 SQRT
102 /
103 STO 20
104 GTO 06
105♦LBL 01
106 STO 20
107 GTO 06
108♦LBL 02
109 RCL 02
110 RCL 04
111 *
112 RCL 05
113 X=0?
114 GTO 05
115 /
116 STO 23
117 1
118 X<>Y
119 X<=Y?
120 GTO 03
121 1
122 STO 23
123♦LBL 03
124 RCL 00
125 SQRT
126 1.9
127 *
128 RCL 03
129 *
130 RCL 04
131 *
132 RCL 08
133 2500
134 *
135 RCL 23
136 *
137 +
138 STO 20
139 RCL 24
140 3.5
141 *
142 RCL 20
143 X<=Y?
144 GTO 06
145 X<>Y
146 STO 20
147 GTO 06
148♦LBL 05
149 RCL 24
150 2
151 *
152 STO 20
153♦LBL 06
154 "Vc="
155 ARCL 20
156 PRA

If $V_u = 0$, set $V_c = 0$

If ρ_w contrib. disregarded go to 05 and use $V_c = 2\sqrt{f_c'}\,bd$ go to step 148

Compute $\dfrac{V_u d}{M_u}$ If $\dfrac{V_u d}{M_u} > 1$, set = 1 Store 23

If ρ_w contribution used $V_c = \left(1.9\sqrt{f_c'} + 2500\rho_w \dfrac{V_u d}{M_u}\right)b_w d$

If $V_c \le 3.5\sqrt{f_c'}\,b_w d$, go to 06 (step 153) If $V_c > (3.5\sqrt{f_c'}\,b_w d)$, store $(3.5\sqrt{f_c'}\,b_w d)$ in 20

Display V_c

157 RCL 21
158 RCL 20
159 2
160 /
161 -
162 X>0?
163 GTO 07
164 0
165 STO 24
166 SF 01
167 "RS REQ"
168 ASTO 26
169 "NO SHEAR BA"
170 ARCL 26
171 PRA
172 "AV/S=0.0000"
173 PRA
174 GTO 08
175♦LBL 07
176 RCL 21
177 RCL 20
178 -
179 X>0?
180 GTO 07
181 0
182 STO 24
183 "AV/S=0.0000"
184 PRA
185 GTO 08
186♦LBL 07
187 RCL 01
188 /
189 RCL 04
190 /
191 STO 24
192 "AV/S="
193 ARCL 24
194 PRA
195♦LBL 08
196 RCL 09
197 0.85
198 /
199 X=0?
200 GTO 10
201 RCL 22
202 -
203 X>0?
204 GTO 09
205♦LBL 10
206 "S REQD"
207 ASTO 26
208 "NO TORS BAR"
209 ARCL 26
210 PRA
211 "2AT/S=0.0000"
212 PRA
213 0
214 GTO 10

If $(V_n - V_c/2) > 0$, go to 07 (step 175) If $(V_n - V_c/2) < 0$, set $V_n - V_c = 0$ and display "no shear bars required"

$\dfrac{A_v}{s} = \dfrac{V_n - V_c}{f_y d}$ If $(V_n - V_c) < 0$, set $\dfrac{A_v}{s} = 0$

If $T_n = 0$, go to 10

If $(T_n - T_c) > 0$, go to 09 (step 215) If $(T_n - T_c) < 0$, no torsion bars required

Figure 13.10 (*cont.*)

215♦LBL 09	If $T_n > 4T_c$, enlarge section
216 RCL 22	
217 4	
218 *	
219 X>Y?	
220 GTO 09	
221 "TS>4Tc ENLA"	
222 "⊦RGE SECTION"	
223 PRA	
224 STOP	
225♦LBL 09	
226 X<>Y	$\dfrac{T_n - T_c}{x_1 y_1 f_y}$ store in 05
227 RCL 01	
228 /	
229 RCL 07	
230 /	
231 RCL 25	
232 /	
233 STO 05	
234 RCL 25	
235 RCL 07	
236 /	
237 0.33	$\alpha_t = 0.66 + 0.33\,\dfrac{y_1}{x_1}$
238 *	
239 0.66	
240 +	
241 STO 08	
242 1.5	If $\alpha_t < 1.5$, go to 12
243 X<>Y	If $\alpha_t > 1.5$, set = 1.5
244 X<=Y?	
245 GTO 12	
246 X<>Y	
247 STO 08	
248♦LBL 12	Compute
249 RCL 05	$\dfrac{2A_t}{s} = 2\dfrac{T_n - T_c}{\alpha_t x_1 y_1 f_y}$
250 RCL 08	
251 /	
252 2	
253 *	
254 STO 05	
255 "2AT/S="	Display
256 ARCL 05	$2A_t/s$
257 PRA	
258♦LBL 10	Compute
259 RCL 24	$\dfrac{A_v}{s} + \dfrac{2A_t}{s}$
260 +	
261 STO 24	
262 FS?C 01	
263 GTO 11	
264 GTO 12	
265♦LBL 11	
266 X=0?	
267 GTO 14	
268♦LBL 12	
269 RCL 24	
270 RCL 03	

271 50	If $\dfrac{A_v}{s} + \dfrac{2A_t}{s} < \dfrac{50b_w}{f_y}$, use $\dfrac{50b_w}{f_y}$ and display $50b_w/f_y$ controls
272 *	
273 RCL 01	
274 /	
275 X<=Y?	
276 GTO 13	
277 STO 24	
278 "TROLS"	
279 ASTO 26	
280 "50bW/FY CON"	
281 ARCL 26	
282 PRA	
283 "AV/S+2AT/S="	
284 ARCL 24	
285 PRA	
286 GTO 11	
287♦LBL 13	
288 X<>Y	Display $\dfrac{A_v}{s} + \dfrac{2A_t}{s}$
289 STO 24	
290 "AV/S+2AT/S="	
291 ARCL 24	
292 PRA	
293♦LBL 11	
294 STOP	
295♦LBL 14	If $V_n - \dfrac{V_c}{2} < 0$ and $T_u = 0$, display "no stirrups required"
296 "STOP NO STI"	
297 "⊦RRUPS REQD."	
298 PRA	
299 STOP	
300♦LBL "SH+TOR2"	$s = \dfrac{A_v}{A_v/s + 2A_t/s}$
301 RCL 10	
302 RCL 24	
303 /	
304 STO 08	
305 RCL 07	
306 RCL 25	
307 +	
308 4	
309 /	
310 12	Select smaller of the two and display as max. "s"
311 X<=Y?	(1) $(x_1 + y_1)/4$
312 GTO 00	(2) 12 in.
313 X<>Y	
314♦LBL 00	
315 STO 26	
316 "MAX. S="	
317 ARCL 26	
318 PRA	
319 RCL 08	
320 X<=Y?	
321 GTO 01	
322 X<>Y	
323♦LBL 01	Display s required
324 STO 08	
325 "S REQD="	
326 ARCL 08	
327 PRA	

Figure 13.10 (*cont.*)

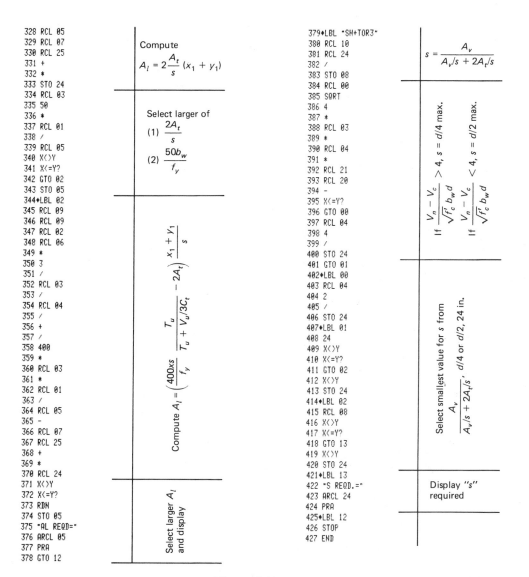

```
328 RCL 05
329 RCL 07
330 RCL 25
331 +
332 *
333 STO 24
334 RCL 03
335 50
336 *
337 RCL 01
338 /
339 RCL 05
340 X<>Y
341 X<=Y?
342 GTO 02
343 STO 05
344*LBL 02
345 RCL 09
346 RCL 09
347 RCL 02
348 RCL 06
349 *
350 3
351 /
352 RCL 03
353 /
354 RCL 04
355 /
356 +
357 /
358 400
359 *
360 RCL 03
361 *
362 RCL 01
363 /
364 RCL 05
365 -
366 RCL 07
367 RCL 25
368 +
369 *
370 RCL 24
371 X<>Y
372 X<=Y?
373 RDN
374 STO 05
375 "AL REQD="
376 ARCL 05
377 PRA
378 GTO 12
```

Compute

$$A_l = 2\frac{A_t}{s}(x_1 + y_1)$$

Select larger of

(1) $\dfrac{2A_t}{s}$

(2) $\dfrac{50b_w}{f_y}$

Compute $A_l = \left(\dfrac{400xs}{f_y}\dfrac{T_u}{T_u + V_u/3C_t} - 2A_t\right)\dfrac{x_1 + y_1}{s}$

Select larger A_l and display

```
379*LBL "SH+TOR3"
380 RCL 10
381 RCL 24
382 /
383 STO 08
384 RCL 00
385 SQRT
386 4
387 *
388 RCL 03
389 *
390 RCL 04
391 *
392 RCL 21
393 RCL 20
394 -
395 X<=Y?
396 GTO 00
397 RCL 04
398 4
399 /
400 STO 24
401 GTO 01
402*LBL 00
403 RCL 04
404 2
405 /
406 STO 24
407*LBL 01
408 24
409 X<>Y
410 X<=Y?
411 GTO 02
412 X<>Y
413 STO 24
414*LBL 02
415 RCL 08
416 X<>Y
417 X<=Y?
418 GTO 13
419 X<>Y
420 STO 24
421*LBL 13
422 "S REQD.="
423 ARCL 24
424 PRA
425*LBL 12
426 STOP
427 END
```

$$s = \frac{A_v}{A_v/s + 2A_t/s}$$

If $\dfrac{V_n - V_c}{\sqrt{f_c}\, b_w d} > 4$, $s = d/4$ max.

If $\dfrac{V_n - V_c}{\sqrt{f_c}\, b_w d} < 4$, $s = d/2$ max.

Select smallest value for s from $\dfrac{A_v}{A_v/s + 2A_t/s}$, $d/4$ or $d/2$, 24 in.

Display "s" required

Figure 13.10 (*cont.*)

13.4.2 Instructions to Run the Program

Step 1: Set at least 27 data storage registers.

(| XEQ | | ALPHA | S I Z E | ALPHA | 027)

Step 2: Load the program from magnetic cards or enter the program steps.

Step 3: Input the following design data.

Variable	Unit	Store in register
f'_c	psi	00
f_y	psi	01
V_u (external shear force at the section being analyzed)	lb	02
b_w	in.	03
d	in.	04
M_u (external moment at the section being analyzed)	in.-lb	05 (store 0.0 if program to use $V_u = 2\sqrt{f'_c}\, b_w d$)
$\sum x^2 y$	in.^{3}	06
x_1	in.	07
y_1	in.	25
*A_s (tension reinforcement)	in.^{2}	08
T_u (external torsion at the section being analyzed)	in.-lb	09
A_v (twice the area of one stirrup leg)	in.^{2}	10

*After every run, A_s must be re-input.

Step 4: Compute the required shear reinforcement area ($|A_t/s|_{\text{total}}$ in.^{2}/two legs/in. spacing) by executing SH + TOR1.

(| XEQ | | ALPHA | SH + TOR1 | ALPHA |)

Step 5: If the section is subjected to shear and torsion or torsion only, execute SH + TOR2.

(| XEQ | | ALPHA | SH + TOR2 | ALPHA |)

The program will calculate and print out the spacing and area of the longitudinal steel, A_l (in.^{2}) required.

If the section is subjected to only shear force, execute SH + TOR3.

(| XEQ | | ALPHA | SH + TOR3 | ALPHA |)

The program will calculate and print out the required spacing for the chosen stirrup size.

13.4.3 Numerical Examples

13.4.3.1 Example 13.10: Design of Web Stirrups

A rectangular beam has an effective span 25 ft and carries a design live load of 8000 lb per linear foot and no external dead load except its self-weight. Design the necessary shear reinforcement. Use the simplified term $(V_c = 2\sqrt{f_c'}\,b_w d)$ for calculating the capacity V_c of the plain concrete web. Given:

f_c' = 4000 psi, normal-weight concrete
f_y = 60,000 psi
b_w = 14 in.
d = 28 in.
h = 30 in.
$\sum x^2 y = (14^2)(30) = 5880$ in.^{3}

Longitudinal tension steel is six No. 9 bars. No axial force acts on the beam.

Solution

Factored shear force

$$\text{beam self-weight} = \frac{14 \times 30}{144} \times 150 = 437.5 \text{ lb/ft}$$

$$\text{factored total load} = 1.7 \times 8000 + 1.4 \times 437.5 = 14{,}212.5 \text{ lb/ft}$$

The factored shear force at the face of the support is

$$V_u = \frac{25}{2} \times 14{,}212.5 = 177{,}656 \text{ lb}$$

The first critical section is at a distance $d = 28$ in. from the face of the support of this beam (half-span = 150 in.).

$$V_u \text{ at } d = \frac{150 - 28}{150} \times 177{,}656 = 144{,}494 \text{ lb}$$

$$V_u \text{ at } 2d = 111{,}331 \text{ lb}$$

$$V_u \text{ at } 3d = 78{,}169 \text{ lb}$$

$$V_u \text{ at } 45.6 \text{ in.} = 123{,}639 \text{ lb}$$

$$V_u \text{ at } 79.96 \text{ in.} = 82{,}954 \text{ lb}$$

These locations (for calculation of stirrup spacing) were chosen to be consistent with the solution of Example 6.1.

At d from support	At 45.6 in. from support	At 2d from support
INPUT	INPUT	INPUT
4,000.0000 STO 00	123,649.0000 STO 02	111,331.0000 STO 02
60,000.0000 STO 01	6.0000 STO 08	6.0000 STO 08
144,494.0000 STO 02		
14.0000 STO 03	OUTPUT	OUTPUT
28.0000 STO 04		
0.0000 STO 05	XEQ "SH+TOR1"	XEQ "SH+TOR1"
5,880.0000 STO 06	VN=145,469.4118	VN=130,977.6471
0.0000 STO 07	Vc=49,584.5137	Vc=49,584.5137
6.0000 STO 08	AV/S=0.0571	AV/S=0.0484
0.0000 STO 09	NO TORS BARS REQD	NO TORS BARS REQD
0.4000 STO 10	2AT/S=0.0000	2AT/S=0.0000
0.0000 STO 25	AV/S+2AT/S=0.0571	AV/S+2AT/S=0.0484
	XEQ "SH+TOR3"	XEQ "SH+TOR3"
OUTPUT	S REQD.=7.0084	S REQD.=8.2562

Say 7.0 in.

XEQ "SH+TOR1"
VN=169,992.9412
Vc=49,584.5137
AV/S=0.0717
NO TORS BARS REQD
2AT/S=0.0000
AV/S+2AT/S=0.0717
   XEQ "SH+TOR3"
S REQD.=5.5810

At 79.96 in. from support	At 3d from support
INPUT	INPUT
82,954.0000 STO 02	78,169.0000 STO 02
6.0000 STO 08	6.0000 STO 08
OUTPUT	OUTPUT
XEQ "SH+TOR1"	XEQ "SH+TOR1"
VN=97,592.9412	VN=91,963.5294
Vc=49,584.5137	Vc=49,584.5137
AV/S=0.0286	AV/S=0.0252
NO TORS BARS REQD	NO TORS BARS REQD
2AT/S=0.0000	2AT/S=0.0000
AV/S+2AT/S=0.0286	AV/S+2AT/S=0.0252
XEQ "SH+TOR3"	XEQ "SH+TOR3"
S REQD.=13.9975	S REQD.=14.0000

Say 14.0 in.

Refer to Fig. 6.10 for reinforcement details.

13.4.3.2 Example 13.11: Alternative Solution to Ex. 13.10

Find the shear force V_c and the change in stirrup spacing for the beam in Ex. 13.10 if the more refined Eq. 6.8 is used where the separate contribution of the main longitudinal steel at the tension side is more accurately reflected.

Solution

$$V_u \text{ at } d \text{ from support} = 144{,}494 \text{ lb} \qquad \text{(Ex. 13.10)}$$

$$M_u \text{ at } d \text{ from support} = \left(14{,}212.5 \times \frac{25}{2}\right) \times 28 - \frac{14{,}212.5(28)^2}{12 \times 2} = 4{,}510{,}100 \text{ lb-in.}$$

$$V_u \text{ at } 2d \text{ from support} = 111{,}331 \text{ lb}$$

$$M_u \text{ at } 2d \text{ from support} = \left(14{,}212.5 \times \frac{25}{2}\right) \times 56 - \frac{14{,}212(52)^2}{12 \times 2}$$

$$= 8{,}347{,}475 \text{ lb}$$

$$V_u \text{ at } 3d \text{ from support} = 78{,}169 \text{ lb}$$

$$M_u \text{ at } 3d \text{ from support} = 10{,}744{,}650 \text{ lb-in.}$$

$$V_u \text{ at } 45.6 \text{ in. from support} = 123{,}649 \text{ lb}$$

$$M_u \text{ at } 45.6 \text{ in. from support} = 6{,}869{,}754 \text{ lb-in.}$$

$$V_u \text{ at } 79.96 \text{ in. from support} = 82{,}954 \text{ lb}$$

$$M_u \text{ at } 79.96 \text{ in. from support} = 9{,}538{,}008 \text{ lb-in.}$$

The program is run the same way as Ex. 13.10 except that the actual value of M_u at the section is input in register 05 instead of 0.0. Note how the required spacing increases slightly at the same section as before. This is expected since the simplified equation $V_c = 2\sqrt{f_c'}\, b_w d$ is conservative.

At d from support

```
        INPUT                        OUTPUT

    4,000.0000 STO 00              XEQ "SH+TOR1"
   60,000.0000 STO 01          VN=169,992.9412
  144,494.0000 STO 02          Vc=60,561.1937
       14.0000 STO 03          AV/S=0.0651
       28.0000 STO 04          NO TORS BARS REQD
4,513,100.000 STO 05           2AT/S=0.0000
    5,880.0000 STO 06          AV/S+2AT/S=0.0651
        0.0000 STO 07                  XEQ "SH+TOR3"
        6.0000 STO 08          S REQD.=6.1408
        0.0000 STO 09
        0.4000 STO 10
        0.0000 STO 25
```

At 45.6 in. from support
INPUT

```
123,649.0000 STO 02
6,869,754.000 STO 05
6.0000 STO 08
```

OUTPUT

```
              XEQ "SH+TOR1"
VN=145,469.4118
Vc=54,664.8862
AV/S=0.0541
NO TORS BARS REQD
2AT/S=0.0000
AV/S+2AT/S=0.0541
              XEQ "SH+TOR3"
S REQD.=7.4005
```

At 2d from support
INPUT

```
111,331.0000 STO 02
8,347,475.000 STO 05
6.0000 STO 08
```

OUTPUT

```
              XEQ "SH+TOR1"
VN=130,977.6471
Vc=52,706.8646
AV/S=0.0466
NO TORS BARS REQD
2AT/S=0.0000
AV/S+2AT/S=0.0466
              XEQ "SH+TOR3"
S REQD.=8.5856
```

At 79.96 in. from support
INPUT

```
82,954.0000 STO 02
9,538,008.000 STO 05
6.0000 STO 08
```

OUTPUT

```
              XEQ "SH+TOR1"
VN=97,592.9412
Vc=50,758.1136
AV/S=0.0279
NO TORS BARS REQD
2AT/S=0.0000
AV/S+2AT/S=0.0279
              XEQ "SH+TOR3"
S REQD.=14.0000
```

At 3d from support
INPUT

```
78,169.0000 STO 02
10,744,650.00 STO 05
6.0000 STO 08
```

OUTPUT

```
              XEQ "SH+TOR1"
VN=91,963.5294
Vc=50,160.8534
AV/S=0.0249
NO TORS BARS REQD
2AT/S=0.0000
AV/S+2AT/S=0.0249
              XEQ "SH+TOR3"
S REQD.=14.0000
```

Note the slightly greater spacing allowed using the more refined equation for V_c.

13.4.3.3 Example 13.12: Design of Web Reinforcement for Combined Torsion and Shear in a T-Beam Section

A T-beam cross section has the geometrical dimensions shown in Fig. 13.11. A factored external shear force acts at the critical section having a value $V_u = 15,000$ lb. It is subjected to the following torques at the critical section: (a) equilibrium factored external

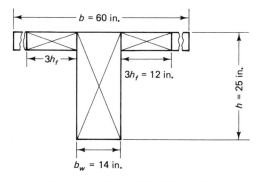

Figure 13.11 Component rectangles of the T beam.

torsional moment T_u = 500,000 in.-lb, (b) compatibility factored T_u = 75,000 in.-lb, and (c) compatibility factored T_u = 300,000 in.-lb. Given:

> Bending reinforcement A_s = 3.4 in.^{2}
>
> f_c' = 4000 psi, normal-weight concrete
>
> f_y = 60,000 psi

Design the web reinforcement needed for this section.

Solution

(a) *Equilibrium Torsion:*

Factored torsional moment: Given equilibrium torsional moment = 500,000 in.-lb. The total torsional moment must be provided for in the design. From Fig. 13.11,

$$\sum x^2 y = 14^2 \times 25 + 4^2 \times 3 \times 4 + 4^2 \times 3 \times 4 = 5284 \text{ in.}^3$$

Assume $1\frac{1}{2}$ in. clear cover and No. 4 closed stirrups.

$$x_1 = 14 - 2(1.5 + 0.25) = 10.5 \text{ in.}$$

$$y_1 = 25 - 2(1.5 + 0.25) = 21.5 \text{ in.}$$

Try No. 3 closed stirrups. Area for two legs = 0.22 in.^{2}.

```
   INPUT                    OUTPUT

  4,000.0000 STO 00              XEQ "SH+TOR1"
 60,000.0000 STO 01       VN=17,647.0588
 15,000.0000 STO 02       Tc=262,094.3281
     14.0000 STO 03       Vc=7,862.8298
     22.5000 STO 04       AV/S=0.0072
      0.0000 STO 05       2AT/S=0.0361
  5,284.0000 STO 06       AV/S+2AT/S=0.0433
     10.5000 STO 07              XEQ "SH+TOR2"
      3.4000 STO 08       MAX. S=8.0000
500,000.0000 STO 09       S REQD=5.0808
      0.2200 STO 10       AL REQD=1.4039
     21.5000 STO 25
```

Use No. 3 closed stirrups at 5 in. center to center.

Distribution of torsion longitudinal bars: Torsional $A_l = 1.41$ in.^{2}. Assume that $\frac{1}{4}A_l$ goes to the top corners and $\frac{1}{4}A_l$ goes to the bottom corners of the stirrups, to be added to the flexural bars. The balance $\frac{1}{2}A_l$ would thus be distributed equally to the vertical faces of the beam cross section at a spacing not to exceed 12 in. center to center.

$$\text{midspan} \sum A_s = \frac{A_l}{4} + A_s = \frac{1.41}{4} + 3.4 = 3.75 \text{ in.}^2$$

Provide five No. 8 bars at the bottom. Provide two No. 4 bars with an area of 0.40 in.^{2} at the top. The required area of $A_l/4$ is 0.35 in.^{2}. The area of steel needed for each vertical face $= 0.35$ in.^{2}. Provide two No. 4 bars on each side. Figure 7.17 shows the geometry of the cross section.

(b) *Compatibility torsion:*

Factored torsional moment: Given $T_u = 75,000$ in.-lb. Using the results of case (a):

```
INPUT

    4,000.0000 STO 00
   60,000.0000 STO 01
   15,000.0000 STO 02
       14.0000 STO 03
       22.5000 STO 04
        0.0000 STO 05
    5,284.0000 STO 06
       10.5000 STO 07
        3.4000 STO 08
   75,000.0000 STO 09
        0.2200 STO 10
       21.5000 STO 25


OUTPUT

        XEQ "SH+TOR1"
VN=17,647.0588
NEGLECT TORSION
Vc=39,844.6985
NO SHEAR BARS REQ
AV/S=0.0000
NO TORS BARS REQD
2AT/S=0.0000
STOP NO STIRRUPS REQD.
```
Neglect torsion and shear. No stirrups required.

(c) *Compatibility torsion:*

Factored torsional moment: Given that $T_u = 300,000$ in.-lb is greater than $\phi(0.5\sqrt{f'_c}\Sigma x^2 y)$. Hence stirrups should be provided. Since this is a compatibility torsion, the section can be designed for a torsional moment of $\phi(4\sqrt{f_c'}\Sigma x^2 y/3)$ if the external torsion exceeds this value.

$$\phi\left(4\sqrt{f_c'}\;\frac{\Sigma x^2 y}{3}\right) = 377,801 \text{ in.-lb} > \text{given } T_u = 300,000 \text{ in.-lb}$$

Hence the section should be designed for $T_u = 300,000$ in.-lb.

```
INPUT                    OUTPUT

  4,000.0000 STO 00           XEQ "SH+TOR1"
 60,000.0000 STO 01     VN=17,647.0588
 15,000.0000 STO 02     Tc=253,467.3716
     14.0000 STO 03     Vc=12,673.3686
     22.5000 STO 04     AV/S=0.0037
      0.0000 STO 05     2AT/S=0.0110
  5,284.0000 STO 06     AV/S+2AT/S=0.0147
     10.5000 STO 07           XEQ "SH+TOR2"
      3.4000 STO 08     MAX. S=8.0000
300,000.0000 STO 09     S REQD=8.0000
      0.2200 STO 10     AL REQD=1.9608
     21.5000 STO 25
```

Provide No. 3 closed stirrups at 8 in. center to center.

Distribution of torsion longitudinal Bars:

$$\text{torsional } A_l = 1.96 \text{ in.}^2 \qquad \frac{A_l}{4} = 0.49 \text{ in.}^2$$

Using the same logic followed in case (a), provide five No. 8 bars at the bottom face. The area required, $A_s + A_l/4 = 3.89$ in.^{2}; the area provided = 3.95 in.^{2}. The required area at the top corners and at each vertical face = $A_l/4 = 0.49$ in.^{2}. Provide two No. 5 bars at the top and at each of the two vertical sides, giving 0.62 in.^{2} in each area. Figure 7.18 shows the geometry of the beam.

13.5 DEEP BEAMS

DEEP BM is a program that computes the necessary reinforcement and its spacing for deep beams ($a/d < 2.5$ for concentrated loading and $l_n/d < 5.0$ for uniform loading) either simply supported or continuous and either uniformly loaded or subjected to concentrated loading. The program designs both flexural and shear reinforcement and uses the same criteria as those discussed in Chapter 6. Because plane sections before bending do not necessarily remain plane after bending in deep beams, the resulting strain distribution is no longer considered as linear and shear deformations that are neglected in normal beams become significant compared to pure flexure.

While the ACI code does specify a rather simple approach to the design of shear reinforcement in deep beams, it does not specify a design procedure for flexural analysis but does require a rigorous nonlinear analysis. As in Chapter 6 for flexure, this program is based on the simplified provisions recommended by the Euro-International Concrete Committee (CEB).

By using the input values of f'_c, f_y, l_n, l, b, d, h, V_u, and M_u at $0.5a \leq d$ for concentrated loading or $0.15l_n < d$ for uniform loading, $M_{u\,max}$, and twice the area of one vertical and one horizontal shear reinforcing bar, the program computes the total area, A_s, of the flexural reinforcement required (both negative and positive for continuous beams); the vertical distance over which it should be distributed, Y_{H1} and Y_{H2} for negative moment regions and Y_{H3} for positive moment regions; the allowable concrete shear capacity, V_c, in psi; and the spacing of both the vertical and horizontal shear reinforcement.

For continuous beams, use the program with the maximum negative moment for the section at the support and then run the program with maximum positive moment for the section at or near the center of the span. The positive flexural reinforcement placed in the bottom of the beam (within Y_{H3}) should be extended through the support region.

13.5.1 Design Equations, Flowchart, and Program Steps

(Refer to Figs. 13.12, 13.13, and 13.14.)

$$\text{Simply supported:} \quad jd = 0.2(l + 2h) \qquad \text{if } l/h \geq 1.0$$

$$jd = 0.6l \qquad \text{if } l/h < 1.0$$

$$\text{Continuous:} \quad jd = 0.2(l + 1.5h) \qquad \text{if } l/h \geq 1.0$$

$$jd = 0.5l \qquad \text{if } l/h < 1.0$$

$$A_s = \frac{M_{u(max)}}{\phi f_y jd} \geq \frac{200bd}{f_y} \qquad \phi = 0.9$$

Simply supported: A_s should be placed within:

$$Y_{H3} = (0.25h - 0.05l) \leq 0.2h$$

Continuous:

$$A_{s1} = 0.5(l/h - 1)A_s \qquad \text{if } l < h \qquad A_{s1} = \frac{200bd}{f_y}$$

$$A_{s2} = (A_s - A_{s1}) \qquad \text{if } l < h \qquad A_{s2} = A_s$$

$$Y_{H1} = 0.2h \qquad\qquad\qquad Y_{H2} = 0.6h$$

If $l_n/d < 2.0$,

$$V_u \leq \phi(8\sqrt{f'_c}\, bd)$$

If $l_n/d \geq 2.0$,

$$V_u \leq \phi \left[\frac{2}{3} \left(10 + \frac{l_n}{d} \right) \sqrt{f_c'}\, bd \right]$$

$$V_c = \left(3.5 - 2.5 \frac{M_u}{V_u d} \right) \left(1.9\sqrt{f_c'} + 2500\rho_w \frac{V_u d}{M_u} \right) bd \leq 6\sqrt{f_c'}\, bd$$

$$1.0 \leq 3.5 - 2.5 \frac{M_u}{V_u d} \leq 2.5$$

$$V_s = \frac{V_u}{\phi} - V_c \qquad V_s = \left(\frac{A_v}{s_v} \frac{1 + l_n/d}{12} + \frac{A_{vh}}{s_h} \frac{11 - l_n/d}{12} \right) f_y d$$

The program solves the equation above assuming that $s_v = s_h$ and then checks the following criteria. If the following criteria result in $s_v \neq s_h$, the program does not recompute s_v and s_h. The result of having $s_v = s_h$ will be slightly conservative.

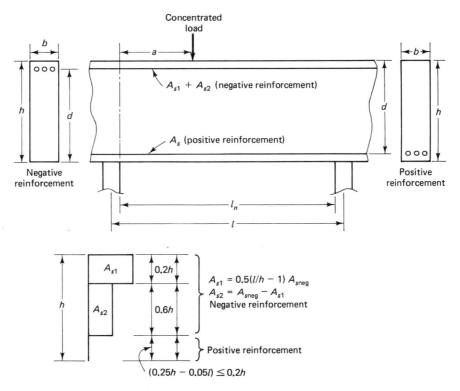

Figure 13.12 Typical continuous deep beam.

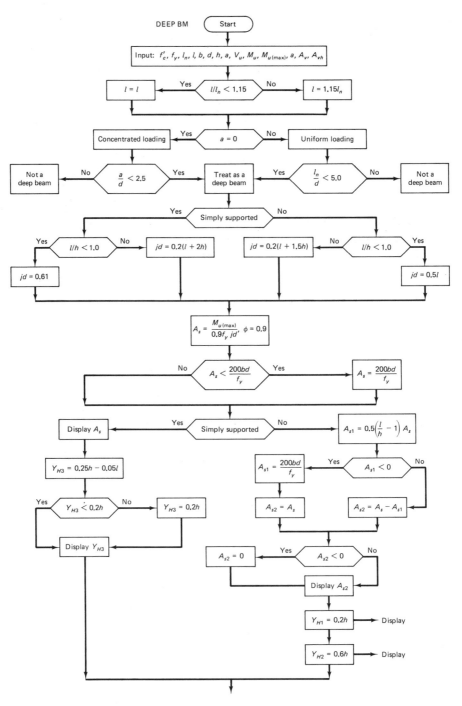

Figure 13.13   Flowchart: deep beams.

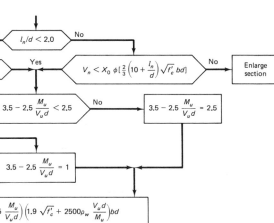

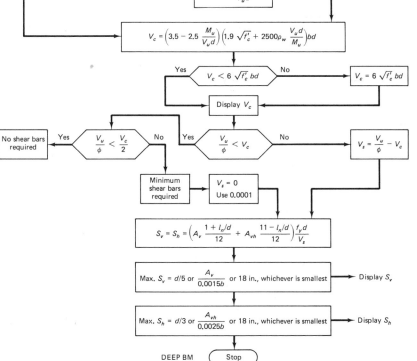

Figure 13.13 (*cont.*)

$$\text{Maximum } s_v = \frac{d}{5} \quad \text{or} \quad \frac{A_v}{0.0015b} \quad \text{or} \quad 18 \text{ in.} \qquad \text{whichever is smallest}$$

$$\text{Maximum } s_h = \frac{d}{3} \quad \text{or} \quad \frac{A_{vh}}{0.0025b} \quad \text{or} \quad 18 \text{ in.} \qquad \text{whichever is smallest}$$

The flowchart and the program steps are presented in Figs. 13.13 and 13.14, respectively.

```
01♦LBL "DEEP RM"
02 FIX 4
03 RCL 03
04 RCL 02
05 /
06 1.15
07 X>Y?
08 X<>Y
09 RCL 02
10 *
11 STO 27
12 RCL 07
13 X=0?
14 GTO 01
15 RCL 05
16 /
17 2.5
18 X<Y?
19 GTO 02
20 GTO 03
21♦LBL 01
22 RCL 02
23 RCL 05
24 /
25 5
26 X>Y?
27 GTO 03
28♦LBL 02
29 "BEAM"
30 ASTO 30
31 "NOT A DEEP "
32 ARCL 30
33 PRA
34 GTO 13
35♦LBL 03
36 RCL 13
37 X=0?
38 GTO 03
39 RCL 27
40 RCL 06
41 /
42 1
43 X>Y?
44 GTO 04
45 RCL 06
46 1.5
47 *
48 RCL 27
49 +
50 .2
```

```
51 *
52 STO 14
53 GTO 05
54♦LBL 04
55 RCL 27
56 .5
57 *
58 STO 14
59 GTO 05
60♦LBL 03
61 RCL 27
62 RCL 06
63 /
64 1
65 X>Y?
66 GTO 04
67 RCL 06
68 2
69 *
70 RCL 27
71 +
72 .2
73 *
74 STO 14
75 GTO 05
76♦LBL 04
77 RCL 27
78 .6
79 *
80 STO 14
81♦LBL 05
82 RCL 10
83 .9
84 /
85 RCL 01
86 /
87 RCL 14
88 /
89 RCL 05
90 RCL 04
91 *
92 200
93 *
94 RCL 01
95 /
96 X<=Y?
97 X<>Y
98 STO 15
99 RCL 13
100 X=0?
```

Annotations (left column):
- If $l/l_n > 1.15$, $l = 1.15 l_n$
- Conc. load: If $a/d > 2.5$, not a deep beam
- Uniform load: If $l_n/d > 5.0$, not a deep beam
- If simply supported, go to 03 (step 60)
- Continuous beams: If $l/h < 1.0$, $jd = 0.5l$; If $l/h > 1.0$, $jd = 0.2(l + 1.5h)$

Annotations (right column):
- Simply supported: If $l/h < 1.0$, $jd = 0.6l$; If $l/h \geq 1.0$, $jd = 0.2(l + 2h)$
- $A_s = M_u/\phi f_y jd$; $A_{s(min)} = (200/f_y)bd$
- If simply supported, go to 06 (step 165)

Figure 13.14 Program steps: deep beams.

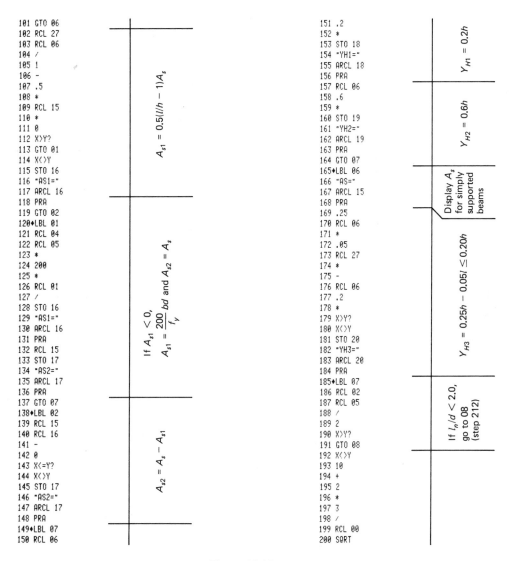

```
101 GTO 06
102 RCL 27
103 RCL 06
104 /
105 1
106 -
107 .5
108 *
109 RCL 15
110 *
111 0
112 X>Y?
113 GTO 01
114 X<>Y
115 STO 16
116 "AS1="
117 ARCL 16
118 PRA
119 GTO 02
120*LBL 01
121 RCL 04
122 RCL 05
123 *
124 200
125 *
126 RCL 01
127 /
128 STO 16
129 "AS1="
130 ARCL 16
131 PRA
132 RCL 15
133 STO 17
134 "AS2="
135 ARCL 17
136 PRA
137 GTO 07
138*LBL 02
139 RCL 15
140 RCL 16
141 -
142 0
143 X<=Y?
144 X<>Y
145 STO 17
146 "AS2="
147 ARCL 17
148 PRA
149*LBL 07
150 RCL 06
```

$A_{s1} = 0.5(l/h - 1)A_s$

If $A_{s1} < 0$,
$A_{s1} = \dfrac{200}{f_y} bd$ and $A_{s2} = A_s$

$A_{s2} = A_s - A_{s1}$

```
151 .2
152 *
153 STO 18
154 "YH1="
155 ARCL 18
156 PRA
157 RCL 06
158 .6
159 *
160 STO 19
161 "YH2="
162 ARCL 19
163 PRA
164 GTO 07
165*LBL 06
166 "AS="
167 ARCL 15
168 PRA
169 .25
170 RCL 06
171 *
172 .05
173 RCL 27
174 *
175 -
176 RCL 06
177 .2
178 *
179 X>Y?
180 X<>Y
181 STO 20
182 "YH3="
183 ARCL 20
184 PRA
185*LBL 07
186 RCL 02
187 RCL 05
188 /
189 2
190 X>Y?
191 GTO 08
192 X<>Y
193 10
194 +
195 2
196 *
197 3
198 /
199 RCL 00
200 SQRT
```

$Y_{H1} = 0.2h$

$Y_{H2} = 0.6h$

Display A_s for simply supported beams

$Y_{H3} = 0.25h - 0.05l \leq 0.20h$

If $l_n/d < 2.0$, go to 08 (step 212)

Figure 13.14 (*cont.*)

```
201 *
202 RCL 04
203 *
204 RCL 05
205 *
206 .85
207 *
208 RCL 08
209 X>Y?
210 GTO 09
211 GTO 10
212◆LBL 08
213 RCL 00
214 SQRT
215 8
216 *
217 RCL 04
218 *
219 RCL 05
220 *
221 .85
222 *
223 RCL 08
224 X<Y?
225 GTO 10
226◆LBL 09
227 "TION"
228 ASTO 30
229 "ENLARGE SEC"
230 ARCL 30
231 PRA
232 GTO 13
233◆LBL 10
234 3.5
235 ENTER↑
236 2.5
237 RCL 09
238 *
239 RCL 08
240 /
241 RCL 05
242 /
243 -
244 2.5
245 X>Y?
246 X<>Y
247 1
248 X<Y?
249 X<>Y
250 STO 21
```

If $V_u > \phi[2/3(10 + l_n/d)\sqrt{f'_c}\,bd]$, enlarge section for $l_n/d > 2.0$

If $V_u > \phi(8\sqrt{f'_c}\,bd)$ enlarge section for $l_n/d < 1.0$

If $3.5 - 2.5\dfrac{M_u}{V_u d} > 2.5$, set = 2.5

If $3.5 - 2.5\dfrac{M_u}{V_u d} < 1.0$, set = 1.0

```
251 RCL 00
252 SQRT
253 1.9
254 *
255 2500
256 RCL 15
257 *
258 RCL 04
259 /
260 RCL 08
261 *
262 RCL 09
263 /
264 +
265 RCL 04
266 *
267 RCL 05
268 *
269 RCL 21
270 *
271 STO 22
272 RCL 00
273 SQRT
274 6
275 *
276 RCL 04
277 *
278 RCL 05
279 *
280 RCL 22
281 X>Y?
282 X<>Y
283 STO 22
284 "Vc="
285 ARCL 22
286 PRA
287 RCL 08
288 .85
289 /
290 X<Y?
291 GTO 11
292 RCL 22
293 -
294 STO 23
295 GTO 12
296◆LBL 11
297 RCL 22
298 2
299 /
300 RCL 08
```

$$V_c = \left(3.5 - 2.5\,\frac{M_u}{V_u d}\right)\left(1.9\sqrt{f'_c} + 2500\rho_w\frac{V_u d}{M_u}\right)bd$$

If $V_c > 6\sqrt{f'_c}\,bd$ set $V_c = 6\sqrt{f'_c}\,bd$ display V_c

If $\dfrac{V_u}{\phi} < V_c$ Check minimum reinforcement

$$V_s = \frac{V_u}{\phi} - V_c$$

Figure 13.14 (*cont.*)

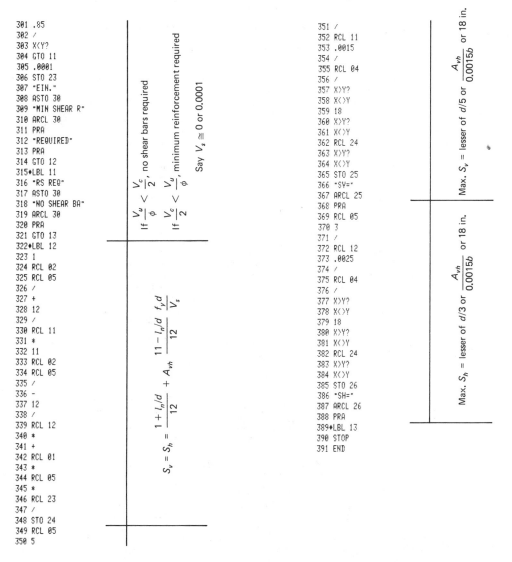

Figure 13.14 (*cont.*)

13.5.2 Instructions to Run the Program

Step 1: Set at least 30 data storage registers.

(☐XEQ☐ ☐ALPHA☐ S̲I̲Z̲E̲ ☐ALPHA☐ 0̲3̲0̲)

Step 2: Load the program from magnetic cards or enter the program steps.
Step 3: Input the following design data.

Variable	Unit	Store in register
f_c'	psi	00
f_y	psi	01
l_n	in.	02
l	in.	03
b	in.	04
d	in.	05
h	in.	06
a (store 0.0 for uniform loading)	in.	07
V_u (at critical section)	lb.	08
M_u (at critical section)	in.-lb	09
$\|M_{u\,max}\|$ (over the support for continuous beam and at or near the midspan for a simply supported beam)	in.-lb	10
A_v (twice area of one bar)	in.^{2}	11
A_{vh} (twice area of one bar)	in.^{2}	12

Simply supported beam: store 0 in 13
Continuous beam: store 1 in 13

The shear force V_u and moment M_u should be calculated at $0.5a \le d$ for concentrated loading and at $0.15l_n \le d$ for uniform loading.

Step 4: Execute DEEP BM.

(☐XEQ☐ ☐ALPHA☐ D̲E̲E̲P̲ ̲B̲M̲ ☐ALPHA☐)

For a simply supported deep beam, the program will calculate and print out the area of required flexural steel, A_s, and the region in which A_s is to be provided, Y_{H3}; the shear capacity of the plain concrete beam V_c; and the spacing of shear reinforcement in vertical s_v and horizontal s_h directions.

For continuous deep beams, the program will print out the area of needed flexural steel area A_{s1} in the region Y_{H1}, A_{s2} in the region Y_{H2}; the shear capacity of the plain concrete beam V_c; and the spacing of shear reinforcement in vertical s_v and horizontal s_h directions. For the positive moment region of a continuous deep beam, use the program as if the beam was simply supported. The complete output will consist of:

Simply supported	Continuous	Unit
A_s	A_{s1}	in.^{2}
	A_{s2}	in.^{2}
Y_{H3}	Y_{H1}	in.
	Y_{H2}	in.
V_c	V_c	lb
s_v	s_v	in.
s_h	s_h	in.

13.5.3 Numerical Examples

13.5.3.1 Example 13.13: Simply Supported Uniformly Loaded Deep Beam

Design the flexural and shear reinforcement for a deep beam given the following details:

$$f_c' = 5000 \text{ psi}$$
$$f_y = 60,000 \text{ psi}$$
$$l_n = 130 \text{ in.}$$
$$l = 155 \text{ in.}$$
$$b = 16 \text{ in.}$$
$$h = 110 \text{ in.}$$
$$V_u = 700,000 \text{ lb at } 0.15l_n$$
$$M = 11,000,000 \text{ in.-lb at } 0.15l_n$$
$$M_{u\,(\max)} = 20,000,000 \text{ in.-lb}$$

Solution

Assume that $d = 0.9h$. Try No. 3 bars for shear reinforcement. $A_v = 2 \times 0.11 = 0.22$ in.^{2}.

INPUT | OUTPUT

```
   5,000.0000 STO 00              XEQ "DEEP BM"
  60,000.0000 STO 01     AS=5.2800
     130.0000 STO 02     YH3=20.0250
     155.0000 STO 03     Vc=672,034.2848
      16.0000 STO 04     SV=8.6260
      99.0000 STO 05     SH=5.5000
     110.0000 STO 06
       0.0000 STO 07
 700,000.0000 STO 08
11,000,000.00 STO 09
20,000,000.00 STO 10
       0.2200 STO 11
       0.2200 STO 12
       0.0000 STO 13
```

Use:

Six No. 9 bars for flexure:
$A_s = 6.0$ in.^{2} > 5.28 in.^{2}
spaced at 20 in./two bars = 10 in.
(three bars on each face)

For shear use:
No. 3 bars at 8 in. vertically
No. 3 bars at 5 in. horizontally

13.5.3.2 Example 13.14: Continuous Uniformly Loaded Beam; Deep Beam

Repeat Ex. 13.13 assuming the beam to be a continuous beam.

Solution   INPUT

```
    5,000.0000 STO 00
   60,000.0000 STO 01
      130.0000 STO 02
      155.0000 STO 03
       16.0000 STO 04
       99.0000 STO 05
      110.0000 STO 06
        0.0000 STO 07
  700,000.0000 STO 08
11,000,000.00 STO 09
20,000,000.00 STO 10
        0.2200 STO 11
        0.2200 STO 12
        1.0000 STO 13
```

OUTPUT

```
        XEQ "DEEP BM"
AS1=1.0572
AS2=4.8310
YH1=22.0000
YH2=66.0000
Vc=672,034.2848
SV=8.6260
SH=5.5000
```

Shear:

Use No. 3 bars at 8 in. vertically.
Use No. 3 bars at 5 in. horizontally.

13.5.3.3 Example 13.15: Simply Supported Deep Beam with a Concentrated Load

Design the flexural and shear reinforcement for a deep beam using the following details.

$f'_c = 4000$ psi
$f_y = 60,000$ psi
$l_n = 60$ in.
$l = 70$ in.
$b = 15$ in.
$h = 50$ in.

$a = 20$ in.

V_u at $0.5a = 260,000$ lb

M_u at $0.5a = 5,700,000$ in.-lb

$M_{u\,(max)} = 10,000,000$ in.-lb.

Try No. 4 bars for shear ($A_S = 2 \times 0.20 = 0.40$ in.^{2}).

```
INPUT

    4,000.0000 STO 00
   60,000.0000 STO 01
       60.0000 STO 02
       70.0000 STO 03
       15.0000 STO 04
       45.0000 STO 05
       50.0000 STO 06
       20.0000 STO 07
  260,000.0000 STO 08
5,700,000.000 STO 09
10,000,000.00 STO 10
        0.4000 STO 11
        0.4000 STO 12
        0.0000 STO 13


OUTPUT

         XEQ "DEEP BM"
AS=5.4789
YH3=9.0500
Vc=249,262.9428
SV=9.0000
SH=10.6667
```

Flexure: Use six No. 9 bars, three on each face.
Shear:

Use No. 4 bars at 9 in. vertically.
Use No. 4 bars at 10 in. (or 9 in.) horizontally.

13.5.3.4 Example 13.16: Design of Shear Reinforcement in Deep Beams

A simply supported beam having a clear span $l_n = 10$ ft and is subjected to a uniformly distributed live load of 86,000 lb/ft on the top. The height h of the beam is 6 ft and its thickness b is 20 in. Given:

$f'_c = 4000$ psi

$f_y = 60,000$ psi

Design the shear and flexural reinforcement for this beam.

Solution

$$\text{beam self-weight} = \frac{20 \times 72}{144} \times 150 = 1500 \text{ lb/ft}$$

$$\text{total factored load} = 1.7 \times 86,000 + 1.4 \times 1500$$

$$= 148,300 \text{ lb/ft}$$

$$\text{distance of the critical section} = 0.15l_n = 0.15 \times 10.0$$

$$= 1.5 \text{ ft} = 18 \text{ in.}$$

The factored shear force V_u at the critical section is

$$V_u = \frac{148,300 \times 10}{2} - 148,300 \times \frac{18}{12} = 519,050 \text{ lb}$$

Try No. 3 bars vertically: $A_s = 2 \times 0.11 = 0.22$, and try No. 4 bars horizontally: $A_v = 2 \times 0.20 = 0.40$.

$$M_{u\,(\text{max})} = \frac{w_u l_n^2}{8} = \frac{148,300(10.0)^2}{8}$$

$$= 2,853,750 \text{ ft-lb} = 22,245,000 \text{ in.-lb}$$

```
        INPUT                        OUTPUT

     4,000.0000 STO 00                    XEQ "DEEP BM"
    60,000.0000 STO 01         AS=7.3040
       120.0000 STO 02         YH3=11.1000
       138.0000 STO 03         Vc=493,315.3150
        20.0000 STO 04         SV=7.3333
        65.0000 STO 05         SH=8.0000
        72.0000 STO 06
         0.0000 STO 07
   519,050.0000 STO 08
 11,344,950.00 STO 09
 22,245,000.00 STO 10
         0.2200 STO 11
         0.4000 STO 12
         0.0000 STO 13
```

Shear: Use No. 3 bars at 7 in. vertically and No. 4 bars at 7 in. or 8 in. horizontally.

Flexure: Use four No. 9 horizontal bars on each face, area = 8.00 in.^{2}. The height over which A_s is to be distributed above the lower beam face is

$$0.25h - 0.05l = 0.25 \times 72 - 0.05 \times 138 = 11.1 \text{ in.}$$

$$\text{spacing of flexural steel} = \frac{11.1}{3} = 3.7 \text{ in.}$$

Space four No. 9 bars at 3.5 in. center to center vertical spacing on each face of the deep beam to be well anchored into the supports. Figure 6.17a and b give, respectively, a sectional elevation and a horizontal cross section showing the details of the vertical and horizontal shear reinforcement as well as the flexural reinforcement concentrated at the lower 10.5 in. of the deep beam.

13.5.3.5 Example 13.17: Reinforcement Design for a Continuous Deep Beam

Design the reinforcement necessary for an interior span of a continuous beam over several supports if the loading and the properties of the beam are the same as those of Ex. 13.16.

Solution

Shear reinforcement

Since the deep beam has a large stiffness, the shear continuity factor for the first interior support is assumed to equal 1.0. Hence use the same vertical and horizontal shear reinforcement as that used in Ex. 13.16.

Flexural reinforcement

The approximate positive factored moment at midspan is

$$M_u = \frac{w_u l_n^2}{24} = \frac{148,300(10)^2}{24} = 617,916.7 \text{ ft-lb}$$

$$= 7,415,000 \text{ in.-lb}$$

The maximum negative factored moment at an interior span is

$$-M_u = \frac{w_u l_n^2}{12} = \frac{148,300(10^2)}{12} = 14,830,000 \text{ in.-lb}$$

Positive moment region

```
INPUT
                                     0.2200 STO 12
    4,000.0000 STO 00                0.4000 STO 12
   60,000.0000 STO 01                0.0000 STO 13
      120.0000 STO 02
      138.0000 STO 03            OUTPUT
       20.0000 STO 04
       65.0000 STO 05                XEQ "DEEP BM"
       72.0000 STO 06        AS=4.3333
        0.0000 STO 07        YH3=11.1000
   519,050.0000 STO 08       Vc=471,083.2303
11,344,950.00 STO 09         SV=7.3333
 7,415,000.000 STO 10        SH=8.0000
```

Negative moment region

```
       INPUT                          OUTPUT

    4,000.0000 STO 00                    XEQ "DEEP BM"
   60,000.0000 STO 01         AS1=2.5584
      120.0000 STO 02         AS2=3.0235
      138.0000 STO 03         YH1=14.4000
       20.0000 STO 04         YH2=43.2000
       65.0000 STO 05         Vc=493,315.3150
       72.0000 STO 06         SV=7.3333
        0.0000 STO 07         SH=8.0000
  519,050.0000 STO 08
 11,344,950.00 STO 09
 14,830,000.00 STO 10
        0.2200 STO 11
        0.4000 STO 12
        1.0000 STO 13
```

Using No. 8 bars:

Zone h_1: two No. 8 bars on each face (3.14 in.^{2} > 2.56 in.^{2})
Zone h_2: three No. 8 bars on each face (4.74 in.^{2} > 3.02 in.^{2})
Zone h_3: three No. 8 bars on each face (4.74 in.^{2} > 4.33 in.^{2})

Figure 6.18 in elevation and cross section shows the arrangement of reinforcement for this beam.

13.6 CORBELS

CORBELS is a program that computes the reinforcement required for a bracket or corbel with a shear span/depth ratio a/d not greater than unity, and subject to a horizontal tensile force N_{uc} not larger than V_u, the factored shear force. The program is based on the provisions of the ACI building code. Because of the large a/d ratio for corbels, there is a tendency for a pure shear failure to occur through essentially vertical planes; ACI recommends a shear-friction approach.

By using the input values of f_c', f_y, b_w, d, h, V_u, N_{uc}, μ, a, and λ, the program checks the allowable shear capacity. If this capacity is exceeded, the program displays "Enlarge section". It then proceeds to calculate A_s, the top horizontal reinforcement, and A_h, the horizontal reinforcement in the side faces of the bracket or corbel. The program ensures that N_{uc}, the horizontal tensile force, is greater than $0.2V_u$. If the input value is less than $0.2V_u$, the program uses $0.2V_u$.

13.6.1 Design Equations, Flowchart, and Program Steps

(Refer to Figs. 13.15, 13.16, and 13.17.)

$$V_n = \frac{V_u}{\phi} \qquad \phi = 0.85 \text{ in all cases}$$

$$V_n \geq 0.20 f_c' bd \text{ and } 800bd\text{—normalweight concrete}$$

$$V_n \geq \left(0.2 - 0.7\frac{a}{d}\right) f_c' bd \text{ and } \left(800 - 280\frac{a}{d}\right)bd\text{—lightweight and sand-light}$$
weight concrete

$$A_{vf} = \frac{V_n}{f_y \mu \lambda} \qquad A_f = \frac{V_u(a) + N_u c(h-d)}{\phi f_y jd} \qquad j \text{ assumed} \simeq 0.85$$

$$N_{uc} \geq 0.20V_u$$

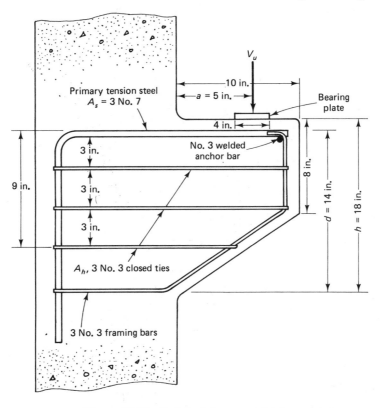

Figure 13.15 Corbel reinforcement details.

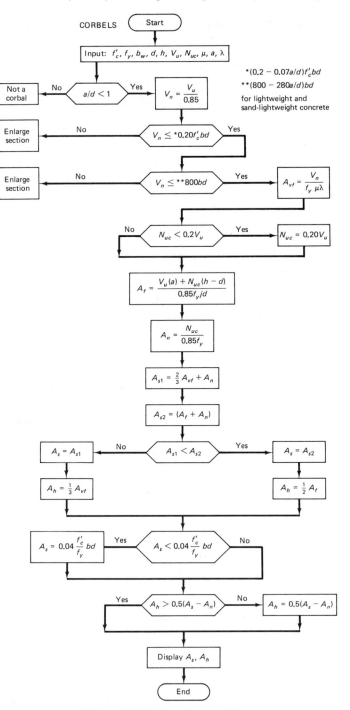

Figure 13.16 Flowchart: corbels.

```
01+LBL "CORBELS"
02 FIX 4
03 RCL 05
04 RCL 03
05 /
06 1
07 X<>Y
08 X<=Y?
09 GTO 05
10 "NO GOOD a>d"
11 PRA
12 GTO 09
13+LBL 05
14 RCL 06
15 .85
16 /
17 STO 11
18 1
19 RCL 09
20 X=Y?
21 GTO 06
22 0.07
23 RCL 05
24 *
25 RCL 03
26 /
27 CHS
28 0.2
29 +
30 RCL 00
31 *
32 RCL 02
33 *
34 RCL 03
35 *
36 RCL 11
37 X<=Y?
38 GTO 07
39 "TION"
40 ASTO 10
41 "ENLARGE SEC"
42 ARCL 10
43 PRA
44 GTO 09
45+LBL 07
46 280
47 RCL 05
48 *
49 RCL 03
50 /
```

```
51 CHS
52 800
53 +
54 RCL 02
55 *
56 RCL 03
57 *
58 RCL 11
59 X<=Y?
60 GTO 02
61 "TION"
62 ASTO 10
63 "ENLARGE SEC"
64 ARCL 10
65 PRA
66 GTO 09
67+LBL 06
68 .2
69 RCL 00
70 *
71 RCL 02
72 *
73 RCL 03
74 *
75 RCL 11
76 X<=Y?
77 GTO 01
78 "TION"
79 ASTO 10
80 "ENLARGE SEC"
81 ARCL 10
82 PRA
83 GTO 09
84+LBL 01
85 800
86 RCL 02
87 *
88 RCL 03
89 *
90 RCL 11
91 X<=Y?
92 GTO 02
93 "TION"
94 ASTO 10
95 "ENLARGE SEC"
96 ARCL 10
97 PRA
98 GTO 09
99+LBL 02
100 RCL 11
```

Annotations (left column):

If $a/d > 1$, not a corbel

$V_n = \dfrac{V_u}{\phi}$ $\phi = 0.85$

If $\lambda = 1.0$, go to 06

If $V_n > *(0.2 - 0.07\, a/d)f'_c bd$, display "enlarge section"

*For lightweight and sand-lightweight concrete

Annotations (right column):

If $V_n > *(800 - 280\, a/d)bd$, display "enlarge section"

If $V_n > 0.20 f'_c bd$, display "enlarge section"

For normal concrete

If $V_n > 800bd$, display "enlarge section"

For normal concrete

Figure 13.17 Program steps: corbels.

$$A_n = \frac{N_{uc}}{\phi f_y}$$

$$A_{s1} = \frac{2}{3}A_{vf} + A_n \qquad A_s = \text{larger of } A_{s1} \text{ or } A_{s2}$$

$$A_{s2} = A_f + A_n \qquad A_s \geq 0.04\,\frac{f'_c}{f_y}\,bd$$

```
101 RCL 01
102 /
103 RCL 08
104 /
105 RCL 09
106 X↑2
107 /
108 STO 12
109 RCL 06
110 0.2
111 *
112 RCL 07
113 X<Y?
114 X<>Y
115 STO 07
116 RCL 06
117 RCL 05
118 *
119 RCL 04
120 RCL 03
121 -
122 RCL 07
123 *
124 +
125 .85
126 /
127 RCL 01
128 /
129 .85
130 /
131 RCL 03
132 /
133 STO 13
134 RCL 07
135 .85
136 /
137 RCL 01
138 /
139 STO 14
140 RCL 12
141 2
142 *
143 3
144 /
145 RCL 14
146 +
147 STO 15
148 RCL 13
149 RCL 14
150 +
```

$$A_{vf} = \dfrac{V_n}{f_y\,\mu\lambda}$$

If $N_{uc} < 0.2V_u$, set $N_{uc} = 0.2V_u$

$$A_f = \dfrac{V_u(a) + N_{uc}(h - d)}{0.85 f_y jd}$$

Assuming $j \cong 0.85$

$$A_{vf} = \dfrac{V_n}{f_y\,\mu\lambda}$$

$$A_{s1} = \dfrac{2}{3}A_{vf} + A_n$$

$$A_{s2} = A_f + A_n$$

```
151 STO 16
152 RCL 15
153 X<=Y?
154 GTO 03
155 STO 17
156 RCL 12
157 3
158 /
159 STO 18
160 GTO 04
161◆LBL 03
162 RCL 16
163 STO 17
164 RCL 13
165 2
166 /
167 STO 18
168◆LBL 04
169 RCL 00
170 RCL 01
171 /
172 .04
173 *
174 RCL 02
175 *
176 RCL 03
177 *
178 RCL 17
179 X<=Y?
180 X<>Y
181 STO 17
182 RCL 17
183 RCL 14
184 -
185 2
186 /
187 RCL 18
188 X<=Y?
189 X<>Y
190 STO 18
191 "AS="
192 ARCL 17
193 PRA
194 "AH="
195 ARCL 18
196 PRA
197◆LBL 09
198 STOP
199 .END.
```

If $A_{s1} < A_{s2}$, $A_s = A_{s2}$
If $A_{s2} > A_{s2}$, $A_s = A_{s1}$
If $A_s = A_{s1}$, $A_h = \dfrac{A_{vf}}{3}$

If $A_s = A_{s2}$, $A_h = \dfrac{A_f}{2}$

If $A_s < 0.04\dfrac{f'_c}{f_y}bd$, set $A_s = 0.04\dfrac{f'_c}{f_y}bd$

If $A_h < 0.5(A_s - A_n)$, $A_h = 0.5(A_s - A_n)$

Display A_s and A_h

Figure 13.17 (*cont.*)

$$A_h = \frac{1}{3}A_{vf} \text{ or } \frac{1}{2}A_f \text{ depending if } A_s = A_{s1} \text{ or } A_{s2}, \text{ respectively}$$

$$A_h > 0.5(A_s - A_n)$$

The flowchart and the program steps are presented in Figs. 13.16 and 13.17, respectively.

13.6.2 Instructions to Run the Program

Step 1: Set at least 19 data storage registers.

$$(\boxed{\text{XEQ}} \quad \boxed{\text{ALPHA}} \; \underline{\text{SIZE}} \; \boxed{\text{ALPHA}} \; \underline{019})$$

Step 2: Load the program from magnetic cards or enter the program steps.

Step 3: Input the following design data.

Variable	Unit	Store in Register
f_c'	psi	00
f_y	psi	01
b	in.	02
d	in.	03
h	in.	04
a	in.	05
V_u	lb	06
N_{uc}	lb	07
μ/λ	—	08
λ	—	09

Step 4: Execute CORBELS.

$$(\boxed{\text{XEQ}} \quad \boxed{\text{ALPHA}} \; \underline{\text{CORBELS}} \; \boxed{\text{ALPHA}})$$

The program will calculate and print out the area of flexural plus tension steel, A_s (in in.^{2}), and the area of horizontal shear steel area, A_h (in in.^{2}). It is assumed that A_s and A_h are perpendicular to the face of the support.

13.6.3 Numerical Examples

13.6.3.1 Example 13.18: Design of a Bracket or Corbel

Design a corbel to support a factored vertical load $V_u = 90{,}000$ lb acting at a distance of $a = 5$ in. from the face of the column. It has a width $b = 10$ in., a total thickness $h = 18$ in., and an effective depth $d = 14$ in. Given:

$f_c' = 5000$ psi, normalweight concrete

$f_y = 60{,}000$ psi

Assume the corbel to be either cast after the supporting column was constructed or both simultaneously cast. Neglect the weight of the corbel.

Solution

Cast simultaneously	*Cast after column*
I N P U T	I N P U T

```
  5,000.0000 STO 00        5,000.0000 STO 00
 60,000.0000 STO 01       60,000.0000 STO 01
     10.0000 STO 02           10.0000 STO 02
     14.0000 STO 03           14.0000 STO 03
     18.0000 STO 04           18.0000 STO 04
      5.0000 STO 05            5.0000 STO 05
 90,000.0000 STO 06       90,000.0000 STO 06
      0.0000 STO 07            0.0000 STO 07
      1.4000 STO 08            1.0000 STO 08
      1.0000 STO 09            1.0000 STO 09
```

O U T P U T	O U T P U T

```
        XEQ "CORBELS"            XEQ "CORBELS"
 AS=1.2130               AS=1.5294
 AH=0.4301               AH=0.5882
```
                                        (Govens)

Select bar sizes:

(a) Required $A_s = 1.529$ in.^{2}; use three No. 7 bars $= 1.80$ in.^{2}.
(b) Required $A_h = 0.588$ in.^{2}; use three No. 3 closed stirrups $= 2 \times 3 \times 0.11 = 0.66$ in.^{2}.

Spread over $\frac{2}{3} d = 9.33$ in. vertical distance; hence use three No. 3 closed stirrups at $4\frac{1}{2}$ in. center to center.

Also use three framing-size No. 3 bars and one welded No. 3 anchor bar. Details of the bracket reinforcement are shown in Figs. 6.24 and 13.15. The bearing area under the load has to be checked and the bearing pad designed such that the bearing stress at the factored load V_u should not exceed $\phi(0.85 f_c' A_1)$, where A_1 is the pad area.

$$V_u = 90,000 \text{ lb} = 0.90(0.85 \times 5,000 A_1)$$

$$A_1 = \frac{90,000}{0.70 \times 0.85 \times 5,000} = 30.25 \text{ in.}^2 \ (15,176 \text{ mm}^2)$$

Use a plate 5 in. $\times$ 6 in. Its thickness has to be designed based on the manner in which V_u is applied.

13.7 RECTANGULAR COLUMNS: ANALYSIS FOR A GIVEN NEUTRAL-AXIS DEPTH c

R/CCOL8 is basically a trial-and-adjustment program that can be used to analyze rectangular columns with steel on two or four faces. This program computes the design load-moment strength of the column using strain compatibility and therefore gives accurate results. The program computes the P_u and M_u for a specific depth c of the neutral axis. The user must, by trial and adjustment, change c until the output eccentricity is equal to the eccentricity of the factored load P_u. When the two eccentricity values are approximately equal, the P_u and M_u output must be compared to the factored values. If they differ significantly, the column size and/or amount of steel should be adjusted as required until a satisfactory and economical result is obtained. The program checks to ensure that $1\% < \rho < 8\%$ as required in the ACI building code.

The user must input f_c', f_y, dimensions of the section, number of bars, the factored axial load, and the first assumed c. By executing RCCOL8, the program immediately outputs c_{bal}, the balanced eccentricity e_b, the capacity reduction factor ϕ, the balanced axial load P_{ub}, and the concentric axial load strength P_{uo}. It then proceeds to calculate the design P_u, M_u, and the eccentricity for the given c. The user must then increase the assumed c if the output eccentricity is less than the actual eccentricity, and vice versa. Since c_b, e_b, ϕ, P_{ub}, and P_{uo} are already computed, the user will save time by pressing $\boxed{\text{XEQ}}$ $\boxed{\text{ALPHA}}$ R/CCOL9 $\boxed{\text{ALPHA}}$ rather than $\boxed{\text{XEQ}}$ $\boxed{\text{ALPHA}}$ R/CCOL8 $\boxed{\text{ALPHA}}$. This will only execute the last portion of R/CCOL8 giving P_u, M_u, and the eccentricity for the new e. Once the eccentricity output is $\simeq$ the actual eccentricity, P_u can be compared to the actual factored axial load at the given eccentricity. If P_u is greater, the capacity is adequate; if not, the column must be enlarged or the amount of reinforcing steel must be increased.

As the user becomes accustomed to the program, it becomes possible to obtain the correct e after only a few trials and the time required is minimal compared to long hand computations.

The program uses a rectangular stress block (i.e., Whitney's approximation to a parabolic stress block) and the corresponding rectangular segment for each value of c. *The stress in each bar that lies in the compression zone is reduced by $0.85f_c'$ to account for the concrete that is displaced.* All moments are referenced to the centroid of the gross concrete section. The ultimate strain in the concrete is assumed to be 0.003 and the modulus of elasticity of the steel is taken as 29×10^6 psi. The program assumes that the steel on the side faces is equally spaced. Execution time is directly proportional to the number of rows of bars in the cross section since the program must calculate the strain in each row.

13.7.1 Design Equations, Flowchart, and Program Steps

(Refer to Figs. 13.18, 13.19, and 13.20.)

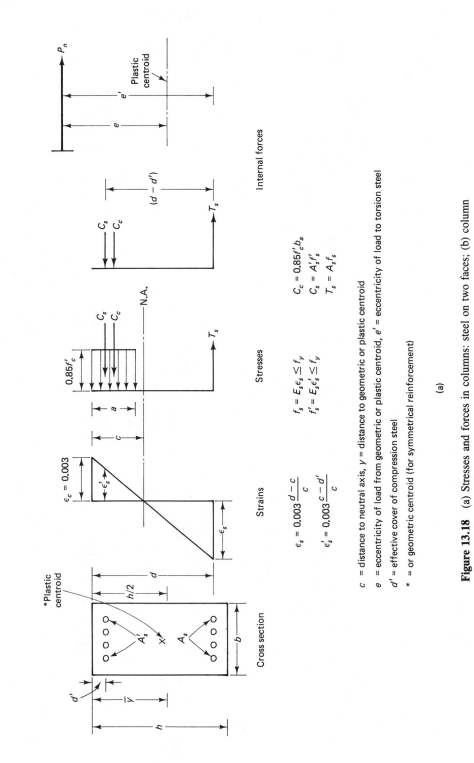

c = distance to neutral axis, y = distance to geometric or plastic centroid

e = eccentricity of load from geometric or plastic centroid, e' = eccentricity of load to torsion steel

d' = effective cover of compression steel

* = or geometric centroid (for symmetrical reinforcement)

Figure 13.18 (a) Stresses and forces in columns: steel on two faces; (b) column details: steel on four faces.

(a)

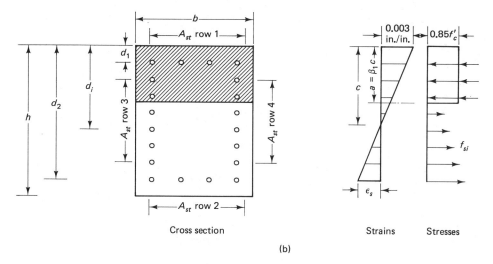

Cross section   Strains   Stresses

(b)

Figure 13.18 (*cont.*)

$$c_b = \frac{87,000}{87,000 + f_y} d_{\max}$$

$$\phi P_{nb} = 0.7 P_{nb} = 0.7(0.85 f_c' a_b b + \Sigma f_{si} A_i)$$

$$\phi M_{nb} = 0.7 M_{nb} = 0.7\left[0.85 f_c' a_b b\left(\frac{h}{2} - \frac{a_b}{2}\right) + \Sigma f_{si} A_i\left(\frac{h}{2} - d_i\right)\right]$$

$$e_b = \frac{M_{nb}}{P_{nb}}$$

$$P_{uo} = 0.7 P_{n(\max)} = (0.7)(0.8)[0.85 f_c', (A_g - A_{st}) + A_s f_y]$$

If $\phi P_{nb} > 0.1 f_c' A_g$,

$$\phi = 0.90 - \frac{2.0 P_u}{f_c' A_g} \geq 0.7$$

If $\phi P_{nb} < 0.1 f_c' A_g$,

$$\phi = 0.90 - \frac{0.20 P_u}{P_{ub}} \geq 0.7$$

$$P_u = \phi(0.85 f_c' a_b + \Sigma f_{si} A_i)$$

$$M_u = \phi\left[0.85 f_c' ab\left(\frac{h}{2} - \frac{a}{2}\right) + \Sigma f_{si} A_i\left(\frac{h}{2} - d_i\right)\right]$$

$$e = \frac{M_u}{P_u}$$

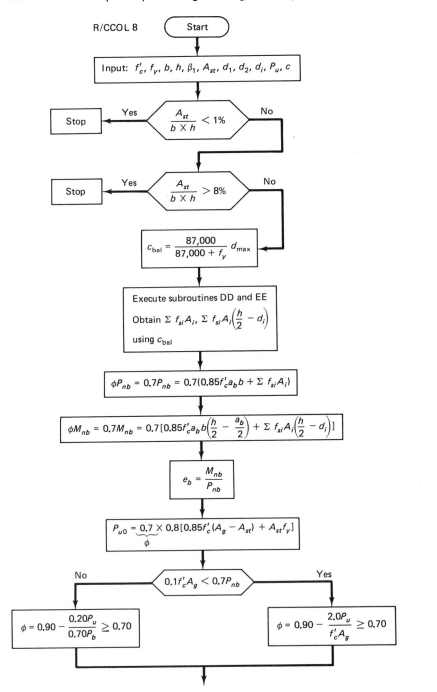

Figure 13.19   Flowchart: rectangular columns—analysis for a given c.

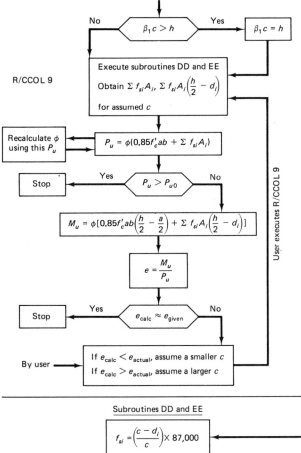

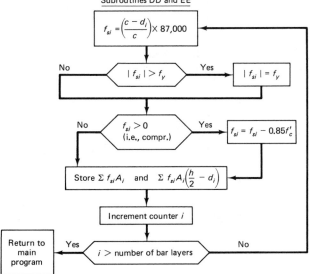

Figure 13.19 (*cont.*)

```
01♦LBL "R/CCOL8"        51 STO 15
02 FIX 4                52 XEQ "DD"
03 RCL 02               53 RCL 15
04 RCL 03               54 RCL 12
05 *                    55 RDN
06 0.01                 56 STO 12
07 *                    57 RDN
08 RCL 05               58 RDN
09 RCL 06               59 RDN
10 +                    60 STO 15
11 RCL 07               61 0.85
12 +                    62 RCL 00
13 X<Y?                 63 *
14 GTO 10               64 RCL 02
15 RCL 02               65 *
16 RCL 03               66 RCL 04
17 *                    67 *
18 0.08                 68 RCL 15
19 *                    69 *
20 X<Y?                 70 RCL 17
21 GTO 11               71 +
22 GTO 12               72 0.7
23♦LBL 10               73 *
24 "ROW GROSS<1%"       74 STO 19
25 PRA                  75 "0.7*PBAL="
26 GTO 07               76 ARCL 19
27♦LBL 11               77 PRA
28 "ROW GROSS>8%"       78 0.85
29 PRA                  79 RCL 00
30 GTO 07               80 *
31♦LBL 12               81 RCL 02
32 87000                82 *
33 RCL 01               83 RCL 04
34 +                    84 *
35 1/X                  85 RCL 15
36 87000                86 *
37 *                    87 RCL 03
38 RCL 10               88 2
39 *                    89 /
40 STO 15               90 RCL 04
41 "c BAL="             91 RCL 15
42 ARCL 15              92 *
43 PRA                  93 2
44 RCL 15               94 /
45 RCL 12               95 -
46 RDN                  96 *
47 STO 12               97 RCL 18
48 RDN                  98 +
49 RDN                  99 0.7
50 RDN                  100 *
```

Annotations (column 1):
- If $\rho_g < 1\%$, stop — not enough steel
- If $\rho_g > 8\%$, stop — excessive steel
- $c_{bal} = \dfrac{87,000}{87,000 + f_y}\, d_{max}$

Annotations (column 2):
- Temporarily exchange c_{bal} and c registers
- $0.7P_{nb} = 0.7[0.85f'_c a_b b + \Sigma f_{si}A_i]$
- $0.7M_{nb} = 0.7\left[0.85f'_c a_b\left(\dfrac{h}{2} - \dfrac{a_b}{2}\right) + \Sigma f_{si}A_i\left(\dfrac{h}{2} - d_i\right)\right]$

Figure 13.20   Program steps: rectangular columns—analysis for a given c.

$$f_{si} = \begin{cases} 87,000\,\dfrac{c - d_i}{c} \le f_y & \text{(tension zone)} \\[2ex] \left(87,000\,\dfrac{c - d_i}{c} \le f_y\right) - 0.85f'_c & \text{(compression zone)} \end{cases}$$

$$A_{st} \ge 0.01bh \qquad A_{st} \le 0.08bh$$

The flowchart and the program steps are presented in Figs. 13.19 and 13.20, respectively.

101 STO 20
102 "0.7*MBAL="
103 ARCL 20
104 PRA
105 RCL 20
106 RCL 19
107 /
108 STO 21
109 "e BAL="
110 ARCL 21
111 PRA
112 RCL 05
113 RCL 06
114 +
115 RCL 07
116 +
117 CHS
118 RCL 02
119 RCL 03
120 *
121 +
122 RCL 00
123 *
124 0.85
125 *
126 RCL 05
127 RCL 06
128 +
129 RCL 07
130 +
131 RCL 01
132 *
133 +
134 0.7
135 *
136 STO 22
137 "0.7*PZERO="
138 ARCL 22
139 PRA
140 RCL 22
141 0.8
142 *
143 STO 23
144 ".7*.8PZERO="
145 ARCL 23
146 PRA
147 0.1
148 RCL 00
149 *
150 RCL 02

$$e_b = \frac{M_{nb}}{P_{nb}}$$

$$0.7P_0 = 0.7[0.85f'_c(A_g - A_{st}) + A_{st}f_y]$$

$$P_{u0} = 0.7 \times 0.8P_0 = 0.7P_{n(max)}$$

151 *
152 RCL 03
153 *
154 RCL 19
155 X>Y?
156 GTO 00
157 0.2
158 RCL 11
159 *
160 RCL 19
161 /
162 CHS
163 0.9
164 +
165 0.7
166 X<Y?
167 X<>Y
168 STO 24
169 GTO 01
170♦LBL 00
171 2.0
172 RCL 11
173 *
174 RCL 00
175 /
176 RCL 02
177 /
178 RCL 03
179 /
180 CHS
181 0.9
182 +
183 0.7
184 X<=Y?
185 X<>Y
186 STO 24
187♦LBL 01
188 "PHI="
189 ARCL 24
190 PRA
191♦LBL "R/CCOL9"
192 RCL 04
193 RCL 12
194 *
195 RCL 03
196 X>Y?
197 GTO 02
198 RCL 04
199 /
200 STO 12

$$0.7P_b > 0.1f'_c A_g \text{ ?}$$
If yes, go to step 170
If no, go to step 157

$$\phi = 0.90 - \frac{0.2P_u}{0.7P_{nb}} \geq 0.7$$
No

$$\phi = 0.9 - \frac{2P_u}{f'_c A_g} \geq 0.70$$
Yes

Display ϕ

If $\beta_1 c > h,$
$\beta_1 c = h,$
i.e. $a > h$

Figure 13.20 (*cont.*)

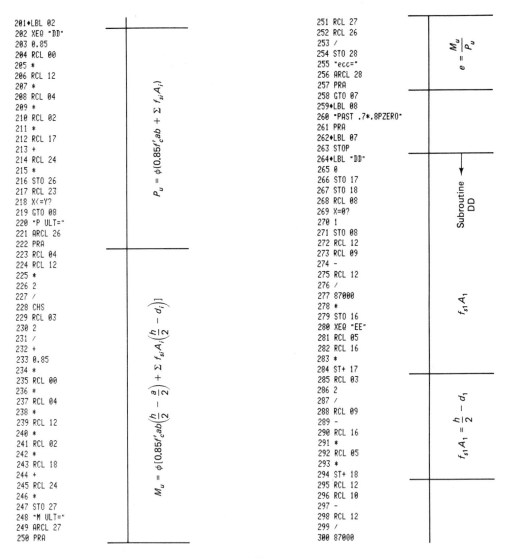

```
201◆LBL 02          251 RCL 27
202 XEQ "DD"        252 RCL 26
203 0.85            253 /
204 RCL 00          254 STO 28
205 *               255 "ecc="
206 RCL 12          256 ARCL 28
207 *               257 PRA
208 RCL 04          258 GTO 07
209 *               259◆LBL 08
210 RCL 02          260 "PAST .7*.8PZERO"
211 *               261 PRA
212 RCL 17          262◆LBL 07
213 +               263 STOP
214 RCL 24          264◆LBL "DD"
215 *               265 0
216 STO 26          266 STO 17
217 RCL 23          267 STO 18
218 X<=Y?           268 RCL 08
219 GTO 08          269 X=0?
220 "P ULT="        270 1
221 ARCL 26         271 STO 08
222 PRA             272 RCL 12
223 RCL 04          273 RCL 09
224 RCL 12          274 -
225 *               275 RCL 12
226 2               276 /
227 /               277 87000
228 CHS             278 *
229 RCL 03          279 STO 16
230 2               280 XEQ "EE"
231 /               281 RCL 05
232 +               282 RCL 16
233 0.85            283 *
234 *               284 ST+ 17
235 RCL 00          285 RCL 03
236 *               286 2
237 RCL 04          287 /
238 *               288 RCL 09
239 RCL 12          289 -
240 *               290 RCL 16
241 RCL 02          291 *
242 *               292 RCL 05
243 RCL 18          293 *
244 +               294 ST+ 18
245 RCL 24          295 RCL 12
246 *               296 RCL 10
247 STO 27          297 -
248 "M ULT="        298 RCL 12
249 ARCL 27         299 /
250 PRA             300 87000
```

$$P_u = \phi(0.85 f'_c ab + \Sigma f_{si} A_i)$$

$$M_u = \phi[0.85 f'_c ab\left(\frac{h}{2} - \frac{a}{2}\right) + \Sigma f_{si} A_i \left(\frac{h}{2} - d_i\right)]$$

$$e = \frac{M_u}{P_u}$$

Subroutine DD

$$f_{s1} A_1$$

$$f_{s1} A_1 = \frac{h}{2} - d_1$$

Figure 13.20 (*cont.*)

```
301 *
302 STO 16
303 XEQ "EE"
304 RCL 06
305 RCL 16
306 *
307 ST+ 17
308 RCL 03
309 2
310 /
311 RCL 10
312 -
313 RCL 16
314 *
315 RCL 06
316 *
317 ST+ 18
318 RCL 08
319 1000
320 /
321 01.00001
322 +
323 STO 25
324 RCL 10
325 RCL 09
326 -
327 RCL 08
328 1
329 +
330 /
331 STO 29
332◆LBL 06
333 RCL 25
334 INT
335 RCL 29
336 *
337 RCL 09
338 +
339 STO 30
340 CHS
341 RCL 12
342 +
343 RCL 12
344 /
345 87000
346 *
347 STO 16
348 XEQ "EE"
349 RCL 07
350 RCL 08
```

$f_{s2}A_2$

$f_{s2}A_2\left(\dfrac{h}{2} - d_2\right)$

Counter

Spacing of row 3 and 4 bars

d_i of row 3 and 4 bars

$f_{si} = \dfrac{c - d_i}{c} \times 87{,}000$

```
351 /
352 RCL 16
353 *
354 ST+ 17
355 RCL 03
356 2
357 /
358 RCL 30
359 -
360 RCL 16
361 *
362 RCL 07
363 *
364 RCL 08
365 /
366 ST+ 18
367 ISG 25
368 GTO 06
369 RTN
370◆LBL "EE"
371 RCL 16
372 ABS
373 RCL 01
374 X>Y?
375 GTO 03
376 RCL 16
377 ABS
378 1/X
379 RCL 16
380 *
381 RCL 01
382 *
383 STO 16
384◆LBL 03
385 0
386 RCL 16
387 X>Y?
388 GTO 04
389 GTO 05
390◆LBL 04
391 0.85
392 RCL 00
393 *
394 CHS
395 RCL 16
396 +
397 STO 16
398◆LBL 05
399 RTN
400 END
```

$f_{si}A_i$

$f_{si}A_i\left(\dfrac{h}{2} - d_i\right)$

Increment counter

Subroutine EE

Tension zone: Check if $f_{si} \leq f_y$

compr. zone: Check if $(f_{si} \leq f_y) - 0.85f'_c$ displaced conc.

Figure 13.20 (*cont.*)

13.7.2 Instructions to Run the Program

Step 1: Set at least 31 data storage registers.

(XEQ ALPHA SIZE ALPHA 031)

Step 2: Load the program from magnetic cards or enter the program steps.

Step 3: Input the following design data.

Variable	Unit	Store in Register
f'_c	psi	00
f_y	psi	01
b	in.	02
h	in.	03
β_1	—	04
A_{st} row 1 (Fig. 13.18b)	in.^{2}	05
A_{st} row 2	in.^{2}	06
A_{st} row 3 + row 4	in.^{2}	07
Number of bars in row 3 or 4	—	08
d_1 (Fig. 13.18b)	in.	09
d_2	in.	10
P_u (external load)	lb.	11
Trial c ($\leq h/\beta_1$)	in.	12

Step 4: Execute R/CCOL8.

(XEQ ALPHA R/CCOL8 ALPHA)

The program will calculate and print out c_b, $0.7P_{nb}$, $0.7M_{nb}$, e_b, the maximum permissible concentric load ($0.7 * 0.8$ PZERO), the ϕ value for the specified P_u, and the values of P_{ultimate} and M_{ultimate} and e for the trial c value.

Compare the P_{ultimate} for the trial value c with the specified external load P_u. If the specified P_u is approximately equal to the calculated P_{ultimate}, proceed to the next step. Otherwise, try other values of c. Input the next trial value of c in register 12 and execute R/CCOL9 (XEQ ALPHA R/CCOL9 ALPHA). If the specified P_u is equal to the calculated P_{ultimate}, compare the calculated eccentricity value e with the external specified eccentricity (M_u/P_u). If the calculated eccentricity is equal to or slightly higher than the external specified eccentricity, the design section is satisfactory. Otherwise, change the dimensions, reinforcement, or both and repeat the entire analysis.

The complete output consists of:

Data	Unit
c_b	in.
$P_{ub} = 0.7P_{nb}$	lb
$M_{ub} = 0.7M_{nb}$	in.-lb
e_b	in.
$0.7P_0$	lb
$P_{u0} = 0.7 \times 0.9P_0$	lb
ϕ	—
P_u } for the specified	
M_u } value of c	in.-lb
e	in.

13.7.3 Numerical Examples

13.7.3.1 Example 13.19: Analysis of a Short Rectangular Tied Column

A short tied column is reinforced with three No. 9 bars on each of the two faces parallel to the axis of bending. Calculate (a) concentric design load P_{u0}, (b) load P_{ub} and e_b corresponding to balanced failure condition, (c) P_u if $e = 14$ in., and (d) P_u if $e = 10$ in. Given:

$$f'_c = 4000 \text{ psi}$$
$$f_y = 60{,}000 \text{ psi}$$
$$b = 12 \text{ in.}$$
$$d = 17.5 \text{ in.}$$
$$d' = 2.5 \text{ in.}$$
$$h = 20 \text{ in.}$$

INPUT

```
    4,000.0000 STO 00
   60,000.0000 STO 01
       12.0000 STO 02
       20.0000 STO 07
        0.8500 STO
        3.0000 STO 05
        3.0000 STO 06
        0.0000 STO 07
        0.0000 STO 08
        2.5000 STO 09
       17.5000 STO 10
  100,000.0000 STO 11
       10.0000 STO 12
```

OUTPUT

```
          XEQ "R/CCOL8"
   c BAL=10.3571
   0.7*PBAL=244,290.0002
   0.7*MBAL=3,244,009.017
   e BAL=13.2793
   0.7*PZERO=808,920.0000
   .7*.8PZERO=647,136.0000
   PHI=0.7000
   P ULT=235,620.0000
   M ULT=3,232,320.000
   ecc=13.7184
```

(a) *First trial:* Assume that $c = 10$ in. and $P_u = 100{,}000$ lb.

```
        INPUT

    230,000.0000 STO 11
        9.8000 STO 12

        OUTPUT

            XEQ "R/CCOL8"
c BAL=10.3571
0.7*PBAL=244,290.0002
0.7*MBAL=3,244,009.017
e BAL=13.2793
0.7*PZERO=808,920.0000
.7*.8PZERO=647,136.0000
PHI=0.7000
P ULT=230,764.8000
M ULT=3,224,624.508
ecc=13.9736
```

(b) *Second trial:* Assume that $c = 9.8$ in. and $P_u = 230,000$ lb. One must execute R/COL8 again since $e = 14$ in. $> e_b$, which indicates that the column is in the tension failure zone and ϕ may not $= 0.70$.

```
        INPUT

        9.7800 STO 12

        OUTPUT

            XEQ "R/CCOL9"
P ULT=230,279.2800
M ULT=3,223,809.562
ecc=13.9996
```

(c) *Third trial:* Refine c; try $c = 9.78$ in.

13.9996 in. $\simeq$ 14 in., say; O.K. Therefore, $P_u = 230,279$ lb. for $e = 14$ in.

(d) Since $e = 10$ in. is $< e_b$, failure will be in compression. There is no need to reinput P_u in register 11, which is used to calculate ϕ since for compression failure $\phi = 0.70$. $\phi = 0.70$ was used in the previous computation and is still in storage.

Trial 1: $c = 11$ in.

```
        INPUT

        11.0000 STO 12

        OUTPUT

            XEQ "R/CCOL9"
P ULT=277,936.9091
M ULT=3,123,109.882
ecc=11.2368
```

Trial 2: $c = 11.5$ in.

```
        INPUT

        11.5000 STO 12

        OUTPUT

            XEQ "R/CCOL9"
P ULT=302,712.2609
M ULT=3,033,640.119
ecc=10.0215
```

Trial 3: c = 11.51 in.

INPUT

11.5100 STO 12

OUTPUT

XEQ "R/CCOL9"
P ULT=303,196.5689
M ULT=3,031,882.098 $e \simeq 10$ in. Conclude that
ecc=9.9997 $P_u = 303,197$ lb.

13.7.3.2 Example 13.20: Analysis of a Column Controlled by Tension Failure; Stress in Compression Steel Less Than Yield Strength

A short rectangular reinforced concrete column is 12 in. × 15 in. and is reinforced with three No. 9 bars on each of the two faces parallel to the axis of bending. Calculate the ultimate design load $P_u = \phi P_n$ if the eccentricity $e = 12$ in. Given:

$f'_c = 4000$ psi
$f_y = 60,000$ psi
$d' = 2.5$ in.

INPUT   OUTPUT

4,000.0000 STO 00   XEQ "R/CCOL8"
60,000.0000 STO 01   c BAL=7.3980
   12.0000 STO 02   0.7*PBAL=167,412.8572
   15.0000 STO 03   0.7*MBAL=1,981,382.662
    0.8500 STO 04   e BAL=11.8353   *Trial 1:*
    3.0000 STO 05   0.7*PZERO=666,120.0000   *c* = 10 in.
    3.0000 STO 06   .7*.8PZERO=532,896.0000 $P_u = 150,000$
    0.0000 STO 07   PHI=0.7000
    0.0000 STO 08   P ULT=315,945.0000
    2.5000 STO 09   M ULT=1,611,645.000
   12.5000 STO 10   ecc=5.1010
150,000.0000 STO 11
   10.0000 STO 12

INPUT

6.5000 STO 12

OUTPUT

XEQ "R/CCOL8"
c BAL=7.3980
0.7*PBAL=167,412.8572
0.7*MBAL=1,981,382.662
e BAL=11.8353
0.7*PZERO=666,120.0000
.7*.8PZERO=532,896.0000
PHI=0.7000
P ULT=137,084.7692
M ULT=1,904,002.921
ecc=13.8892

Trial 2:
c = 6.5 in.

INPUT

7.2000 STO 12

OUTPUT

XEQ "R/CCOL9"
P ULT=160,909.7000
M ULT=1,966,667.668
ecc=12.2222

Trial 3
c = 7.2 in.

INPUT

7.3000 STO 12

OUTPUT

XEQ "R/CCOL9"
P ULT=164,206.3069
M ULT=1,974,259.617
ecc=12.0230

Trial 4:
c = 7.3 in.

INPUT

7.3100 STO 12

OUTPUT

XEQ "R/CCOL9"
P ULT=164,534.6599
M ULT=1,975,000.925
ecc=12.0036

Trial 5:
c = 7.31 in.

Conclude that $P_u = 164,535$ lb for $e = 12$ in.
$P_u = 164,535$ lb $< \phi P_{nb}$;
therefore, failure will be in
tension, or alternatively, $c = 7.31$ in. $< c_b$.

13.7.3.3 Example 13.21: Design of a Column with Large Eccentricity; Initial Tension Failure

A tied reinforced concrete column is subjected to a service axial force due to dead load = 65,000 lb and a service axial force due to live load = 125,000 lb. Eccentricity to the plastic and geometric centroid is $e = 16$ in.

Design the longitudinal and lateral reinforcement for this column, assuming a nonslender column with a total reinforcement ratio between 2 and 3%. Given:

f'_c = 4000 psi, normalweight concrete
f_y = 60,000 psi

Solution

Calculate the factored external load and moment.

$$P_u = 1.4D + 1.7L = 1.4 \times 65,000 + 1.7 \times 125,000 = 303,500 \text{ lb}$$

$$P_u e = 303,500 \times 16 = 4,856,000 \text{ in.-lb}$$

Assume a section 20 in. × 20 in. and a total reinforcement ratio of 3%

Assume that $\rho = \rho' = A_s/bd = 0.015$ and $d' = 2.5$ in.

$$A_s = A'_s = 0.015 \times 20(20 - 2.5) = 5.25 \text{ in.}^2$$

Try five No. 9 bars—5.00 in.^{2} on each face (3225 mm^2).

$$\rho = \frac{5.00}{20 \times 17.5} = 0.0143$$

Trial 1:
$c = 9$ in.

INPUT   OUTPUT

```
    4,000.0000 STO 00              XEQ "R/CCOLS"
   60,000.0000 STO 01        c BAL=10.3571
       20.0000 STO 02        0.7*PBAL=407,150.0003
       20.0000 STO 03        0.7*MBAL=5,406,681.696
        0.8500 STO 04        ε BAL=13.2793
        5.0000 STO 05        0.7*PZERO=1,348,200.000
        5.0000 STO 06        .7*.8PZERO=1,078,560.000
        0.0000 STO 07        PHI=0.7000
        0.0000 STO 08        P ULT=352,239.9999
        2.5000 STO 09        M ULT=5,309,314.500
       17.5000 STO 10        ecc=15.0730
  303,500.0000 STO 11
        9.0000 STO 12
```

Trial 2:	*Trial 3:*
c = 8.5 in.	c = 8.4 in.
INPUT	INPUT

8.5000 STO 12	8.4000 STO 12

OUTPUT   OUTPUT

XEQ "R/CCOL8"   XEQ "R/CCOL9"
```
c BAL=10.3571                    P ULT=327,964.0000
0.7*PBAL=407,150.0003            M ULT=5,246,075.520
0.7*MBAL=5,406,681.696           ecc=15.9959
e BAL=13.2793
0.7*PZERO=1,348,200.000
.7*.8PZERO=1,078,560.000
PHI=0.7000
P ULT=332,010.0001
M ULT=5,257,475.125
ecc=15.8353
```

15.9959 in. $\simeq$ 16.00 in., say; O.K.
P_u = 327,964 lb > 303,500 lb
Therefore, adopt section 20 in. $\times$ 20 in.
with five No. 9 bars on each face
$\phi P_{nb} > P_u$. Therefore, tension failure.

13.7.3.4 Example 13.22: Design of a Column with Small Eccentricity; Initial Compression Failure

A nonslender column is subjected to a factored P_u = 365,000 lb (1620 kN) and a factored M_u = 1,640,000 in.-lb (185 kN-m). Assume that the gross reinforcement ratio ρ_g = 1.5 to 2% and that the effective cover to the center of the longitudinal steel is d' = $2\frac{1}{2}$ in. (63.5 mm). Design the column section and the necessary longitudinal and transverse reinforcement. Given:

f'_c = 4500 psi (31.03 MPa), normalweight concrete
f_y = 60,000 psi (414 MPa)

Solution

Calculation of factored design loads

$$P_u = 365,000 \text{ lb} \qquad e = \frac{1,640,000}{365,000} = 4.5 \text{ in. (114 mm)}$$

Assume a 15 in. $\times$ 15 in. (d = 12.5 in.) section

Assume a reinforcement ratio $\rho = \rho' = 0.01$.

$$A_s = A'_s \simeq 0.01 \times 15 \times 15 = 4.5 \text{ in.}^2$$

Provide two No. 9 bars on each side.

$$A_s = A'_s = 2.0 \text{ in.}^2 \text{ (1290 mm}^2\text{)}$$

Trial 1:
$c = 9$ in.

INPUT	OUTPUT
4,500.0000 STO 00	XEQ "R/CCOL8"
60,000.0000 STO 01	c BAL=7.3980
15.0000 STO 02	0.7*PBAL=236,409.4420
15.0000 STO 03	0.7*MBAL=1,886,822.310
0.8250 STO 04	e BAL=7.9812
2.0000 STO 05	0.7*PZERO=759,727.5000
2.0000 STO 06	.7*.8PZERO=607,782.0000
0.0000 STO 07	PHI=0.7000
0.0000 STO 08	P ULT=329,484.8958
2.5000 STO 09	M ULT=1,759,515.689
12.5000 STO 10	ecc=5.3402
365,000.0000 STO 11	
9.0000 STO 12	

Trial 2:
$c = 10$ in.

INPUT

10.0000 STO 12

OUTPUT

XEQ "R/CCOL9"
P ULT=379,535.6250
M ULT=1,663,749.610
ecc=4.3836

Trial 3:
$c = 9.8$ in.

INPUT

9.8000 STO 12

OUTPUT

XEQ "R/CCOL9"
P ULT=369,801.6697
M ULT=1,683,708.722
ecc=4.5530

Trial 4:
$c = 9.85$ in.

INPUT

9.8500 STO 12

OUTPUT

XEQ "R/CCOL9"
P ULT=372,246.9878
M ULT=1,678,762.307
ecc=4.5098

$\phi P_{nb} < P_u$. Therefore, compression failure.
4.5098 in. $\simeq$ 4.50 in., say; O.K.
$P_u = 372,247$ lb $> 365,000$ lb
Therefore, adopt 15 in. $\times$ 15 in. section with
two No. 9 bars on each face.

13.7.3.5 Example 13.23: Construction of a Load–Moment Design Strength Interaction Diagram for a Rectangular Column

Construct a load–moment design strength interaction diagram for a rectangular column having the following properties:

$f'_c = 6000$ psi

$f_y = 60,000$ psi

$b = 12$ in.

$h = 14$ in.

$d' = 3$ in.

$A_s = A'_s = 3.12$ in.^{2} (two No. 11 bars)

Solution

This problem can be solved using the R/CCOL8 program. The program automatically finds the balanced failure point (ϕP_{nb}, ϕM_{nb}) as well as the maximum concentric load $P_{u0} = \phi P_{n(max)} = (0.7 \times 0.8 P_0)$. To find enough points to construct the entire interaction diagram, one has only to vary the value of c in register 12. For points above the balanced point, it does not matter what value of P_u is placed in register 11 as long as it is greater than ϕP_{nb} since this value is used to compute ϕ and one knows that ϕ is always equal to 0.7 for a rectangular column in the compression failure zone. However, in the tension zone, ϕ will vary from 0.7 up to 0.9 at M_{u0}. Therefore, for a given value of c in the tension zone, one must revise the value in register 11 until it approximately equals P_u output for that c.

As a series of coordinates of load–moment design strength values of P_u–M_u are obtained, an interaction diagram for each f'_c/f_y and A_s/A'_s combination can be constructed. This example gives four coordinate runs for $c = 10$ in., 7 in., 6 in., and 4 in.. Other points can be similarly obtained and a series of P_u–M_u or P_n–M_n interaction curves can be constructed similar to the plot in Fig. 9.19.

First run:

Try $P_u = 200,000$ lb.

$c = 10$ in.

INPUT	OUTPUT
	XEQ "R/CCOL8"
6,000.0000 STO 00	c BAL=6.5102
60,000.0000 STO 01	0.7*PBAL=169,443.9117
12.0000 STO 02	0.7*MBAL=1,842,954.973
14.0000 STO 03	e BAL=10.8765
0.7500 STO 04	0.7*PZERO=839,563.2000
3.1200 STO 05	.7*.8PZERO=671,650.5600
3.1200 STO 06	PHI=0.7000
0.0000 STO 07	P ULT=422,200.8000
0.0000 STO 08	M ULT=1,599,834.600
3.0000 STO 09	ecc=3.7893
11.0000 STO 10	
200,000.0000 STO 11	
10.0000 STO 12	

Second run (compression failure zone):

Leave $P_u = 200,000$ lb.
Try $c = 7$ in. $> C_{bal} = 6.51$ in.
Use the R/CCOL9 portion of the program to
find P_{u2} and M_{u2}.

INPUT

7.0000 STO 12

OUTPUT

XEQ "R/CCOL9"
P ULT=213,771.6000
M ULT=1,808,035.450
ecc=8.4578

$P_{u2} = 213,772$ lb
$M_{u2} = 1,808,036$ in.-lb

$P_{u0} = 671,651$ lb $P_{u1} = 422,210$ lb

$P_{ub} = 169,444$ lb $M_{u1} = 1,599,835$ in.-lb

$M_{ub} = 1,842,955$ in.-lb $c = 10$ in. $> c_{bal}$

Therefore, compression failure zone.

Third Run (tension failure zone):

Try $P_u = 140,000$ lb.
$c = 6.0$ in. $< c_{bal} = 6.57$ in.

INPUT

140,000.0000 STO 11
6.0000 STO 12

OUTPUT

XEQ "R/CCOL8"
c BAL=6.5102
0.7*PBAL=169,443.9117
0.7*MBAL=1,842,954.973
e BAL=10.8765
0.7*PZERO=839,563.2000
.7*.8PZERO=671,650.5600
PHI=0.7000
P ULT=145,605.6000
M ULT=1,775,327.400
ecc=12.1927

$\phi = 0.70$ and register 11 is less
than $P_u = 145,606$. Therefore,
revising register 11 will not change
ϕ and therefore not P_{u3} and M_{u3}.

$P_{u3} = 145,606$ lb
$M_{u3} = 1,775,327$ in.-lb

$39,700 \simeq 39,704$ lb, say; O.K.

Fourth Run (tension failure zone):

Try $P_u = 50,000$ lb.
$c = 4.0$ in. $< c_{bal} = 6.51$ in.

INPUT

50,000.0000 STO 11
4.0000 STO 12

OUTPUT

XEQ "R/CCOL9"
c BAL=6.5102
0.7*PBAL=169,443.9117
0.7*MBAL=1,842,954.973
e BAL=10.8765
0.7*PZERO=839,563.2000
.7*.8PZERO=671,650.5600
PHI=0.8008
P ULT=38,716.7714
M ULT=1,574,674.229
ecc=40.6716

$\phi > 0.70$. Revise P_u.

Try 39,700.

```
INPUT

39,700.0000 STO 11

OUTPUT

                 XEQ "R/CCOL8"
c BAL=6.5102
0.7*PBAL=169,443.9117
0.7*MBAL=1,842,954.973
e BAL=10.8765
0.7*PZERO=839,563.2000
.7*.8PZERO=671,650.5600
PHI=0.8212
P ULT=39,704.8357
M ULT=1,614,860.414
ecc=40.6716
```

$P_{u4} = 39{,}704$ lb
$M_{u4} = 1{,}614{,}860$ in.-lb

Additional points can be found in a similar manner.

13.8 CIRCULAR COLUMNS: ANALYSIS FOR A GIVEN NEUTRAL-AXIS DEPTH c

R/CCOL5 is basically a trial-and-adjustment program that can be used to analyze circular columns. This program functions similar to R/CCOL8 in that it computes the design load-moment strength of the column using strain compatibility and therefore gives accurate results. It computes the P_u and M_u for a specific depth c of the neutral axis used in the first input cycle. The user must change c until the eccentricity output is equal to the eccentricity of the factored load P_u. When the two eccentricity values are approximately equal, the P_u and M_u output must be compared to the factored values. If they differ significantly, the column size and/or amount of steel should be adjusted as required until a satisfactory and economical result is obtained. The program will check to ensure that $1\% < \rho < 8\%$.

The user must input f'_c, f_y, h, D_s, β_1, $A_{st\ total}$, number of bars, d_{max}, the factored axial load, and the first assumed c. By executing R/CCOL5 the program first outputs c_b, the balanced eccentricity e_b, the capacity reduction factor ϕ, the balanced axial load P_{ub}, and the concentric axial load capacity P_{uo}. It then proceeds to calculate P_u, M_u, and the eccentricity for the given c. The user must then increase the assumed c if the eccentricity is less than the actual eccentricity, and vice versa. Since c_b, e_b, ϕ, P_{ub}, and

P_{uo} are already computed, the user saves time by pressing $\boxed{\text{XEQ}}$ $\boxed{\text{ALPHA}}$ R/CCOL6 $\boxed{\text{ALPHA}}$ rather than $\boxed{\text{XEQ}}$ $\boxed{\text{ALPHA}}$ R/CCOL5 $\boxed{\text{ALPHA}}$. This will only execute the last portion of R/CCOL5 giving P_u, M_u, and the eccentricity for the new c. Once the eccentricity output is $\approx$ the actual eccentricity, P_u can be compared to the actual factored axial load at the given eccentricity. If P_u is greater, the capacity is adequate; if not, the column must be enlarged or the amount of reinforcing steel must be increased. As the user becomes accustomed to the program it becomes possible to obtain the correct e after only a few trials.

The program uses a rectangular stress block (i.e., Whitney's approximation to a parabolic stress block) and the corresponding segment of a circle for each value of c. The stress in each bar that lies in the compression zone is reduced by $0.85f'_c$ to account for the concrete that is displaced. All moments are referenced to the centroid of the gross concrete section. The ultimate strain in the concrete is assumed to be 0.003 and the modulus of elasticity of the steel is taken as 29×10^6 psi. The bars are assumed to be equally spaced. Execution time is directly proportional to the number of bars in the cross section since the program must calculate the strain in each bar.

13.8.1 Design Equations, Flowchart, and Program Steps

(Refer to Figs. 13.21, 13.22, and 13.23.)

$$c_{\text{bal}} = \frac{87,000}{87,000 + f_y} d_{\max} \qquad \alpha = \cos^{-1} \frac{h/2 - \beta_1 c}{h/2}$$

$$\text{compression area} = h^2 \frac{\alpha(\pi/180) - \sin\alpha\cos\alpha}{4}$$

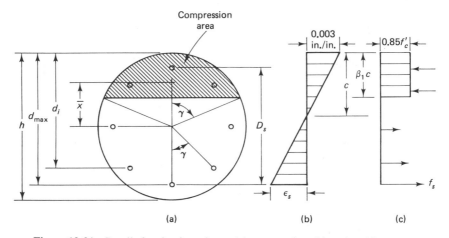

Figure 13.21   Details for circular column: (a) cross section; (b) strains; (c) stresses.

$$\bar{x} = \frac{h^3 \sin^3 \alpha}{12 \times \text{compression area}} \qquad d_i = \frac{h}{2} - \frac{D_s}{2} \cos[(\text{bar no. } i)(\gamma) - \gamma]$$

$$f_{si} = \begin{cases} \dfrac{c - di}{c} \times 87{,}000 \leq f_y & \text{tension zone} \\[3mm] \dfrac{c - di}{c} \times 87{,}000 \leq f_y - 0.85 f_c' & \text{compression zone} \end{cases}$$

$$\rho_{\min} = 1\% \qquad \rho_{\max} = 8\%$$

$$P_{n\text{bal}} = 0.85 f_c' \,(\text{compression area}) + \Sigma f_{si} A_i \text{ for } C_{\text{bal}}$$

$$M_{n\text{bal}} = 0.85 f_c' \,(\text{compression area})\,(\bar{x}) + \Sigma f_{si} A_i \left(\frac{h}{2} - d_i\right) \text{ for } c_{\text{bal}}$$

$$e_b = \frac{M_b}{P_b}$$

$$\text{If } 0.1 f_c' A_g < \left. \begin{array}{l} 0.7 P_{nb} \text{ (tied)} \\ 0.75 P_{nb} \text{ (spiral)} \end{array} \right\} \qquad \begin{array}{l} \phi = 0.90 - \dfrac{1.5 P_u}{f_c' A_g} \geq 0.75 \text{ (spiral)} \\[4mm] \phi = 0.90 - \dfrac{2.0 P_u}{f_c' A_g} \geq 0.70 \text{ (tied)} \end{array}$$

$$\text{If } 0.1 f_c' A_g > \left. \begin{array}{l} 0.7 P_{nb} \text{ (tied)} \\ 0.75 P_{nb} \text{ (spiral)} \end{array} \right\} \qquad \begin{array}{l} \phi = 0.90 - \dfrac{0.15 P_u}{0.75 P_{nb}} \geq 0.75 \text{ (spiral)} \\[4mm] \phi = 0.90 - \dfrac{0.20 P_u}{0.70 P_{nb}} \geq 0.70 \text{ (tied)} \end{array}$$

$$P_{u\max} = \begin{cases} P_{uo} = 0.7 \times 0.8[0.85 f_c'(A_g - A_{st}) + A_s f_y] & \text{tied} \\ P_{uo} = 0.75 \times 0.85[0.85 f_c'(A_g - A_{st}) + A_s f_y] & \text{spiral} \end{cases}$$

$$P_u = \phi[0.85 f_c' \,(\text{compression area}) + \Sigma f_{si} A_i]$$

$$M_u = \phi\left[0.85 f_c' \,(\text{compression area})\,\bar{x} + \Sigma f_{si} A_i \left(\frac{h}{2} - d_i\right)\right]$$

$$e = \frac{M_u}{P_u}$$

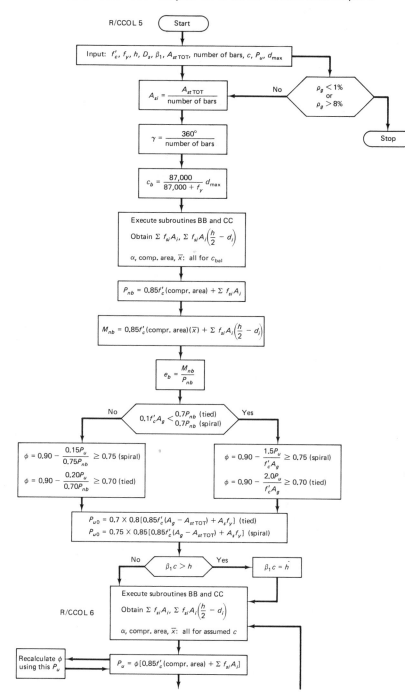

Figure 13.22 Flowchart: circular columns—analysis for a given c.

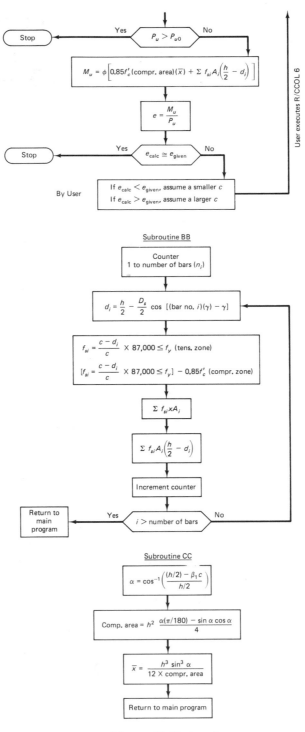

Figure 13.22 *(cont.)*

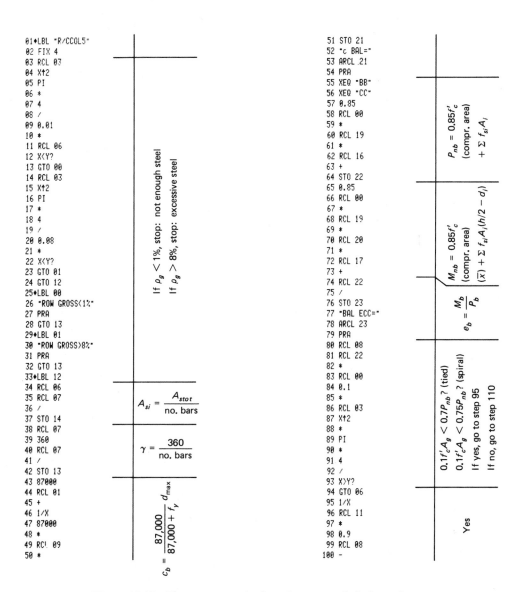

Figure 13.23   Program steps: circular columns—analysis for a given c.

```
101 *
102 CHS
103 0.9
104 +
105 RCL 08
106 X<=Y?
107 GTO 07
108 X<>Y
109 GTO 07
110*LBL 06
111 0.9
112 RCL 08
113 -
114 RCL 11
115 *
116 RCL 08
117 /
118 RCL 22
119 /
120 CHS
121 0.9
122 +
123 RCL 08
124 X<=Y?
125 GTO 07
126 X<>Y
127*LBL 07
128 X<>Y
129 STO 25
130 "PHI="
131 ARCL 25
132 PRA
133 RCL 08
134 RCL 22
135 *
136 STO 22
137 "PULT BAL="
138 ARCL 22
139 PRA
140 RCL 03
141 X↑2
142 PI
143 *
144 4
145 /
146 RCL 06
147 -
148 0.85
149 *
150 RCL 00
```

$\phi = 0.90 - \dfrac{2.0P_u}{f'_c A_g}$ ≥ 0.70 (tied)

$\phi = 0.90 - \dfrac{1.5P_u}{f'_c A_g}$ ≥ 0.75 (spiral)

$\phi = 0.90 - \dfrac{0.15P_u}{0.75P_b} \geq 0.75$ (spiral)

$\phi = 0.90 - \dfrac{0.20P_u}{0.70P_b} \geq 0.70$ (tied)

No

Display ϕ

Display 0.70P_b (tied)
Display 0.75P_b (spiral)

```
151 *
152 RCL 06
153 RCL 01
154 *
155 +
156 RCL 12
157 *
158 RCL 08
159 *
160 STO 27
161 "PULT ZERO="
162 ARCL 27
163 PRA
164*LBL "R/CCOL6"
165 ADV
166 RCL 10
167 STO 21
168 RCL 05
169 *
170 RCL 03
171 X<=Y?
172 GTO 10
173 GTO 11
174*LBL 10
175 RCL 03
176 RCL 05
177 /
178 STO 21
179*LBL 11
180 XEQ "BB"
181 XEQ "CC"
182 0.85
183 RCL 00
184 *
185 RCL 19
186 *
187 RCL 16
188 +
189 RCL 25
190 *
191 STO 28
192 RCL 27
193 X>Y?
194 GTO 08
195 GTO 09
196*LBL 08
197 "P ULT="
198 ARCL 28
199 PRA
200 0.85
```

Tied: $P_{uO} = 0.7 \times 0.8[0.85f'_c$
$(A_g - A_{st}) + A_s f_y]$
Spiral: $P_{uO} = 0.75 \times 0.85[0.85f'_c$
$(A_g - A_{st}) + A_s f_y]$

If $\beta_1 c > h$,
set $\beta_1 c = h$
i.e. $a > h$

$P_u = \phi\{0.85f'_c(\text{compr. area}) + \Sigma f_{si}A_i\}$
If $P_u > P_{uO}$, stop

Figure 13.23 (cont.)

201 RCL 00		
202 *		
203 RCL 19		
204 *		
205 RCL 20		
206 *		
207 RCL 17		$M_u = \phi[0.85f'_c(\text{comp. area})(\overline{x})$
208 +		$+ \Sigma f_{si}A_i\left(\frac{h}{2} - d_i\right)]$
209 RCL 25		
210 *		
211 STO 29		
212 "M ULT="		
213 ARCL 29		
214 PRA		
215 RCL 29		
216 RCL 28		
217 /		$e = \dfrac{M_u}{P_u}$
218 STO 30		
219 "ecc="		
220 ARCL 30		
221 PRA		
222 GTO 13		
223♦LBL 09		
224 "PAST PUO N.G."		Display if
225 PRA		$P_u > P_{uO}$
226♦LBL 13		
227 STOP		
228♦LBL "BB"		
229 0		
230 STO 16		
231 STO 17		
232 RCL 07		
233 1000		
234 /		Counter
235 01.00001		
236 +		
237 STO 26		
238♦LBL 02		
239 RCL 26		
240 INT		
241 RCL 13		
242 *		
243 RCL 13		
244 -		$d_i = \dfrac{h}{2} - \dfrac{D_s}{2}\cos(i\gamma - \gamma)$
245 COS		
246 RCL 04		
247 2		
248 /		
249 *		
250 CHS		

251 RCL 03		
252 2		
253 /		
254 +		
255 STO 24		
256 RCL 21		
257 RCL 24		
258 -		
259 RCL 21		
260 /		
261 87000		
262 *		
263 STO 15		
264 ABS		
265 RCL 01		
266 X>Y?		
267 GTO 03		
268 RCL 15		Tension zone: $f_{si} = \dfrac{c - d_i}{c} \times 87{,}000 \le f_y$
269 ABS		
270 1/X		
271 RCL 15		
272 *		
273 RCL 01		
274 *		
275 STO 15		
276♦LBL 03		
277 0		
278 ENTER↑		
279 RCL 15		
280 X>Y?		
281 GTO 04		
282 GTO 05		
283♦LBL 04		Compr. zone: $f_{si} = \left(\dfrac{c - d_i}{c} \times 87{,}000 \le f_y\right) - 0.85f'_c$
284 0.85		
285 RCL 00		
286 *		
287 CHS		
288 RCL 15		
289 +		
290 STO 15		
291♦LBL 05		
292 RCL 15		
293 RCL 14		
294 *		
295 ST+ 16		$\Sigma f_{si}A_i$
296 RCL 03		
297 2		
298 /		
299 RCL 24		
300 -		

Figure 13.23   (*cont.*)

```
301 RCL 15
302 *
303 RCL 14
304 *
305 ST+ 17
306 ISG 26
307 GTO 02
308 RTN
309*LBL "CC"
310 RCL 03
311 2
312 /
313 RCL 05
314 RCL 21
315 *
316 -
317 RCL 03
318 /
319 2
320 *
321 ACOS
322 STO 18
323 PI
324 *
325 180
326 /
327 RCL 18
328 SIN
329 RCL 18
330 COS
331 *
332 -
333 4
334 /
335 RCL 03
336 X↑2
337 *
338 STO 19
339 RCL 18
340 SIN
341 ENTER↑
342 3
343 Y↑X
344 ENTER↑
345 RCL 03
346 ENTER↑
347 3
348 Y↑X
349 *
350 12
```

```
351 /
352 RCL 19
353 /
354 STO 20
355 RTN
356 END
```

$$\Sigma f_{si} A_i \frac{h}{2} - d_i$$

Increment counter

$$\alpha = \cos^{-1} \frac{(h/2) - \beta_1 c}{h/2}$$

$$\text{Compr. area} = h^2 \frac{(\pi/180)\,\alpha - \sin\alpha\cos\alpha}{4}$$

$$\bar{x} = \frac{h^3 \sin^3\alpha}{12\ \text{compr. area}}$$

Figure 13.23   (*cont.*)

13.8.2 Instructions to Run the Program

Step 1: Set at least 31 data storage registers.

(XEQ   ALPHA S I Z E ALPHA 0 3 1)

Step 2: Load the program from magnetic cards or enter the program steps.
Step 3: Input the following design data.

Variable	Unit	Store in register
f_c'	psi	00
f_y	psi	01
h	in.	03
D_s	in.	04
β_1	—	05
A_{st} total	in.^{2}	06
Number of bars	—	07
Store 0.70 for tied column and 0.75 for spiral column	—	08
d_{max} (Fig. 13.21)	in.	09
Trial c	in.	10
P_u (external load)	lb	11
0.8 for tied column and 0.85 for spiral column	—	12

Step 4: Execute R/CCOL5.

(XEQ   ALPHA R / C C O L 5 ALPHA)

The program calculates and prints out the values of c_b and ϕ for the specified P_u, $\phi P_{nb} = P_{ub}$, e_b, P_{uo}, P_u, M_u, and e for the chosen trial value c.

Compare the P_u with the external load P_u. If they are equal, proceed to the next step; otherwise, input a new trial value for c in register 10 and execute R/CCOL6. Repeat the trials until the calculated and specified ultimate loads are approximately equal.

Step 5: If P_u calculated is approximately equal to P_u specified, compare the calculated e value with the specified e value (M_u/P_u). If the calculated e is equal or slightly greater than the specified e value, the design is satisfactory. Otherwise, revise the column dimensions, the reinforcement or both, and repeat the analysis.

The complete output is:

Data	Unit
c_b	in.
ϕ	—
$P_{ub} = \phi P_n b$	lb
e_b	in.
P_{uo}	lb
P_u	lb
M_u	in.-lb
e	in.

13.8.3 Numerical Examples

13.8.3.1 Example 13.24: Analysis of a Circular Column

A circular column 20 in. in diameter is reinforced with six No. 8 equally spaced bars. Calculate (a) design load and eccentricity for the balanced failure condition, (b) the load P_u for $e = 16.0$ in., and (c) the load P_u for $e = 5.0$ in. Assume the column to be nonslender (short) and spirally reinforced. Given:

$$f_c' = 4000 \text{ psi}$$
$$f_y = 60,000 \text{ psi}$$

(a, b)   *Trial 1:*   *Trial 2:*
         $c = 10$ in. $c = 7$ in.
         $P_u = 150,000$

```
    INPUT                    OUTPUT                    INPUT

  4,000.0000 STO 00           XEQ "R/CCOL5"          7.0000 STO 10
 60,000.0000 STO 01     c BAL=10.3571
    20.0000 STO 03       BAL ECC=7.1801                OUTPUT
    15.0000 STO 04       PHI=0.7500
     0.8500 STO 05       PULT BAL=340,747.5929         XEQ "R/CCOL6"
     4.7400 STO 06       PULT ZERO=851,971.2585
     6.0000 STO 07                              P ULT=131,287.4774
     0.7500 STO 08     P ULT=318,297.4145       M ULT=2,092,375.202
    17.5000 STO 09     M ULT=2,435,936.142      ecc=15.9374
    10.0000 STO 10     ecc=7.6530
150,000.0000 STO 11
     0.8500 STO 12
```

Trial 3:   *Trial 4:*
$c = 6.97$ in. $c = 6.985$ in.
$P_u = 130,000$
    INPUT   INPUT

        6.9700 STO 10   6.9850 STO 10
    130,000.0000 STO 11
                                               OUTPUT
    OUTPUT
                                                   XEQ "R/CCOL6"
        XEQ "R/CCOL5"
    c BAL=10.3571   P ULT=130,455.9209
    BAL ECC=7.1801   M ULT=2,088,926.516
    PHI=0.7500   ecc=16.0125
    PULT BAL=340,747.5929
    PULT ZERO=851,971.2585

                                               Conclude $P_u = 130,460$ lb
    P ULT=129,623.7119   for $e = 16.0$ in.
    M ULT=2,085,467.287 $\phi P_{nb} = 340,748$ lb
    ecc=16.0886 $e_b = 7.18$ in.

(c)
Trial 1:   *Trial 2:*   *Trial 3:*
$c = 11$ in. $c = 12$ in. $c = 12.24$ in.
INPUT   INPUT   INPUT

    12.0000 STO 10   11.0000 STO 10   12.2400 STO 10

OUTPUT   OUTPUT   OUTPUT

    XEQ "R/CCOL6"   XEQ "R/CCOL6"   XEQ "R/CCOL6"

P ULT=450,997.9460   P ULT=385,217.5805   P ULT=466,240.1264
M ULT=2,354,211.039   M ULT=2,417,689.456   M ULT=2,335,277.890
ecc=5.2200   ecc=6.2762   ecc=5.0087

Conclude that $P_u = \phi P_n = 466,240$ lb for $e = 5$ in.

13.8.3.2 Example 13.25: Circular Spirally Reinforced Column

A spirally reinforced circular column is subjected to an external factored load $P_u = 110,000$ lb acting at an eccentricity to the plastic and geometric centroid of magnitude $e = 16$ in. Design the column cross section and the longitudinal and spiral reinforcement necessary, assuming a nonslender column with a total reinforcement ratio of about 2%. Given:

$f_c' = 4000$ psi, normalweight concrete
$f_y = 60,000$ psi

Solution

Calculation of factored external loads

Given:

$P_u = 110,000$ lb
$e = 16$ in.

*Try a 20-in. circular column with six No. 8 bars
(area = 4.74 in.^{2})*

Provide a clear cover of 1.5 in. and effective cover (to the center of the bar) of 2.5 in.

| | Trial 1: | | Trial 2: |
| | c = 7 in. | | c = 6.986 in. |

INPUT	OUTPUT	INPUT
4,000.0000 STO 00	XEQ "R/CCOL5"	6.9860 STO 10
60,000.0000 STO 01	c BAL=10.3571	
20.0000 STO 03	BAL ECC=7.1801	OUTPUT
15.0000 STO 04	PHI=0.7687	
0.8500 STO 05	PULT BAL=340,747.5929	XEQ "R/CCOL6"
4.7400 STO 06	PULT ZERO=851,971.2585	
6.0000 STO 07		P ULT=133,764.9698
0.7500 STO 08	P ULT=134,560.4167	M ULT=2,141,238.522
17.5000 STO 09	M ULT=2,144,537.201	ecc=16.0075
7.0000 STO 10	ecc=15.9374	
110,000.0000 STO 11		
0.8500 STO 12		

$\phi P_n = P_u = 133,765$ lb $> 110,000$ lb   O.K.

Therefore, adopt 20-in. round column with six No. 8 bars equally spaced.

13.8.3.3 Example 13.26: Construction of a Load–Moment Design Strength Interaction Diagram Coordinates for a Round Column

Calculate the coordinates of P_u and M_u that can be used to construct a load–moment interaction diagram for a round column having the following properties: $f_c' = 6000$ psi, $f_y = 60,000$ psi, $h = 30$ in., $D_s = 24.5$ in., $A_s = 10.16$ in.^{2} (eight No. 10 bars), and spiral ties.

Solution

This problem can be solved in the same way as Ex. 13.23 except that one must use R/CCOL5 rather than R/CCOL8. In the tension zone, ϕ will vary from 0.75 up to 0.90 at M_{uo}.

First Run (compression failure zone): Try $P_u = 1,500,000$ lb and $c = 25$ in.

INPUT   OUTPUT

```
        6,000.0000 STO 00              XEQ "R/CCOL5"
       60,000.0000 STO 01         c BAL=16.1276
          30.0000 STO 03        BAL ECC=10.5953
          24.5000 STO 04        PHI=0.7500
           0.7500 STO 05        PULT BAL=1,031,161.328
          10.1600 STO 06        PULT ZERO=2,653,760.501
           8.0000 STO 07
           0.7500 STO 08        P ULT=1,982,805.104
          27.2500 STO 09        M ULT=9,457,339.778
          25.0000 STO 10        ecc=4.7697
     1,500,000.000 STO 11
           0.8500 STO 12
```

$P_{uo} = 2,653,761$ lb

$P_{ub} = 1,031,161$ lb

$M_{ub} = P_{ub} \times 10.595$ in

$= 10,925,151$ in.-lb

$c = 25$ in. $> c_b = 16.13$ in. Therefore, compression failure zone.

$$P_{u1} = 1,982,805 \text{ lb}$$

$$M_{u1} = 9,457,340 \text{ in.-lb}$$

Second Run (compression failure zone):   *Third run* (tension failure zone):
Try $e = 18$ in.   Try $c = 12$ in.
No need to revise register 11. $P_u = 500,000$ lb

INPUT   INPUT

```
        18.0000 STO 10                    12.0000 STO 10
                                     500,000.0000 STO 11
```

OUTPUT
                                             OUTPUT

```
    XEQ "R/CCOL6"                          XEQ "R/CCOL5"

P ULT=1,253,172.959              c BAL=16.1276
M ULT=10,937,266.25              BAL ECC=10.5953
ecc=8.7277                       PHI=0.7500
                                 PULT BAL=1,031,161.328
                                 PULT ZERO=2,653,760.501

                                 P ULT=590,087.6666
                                 M ULT=9,549,675.450
                                 ecc=16.1835
```

$500,000 < P_u = 590,088$ and $\phi = 0.75$.

Therefore, revising register 11 to equal P_u output will not change P_u and M_u. Therefore,

$$P_{u3} = 590,088 \text{ lb}$$

$$M_{u3} = 9,549,675 \text{ in.-lb}$$

Fourth run (tension failure zone):

Try $c = 8$ in.

$P_u = 200,000$ lb

INPUT

```
      8.0000 STO 10
200,000.0000 STO 11
```

OUTPUT

```
        XEQ "R/CCOL5"
c BAL=16.1276
BAL ECC=10.5953
PHI=0.8293
PULT BAL=1,031,161.328
PULT ZERO=2,653,760.501

P ULT=191,775.3065
M ULT=7,648,396.535
ecc=39.8821
```

$\phi > 0.75$

Therefore, revise P_u.

Try 192,000.

Revise P_u again; try 192,400 lb.

INPUT

```
192,000.0000 STO 11
```

OUTPUT

```
        XEQ "R/CCOL5"
c BAL=16.1276
BAL ECC=10.5953
PHI=0.8321
PULT BAL=1,031,161.328
PULT ZERO=2,653,760.501

P ULT=192,429.6371
M ULT=7,674,492.595
ecc=39.8821
```

INPUT

```
192,400.0000 STO 11
```

OUTPUT

```
        XEQ "R/CCOL5"
c BAL=16.1276
BAL ECC=10.5953
PHI=0.8320
PULT BAL=1,031,161.328
PULT ZERO=2,653,760.501

P ULT=192,396.9205
M ULT=7,673,187.792
ecc=39.8821
```

Say O.K.

$$P_{u4} = 192,397 \text{ lb}$$

$$M_{u4} = 7,673,188 \text{ in.-lb.}$$

$$\phi = 0.83$$

13.9 RECTANGULAR COLUMNS UNDER BIAXIAL BENDING

This computer program uses the method proposed by Parme et al. (see Section 9.16.2 for more details) to determine approximately an equivalent uniaxial moment for a biaxially loaded column. The program applies only to square or rectangular columns since round columns have the same geometry and capacity about any axis through the centroid. The equivalent uniaxial moment for the rectangular column shape and the equivalent eccentricity can then be used to select a column section that can carry the higher eccentric load using the programs already presented in the preceding sections for uniaxial loading.

13.9.1 Design Equations, Flowchart, and Program Steps
(Refer to Figs. 13.24, 13.25, 13.26, and 13.27)

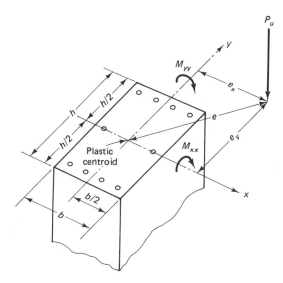

Figure 13.24   Biaxially stressed column cross section.

The governing equations are:

$$M_{nx} + M_{ny}\frac{h}{b}\frac{1-\beta}{\beta} = M_{0x} \qquad \text{if } \frac{M_{ny}}{M_{nx}} \le \frac{b}{h}$$

$$M_{ny} + M_{nx}\frac{b}{h}\frac{1-\beta}{\beta} = M_{0y} \qquad \text{if } \frac{M_{ny}}{M_{nx}} \ge \frac{b}{h}$$

where

$$M_{nx} = \frac{M_{ux}}{\phi} = \text{moment resistance about } x \text{ axis}$$

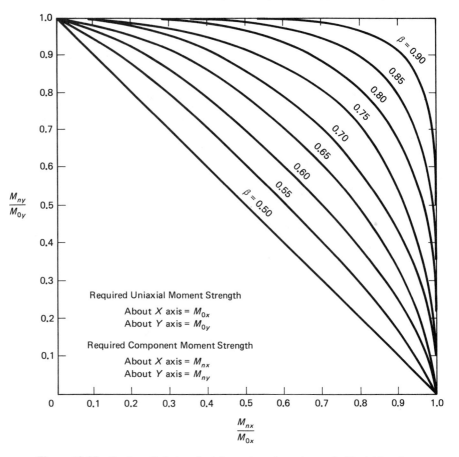

Figure 13.25   Contour β-factor chart for rectangular columns in biaxial bending.

$$M_{ny} = \frac{M_{uy}}{\phi} = \text{moment resistance about } y \text{ axis}$$

$$M_{0x} = M_{nx} \text{ at such axial load } P_n \text{ where } M_{uy} \text{ or } e_x = 0$$

$$M_{0y} = M_{ny} \text{ at such an axial load } P_n \text{ where } M_{ux} \text{ or } e_y = 0$$

β = Modifying contour factor such that the M_{nx}/M_{ny} ratio would have the same value as M_{0x}/M_{0y}

β can be obtained from the chart Fig. 13.25 as well as Fig. 9.35.

The flow chart and the program steps are presented in Figs. 13.26 and 13.27, respectively.

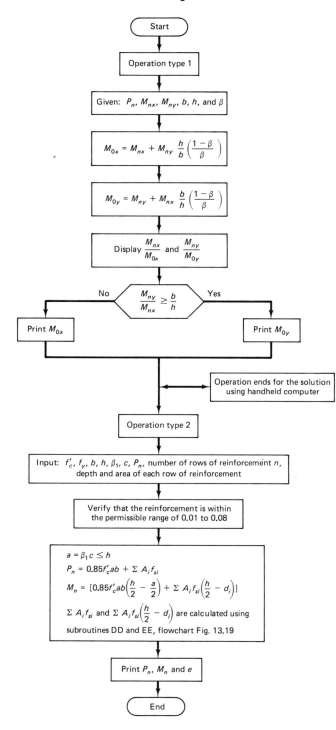

Figure 13.26 Flowchart for rectangular columns under biaxial bending.

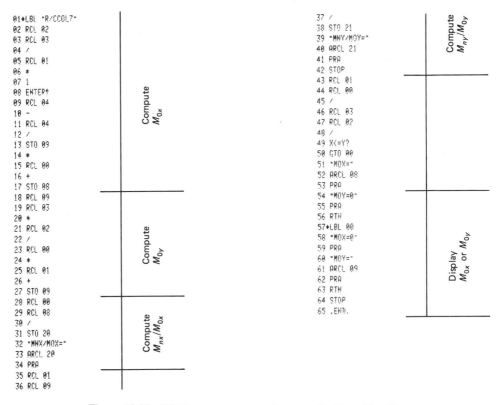

Figure 13.27   HP41 program steps: columns under biaxial bending.

13.9.2 Instructions to Run the Program

Step 1: Set at least 22 data storage registers.

(XEQ   ALPHA   S I Z E   ALPHA 022)

Step 2: Load the program from magnetic cards or enter the program steps.

Step 3: Input the following design data.

Variable	Unit	Store in register
M_{nx}	in.-lb	00
M_{ny}	in.-lb	01
h	in.	02
b	in.	03
Trial β value	—	04

Step 4: Execute R/CCOL7.

$$(\boxed{\text{XEQ}}\quad\boxed{\text{ALPHA}}\quad \underline{R\,/\,C\,C\,O\,L\,7}\quad \boxed{\text{ALPHA}}\,)$$

If $M_{ny}/M_{nx} \leq b/h$, the program will calculate the value of equivalent uniaxial moment, M_{0x} for which the column has to be designed. If $M_{ny}/M_{nx} \geq b/h$, the program will calculate and print out M_{0y}.

The following design steps can be used for the design (analysis) of rectangular columns subjected to biaxial bending.

Step 1: Choose a trial section. Assume a β value.

Step 2: Use the R/CCOL7 program to obtain the equivalent uniaxial moment.

Step 3: Using R/CCOL8, check if the assumed section can safely carry the equivalent uniaxial moment calculated in step 2. Otherwise, revise the cross section, reinforcement, or both and repeat steps 2 and 3 until the capacity of the assumed section is approximately equal to the external (equivalent uniaxial) moment.

Step 4: For the assumed section and the given external axial load, P_u, calculate M_{nx} and M_{ny} using R/CCOL8 and obtain the ratio M_{nx}/M_{0x} and M_{ny}/M_{0y}.

Step 5: Using the ratios M_{nx}/M_{0x} and M_{ny}/M_{0y}, obtain β from the graph, Fig. 13.25.

If this β value is approximately equal to the assumed β value, the design is satisfactory. Otherwise, assume a different value for β and repeat steps 2 to 5.

13.9.3 Numerical Examples

13.9.3.1 Example 13.27: Design of a Biaxially Loaded Column

A corner column is subjected to a factored compression axial load $P_u = 210,000$ lb, a factored bending moment $M_{ux} = 1,680,000$ in.-lb about the x axis, and a factored bending moment $M_{uy} = 980,000$ in.-lb about the y-axis. Given:

$f_c' = 4000$ psi, normal-weight concrete
$f_y = 60,000$ psi

Design a rectangular tied column section to resist the factored biaxial bending moments resulting from the given factored eccentric compressive load.

Solution

Calculate the equivalent uniaxial bending moments
assuming equal numbers of bars on all faces

Assume that $\phi = 0.70$ for tied columns.

$$\text{required nominal } P_n = \frac{210,000}{0.70} = 300,000 \text{ lb}$$

$$\text{required } M_{nx} = \frac{1,680,000}{0.70} = 2,400,000 \text{ in.-lb}$$

$$\text{required nominal } M_{ny} = \frac{980,000}{0.70} = 1,400,000 \text{ in.-lb}$$

Since the column dimensions are proportional to the applied moments, assume that $h/b = 1.71$ or $b = 12$ in. and $h = 20$ in. to give $h/b = 1.67$. Assume that interaction contour factor $\beta = 0.61$.

$$e_y = \frac{1,680,000}{210,000} = 8.0 \text{ in.}$$

$$e_x = \frac{980,000}{210,000} = 4.67 \text{ in.}$$

INPUT

```
2,400,000.000 STO 00
1,400,000.000 STO 01
      20.0000 STO 02
      12.0000 STO 03
       0.6100 STO 04
```

OUTPUT

```
              XEQ "R/CCOL7"
MNX/MOX=0.6167
MNY/MOY=0.6033
                       RUN
MOX=3,891,803.279
MOY=0
```

Trial 1: $\beta = 0.61$

req'd $M_{0xn} = 3,891,803$

$$\text{req'd } M_{0yn} = \frac{1,400,000}{0.6033} = 2,320,570$$

$$e_{0y} = \frac{2,724,262}{210,000} = 12.97 \text{ in.}$$

$$e_{0x} = 2,320,570 \times \frac{0.70}{210,000} = 7.73 \text{ in.}$$

Try the ratio $\rho = \rho' = 0.012$ and $d' = 2.5$ in., $d = 20.0 - 2.5 = 17.5$ in.

$$A_s = A_s' = 0.012 \times 12(20.0 - 2.5) = 2.52 \text{ in.}^2$$

Try $A_s = A_s' = 2.37$ in.^{2} on each face.

$$\rho_g = \frac{12 \times 0.79}{12 \times 20} = 0.4$$

Calculation of actual M_{0xn} (c = 9 in.)   *Calculation of actual M_{0yn} (c = 8 in.)*

INPUT   INPUT

```
    4,000.0000 STO 00                4,000.0000 STO 00
   60,000.0000 STO 01               60,000.0000 STO 01
       12.0000 STO 02                   20.0000 STO 02
       20.0000 STO 03                   12.0000 STO 03
        0.8500 STO 04                    0.8500 STO 04
        2.3700 STO 05                    2.3700 STO 05
        2.3700 STO 06                    2.3700 STO 06
        0.0000 STO 07                    0.0000 STO 07
        0.0000 STO 08                    0.0000 STO 08
        2.5000 STO 09                    2.5000 STO 09
       17.5000 STO 10                    9.5000 STO 10
  210,000.0000 STO 11              210,000.0000 STO 11
        9.0000 STO 12                    8.0000 STO 12
```

OUTPUT   OUTPUT

```
        XEQ "R/CCOL8"                    XEQ "R/CCOL8"
c BAL=10.3571                    c BAL=5.6224
0.7*PBAL=245,789.4002            0.7*PBAL=202,459.5804
0.7*MBAL=2,858,354.517           0.7*MBAL=1,430,516.260
e BAL=11.6293                    e BAL=7.0657
0.7*PZERO=758,998.8000           0.7*PZERO=758,998.8000
.7*.8PZERO=607,199.0400          .7*.8PZERO=607,199.0400
PHI=0.7000                       PHI=0.7000
P ULT=212,843.4000               P ULT=390,205.9000
M ULT=2,799,934.200              M ULT=1,263,845.713
ecc=13.1549                      ecc=3.2389
```

Trial 2:   *Trial* 2:
$c = 9.18$ in. $c = 5.18$ in.

INPUT   INPUT

```
        9.1800 STO 12                    5.1800 STO 12
```

OUTPUT   OUTPUT

```
        XEQ "R/CCOL9"                    XEQ "R/CCOL9"
P ULT=217,213.0800              P ULT=179,076.4162
M ULT=2,809,868.667             M ULT=1,386,107.923
ecc=12.9360                     ecc=7.7403
```

Trial 3:

$c = 9.16$ in.

```
INPUT

    9.1600 STO 12
```

$$\text{actual } M_{0yn} = \frac{1,386,108}{0.7} = 1,980,153$$

```
OUTPUT
```

$$\text{actual } M_{0xn} = \frac{2,808,798}{0.7} = 4,012,569$$

```
    XEQ "R/CCOL9"
P ULT=216,727.5600
M ULT=2,808,797.853
ecc=12.9600
```

$$> \text{req'd } M_{0xn} = 3,891,803$$

Say O.K. so far.

$$\frac{M_{nx}}{M_{0xn}} = \frac{2,400,000}{4,012,569} = 0.598$$

$$\frac{M_{ny}}{M_{0yn}} = \frac{1,400,000}{1,980,153} = 0.707$$

From Fig. 13.25, $\beta = 0.65$. It was assumed that $\beta = 0.61$; hence make another trial assuming that $\beta = 0.64$.

```
        INPUT                      OUTPUT

  2,400,000.000 STO 00               XEQ "R/CCOL7"
  1,400,000.000 STO 01       MNX/MOX=0.6465
        20.0000 STO 02       MNY/MOY=0.6335
        12.0000 STO 03                       RUN
         0.6400 STO 04       MOX=3,712,500.000
                             MOY=0
```

From previous output:

actual $M_{0xn} = 4,012,569 > 3,712,500$ O.K.

$$\text{req'd } e_{0y} = \frac{3,712,500}{300,000} = 12.38 \text{ in.}$$

$$\text{req'd } e_{0x} = \frac{2,209,945}{300,000} = 7.37 \text{ in.}$$

	INPUT		OUTPUT
	4,000.0000 STO 00		XEQ "R/CCOL8"

```
                  INPUT                                    OUTPUT

              4,000.0000 STO 00                        XEQ "R/CCOL8"
             60,000.0000 STO 01           c BAL=10.3571
                 12.0000 STO 02           0.7*PBAL=245,789.4002
                 20.0000 STO 03           0.7*MBAL=2,858,354.517
                  0.8500 STO 04           e BAL=11.6293
                  2.3700 STO 05           0.7*PZERO=758,998.8000
                  2.3700 STO 06           .7*.8PZERO=607,199.0400
                  0.0000 STO 07           PHI=0.7000
                  0.0000 STO 08           P ULT=224,981.4000
                  2.5000 STO 09           M ULT=2,825,879.175
                 17.5000 STO 10           ecc=12.5605
Trial 1:    210,000.0000 STO 11
c = 9.5 in.       9.5000 STO 12
```

```
                  INPUT                                    OUTPUT

              4,000.0000 STO 00                        XEQ "R/CCOL8"
             60,000.0000 STO 01           c BAL=5.6224
                 20.0000 STO 02           0.7*PBAL=202,459.5804
                 12.0000 STO 03           0.7*MBAL=1,430,516.260
                  0.8500 STO 04           e BAL=7.0657
                  2.3700 STO 05           0.7*PZERO=758,998.8000
                  2.3700 STO 06           .7*.8PZERO=607,199.0400
                  0.0000 STO 07           PHI=0.7000
                  0.0000 STO 08           P ULT=185,508.7962
                  2.5000 STO 09           M ULT=1,399,134.192
                  9.5000 STO 10           ecc=7.5421
Trial 1:    210,000.0000 STO 11
c = 5.3 in.       5.3000 STO 12
```

```
              INPUT                                    INPUT

          9.7000 STO 12                             5.5000 STO 12

              OUTPUT                                   OUTPUT

                  XEQ "R/CCOL9"                            XEQ "R/CCOL9"
          P ULT=229,836.6000                     P ULT=196,076.4909
Trial 2:  M ULT=2,834,812.743        Trial 2:    M ULT=1,419,208.844
c = 9.7 in. ecc=12.3340              c = 5.5 in.  ecc=7.2380
```

Trial 3:   INPUT

$c = 9.67$ in.

$\qquad\qquad$ 9.6700 STO 12

$\qquad$ OUTPUT

$\qquad\qquad$ XEQ "R/CCOL9"

P ULT=229,108.3200

M ULT=2,833,525.326

ecc=12.3676

Trial 3:   INPUT

$c = 5.4$ in.

$\qquad\qquad$ 5.4000 STO 12

$\qquad$ OUTPUT

$\qquad\qquad$ XEQ "R/CCOL9"

P ULT=190,815.5667

M ULT=1,409,423.703

ecc=7.3863

$$\text{actual } M_{n0x} = \frac{2,833,525}{0.7}$$

$$= 4,047,893$$

$$\text{actual } M_{n0y} = \frac{1,409,424}{0.7}$$

$$= 2,013,463$$

$$\frac{M_{nx}}{M_{0xn}} = \frac{2,400,000}{4,047,893} = 0.593$$

$$\frac{M_{ny}}{M_{n0y}} = \frac{1,400,000}{2,013,463} = 0.695$$

From Fig. 13.25, $\beta \simeq 0.64$, which is equal to the assumed β value. Hence adopt the design. The details of the reinforcement are presented in Fig. 9.38.

13.10 USE OF DESKTOP AND TRANSPORTABLE PERSONAL COMPUTERS

This section discusses the use of desktop and transportable personal computers. Its main objective is to demonstrate that the appropriate analysis and design equations and the logic flowcharts presented in earlier chapters of this book can be effectively used to develop program steps in such computer languages as BASIC. Desktop and transportable computers are personal computers with generally larger memory capacity than that of handheld computers and are compatible with mainframe computers. Programs and input data are stored on $5\frac{1}{4}$-in. floppy diskettes as compared to the magnetic cards, cassettes, or modules used in the handheld computer calculators.

The Apple IIe desktop personal computer is chosen to represent the numerous types of desktop computers available in the market. The program steps, with minor modifications, can be used in the other type of personal computers, such as the IBM PC or HP150.

Program steps are provided for such topics as the analysis (design) of beams in flexure, diagonal tension, and torsion or the design of columns. The fundamental differences between the BASIC language used in the desktop computer and the language used in the HP41C/ or HP41CV/41CX are discussed first, followed by the programming steps needed in the analysis and design of sections.

13.10.1 Programming in BASIC Language

The major differences between the programming procedure used in HP41C/ or HP41CV/41CX and the BASIC language procedure presented in this section are: (1) in BASIC more than one operation can be done in a single statement or program line, and (2) each parameter such as f_c', the compressive strength of concrete, is assigned to a variable F. This variable F will be used during the input process whenever f_c' is needed instead of recalling the total item from memory as in the case of HP41C. The differences can be better explained using the example for the calculation of bending moment in a simply supported beam.

As shown in Section 13.1.2.3, the variables l, x, and w were stored in storage registers 00, 01, and 02. These variables were recalled to compute the required expression, $W/2(lx - x^2)$.

In BASIC, the variables l, x, and w can be assigned to letters, say L, X, and W. The program steps can be written in such a manner that the computer will ask for these variables during the input process. The mathematical expression can be computed in a single step, as shown below.

The following program steps can be compared with the program steps presented in Section 13.1.2.3 for HP41C or HP41CV/41CX:

1. PRINT "L=?": INPUT L: PRINT "x="?: INPUT x; PRINT "W=?": INPUT W
2. M = W* (L*x − x*x)/2
3. PRINT "MOMENT" = ", M
4. END

Line number 1 covers the input. Line 2 does all the calculations and line 3 prints the moment value.

All the program lines have to be entered in the computer (as in HP41CV or 41CX). Then the program can be executed by entering the command RUN. The computer will respond by printing L=?. The variable L has then to be specified. The machine will then ask for X and W in a similar fashion. Using the specified input values of L, X and W, the computer proceeds to calculate the moment and print out its value.

The similarities between handheld and desktop computers are more pronounced than the differences in developing the program steps. The input variables, the necessary equations and the sequence of steps (presented in the form of a flowchart) are essentially the same for both types of instruments. Consequently, the equations and flowcharts presented in this chapter can be used to write programs in any computer language, including BASIC, for use in the pocket, transportable, and desktop personal computers.

13.11 PROGRAMS FOR APPLE IIe DESKTOP PERSONAL COMPUTERS: RECTANGULAR BEAMS IN FLEXURE, SHEAR AND TORSION

A modified computer program in BASIC language, EGNAWY1, is presented for the analysis (design) of beams. It is divided into three major segments or operations. The *first* covers flexure for singly-reinforced and doubly-reinforced beams, and can be applied to T-beams. The *second* deals with diagonal tension and the *third* with combined diagonal tension and torsion for normal beams including the design of web steel spacing. In effect, the program can be considered as three separate programs in BASIC language for proportioning rectangular beams, but combined for efficiency in programming the input and output steps.

Figure 13.28 presents the program steps for the analysis of beams in flexure diagonal tension and torsion. The necessary equations and the flow charts were taken from Sections 13.2.1 and 13.4.1

In order to use the program, one has to key-in (enter) all the program steps into the computer. Magnetic discs (diskettes) can be used to store and retrieve programs.

Once the program steps are entered, they can be executed by entering the command RUN.

The program will respond with "ENTER YOUR SELECTION". If the input is operation "1," the program will go to the flexural analysis part. An input "2" takes the program to the diagonal tension operation and an input "3" takes it to the combined diagonal tension and torsion operation.

Once the program selects the appropriate analysis section, the necessary input variables are requested.

The following symbols for the variables are used in the program.

FC′ = compressive strength of concrete, psi
FY = yield strength of steel, psi
B or b = width of the beam, in.
D or d = depth of the beam, in.
AS or As = area of tension steel, in^2
AS′ = area of compression steel, in^2
MU or Mu = design moment, in-lb.
A = depth of the rectangular stress block, in.
Vu = design shear force, lb.
At, Av = area of two legs of the stirrup, in.^{2}
x1, y1 = shorter and longer dimensions of the stirrup, in.
Tu = torsional moment, in-lb.
Sum (x2y) = $\Sigma\, x^2 y$, in.^{3}

In flexural analysis, program operation 1 calculates and prints the design moment M_u, if the given reinforcement satisfies the maximum and minimum permissible

```
1    REM   EGNAWY1- BEAMS IN FLEXURE
     ,DIAGONAL TENSION AND TORSIO
d    N
10   HOME : CLEAR
20   REM    PROGRAM ENGINEERING CON
]    CRETE DESIGN
25   S1$ = "----------------------
     ----------------"
50   T2$ = "**********************
     ****************"
70   HOME : PRINT T2$: VTAB 10: PRINT
     " 1-RECTANGULAR BEAMS UNDER
     FLEXURE": PRINT " 2-BEAMS UN
     DER SHEAR AND DIAGONAL TENSI
     ON"
80   PRINT "  3- BEAMS UNDER SHEAR
      AND TORSION"
150  VTAB 20: HTAB 8: INPUT "ENTE
     R YOUR SELECTION >  ";AN$
152  AN =  VAL (AN$)
153  IF AN < 1 OR AN > 3 THEN   VTAB
     20: HTAB 8: PRINT "
                        "
155  IF AN < 1 OR AN > 3 THEN   GOTO
     150
200  IF AN = 1 THEN   GOSUB 1000
210  IF AN = 2 THEN   GOTO 2000
220  IF AN = 3 THEN   GOTO 3000
1000 REM   TRIAL PROG. **BEAM**
1005 REM   **ANALYSIS OF DOUBLY R
     EINFORCED CONCRETE BEAM USIN
     G 318-83
1205 HOME : PRINT T2$: PRINT : PRINT
     "**INPUT DATA:**": PRINT "--
     ---------------": VTAB 8
1210 INPUT "Fc'= ";F: VTAB 8: HTAB
     15: INPUT " Fy= ";S
1211 PRINT
1212 INPUT "B  = ";B: VTAB 10: HTAB
     15: INPUT " D = ";D: VTAB 10
     : HTAB 30: INPUT " D' = ";D1

1214 PRINT : INPUT "AS = ";A1: VTAB
     12: HTAB 15: INPUT " AS'= ";
     A2
1220 PRINT : PRINT "** VERIFICAT
     ION OF INPUT DATA **": PRINT

1225 PRINT "FC'= ";F: VTAB 16: HTAB
     15: PRINT " FY= ";S
1228 PRINT : PRINT "B  = ";B: VTAB
     18: HTAB 15: PRINT " D = ";D
     : VTAB 18: HTAB 30: PRINT "
     D' = ";D1
1230 PRINT : PRINT "AS = ";A1: VTAB
     20: HTAB 15: PRINT " AS'= ";
     A2
1233 PRINT
1234 INPUT "ARE THE DATA CORRECT
     ? Y/N ";Z$
1235 IF Z$ = "N" THEN   GOTO 10
1238 HOME : PRINT T2$: PRINT : PRINT
     : PRINT : PRINT : INVERSE : FLASH
     : HTAB 8: PRINT "PLEASE WAIT
     FOR CALCULATIONS": NORMAL
```

Selection of the type of analysis:

1. Flexural analysis
2. Shear and diagonal tension
3. Shear and torsion

Flexural Analysis

Input the design variables:
f'_c, f_y, b, d, d', A_s, and A'_s

Verification of input data

If the input data are correct, type Y. Otherwise, type N and re-input the correct data.

Figure 13.28 Program steps: analysis of rectangular beams.

```
1240   IF F < 4000. THEN 1270
1250   IF F > 8000. THEN 1280
1260   BE = .85 - (.05 * (F - 4000.
       ) / 1000): GOTO 1290
1270   BETA = .85: GOTO 1290
1280   BETA = .65: GOTO 1290
1290   REM  CONTINUE
1300   PM = 200 / S
1310   R1 = A1 / (B * D)
1325   F1 = S
1330   A = ((A1 * S) - (A2 * F1)) /
       (.85 * F * B)
1332   E1 = .003 * (A - (BE * D1)) /
       A
1334   IF E1 > = S / 29000000 THEN
       F1 = S: GOTO 1350
1336   F1 = 29000000 * E1
1340   IF F1 < 0 THEN F1 = 0.: GOTO
       1350
1344   A = (A1 * S - A2 * F1) / (.8
       5 * F * B)
1350   E2 = .003 * ((BE * D) - A) /
       A
1355   M1$ = " "
1365   PB = .75 * ((.85 * F * BE /
       S) * (.003 * 29000000 / ((.0
       03 * 29000000) + S))) + (A2 *
       F1 / (S * B * D))
1366   IF E2 < = (S / 29000000) THEN
       M1$ = "SECTION IS OVER REINF
       ORCED": GOTO 1450
1370   R1 = A1 / (B * D):PM = 200 /
       S
1375   IF R1 > = PB THEN M1$ = "S
       ECTION IS OVER REINFORCED": GOTO
       1450
1380   IF R1 < = PM THEN M1$ = "A
       REA OF STEEL IS LESS THAN MI
       NIMUM": GOTO 1450
1395   HOME : PRINT T2$: PRINT : PRINT
       "** FINAL RESULTS **": PRINT
1430   M = .9 * (.85 * F * A * B *
       (D - A / 2) + A2 * F1 * (D -
       D1))
1440   PRINT "MU (LB-IN)=      ";M
1450   PRINT "BETA1=           ";BET
       A
1460   PRINT "A=               ";A
1470   PRINT ".75*ROW.BAL=     ";PB:
       PRINT "ROW=             ";R1:
       PRINT "ROW MINIMUM=     ";PM:
       PRINT
1480   PRINT "** MESSAGES **"
1490   PRINT M1$: PRINT
1500   INPUT "ANOTHER SECTION ? Y/
       N ";AN$
1510   IF AN$ = "Y" THEN  GOTO 100
       0
1900   RETURN
```

If $f'_c \leq 4000$ psi, $\beta_1 = 0.85$.

If $f'_c \leq 8000$ psi, $\beta_1 = 0.65$.

If 4000 psi $< f'_c < 8000$ psi.

$$\beta_1 = 0.85 - 0.05 \frac{f'_c - 4000}{1000}$$

$$\rho_{min} = \frac{200}{f_y}, \ \rho = \frac{A_s}{bd}$$

$$a = \frac{A_s f_y - A'_s f'_s}{0.85 b f'_c}$$

$$f'_s = 29 \times 10^6 \left(0.003 \frac{a - \beta_1 d'}{a}\right) \leq f_y$$

$$\rho_{max} = 0.75 \frac{87,000}{87,000 + f_y} \beta_1 \frac{0.85 f'_c}{f_y}$$

$$+ \frac{A'_s f'_s}{bd f_y}$$

If $\rho > \rho_{max}$, print "section is over-reinforced".

If $\rho < \rho_{min}$, print "area of steel is less than minimum".

$$M_u = 0.9 \left[0.85 f'_c ba \left(d - \frac{a}{2}\right) + A'_s f'_s (d - d')\right]$$

Print: M_u, $0.75\rho_b$, ρ, ρ_{min}.
  Another section ?
Type Y to analyze another section.

Figure 13.28 (*cont.*)

```
2000   REM     SHEAR AND DIAGONAL TE
       NSION
2100   HOME : PRINT T1$: PRINT T2$
       : PRINT "***** SHEAR AND DIA
       GONAL TENSION *****"
2300   PRINT : PRINT
2305   CLEAR
2310   INPUT "Fc'      = ";FC
2315   INPUT "FY       = ";FY
2317   INPUT "b        = ";B
2319   INPUT "d        = ";D
2322   INPUT "Vu       = ";VU
2325   INPUT "Mu       = ";M
2328   INPUT "As       = ";A
2350   INPUT "At       = ";AO
2400   INPUT "ARE DATA CORRECT Y/N
              ? ";AX$: IF A
       X$ = "N" THEN  VTAB 6: GOTO
       2310
2430   V1 = VU / 0.85:Q = M
2440    IF Q = 0 THEN 2530
2450   T1 = VU * D / M
2460    IF T1 <  = 1 THEN 2480
2470   T1 = 1
2480   C1 = 1.9 *  SQR (FC) * B * D
       + 2500 * A * T1
2490   C2 = 3.5 *  SQR (FC) * B * D

2500    IF C1 < C2 THEN 2520
2510   C = C2: GOTO 2540
2520   C = C1: GOTO 2540
2530   C = 2. *  SQR (FC) * B * D
2540   T = V1 - C
2550   C3 = 0.5 * C
2555    IF C3 > V1 THEN 2750
2560   T1 = 8 *  SQR (FC) * B * D
2570    IF T > T1 THEN 2590
2580    GOTO 2600
2590    PRINT "THE CROSS SECTION IS
       TOO SMALL,CHANGE THE DIMENS
       IOS"
2595    GOTO 2760
2600   A2 = T / (FY * D)
2610   A3 = 50 * B / FY
2620    IF A2 < A3 THEN 2640
2630   A4 = A2: GOTO 2650
2640   A4 = A3
2650   S1 = AO / A4
2660   T2 = T1 / 2
2670    IF T2 >  = T THEN 2690
2680   S2 = D / 4: GOTO 2700
2690   S2 = D / 2
2700    IF S1 < S2 THEN 2720
2710   S1 = S2
2720    IF S1 <  = 24 THEN 2740
2740    PRINT "SPACING OF STIRRUP=
       ";S1
2741    GOTO 2760
2750    PRINT "SHEAR REINFORCEMENT
       IS NOT NECESSARY"
2760    INPUT " HIT RETURN TO CONTI
       NUE";SS$
2770    GOTO 10
```

Shear and Diagonal Tension

Input the design variables:
f_c', f_y, b, d, V_u, M_u, A_s, and
area of two legs of the stirrup.

Verification of input data. If the input
data are correct, type Y. Otherwise, type
N, and re-input the correct data.

$$V_n = \frac{V_u}{0.85}$$

$$V_c = \left(1.9 f_c' + 2{,}500\, \rho_w\, \frac{V_u d}{M_u}\right) b_w d$$

$$\leq 3.5\, \sqrt{f_c'}\, b_w d$$

If contribution $\rho_w\, \dfrac{V_u d}{M_u}$ is disregarded,

V_c is taken $= 2\, \sqrt{f_c'}\,(b_w d)$

(Command $Q = 0$ is step 2440)

If $V_n - V_c > 8\, \sqrt{f_c'}\, b_w d$,
print "the cross section is too small,
change the dimensions".

$$\frac{A_t}{s} = \frac{V_n - V_c}{f_y d} \geq \frac{50 b_w}{f_y}$$

$$s = \frac{A_t/s}{A_v} \leq \frac{d}{2} \leq 24 \text{ in.}$$

$$s \leq \frac{d}{4} \text{ if } V_n - V_c \geq 4\, \sqrt{f_c'}\, b_w d$$

Print s.

If $\dfrac{V_c}{2} > V_n$, print "shear reinforcement is
not required.

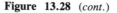

Figure 13.28 (*cont.*)

```
3000  REM  BEAM ANALYSIS UNDER SH
      EAR AND TORSION
3050  HOME
3100  PRINT T2$: PRINT "**** BEAM
      S UNDER SHEAR AND TORSION **
      *"
3105  PRINT : PRINT
3106  CLEAR
3110  INPUT "Fc'      =";FC
3120  INPUT "Fy       =";FY
3130  INPUT "Vu       =";VU
3140  INPUT "bw       =";BW
3150  INPUT "d        =";D
3160  INPUT "Mu       =";MU
3170  INPUT "SUM(X^2*Y)=";SM
3180  INPUT "x1       =";X1
3190  INPUT "y1       =";Y1
3200  INPUT "As       =";AS
3210  INPUT "Tu       =";TU
3230  INPUT "Av       =";AV
3300  PRINT : PRINT : INPUT "ARE
      DATA CORRECT Y/N ? ";AN$
3310  IF AN$ = "N" THEN  GOTO 300
      0
3315  PRINT : PRINT : PRINT "****
      OUTPUT RESULTS ***"
3320 VN = VU / 0.85:Q = MU
3330 V1 = 10 * SQR (FC) * BW * D

3335  PRINT "Vn        = ";VN
3340  IF VN > V1 THEN  PRINT "ENL
      ARGE SECTION Vn > 10*SQR(Fc'
      )*bw*d": INPUT X$: GOTO 10
3350  IF TU = 0. THEN 3500
3360 T1 = .85 * (.5 *  SQR (FC) *
      SM)
3370  IF TU <  = T1 THEN  PRINT "
      TORSION CAN BE NEGLECTED":TU
      = 0: GOTO 3500
3380 CT = BW * D / SM
3390 TC = .8 * SM *  SQR (FC) / (
      SQR ((((.4 * VU / (CT * TU))
      ^ 2) + 1))
3395  PRINT " TC =        ";TC
3400 VC = 2 *  SQR (FC) * BW * D /
      ( SQR (1 + ((2.5 * CT * TU /
      VU) ^ 2)))
3410  IF VU = 0 THEN VC = 0.
3420  GOTO 3600
3500  IF Q = 0 THEN VC = 2 *  SQR
      (FC) * BW * D: GOTO 3600
3510 V2 = VU * D / MU
3520  IF V2 > 1 THEN V2 = 1
3530 VC = (1.9 *  SQR (FC) * BW *
      D) + (2500 * AS * V2)
3535 V3 = 3.5 * BW * D * ( SQR (F
      C))
3540  IF VC <  = V3 THEN 3600
3550 VC = V3
3600  PRINT "Vc        = ";VC
3610  IF (VN - (VC / 2)) > 0 THEN
      3650
```

Shear and Torsion

Input the design variables:
f_c', f_y, V_u, b_w, d, M_u, $\Sigma\, x^2 y$, x_1, y_1, A_s, T_u, A_v.

Verification of input data. If the input data are correct, type Y. Otherwise, type N, and re-input the correct data.

$V_n = \dfrac{V_u}{0.85}$ If $V_n > 10\, \sqrt{f_c'}\, b_w d$, print "enlarge section". If $T_u < 0.85\, (0.5\, \sqrt{f_c'}\, \Sigma\, x^2 y)$, print "torsion can be neglected". Go to shear and diagonal tension section, steps 3500–3550.

$c_t = \dfrac{b_w d}{\Sigma\, x^2 y}$, $T_c = \dfrac{0.8\, \sqrt{f_c'}\, \Sigma\, x^2 y}{\sqrt{1 + (0.4 V_u / C_t T_u)^2}}$, print T_c.

$V_c = \dfrac{2\, \sqrt{f_c'}\, b_w d}{\sqrt{1 + (2.5 C_t T_u / V_u)^2}}$

$V_c = (1.9\, \sqrt{f_c'} + 2500\, \rho_w\, \dfrac{V_u d}{M_u}\, b_w d\,) \leqslant 3.5\, \sqrt{f_c'}\, b_w d$

If contribution $\rho_w\, \dfrac{V_u d}{M_u}$ is disregarded,

V_c is taken $= 2\, \sqrt{f_c'}\, (b_w d)$

Figure 13.28 (*cont.*)

```
3620  IF (VN - (VC / 2)) < O THEN
      V4 = O: PRINT " NO SHEAR BAR
      S REQUIRED ": INPUT X$: GOTO
      10
3650  SS = (VN - VC) / (FY * D)
3655  TN = TU / .85
3660  IF VN - VC < O THEN SS = O
3665  IF TU = O THEN  PRINT "NO T
      ORSION BARS REQUIRED": GOTO
      3800
3668  TS = TN - TC
3670  IF TS < O THEN  PRINT "NO T
      ORSION BARS REQUIRED":TS = O
      : GOTO 3800
3680  IF TS > 4 * TC THEN  PRINT
      "ENLARGE SECTION DUE TO TORS
      INAL SHEAR": INPUT X$: GOTO
      10
3690  S5 = TS / (X1 * Y1 * FY)
3695  AA' = 0.66 + .33 * Y1 / X1
3700  IF AA < 1.5 THEN 3750
3710  AA = 1.5: GOTO 3750
3750  A6 = 2 * S5 / (AA)
3760   PRINT "2*At/s      = ";A6
3800  A7 = A6 + SS
3810  AP = 50 * BW / FY
3820  IF A7 > AP THEN 4000
3890  A7 = AP: PRINT "Av/s + 2*At/
      s    = ";A7
3900  IF TU = O. AND (VN - (VC /
      2)) < O THEN A7 = O: PRINT "
      NO STIRRUPS REQUIRED": INPUT
      X$: GOTO 10
4000   PRINT "50 * bw/Fy  = ";AP
4010   PRINT "AV/S + 2*AT/S ";A7
4050   IF TU = O THEN  GOTO 4500
4100   REM  SHEAR AND TORSION PART
      2
4200  S = AV / A7
4210   IF (X1 + Y1) / 4 >  = 12 THEN
      S1 = 12: GOTO 4250
4220  S1 = (X1 + Y1) / 4
4250  IF S > S1 THEN S = S1
4260   PRINT "SPACING OF STIRRUPS
      = ";S
4270  AL = A6 * (X1 + Y1)
4280  AP = 50 * BW / FY
4290   IF AP > (A6) THEN AX = AP *
      (S)
4300   IF AP <  = A6 THEN AX = A6 *
      S
4310  X = BW
4320   IF D < BW THEN X = D
4330  LL = ((400 * S * X / FY) * (
      TU / ((VU / (3 * CT)) + TU))
      - AX) * ((X1 + Y1) / S)
4340   IF LL > AL THEN AL = LL
4350   PRINT " AL =           ";AL
4400   INPUT "HIT   (RETURN) TO CON
      TINUE ";X$: GOTO 10
4500   REM  SHEAR AND TORSION MODU
      LE #3
4510  S = AV / A7
```

If $V_n - \dfrac{V_c}{2} \leqslant 0$, print "no shear bars

required". $T_s = \dfrac{T_u}{0.85} - T_c$, $\dfrac{A_v}{s} = \dfrac{V_n - V_c}{f_y d}$

If T_u or $T_s \leq 0$, print "no torsion bars required".

If $T_s > 4T_c$, print "enlarge section due to torsional shear".

$\alpha_t = 0.66 + 0.33\dfrac{y_1}{x_1} \leq 1.5$

$\dfrac{2A_t}{s} = 2\dfrac{T_n - T_c}{\alpha_t x_1 y_1 f_y}$

Compute and print $A_v + \dfrac{2A_t}{s}$.

If $T_u = 0$ and $V_n - \dfrac{V_c}{2} \leq 0$,

print "no stirrups required".

If $\dfrac{A_v}{s} + 2\dfrac{A_t}{s} < \dfrac{50b_w}{f_y}$, set $\dfrac{A_v}{s} + \dfrac{2A_t}{s}$

$\quad = 50\dfrac{b_w}{f_y}$

$s = \left(\dfrac{\text{area of two legs of stirrup}}{A_v/s + 2(A_t/s)}\right)$

$\quad \leq \dfrac{x_1 + y_1}{4} \leq 12$ in., print s.

$A_l = 2\dfrac{A_t}{s}(x_1 + y_1)$

$\quad \geq \left(\dfrac{400xs}{f_y}\dfrac{T_u}{T_u + V_u/3C_t} - 2A_t\right)\dfrac{x_1 + y_1}{s}$

If $\dfrac{50b_w s}{f_y} > 2A_t$ replace $2A_t$ with $\dfrac{50b_w s}{f_y}$.

Print A_l.

Figure 13.28 (*cont.*)

```
4520  VS = 4 * BW * D * ( SQR (FC)
      )
4530  SZ = D / 2
4540  IF (VN - VC) > VS THEN SZ =
      D / 4
4550  IF SZ > 24 THEN SZ = 24
4560  IF S > SZ THEN S = SZ
4570  PRINT " FINAL SPACING = S =
      ";S
4580  PRINT
4600  INPUT "HIT (RETURN) TO CONT
      INUE ";X$: GOTO 10
40000 END
```

$$s \le \frac{d}{4} \text{ if } V_n - V_c > 4\sqrt{f_c'}\, b_w d$$

$s \le 24$ in. Print s.

Figure 13.28 (*cont.*)

reinforcement ratios. Otherwise the program prints the appropriate error message, namely, "Section is over-reinforced" or "Area of steel is less than minimum".

For diagonal tension in beams, program operation 2 calculates and prints the spacing of stirrups. If necessary, it will also print one of the two following messages: "The cross-section is too small, change the dimensions" or "Shear reinforcement is not required."

For beams subjected to combined diagonal tension and torsion, program operation 3 calculates and prints the spacing of stirrups and the area of the longitudinal reinforcement. It will also print the following messages in the appropriate cases: "Enlarge section due to torsional shear" or "No torsion bars required".

When the beam section is not rectangular such as a T-beam or an L-beam with the neutral axis falling outside the flange, flexural analysis operation "1" of program EGNAWY 1 can be used for the analysis (design). Minor modifications have to be made in the input values of b, A_s' and d' such that $b = $ width b_w of the web, $A_s' = A_{sf}$ as described in flowchart Fig. 13.5 and $d' = h_f/2$. A check has to be made that $\rho \le 0.75\rho_b$ of the flanged section.

If the neutral axis falls within the flange, the beam can be treated in the flexural analysis as a singly reinforced rectangular section using input values for $b = $ width of the compression flange and $A_s' = 0$.

13.11.1 Example 13.28: Flexural Analysis of Singly and Doubly Reinforced Beams

Solve Ex. 13.1 and 13.5 using desktop personal computer Apple IIe.

Solution

(a) *Solution to Ex. 13.1(a) and (b):*

Example 13.1(a): The input variables are: $f_c' = 4000$ psi, $f_y = 60,000$ psi, $b = 10$ in., $d = 18$ in., $d' = 0$, $A_s = 6.0$ in.^{2}, and $A_s' = 0.0$ in.^{2}.

Example 13.1(b): The input variables are: $f_c' = 4000$ psi, $f_y = 40,000$ psi, $b = 10$ in., $d = 18$ in., $d' = 0$ in., $A_s = 6.0$ in.^{2}, and $A_s' = 0.0$ in.^{2}.

Example 13.1(a):

```
**INPUT DATA:**
-----------------
Fc'= 4000
              Fy= 60000

B  = 10
              D = 18
                          D' = 0

AS = 6
              AS'= 0

** VERIFICATION OF INPUT DATA **

FC'= 4000
              FY= 60000

B  = 10
              D = 18
                          D' = 0

AS = 6
              AS'= 0

ARE THE DATA CORRECT  ? Y/N Y
****************************************

        PLEASE`WAIT`FOR`CALCULATIONS
BETA1=          .85
A=              10.5882353
.75*ROW.BAL=    .021380102
ROW=            .0333333333
ROW MINIMUM=    3.33333333E-03

** MESSAGES **
SECTION IS OVER REINFORCED
```

Example 13.1(b):

```
**INPUT DATA:**
-----------------
Fc'= 4000
              Fy= 40000

B  = 10
              D = 18
                          D' = 0

AS = 6
              AS'= 0

** VERIFICATION OF INPUT DATA **

FC'= 4000
              FY= 40000

B  = 10
              D = 18
                          D' = 0

AS = 6
              AS'= 0

ARE THE DATA CORRECT  ? Y/N Y
****************************************

        PLEASE`WAIT`FOR`CALCULATIONS
****************************************

** FINAL RESULTS **

MU (LB-IN)=     3125647.06
BETA1=          .85
A=              7.05882353
.75*ROW.BAL=    .0371205709
ROW=            .0333333333
ROW MINIMUM=    5E-03
```

(b) *Solution to Ex. 13.5:* The input variables are: $f'_c = 5000$ psi, $f_y = 60,000$ psi, $b = 14$ in., $d = 21$ in., $d' = 2.5$ in., $A_s = 5.08$ in.^{2}, and $A'_s = 1.20$ in.^{2}.

```
**INPUT DATA:**
-----------------
Fc'= 5000
              Fy= 60000

B  = 14
              D = 21
                          D' = 2.5

AS = 5.08
              AS'= 1.2

** VERIFICATION OF INPUT DATA **

FC'= 5000
              FY= 60000

B  = 14
              D = 21
                          D' = 2.5

AS = 5.08
              AS'= 1.2

ARE THE DATA CORRECT  ? Y/N Y
****************************************
```

```
        PLEASE`WAIT`FOR`CALCULATIONS
****************************************

** FINAL RESULTS **

MU (LB-IN)=     5158856.01
BETA1=          .8
A=              4.26497444
.75*ROW.BAL=    .0280461463
ROW=            .0172789116
ROW MINIMUM=    3.33333333E-03
```

13.11.2 Example 13.29: Design of Web Stirrups for Shear

Solve Exs. 13.10 and 13.11 for V_u (at d) = 144,499 lb and V_u (at $2d$) = 111,331 lb.

Solution

(a) *Solution to Ex. 13.10:* The input variables are: f'_c = 4000 psi, f_y = 60,000 psi, b = 14 in., d = 28 in., V_u = 144,499 lb and 111,331 lb, M_u = 0, A_s = 6 in.^{2}, and A_v = 0.4 in.^{2}.

```
***** SHEAR AND DIAGONAL TENSION *****


Fc'      = 4000
FY       = 60000
b        = 14
d        = 28
Vu       = 144499
Mu       = 0
As       = 6
At       = 0.4
ARE DATA CORRECT Y/N              ? Y
SPACING OF STIRRUP= 5.58073207
```

```
****************************************
***** SHEAR AND DIAGONAL TENSION *****


Fc'      = 4000
FY       = 60000
b        = 14
d        = 28
Vu       = 111331
Mu       = 0
As       = 6
At       = .4
ARE DATA CORRECT Y/N              ? Y
SPACING OF STIRRUP= 8.25622473
```

(b) *Solution to Ex. 13.11:* The input variables are the same as those in Ex. 13.10 except that the value of the factored moment M_u is given, justifying use of the more precise diagonal tension expression. M_u = 4,510,100 in.-lb and 8,347,475 in.-lb

```
***** SHEAR AND DIAGONAL TENSION *****


Fc'      = 4000
FY       = 60000
b        = 14
d        = 28
Vu       = 144499
Mu       = 4510100
As       = 6
At       = .4
ARE DATA CORRECT Y/N              ? Y
SPACING OF STIRRUP= 6.14050999
```

```
****************************************
***** SHEAR AND DIAGONAL TENSION *****


Fc'      = 4000
FY       = 60000
b        = 14
d        = 28
Vu       = 111331
Mu       = 8347475
As       = 6
At       = .4
ARE DATA CORRECT Y/N              ? Y
SPACING OF STIRRUP= 8.58557918
```

13.11.3 Example 13.30: Design of Web Stirrups for Combined Shear and Torsion

Solve Ex. 13.12 using the desktop personal computer program in BASIC.

Solution

(a) *Solution to Ex. 13.12(a):* The input variables are: $f'_c = 4000$ psi, $f_y = 60,000$ psi,, $V_u = 15,000$ lb, $b_w = 14$ in., $d = 22.5$ in., $M_u = 0$ in.-lb, $\Sigma\, x^2 y = 5384$ in.^{3}, $x_1 = 10.5$ in., $y_1 = 21.5$ in., $A_s = 3.4$ in.^{2}, $T_u = 75,000$ in.-lb, and $A_v = 0.22$ in.^{2}.

```
**** BEAMS UNDER SHEAR AND TORSION ***          **** OUTPUT RESULTS ***
                                                Vn          = 17647.0588
                                                TORSION CAN BE NEGLECTED
Fc'          =4000                              Vc          = 39844.6985
Fy           =60000                                NO SHEAR BARS REQUIRED
Vu           =15000
bw           =14
d            =22.5
Mu           =0
SUM(X^2*Y)=5284
x1           =10.5
y1           =21.5
As           =3.4
Tu           =75000
Av           =.22


ARE DATA CORRECT Y/N ? Y
```

(b) *Solution to Ex. 13.12(b):* The input variables are the same as those of Ex. 13.12(a), except that $T_u = 300,000$ in.-lb.

```
                    Fc'          =4000
                    Fy           =60000
                    Vu           =15000
                    bw           =14
                    d            =22.5
                    Mu           =0
                    SUM(X^2*Y)=5284
                    x1           =10.5
                    y1           =21.5
                    As           =3.4
                    Tu           =300000
                    Av           =.22


                    ARE DATA CORRECT Y/N ? Y


                    **** OUTPUT RESULTS ***
                    Vn          = 17647.0588
                      TC =          253467.372
                    Vc          = 12673.3686
                    2*At/s      = .010996289
                    50 * bw/Fy  = .0116666667
                    AV/S + 2*AT/S .014680504
                    SPACING OF STIRRUPS = 8
                      AL =          1.96077186
```

(c) *Solution to Ex. 13.12(c):* The input variables are the same as those of Ex. 13.12(a) except that $T_u = 500{,}000$ lb.

```
Fc            =4000
Fy            =60000
Vu            =15000
bw            =14
d             =22.5
Mu            =0
SUM(X^2*Y)=5284
x1            =10.5
y1            =21.5
As            =3.4
Tu            =500000
Av            =.22


ARE DATA CORRECT Y/N ? Y


**** OUTPUT RESULTS ***
Vn           =  17647.0588
 TC =                262094.328
Vc           =  7862.82985
2*At/s        = .0360531129
50 * bw/Fy   = .0116666667
AV/S + 2*AT/S .0433006899
SPACING OF STIRRUPS = 5.08075046
   AL =             1.40393414
```

13.12 PROGRAMS FOR APPLE IIe DESKTOP PERSONAL COMPUTERS: COMPRESSION MEMBERS

Three separate computer programs in BASIC language have been written for the Apple IIe computer with subroutines to enable analyzing and designing columns and constructing load-moment strength interaction diagrams.

The *first* program, EGNAWY2, is for the analysis and design of rectangular columns, the *second,* EGNAWY3, is for the analysis and design of circular columns and the *third,* EGNAWY4, is for the design of biaxially loaded rectangular columns.

Uniaxial Loading

Figs. 13.29 and 13.30 present the program steps for the analysis of rectangular and circular columns respectively. The necessary equations and the flow charts were taken from Sections 13.7.1 and 13.8.1.

In order to use the programs, one has to key-in (enter) all the program steps into the computer. Magnetic diskettes are used to store and retrieve the programs. Once the program steps are entered, they can be executed by entering the command "RUN".

The program executes two types of operations: Type 1 is for analyzing a particular column sections by carrying out the analysis for a given neutral-axis depth c. This value of c is assumed by the user at the outset for the first input cycle, to be adjusted subsequently. Type 2 is for evaluating coordinate values for constructing load-moment strength interaction diagrams.

Operation Input 1: Analysis of Column Sections

The program asks for the necessary input variables. The following symbols for the variables are used in the program.

FC' = compressive strength of concrete, psi

FY = yield strength of steel, psi

B   = width of the column, in.

H   = total thickness of the column or diameter of the column

β_1 = adjustment factor β_1 for f'_c

C   = depth of neutral axis, in.

PU = specified factored axial load

N   = number of rows of bars in a rectangular column section or number of bars in a circular column section

DM = distance of the tension bar closest to the tension face measured from the extreme compression fiber

D   = depth of a specific layer of bars from the compression fibers

A   = area of the bars located at depth D

When analyzing a section for the entered neutral-axis depth c, the program calculates and prints the values of c_b, P_{ub}, M_{ub}, and e_b, the depth of neutral axis, design axial load, design moment, and eccentricity, respectively, for the balanced condition and the maximum permissible axial load P_{u0}.

For the first cycle, the computer calculates the strength reduction factor ϕ value corresponding to the factored axial load input. If the design load strength P_u differs from the applied factored load P_u, the computer goes through additional computational cycles using the *design* P_u value in the strength reduction factor ϕ equation. Convergence results in applying the correct ϕ value for the computation of P_u and moment strength M_u of the particular section which is being analysed. This is necessary since ϕ increases in the tension type failure from $\phi = 0.70$ for tied columns and $\phi = 0.75$ for spirally reinforced columns up to a maximum $\phi = 0.90$ for the pure bending state.

The output printout lists the actual permissible design load P_u, permissible design moment M_u and the corresponding eccentricity e. Based on the output values of these three last parameters and the c values used in the input transfer, the user adjusts the geometrical properties of the first section. The adjustment involves reducing or increasing the dimensions and/or the reinforcement area. This step can also involve using another trial "c" value in order to arrive at the proper column section where the output eccentricity is *almost equal* to the eccentricity of the actual factored load P_u applied to the column.

Operation Input 2: Coordinate Values
for Load-Moment Strength Interaction Plots

In order to execute the program, a choice of the dimensions B and H of a hypothetical cross-section for input is to be based on a predetermined γ value for the chart's set of diagrams and the appropriate β_1. Hypothetical reasonable values for a reinforcement area A_s, a design load P_u and a neutral axis depth c are used in the input in order to maintain the flow of calculations for the $(\phi P_n/A_g)/(\phi M_n/A_g h)$ coordinates.

The program calculates and prints the coordinate values of $\phi P_n/A_g$ (psi) and $\phi M_n/A_g h$ (psi) for a range of gross reinforcement ratios from $\rho_g = 1\%$ to $\rho_g = 8\%$ for the particular input value of β_1 corresponding to the concrete compressive strength f_c' used. It gives the maximum permissible axial load value $\phi P_{no(max)}/A_g$ and indicates the co-ordinates of the point for $f_s = 0$, namely, when the stress in the extreme tension bar changes from compression to tension and $f_s = f_y$, namely, the balanced condition.

It also prints the following input data values preceeding the tabulation of the coordinate values:

$f_c' = $ Compressive strength of concrete, psi

$f_y = $ Reinforcement yield strength, psi

$\rho_g = $ Gross reinforcement ratio $\dfrac{A_{st}}{A_g}$

$\gamma = $ Ratio $\dfrac{\text{center-line distance between extreme bars}}{\text{total thickness } h \text{ of the rectangular column}}$

or

$\gamma = $ Ratio $\dfrac{\text{center line distance between extreme bars}}{\text{External Diameter } h \text{ of the circular column}}$

The program checks whether the minimum and maximum limits of the total permissible reinforcement ratio ρ_g are followed then prints the appropriate error messages, namely, "Reinforcement ratio is less than 1%" or "Reinforcement ratio is greater than 8%."

Biaxial Loading

The third computer program, EGNAWY4, as documented in Fig. 13.31, presents steps for the analysis of rectangular columns subjected to biaxial bending. The applicable equations and the flow chart are given in Sections 13.7 and 13.9. A contour β-factor chart for moment ratios in columns subjected to biaxial bending is given in Fig. 13.26. The flow chart outlining the principal operational steps is shown in Fig. 13.27.

EGNAWY2-RECTANGULAR COLUMNS

```
1   REM
10  HOME : CLEAR
20  REM   PROGRAM ENGINEERING CONC
    RETE DESIGN
30  T2$ = "*******ENGINEERING PROG
    RAM*******"
40  HOME : PRINT T2$: PRINT "**IN
    PUT DATA:**": PRINT "-------
    -------": VTAB 8
50  DIM D(20),A(20),FS(20),P1(100
    ),M1(100)
51  WW = 0.0:ZK = 1
52  XX = 0.0
53  FL = 0.0
55  J = 1
56  INPUT "TYPE OF PROBLEM=";TY
60  INPUT "FC'=";FC: INPUT "FY=";
    FY
70  INPUT "B=";B: INPUT "H=";H: INPUT
    "B1=";B1
100 INPUT "C=";C: INPUT "PU=";PU

120 INPUT "NO. OF ROWS N=";N
140 FOR I = 1 TO N
150 PRINT "INPUT DEPTH AND AREA
    OF ROW NO.";I
170 INPUT "D=";D(I): INPUT "A=";
    A(I)
180 NEXT I
190 HOME : PRINT "**VERIFICATION
    OF INPUT DATA**"
200 PRINT "FC'=";FC: PRINT "FY="
    ;FY
210 PRINT "B=";B: PRINT "H=";H: PRINT
    "B1=";B1
220 PRINT "C=";C: PRINT "PU=";PU

230 PRINT "N=";N
250 AST = 0.0
260 FOR I = 1 TO N
270 PRINT "D AND A OF ROW NO.";TAB(
    20);I; TAB( 23);"ARE"; TAB(
    28);D(I); TAB( 31);"AND"; TAB(
    36);A(I)
280 AST = AST + A(I)
290 NEXT I
300 INPUT "ARE THE DATA CORRECT?
    Y/N";Z$
320 IF Z$ = "N" THEN  GOTO 10
330 HOME : PRINT T2$: PRINT : PRINT
    : PRINT : PRINT : INVERSE : FLASH
    : HTAB 8: PRINT "PLEASE WAIT
    FOR CALCULATION": NORMAL
331 IF TY = 2 THEN  GOTO 1021
332 IF TY = 2 THEN C = 1.075 * H
    / B1
```

Selection of the type of analysis

1. Analysis for a given c, the depth of neutral axis
2. Calculate a set of values for P_u and M_u for interaction diagram.

Input the design variables: f'_c, f_y, b, h, β_1, c, P_u, N, d_i, A_{si}.

Verification of input data

If the input data are correct, type Y. Otherwise, type N and re-input the correct data.

Figure 13.29 Program steps: analysis of rectangular columns under uniaxial bending.

```
340 A1 = AST / (B * H)
350 IF A1 < 0.0095 THEN GOTO 66
    0
360 IF A1 > 0.0801 THEN GOTO 69
    0
370 CBAL = 87000.0 * D(N) / (8700
    0.0 + FY)
380 CC = CBAL
390 GOSUB 750
400 PB = 0.7 * (0.85 * FC * B1 *
    CBAL * B + S1)
410 MB = 0.7 * (0.85 * FC * B1 *
    CBAL * B * (H / 2 - CBAL * B
    1 / 2) + S2)
420 EB = MB / PB
430 PO = 0.7 * 0.8 * (0.85 * FC *
    (B * H - AST) + AST * FY)
435 IF TY = 2 THEN PU = PB
440 W = 0.1 * FC * B * H
441 ZK = ZK + 1
450 IF W < PB THEN PI = 0.9 - 2.
    0 * PU / (FC * B * H)
460 IF W > = PB THEN PI = 0.9 -
    0.2 * PU / PB
465 IF PI < 0.7 THEN PI = 0.7
470 IF B1 * C < = H THEN AA = B
    1 * C
480 IF B1 * C > H THEN AA = H
490 CC = C
500 GOSUB 750
510 PU = PI * (0.85 * FC * AA * B
    + S1)
511 IF TY = 1 AND ZK = 2 THEN GOTO
    440
512 IF TY = 2 THEN GOTO 530
520 IF PU > PO THEN GOTO 720
530 MU = PI * (0.85 * FC * AA * B
    * (H / 2 - AA / 2) + S2)
540 EC = MU / PU
545 IF TY = 2 THEN GOTO 900
550 HOME : PRINT : PRINT "**FINA
    L RESULTS**"
560 PRINT "CB (IN)    =";CBAL
570 PRINT "PUB (LB)   =";PB
580 PRINT "MUB (IN-LB) =";MB
590 PRINT "EB (IN)    =";EB
600 PRINT "PUO (LB)   =";PO
610 PRINT "PI         =";PI
620 PRINT "PU (LB)    =";PU
630 PRINT "MU (IN-LB) =";MU
640 PRINT "E (IN)     =";EC
650 GOTO 740
660 HOME : PRINT : PRINT "**FINA
    L RESULTS**"
670 PRINT "REINFORCEMENT RATIO I
    S LESS THAN 1 %"
680 GOTO 740
690 HOME : PRINT : PRINT "**FINA
    L RESULTS**"
700 PRINT "REINFORCEMENT RATIO I
    S GREATER THAN 8 %"
```

If ρ_g is <1% or ρ_g is >8% go to step 660 or 690, respectively.

$$c_b = \frac{87,000}{87,000 + f_y} d_{max}$$

$$P_{ub} = 0.7(0.85f'_c\beta_1 c_b b + \Sigma f_{si}A_i)$$

$$M_{ub} = 0.7[0.85f'_c\beta_1 c_b b \frac{h}{2} - \frac{\beta_1 c_b}{2}$$

$$+ \Sigma f_{si}A_i \frac{h}{2} - d_i]$$

Contributions $\Sigma f_{si}A_i$ and $\Sigma f_{si}A_i\left(\frac{h}{2} - d_i\right)$ are calculated using the subroutine starting with statement 750.

$$e_b = \frac{M_{ub}}{P_{ub}}$$

$$P_{u0} = 0.7 \times 0.8[0.85f'_c(A_g - A_{st}) + A_{st}f_y]$$

If $0.7P_{nb} \geq 0.1f'_cA_g$, $\phi = 0.9 - \frac{2.0P_u}{f'_cA_g} \geq 0.7$

If $0.7P_{nb} < 0.1f'_cA_g$, $\phi = 0.9 - \frac{0.2P_u}{0.7P_{nb}} \geq 0.7$

$$P_u = \phi(0.85f'_cba + \Sigma f_{si}A_i)$$

$$M_u = \phi[0.85f'_cba\left(\frac{h}{2} - \frac{a}{2}\right) + \Sigma f_{si}A_i\left(\frac{h}{2} - d_i\right)]$$

If $P_u > P_{u0}$, go to step 720 and stop.

Print the results (for the analysis for a given c):
$c_b, P_{ub}, M_{ub}, e_b, P_{u0}, \phi, P_u, M_u$, and e.

Print the message "reinforcement ratio is less than 1%".

Print the message "reinforcement ratio is greater than 8%".

Figure 13.29 (*cont.*)

```
710    GOTO 740
720    HOME : PRINT : PRINT "**FINA
       L RESULTS**"
730    PRINT "SPECIFIED DESIGN LOAD
        PU IS GREATER THAN MAXIMUM
        POSSIBLE AXIAL FORCEPUO"
740    END
750    REM   SUBROUTINE DD AND EE
760    S1 = 0.0
770    S2 = 0.0
780    FOR I = 1 TO N
790    FS(I) = (CC - D(I)) * 87000.0
        / CC
800    IF   ABS (FS(I)) <  = FY THEN
        GOTO 830
810    IF FS(I) < 0.0 THEN FS(I) =
        - 1 * FY
820    IF FS(I) > 0.0 THEN FS(I) =
       FY
830    IF FS(I) < 0.0 THEN   GOTO 85
       0
840    FS(I) = FS(I) - 0.85 * FC
850    S1 = S1 + FS(I) * A(I)
860    S2 = S2 + FS(I) * A(I) * (H /
        2 - D(I))
870    NEXT I
880    RETURN
900    P1(J) = PU / (H * B)
910    M1(J) = MU / (H ^ 2 * B)
911    IF C = D(N) THEN YY = P1(J)
920    J = J + 1
930    IF PU <  = 0.1 * FC * B * H THEN
       C = C - 0.005 * H
931    IF PU > 0.1 * FC * B * H THEN
       C = C - 0.03 * H
932    IF C > D(N) THEN C = C - 0.0
       3 * H
933    IF C <  = D(N) AND FL = 0.0 THEN
       C = D(N)
934    IF C = D(N) THEN FL = 1.0
940    IF PU > 0.0 THEN   GOTO 440
951    PRINT : PRINT : PRINT "*****
       *****************************
       *********"
980    PRINT "FC'      =";FC: PRINT
       "FY       =";FY
981    PRINT "RG       =";A1
982    PRINT "GAMMA       =";(D(N) -
       D(1)) / H
983    PRINT : PRINT : PRINT "*****
       ***COORDINATES OF INTERACTIO
       N DIAGRAM FOR RG=";A1;"*****
       ***"
985    PRINT "PU/AG(PSI)
           MU/(AG*H)(PSI)"
986    PRINT PO / (H * B); TAB( 25)
       ;"0.0000"
990    FOR K = 1 TO J - 1
991    IF XX = 1.0 THEN   GOTO 995
992    IF PB / (H * B) < P1(K) THEN
        GOTO 995
993    PRINT PB / (H * B); TAB( 25)
       ;MB / (H ^ 2 * B);"FS=FY"
994    PRINT :XX = 1.0
```

Print the message "specified design load P_u is greater than maximum possible axial force P_{u0}".

$$f_{si} = 87{,}000\,\frac{c - d_i}{c}$$

If $f_{si} > f_y$, $f_{si} = f_y$, $f_{si} = f_{si} - 0.85f'_c$

If $f_{si} < f_y$ and $|f_{si}| > f_y$, $f_{si} = -f_y$

$s_1 = \Sigma\, f_{si}A_i$

$s_2 = \Sigma\, f_{si}A_i\,\dfrac{h}{2} - d_i$

Decrease c from $1.075h/\beta_1$
in steps of $0.03h$ and calculate
a set of values for P_u/bh and
M_u/bh^2.
If $P_u \leq 0.1f'_c bh$, decrease c by $0.005h$ instead of $0.03h$.

Print the results: f'_c, f_y, ρ_g, γ
and a set of values of
P_u/bh and M_u/bh^2.

Figure 13.29 *(cont.)*

```
995   IF P1(K) = YY THEN  GOTO 100
      2
1000  PRINT P1(K); TAB( 25);M1(K)

1001  GOTO 1005
1002  PRINT P1(K); TAB( 25);M1(K)
      ; TAB( 35);"FS=0.0"
1005  PRINT " "
1010  NEXT K
1020  IF WW = 0.08 GOTO 740
1021  XX = 0.0
1022  FL = 0.0
1030  J = 1
1050  WW = WW + 0.01
1060  AST = WW * H * B
1070  N = 2
1080  A(1) = AST / 2.0
1090  A(2) = AST / 2.0
1100  GOTO 332
```

Identify the values when
$f_s = 0$ and $f_s = f_y$

Increase ρ_g from 0.01 in steps
of 0.01 up to 0.08

Figure 13.29 (*cont.*)

```
      EGNAWY3-CIRCULAR COLUMNS
1     REM
10    HOME : CLEAR
20    REM   PROGRAM ENGINEERING CONC
      RETE DESIGN
30    T2$ = "*******ENGINEERING PROG
      RAM*******"
40    HOME : PRINT T2$: PRINT "**IN
      PUT DATA:**": PRINT "-------
      ------": VTAB 8
1010  DIM D(30),FS(30),P2(100),M2
      (100)
1012  FL = 0.0
1013  WW = 0.0:ZK = 1
1014  XX = 0.0
1015  J = 1
1020  PRINT "INPUT TYPE OF COLUMN
      SPIRAL OR TIED": INPUT Z$
1021  INPUT "TYPE OF PROBLEM=";TY
      : INPUT "FC=";FC: INPUT "FY=
      ";FY
1022  INPUT "H=";H: INPUT "DS=";D
      S: INPUT "B1=";B1
1023  INPUT "AS=";AS: INPUT "NO.
      OF BARS N=";N
1024  INPUT "C=";C: INPUT "PU=";P
      U: INPUT "DMAX=";DM
1025  HOME : PRINT "**VERIFICATIO
      N OF INPUT DATA**": PRINT
1026  PRINT "THE COLUMN IS"; TAB(
      30);Z$
1027  PRINT "FC'=";FC: PRINT "FY=
      ";FY
1028  PRINT "H=";H: PRINT "DS=";D
      S: PRINT "B1=";B1
1029  PRINT "AS=";AS: PRINT "NO.
      OF BARS N=";N
1030  PRINT "C=";C: PRINT "PU=";P
      U: PRINT "DMAX=";DM
1031  INPUT "ARE THE DATA CORRECT
      ? Y/N";Y$
1032  IF Y$ = "N" THEN  GOTO 10
```

Selection of the type of analysis

1. Analysis for a given c, the depth of neutral axis.
2. Calculate a set of values for P_u and M_u for interaction diagram.

Input the design variables.
Type of column: spiral or tied,
f'_c, f_y, b, h, D_s, β_1, c, P_u, N,
A_{st}, d_{max}.

Verification of input data. If the input data are correct, type Y. Otherwise, type N and re-input the correct data.

Figure 13.30 Program steps: analysis of circular columns.

```
1033  HOME : PRINT T2$: PRINT : PRINT
      : PRINT : PRINT : INVERSE : FLASH
      : HTAB 8: PRINT "PLEASE WAIT
      FOR CALCULATIONS": NORMAL
1034  IF TY = 2 THEN N = 12
1035  IF TY = 2 THEN  GOTO 1850
1070 DG = AS * 28 / (22 * H ^ 2)
1080  IF DG < 0.0095 THEN  GOTO 1
      480
1090  IF DG > 0.0801 THEN  GOTO 1
      510
1100 AI = AS / N
1120 CB = 87000 * DM / (87000 + F
      Y)
1130 CC = CB
1131 GA = 360 / N


1140  GOSUB 1570
1150 PN = 0.85 * FC * CA + S1
1151  IF Z$ = "TIED" THEN P1 = 0.
      7 * PN
1152  IF Z$ = "SPIRAL" THEN P1 =
      0.75 * PN
1154  IF Z$ = "SPIRAL" THEN M1 =
      0.75 * MN
1160 MN = 0.85 * FC * CA * X + S2

1163  IF Z$ = "TIED" THEN M1 = 0.
      7 * MN
1164  IF Z$ = "SPIRAL" THEN M1 =
      0.75 * MN
1170 EB = MN / PN
1176  IF TY = 2 THEN PU = P1
1177  IF TY = 2 THEN CC = 1.04 *
      H / B1
1180 W = 0.1 * FC * 22 * H ^ 2 /
      28
1181  IF TY = 1 THEN ZK = ZK + 1
1190  IF Z$ = "TIED" THEN  GOTO 1
      250
1200  IF W < PN THEN PI = 0.9 - 1
      .5 * PU / (FC * H ^ 2 * 22 /
      28)
1210  IF W >  = PN THEN PI = 0.9 -
      0.15 * PU / PN
1220  IF PI < 0.75 THEN PI = 0.75

1230 PO = 0.75 * 0.85 * (0.85 * F
      C * (22 * H ^ 2 / 28 - AS) +
      AS * FY)
1240  GOTO 1290
1250  IF W < PN THEN PI = 0.9 - 2
      * PU / (FC * 22 * H ^ 2 / 2
      8)
1260  IF W >  = PN THEN PI = 0.9 -
      0.2 * PU / PN
1270  IF PI < 0.7 THEN PI = 0.7
1280 PO = 0.7 * 0.8 * (0.85 * FC *
      (22 * H ^ 2 / 28 - AS) + AS *
      FY)
1290  IF B1 * C > H THEN AA = H
1300  IF B1 * C <  = H THEN AA =
      B1 * C
1310  IF TY = 1 THEN CC = C
1320  GOSUB 1570
```

If ρ_g is <1% or ρ_g >8%, go to steps 1480 or 1510, respectively.

$$c_b = \frac{87{,}000}{87{,}000 + f_y}\, d_{max}$$

$$\gamma = \frac{360}{N}$$

$P_{nb} = [0.85 f'_c \text{ (compression area)} + \Sigma\, f_{si} A_i]$

M_{nb} = moment contribution of concrete

$$+ \Sigma\, f_{si} A_i \left(\frac{h}{2} - d_i\right)$$

The compression area and moment contribution from concrete, $\Sigma\, f_{si} A_i$ and $\Sigma\, f_{si} A_i \left(\frac{h}{2} - d_i\right)$ are calculated in the subroutine starting with statement 1570.

For tied column: $F_{ub} = 0.7 P_{nb}$, $M_{ub} = 0.7 M_{nb}$
For spiral column: $P_{ub} = 0.75 P_{nb}$, $M_{ub} = 0.75 M_{nb}$

If $\phi P_{nb} \geq 0.1 f'_c A_g$: $\phi = 0.9 - \dfrac{1.5 P_u}{f'_c A_g}$

≥ 0.75 for spiral column, or

$\phi = 0.9 - \dfrac{2.0 P_u}{f'_c A_g} \geq 0.70$ for tied column.

If $\phi P_{nb} < 0.1 f'_c A_g$: $\phi = 0.9 - \dfrac{0.15 P_u}{0.75 P_{nb}}$

≥ 0.75 for spiral column, or

$\phi = 0.9 - \dfrac{0.2 P_u}{0.7 P_{nb}} \geq 0.70$ for tied column

$P_u = \phi P_n$, $M_u = \phi M_n$

$P_{u0} = 0.75 \times 0.85 [0.85 f'_c (A_g - A_{st}) + A_{st} f_y]$
   for spiral column,

$P_{u0} = 0.7 \times 0.8 [0.85 f'_c (A_g - A_{st}) + A_{st} f_y]$
   for tied column.

If $P_u > P_{u0}$, go to step 1540 and stop.

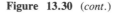

Figure 13.30 (*cont.*)

```
1330 PU = PI * (0.85 * FC * CA +
     S1)
1331  IF TY = 2 THEN  GOTO 1350
1332  IF TY = 1 AND ZK = 2 THEN  GOTO
     1180
1340  IF PU > PO + 0.00001 THEN  GOTO
     1540
1350 MU = PI * (0.85 * FC * CA *
     X + S2)
1360 E = MU / PU
1365  IF TY = 2 THEN  GOTO 1720
1370  HOME : PRINT : PRINT "**FIN
     AL RESULTS**"
1380  PRINT "CB (IN)     =";CB
1390  PRINT "PUB (LB)    =";P1
1400  PRINT "MUB (IN-LB)=";M1
1410  PRINT "EB (IN)     =";EB
1420  PRINT "PUO (LB)    =
     ";PO
1430  PRINT "PI          =";PI
1440  PRINT "PU (LB)     =";PU
1450  PRINT "MU(IN-LB)   =";MU
1460  PRINT "E (IN)      =";E
1470  GOTO 1560
1480  HOME : PRINT : PRINT "**FIN
     AL RESULTS**": PRINT
1490  PRINT "REINFORCEMENT RATIO
     IS LESS THAN 1%"
1500  GOTO 1560
1510  HOME : PRINT : PRINT "**FIN
     AL RESULTS**": PRINT
1520  PRINT "REINFORCEMENT RATIO
     IS GREATER THAN 8%"
1530  GOTO 1560
1540  HOME : PRINT : PRINT "**FIN
     AL RESULTS**": PRINT
1550  PRINT "SPECIFIED DESIGN LOA
     D PU IS GREATER THAN MAXIMUM
     POSSIBLE AXIAL FORCE PUO"
1560  END
1570  REM  SUBROUTINE BB AND CC
1580 S1 = 0.0
1590 S2 = 0.0
1600  FOR I = 1 TO N
1610 D(I) = H / 2 - 0.5 * DS * ( COS
     ((I * GA - GA) * 3.14159 / 1
     80.))
1620 FS(I) = (CC - D(I)) * 87000 /
     CC
1621  IF FS(I) > 0.0 AND FS(I) >
     FY THEN FS(I) = FY
1622  IF FS(I) < 0. AND  ABS (FS(
     I)) > FY THEN FS(I) = - FY
1623  IF FS(I) > 0. THEN FS(I) =
     FS(I) - 0.85 * FC
1630 S1 = S1 + FS(I) * AI
1640 S2 = S2 + FS(I) * AI * (H /
     2 - D(I))
1650  NEXT I
1660 A9 = B1 * CC: IF A9 > H THEN
     A9 = H
```

$P_u = \phi[0.85f_c'\ (\text{compression area}) + \Sigma\ f_{si}A_i]$

$M_u = \phi[\text{moment contribution of concrete}$
$\quad + \Sigma\ f_{si}A_i\ \dfrac{h}{2} - d_i\ ]\ e = \dfrac{M_u}{P_u}$

Print the results (analysis for a given c):
c_b, P_{ub}, M_{ub}, e_b, P_{u0}, ϕ, P_u, M_u, and e.

Print the message "reinforcement ratio is less than 1%".

Print the message "reinforcement ratio is greater than 8%".

Print the message "specified design load is greater than maximum permissible axial force P_{u0}".

$d_i = \dfrac{h}{2} - 0.5D_s\ [\cos\ (i\gamma_i - \gamma)]$

$f_{si} = 87{,}000\ \dfrac{c - d_i}{c}$

If $f_{si} > f_y$, $f_{si} = f_y - 0.85f_c'$

If $f_{si} > 0$, $f_{si} = f_{si} - 0.85f_c'$

If $f_{si} < 0$ and $|f_{si}| > f_y$, $f_{si} = -f_y$

$s_1 = \Sigma\ f_{si}A_i$

$s_2 = \Sigma\ f_{si}A_i\left(\dfrac{h}{2} - d_i\right)$

Figure 13.30 (*cont.*)

```
1661 H9 = H / 2
1662  IF A9 < H9 THEN Z1 =  SQR (
     0.25 * H ^ 2 - (H / 2 - A9) ^
     2)
1663  IF A9 > H9 THEN Z1 =  SQR (
     0.25 * H ^ 2 - (A9 - H / 2) ^
     2)
1664  IF A9 < H9 THEN AL =  ATN (
     Z1 / (H / 2 - A9))
1665  IF A9 > H9 THEN AL =  ATN (
     Z1 / (A9 - H / 2))
1666  IF A9 > H9 THEN AL = 3.1415
     9 - AL
1690 CA = H ^ 2 * (AL -  SIN (AL)
     *  COS (AL)) / 4
1700 X = H ^ 3 *  SIN (AL) ^ 3 /
     (12 * CA)
1710  RETURN
1720 P2(J) = PU / (22 * H ^ 2 / 2
     8)
1730 M2(J) = MU / (H * 22 * H ^ 2
     / 28)
1731  IF CC = DM THEN YY = P2(J)
1740 J = J + 1
1741  IF PU >  = 0.1 * FC * 22 *
     H ^ 2 / 28 THEN CC = CC - 0.
     03 * H
1742  IF PU < 0.1 * FC * 22 * H ^
     2 / 28 THEN CC = CC - 0.005 *
     H
1743  IF CC > DM THEN CC = CC - 0
     .03 * H
1744  IF CC <  = DM AND FL = 0.0 THEN
     CC = DM
1745  IF CC = DM THEN FL = 1.0
1751  IF CC < 1 GOTO 1780
1760  IF PU > 0 THEN  GOTO 1180
1780  PRINT :  PRINT :  PRINT "****
     ****************************
     ****************"
1781  PRINT :  PRINT :  PRINT :  PRINT
     "FC '        =";FC:  PRINT "FY
             =";FY
1782  PRINT "RG         =";DG:  PRINT
     "GAMMA      =";DS / H
1801  PRINT " ":  PRINT " "
1809  PRINT :  PRINT :  PRINT "****.
     ****COORDINATES OF INTERACTI
     ON DIAGRAM FOR RG=";DG;"****
     *****"
1810  PRINT "PU/AG(PSI)      MU/(A
     G*H)(PSI)"
1811  PRINT :  PRINT PO / (22 * H ^
     2 / 28);  TAB( 15);"0.0000"
1820  FOR K = 1 TO J - 1
1821  IF XX = 1.0 THEN  GOTO 1825
1822 ZX = P1 / (22 * H ^ 2 / 28)
1823  IF ZX > P2(K) THEN  PRINT Z
     X;  TAB( 15);M1 / (22 * H ^ 3
     / 28);"FS=FY"
1824  IF ZX > P2(K) THEN XX = 1.0
```

Compression area = $\dfrac{h^2}{4}$ $(\alpha - \sin\alpha\cos\alpha)$

Moment contribution of concrete

$= \dfrac{h^3\,(\sin^3\alpha)}{12}$

$\alpha = \tan^{-1}\dfrac{\sqrt{(h^2/4) - [(h/2) - a]^2}}{(h/2) - a}$ for $a < \dfrac{h}{2}$

$= \pi - \tan^{-1}\dfrac{\sqrt{(h^2/4) - (a - h/2)^2}}{a - h/2}$ for $a > \dfrac{h}{2}$

Decrease c from $1.04h/\beta_1$ in stops of $0.03h$ and calculate a set of values for $P_u/\pi h^2/4$ and $M_u/\pi h^3/4$.

If $P_u \leq 0.1f'_c\,(\pi h^2/4)$, decrease c by $0.005h$ instead of $0.03h$.

Print the results: f'_c, f_y, ρ_g, γ and a set of values of $P_u/\pi h^2/4$ and $M_u/\pi h^3/4$.

Identify the values when $f_s = 0$ and $f_s = f_y$.

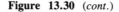

Figure 13.30 (*cont.*)

```
1825   IF P2(K) = YY THEN  GOTO 18
       32
1830   PRINT P2(K); TAB( 15);M2(K)

1831   GOTO 1840
1832   PRINT P2(K); TAB( 15);M2(K)
       ; TAB( 25);"FS=0.0"
1840   NEXT K
1850   IF WW = 0.08 THEN  GOTO 156
       0
1860   XX = 0.0
1861   FL = 0.0
1870   J = 1
1880   WW = WW + 0.01
1890   AS = WW * H ^ 2 * 22 / 28
1900   GOTO 1070
```

Increase ρ_g from 0.01 in steps of 0.01
up to 0.08.

Figure 13.30   (*cont.*)

EGNAWY4—BIAXIAL BENDING

```
1    REM
10   HOME : CLEAR
20   REM  PROGRAM ENGINEERING CONC
     RETE DESIGN
30   T2$ = "*******ENGINEERING PROG
     RAM*******"
40   HOME : PRINT T2$: PRINT "**IN
     PUT DATA:**": PRINT "-------
     ------": VTAB 8
41   INPUT "TYPE OF PROBLEM=";TY
42   IF TY = 2 THEN  GOTO 160
50   INPUT "PN=";PN: INPUT "MNX=";
     MX: INPUT "MNY=";MY: INPUT "
     B=";B: INPUT "H=";H: INPUT "
     BE=";BE
60   HOME : PRINT "**VERIFICATION
     OF INPUT DATA**": PRINT
70   PRINT "PN=";PN: PRINT "MNX=";
     MX: PRINT "MNY=";MY: PRINT "
     B=";B: PRINT "H=";H: PRINT "
     BE=";BE
71   INPUT "ARE THE DATA CORRECT?
     Y/N";Z$
72   IF Z$ = "N" THEN  GOTO 10
73   HOME : PRINT T2$: PRINT : PRINT
     : PRINT : PRINT : INVERSE : FLASH
     : HTAB 8: PRINT "PLEASE WAIT
     FOR CALCULATION": NORMAL
80   M1 = MX + MY * (H / B) * (1 -
     BE) / BE
90   M2 = MY + MX * (B / H) * (1 -
     BE) / BE
100  A1 = MX / M1
110  B1 = MY / M2
120  PRINT "MNX/MOX =";A1
130  PRINT "MNY/MOY =";B1
140  IF MY / MX > B / H THEN  PRINT
     "MOY=";M2
150  IF MY / MX < B / H THEN  PRINT
     "MOX=";M1
151  IF TY = 1 THEN  GOTO 740
```

Selection of the type of analysis:

1. Calculate the controlling equivalent uniaxial
   moment M_{0x} or M_{0y}.
2. Calculate the nominal load strength and moment
   strength of the section for a given neutral axis
   depth C

Input the design variables:
P_n, M_{nx}, M_{ny}, b, h, β.

Verification of input data. If the
input data are correct, type *Y*.
Otherwise, type *N* and re-input
the correct data.

$$M_{0x} = M_{nx} + M_{ny}\frac{h}{b}\left(\frac{1-\beta}{\beta}\right)$$

$$M_{0y} = M_{ny} + M_{0x}\frac{b}{h}\left(\frac{1-\beta}{\beta}\right)$$

Print: $\dfrac{M_{nx}}{M_{0x}}$, $\dfrac{M_{ny}}{M_{0y}}$,

M_{0y} if $\dfrac{M_{ny}}{M_{nx}} > \dfrac{b}{h}$

M_{0x} if $\dfrac{M_{ny}}{M_{nx}} < \dfrac{b}{h}$

Figure 13.31   Program steps: analysis of rectangular columns under biaxial
bending.

```
160   INPUT "FC'=";FC: INPUT "FY="
      ;FY
161   INPUT "B=";B: INPUT "H=";H: INPUT
      "B1=";B1
162   INPUT "C=";C: INPUT "PN=";PU

163   INPUT "NO.OF ROWS N=";N
164   FOR I = 1 TO N
165   PRINT "INPUT DEPTH AND AREA
      OF ROW NO.";I
170   INPUT "D=";D(I): INPUT "A=";
      A(I)
180   NEXT I
190   HOME : PRINT "**VERIFICATION
      OF INPUT DATA**"
200   PRINT "FC'=";FC: PRINT "FY="
      ;FY
210   PRINT "B=";B: PRINT "H=";H: PRINT
      "B1=";B1
220   PRINT "C=";C: PRINT "PU=";PU

230   PRINT "N=";N
250 AST = 0.0
260   FOR I = 1 TO N
270   PRINT "D AND A OF ROW NO."; TAB(
      20);I; TAB( 23);"ARE"; TAB(
      28);D(I); TAB( 31);"AND"; TAB(
      36);A(I)
280 AST = AST + A(I)
290   NEXT I
300   INPUT "ARE THE DATA CORRECT?
      Y/N";Z$
320   IF Z$ = "N" THEN   GOTO 10
330   HOME : PRINT T2$: PRINT : PRINT
      : PRINT : PRINT : INVERSE : FLASH
      : HTAB 8: PRINT "PLEASE WAIT
      FOR CALCULATION": NORMAL
340 A1 = AST / (B * H)
350   IF A1 < 0.0095 THEN   GOTO 66
      0
360   IF A1 > 0.0801 THEN   GOTO 69
      0
470 AA = B1 * C
480   IF AA > H THEN AA = H
490 CC = C
500   GOSUB 750
510 PU = 0.85 * FC * AA * B + S1
530 MU = 0.85 * FC * AA * B * (H /
      2 - AA / 2) + S2
540 EC = MU / PU
550   HOME : PRINT : PRINT "**FINA
      L RESULTS**"
570   PRINT "PN(LB)      =";PU
580   PRINT "MN(IN-LB)   =";MU
590   PRINT "E(IN)       =";EC
600   GOTO 740
660   HOME : PRINT : PRINT "**FINA
      L RESULTS**"
670   PRINT "REINFORCEMENT RATIO I
      S LESS THAN 1 %"
680   GOTO 740
690   HOME : PRINT : PRINT "**FINA
      L RESULTS**"
```

Input the design variables:
$f'_c, f_y, b, h, \beta_1, c, P_n, N, d_i, A_{si}$

Verification of input data

If the input data is correct, type *Y*. Otherwise, type *N* and re-input the correct data.

If ρ_g is <1% or ρ_g is >8% go to steps 660 or 690 respectively.

$a = \beta_1 c$

If $a > h$, $a = h$.

$P_n = 0.85 f'_c\, ab + s_1$

$M_n = 0.85 f'_c\, ab\left(\dfrac{h}{2} - \dfrac{a}{2}\right) + s_2$

Print the results: (for the analysis for a given *c*)

Print the message "reinforcement ratio is less than 1%".

Print the message "reinforcement ratio is greater than 8%".

Figure 13.31 *(cont.)*

```
700  PRINT "REINFORCEMENT RATIO I
     S GREATER THAN 8 %"
710  GOTO 740
720  HOME : PRINT : PRINT "**FINA
     L RESULTS**"
740  END
750  REM    SUBROUTINE DD AND EE
760  S1 = 0.0
770  S2 = 0.0
780  FOR I = 1 TO N
790  FS(I) = (CC - D(I)) * 87000.0
     / CC
800  IF   ABS (FS(I)) <  = FY THEN
     GOTO 830
810  IF FS(I) < 0.0 THEN FS(I) =
     - 1 * FY
820  IF FS(I) > 0.0 THEN FS(I) =
     FY
830  IF FS(I) < 0.0 THEN   GOTO 85
     0
840  FS(I) = FS(I) - 0.85 * FC
850  S1 = S1 + FS(I) * A(I)
860  S2 = S2 + FS(I) * A(I) * (H /
     2 - D(I))
870  NEXT I
880  RETURN
```

$$f_{si} = 87{,}000\left(\frac{c - d_i}{c}\right),$$

If $f_{si} > 0$, $f_{si} = (f_{si} - 0.85 f'_c)$

If $f_{si} > f_y$, $f_{si} = (f_y - 0.85 f'_c)$

If $f_{si} < f_y$ and $|f_{si}| > f_y$, $f_{si} = -f_y$

$s_1 = \Sigma\, f_{si} A_i$

$s_2 = \Sigma\, f_{si} A_i \left(\dfrac{h}{2} - d_i\right)$

Figure 13.31 (*cont.*)

The program executes two types of operations.

Operation type 1: The input consists of the required axial load P_u/ϕ designated as P_n (lb.) and the required uniaxial moments M_{ux}/ϕ and M_{uy}/ϕ (in.-lb.) designated as M_{nx} and M_{ny} respectively in the program, a trial shorter total dimension b (in.) and longer total dimension h (in.) of the column section, and a trial contour β-factor for moment ratios. The computer evaluates and prints out the *required controlling* equivalent uniaxial moment strength M_{ox} and M_{oy} and the moment ratios M_{nx}/M_{ox} and M_{ny}/M_{oy}. M_{ox} controls in designing the column in operation type 2 when $M_{ny}/M_{nx} < b/h$ and M_{oy} controls when $M_{ny}/M_{oy} > b/h$ as discussed in Sec. 9.16.2 and shown in the flow chart.

Operation type 2: The input consists of the concrete compressive strength f'_c (psi), the trial column cross-sectional dimensions b (in.) and h (in.), an assumed neutral axis depth c value, the required axial load P_n, the number of rows of bars, their area (in.^{2}) and distance (in.) from the compression face of the column section. The computer evaluates and prints out the *actual* axial load strength P_n, the actual nominal moment strength M_n (in.-lb), and the corresponding eccentricity e (in.) of the assumed section. The output nominal moment strength M_n represents M_{oxn} or M_{oyn} based on the input cross-sectional column dimensions, namely, whether the larger dimension is entered as perpendicular to the bending axis of the equivalent uniaxially loaded column (H in the program) or the shorter dimension is entered as such. Computation of both M_{oxn} and M_{oyn} is needed for the solution as seen in Ex. 13.35. The resulting moment ratios M_{nx}/M_{oxn} and M_{ny}/M_{oyn} are entered into the β-factor chart of Fig. 13.26 in order to verify if the initially assumed β value for the trial section is close to the β value obtained on the basis of the computer calculated M_{oxn} and M_{oyn}.

If not, another trial and adjustment cycle is executed since a new eccentricity results from each new β-factor value used as outlined in the step-by-step operational procedure in Sec. 9.16.3. The process is repeated until the β-factor value converges for the final nominal equivalent moment strengths M_{oxn} and M_{oyn} about the x and y axes respectively. These moment values have to be at least equal to the *required* controlling equivalent moment strengths M_{ox} or M_{oy} about the respective axes.

13.12.1 Example 13.31: Analysis of a Rectangular Tied Column

Select the appropriate section for a rectangular tied column to carry a service axial dead load $P_{wd} = 120,000$ lb and a service axial live load $P_{wl} = 360,000$ lb acting at an eccentricity (a) $e = 19.5$ in., and (b) $e = 4.0$ in. Given:

$$f_c' = 4000 \text{ psi}$$
$$f_y = 60,000 \text{ psi}$$
$$d' = 3.0 \text{ in.}$$

Solution

(a) *Eccentricity e = 19.5 in:*

factored axial load $P_u = 1.4 \times 120,000 + 1.7 \times 360,000 = 780,000$ lb

Assume a section 27 in. $\times$ 32 in. ($d = 29$). Assume that $\rho_g \simeq 0.015$.

$$A_s = A_s' = 0.0075 \times 27 \times 29$$

$$= 5.87 \text{ in.}^2$$

Try six No. 9 bars on each face $= 6.00$ in.^{2}. Assume that neutral-axis depth $c = 13$ in. and input all the data using the Problem Type 1 Program. The input and output for this cycle are as follows:

```
]PR#1
]RUN
******ENGINEERING PROGRAM*******
**INPUT DATA:**
---------------
TYPE OF PROBLEM=1
FC'=4000
FY=60000
B=27
H=32
B1=0.85
C=13
PU=780000
NO. OF ROWS N=2
INPUT DEPTH AND AREA OF ROW NO.1
D=3
A=6
INPUT DEPTH AND AREA OF ROW NO.2
D=29
A=6
**VERIFICATION OF INPUT DATA**
FC'=4000
FY=60000
B=27
H=32
```

```
B1=.85
C=13
PU=780000
N=2
D AND A OF ROW NO. 1   ARE   3  AND  6
D AND A OF ROW NO. 2   ARE  29 AND  6
ARE THE DATA CORRECT? Y/NY


              PLEASE`WAIT`FOR`CALCULATION

**FINAL RESULTS**
CB (IN)      =17.1632653
PUB (LB)     =923194.714
MUB (IN-LB)  =14527651.4
EB (IN)      =15.7362809
PUO (LB)     =2025408
PI           =.7
PU (LB)      =695793
MU (IN-LB)   =13804374.7
E (IN)       =19.8397723
```

Output $e = 19.84$ in. is approximately equal to the given eccentricity of 19.5 in., while output $P_u = 695,793$ lb is less than 780,000 lb. Hence a new trial c value for the same section or same reinforcement area would not suffice.

Try in the next cycle a reinforcement area $A_s = A'_s = 7.62$ (six No. 10 bars) and a trial value of $c = 14.9$ in.

The input and output values are as follows:

```
*******ENGINEERING PROGRAM*******
**INPUT DATA:**
---------------
TYPE OF PROBLEM=1
FC'=4000
FY=60000
B=27
H=32
B1=.85
C=14.9
PU=780000
NO. OF ROWS N=2
INPUT DEPTH AND AREA OF ROW NO.1
D=3
A=7.62
INPUT DEPTH AND AREA OF ROW NO.2
D=29
A=7.62
**VERIFICATION OF INPUT DATA**
FC'=4000
FY=60000
B=27
H=32
B1=.85
C=14.9
PU=780000
N=2
D AND A OF ROW NO. 1   ARE   3   AND   7.62
D AND A OF ROW NO. 2   ARE   29 AND   7.62
ARE THE DATA CORRECT? Y/NY
*******ENGINEERING PROGRAM*******


          PLEASE`WAIT`FOR`CALCULATION

**FINAL RESULTS**
CB (IN)      =17.1632653
PUB (LB)     =919339.114
MUB (IN-LB) =16246568.6
EB (IN)      =17.6720084
PUO (LB)     =2128103.04
PI           =.7
PU (LB)      =795717.3
MU (IN-LB)   =15953200.1
E (IN)       =20.048829
```

The computer output gives a design $P_u = 795,717$ lb $\simeq$ factored $P_u = 780,000$ lb and output eccentricity $e = 20.05$ in., which is close to the actual eccentricity of 19.5 in. Adopt the design.

(b) *Eccentricity e = 4.0 in:* Assume a section 18 in. $\times$ 24 in. ($d = 21$ in.). Try $A_s = A'_s = 5.0$ in.^{2}. Assume that $c = 18$ in. in the first cycle.

Call the Problem Type 1 Program and input all the data. The computer input and output data are as follows:

```
*******ENGINEERING PROGRAM*******
**INPUT DATA:**
---------------
TYPE OF PROBLEM=1
FC'=4000
FY=60000
B=18
H=24
B1=.85
C=18
PU=780000
NO. OF ROWS N=2
INPUT DEPTH AND AREA OF ROW NO.1
D=3
A=5
INPUT DEPTH AND AREA OF ROW NO.2
D=21
A=5
**VERIFICATION OF INPUT DATA**
FC'=4000
FY=60000
B=18
H=24
B1=.85
C=18
PU=780000
N=2
D AND A OF ROW NO. 1   ARE   3   AND   5
D AND A OF ROW NO. 2   ARE  21   AND   5
ARE THE DATA CORRECT? Y/NY
*******ENGINEERING PROGRAM*******



          PLEASE'WAIT'FOR'CALCULATION

**FINAL RESULTS**
CB (IN)      =12.4285714
PUB (LB)     =440674
MUB (IN-LB) =6713227.48
EB (IN)      =15.2339995
PUO (LB)     =1139488
PI           =.7
PU (LB)      =802802
MU (IN-LB)  =5090866.2
E (IN)       =6.3413721
```

The output gives $e = 6.34$ in. and $P_u = 802,802$ lb. Since both e and P_u are larger than the actual values, decrease the size of the section.

Assume a section size 18 in. $\times$ 20 in. and a value of $c = 16.5$ in for the new input. The computer output as shown on following page gives design $P_u = 789,703$ lb $\sim$ factored $P_u = 780,000$ lb and output eccentricity $e = 4.11$ in., which is close to the actual eccentricity. Adopt the design.

```
******ENGINEERING PROGRAM*******
**INPUT DATA:**
---------------
TYPE OF PROBLEM=1
FC'=4000
FY=60000
B=18
H=20
B1=.85
C=16.5
PU=780000
NO. OF ROWS N=2
INPUT DEPTH AND AREA OF ROW NO.1
D=3
A=5
INPUT DEPTH AND AREA OF ROW NO.2
D=17
A=5
**VERIFICATION OF INPUT DATA**
FC'=4000
FY=60000
B=18
H=20
B1=.85
C=16.5
PU=780000
N=2
D AND A OF ROW NO. 1   ARE   3   AND   5
D AND A OF ROW NO. 2   ARE  17   AND   5
ARE THE DATA CORRECT? Y/NY
******ENGINEERING PROGRAM*******



          PLEASE'WAIT'FOR'CALCULATION

**FINAL RESULTS**
CB (IN)      =10.0612245
PUB (LB)     =354469.429
MUB (IN-LB) =4953791.13
EB (IN)      =13.9752281
PUO (LB)     =1002400
PI           =.7
PU (LB)      =789703.728
MU (IN-LB)  =3246273.52
E (IN)       =4.11074864
```

13.12.2 Example 13.32: Analysis of a Spirally Reinforced Circular Column

Solve Ex. 13.31 using a circular spirally reinforced column with a load eccentricity $e = 19.5$ in.

Solution

The factored axial load $P_u = 780,000$ lb. Assume a 30.0-in.-diameter circular section reinforced with 12 No. 10 bars giving a total reinforcement area of 15.24 in.^{2} and a trial neutral-axis depth $c = 18$ in.

The computer input and output for this cycle are as follows:

```
******ENGINEERING PROGRAM*******              PLEASE`WAIT`FOR`CALCULATIONS
**INPUT DATA:**
-------------                                 **FINAL RESULTS**
INPUT TYPE OF COLUMN SPIRAL OR TIED           CB (IN)     =15.9795918
?SPIRAL                                       PUB (LB)    =817463.413
TYPE OF PROBLEM=1                             MUB (IN-LB)=9836549.82
FC=4000                                       EB (IN)     =12.0330154
FY=60000                                      PUO (LB)    =
H=30                                          2082629.44
DS=24                                         PI          =.75
B1=.85                                        PU (LB)     =1046339
AS=15.24                                      MU(IN-LB)   =9330154.81
NO. OF BARS N=12                              E (IN)      =8.91695218
C=18
PU=780000
DMAX=27
**VERIFICATION OF INPUT DATA**

THE COLUMN IS                SPIRAL
FC´=4000
FY=60000
H=30
DS=24
B1=.85
AS=15.24
NO. OF BARS N=12
C=18
PU=780000
DMAX=27
ARE THE DATA CORRECT ? Y/NY
******ENGINEERING PROGRAM*******
```

The output gives a design axial load P_u = 1,046,399 lb, which is larger than the factored P_u = 780,000 lb. But the output eccentricity of 8.91 in. is too low compared to given eccentricity of 19.5 in. Hence revise the section.

In the last trial cycle a circular column section having a diameter of 36.0 in. and reinforced with 14 No. 10 bars is used as adequate. Following are the computer input and output details.

```
******ENGINEERING PROGRAM*******              C=16
**INPUT DATA:**                               PU=780000
-------------                                 DMAX=33
INPUT TYPE OF COLUMN SPIRAL OR TIED           ARE THE DATA CORRECT ? Y/NY
?SPIRAL                                       ******ENGINEERING PROGRAM*******
TYPE OF PROBLEM=1
FC=4000
FY=60000
H=36
DS=30                                                 PLEASE`WAIT`FOR`CALCULATIONS
B1=.85
AS=17.78                                      **FINAL RESULTS**
NO. OF BARS N=14                              CB (IN)     =19.5306123
C=16                                          PUB (LB)    =1211384.98
PU=780000                                     MUB (IN-LB)=15913504.6
DMAX=33                                       EB (IN)     =13.1366204
**VERIFICATION OF INPUT DATA**                PUO (LB)    =
                                             2848681.14
THE COLUMN IS                SPIRAL           PI          =.75
FC´=4000                                      PU (LB)     =795686.296
FY=60000                                      MU(IN-LB)   =15636843.1
H=36                                          E (IN)      =19.6520201
DS=30
B1=.85
AS=17.78
NO. OF BARS N=14
```

13.12.3 Example 13.33: Load–Moment Design Strength Interaction Plots for Rectangular Columns

Using Section 13.12 computer program in BASIC, construct a set of Load–moment strength interaction diagrams for rectangular columns for the two cases (a) $\gamma = 0.75$, and (b) $\gamma = 0.90$ from the sets of printout coordinate values $\phi P_n/A_g$ and $\phi M_n/A_g h$. Give also an output set of

$$\frac{\phi P_n}{A_g} \bigg/ \frac{\phi M_n}{A_g h}$$

values for $\rho_g = 0.04$ and $\gamma = 0.90$. Given:

$$f'_c = 4000 \text{ psi}$$
$$f_y = 60,000 \text{ psi}$$

Solution

(a) *Case $\gamma = 0.75$:* The input variables are: type of problem $= 2$, $f'_c = 4000$ psi, $f_y = 60,000$ psi, $b = 12$ in., $h = 20$ in. $\beta_1 = 0.85$, $c = 10$ in., $P_u = 200,000$ lb, number of rows $= 2$, $d_1 = 2.5$ in., $A_1 = 3.0$ in.^{2}, $d_2 = 17.5$ in., and $A_2 = 3.0$ in.^{2}. The output is given in five plots in Fig. 13.32 for values of $\rho_g = 0.01, 0.02, 0.04, 0.06$, and 0.08.

Coordinates of the first point on each plot define the maximum allowable axial load strength ϕP_n, namely $\phi P_{n(max)}$ or $P_{u(max)} = 0.80\phi[0.85f'_c(A_g - A_{st}) + f_y A_{st}$ and $\phi M_n = 0$. The output plots were obtained by Tektronix Easygraphing with a peripheral interactive digital plotter.

The second point on the interaction diagram in Fig. 13.32 gives the load-moment strength coordinate values for reinforcement tensile stress $f_s = 0$ in the bar closest to the extreme tension fibers, that is, at the load level where the stress in the tension bar changes from compression to tension. The horizontal plateau of the diagrams at $\phi P_{n(max)}$ is manually drawn to intersect with the load-moment strength interaction curve extended upward beyond the point at $f_s = 0$.

The coordinate values in the plots at $f_s = f_y$ represent the balanced condition, namely the ϕP_{nb}–ϕM_{nb} strength levels. The line $e/H = 0.10$ denotes the minimum eccentricity ratio allowed in rectangular columns in previous codes. The eccentricity ratio lines are hand-drawn to aid in speedy use of the charts as a design tool.

(b) *Case $\gamma = 0.90$:* The input variables are: type of problem $= 2$, $f'_c = 4000$ psi, $f_y = 60,000$ psi, $b = 36$ in., $h = 50$ in., $\beta_1 = 0.85$, $c = 15$ in., $P_u = 600,000$ lb, number of rows $= 2$, $d_1 = 2.5$ in., $A_1 = 18.0$ in.^{2}, $d_2 = 47.5$ in., and $A_2 = 18.0$ in.^{2}.

The output is given in five plots in Fig. 13.33 for values of $\rho_g = 0.01, 0.02, 0.04$, 0.06, and 0.08. Coordinates for other ρ_g values can be executed similarly.

Figure 13.34 gives a typical computer printout of the load-moment strength coordinate values for the case of $\rho_g = 0.04$ and $\gamma = 0.90$. Note the starting maximum permissible axial load value $P_{u(max)}/A_g$ at $M_u/A_g h = 0$, which is the first point on the horizontal plateau of diagram in Fig. 13.32 for the gross reinforcement percentage $\rho_g = 0.04$.

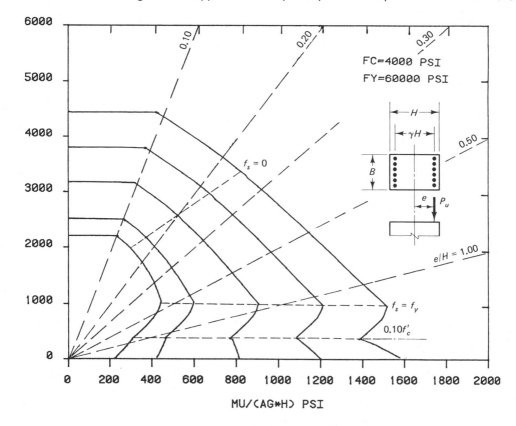

LOAD—MOMENT STRENGTH INTERACTION DIAGRAM - RECTANGULAR COLUMNS
GAMMA=0.75

Figure 13.32 Load—moment strength interaction diagram: rectangular columns.
$\gamma = 0.75$.

13.12.4 Example 13.34: Load—Moment Design Strength Interaction Plots for Circular Columns

Using the Section 13.12 computer program in BASIC, construct a set of load—moment strength interaction diagrams for spiral circular columns for the two cases (a) $\gamma = 0.75$, and (b) $\gamma = 0.90$ from the sets of printout coordinate values $\phi P_n / A_g$ and $\phi M_n / A_g h$. Given:

$f'_c = 4000$ psi
$f_y = 60,000$ psi

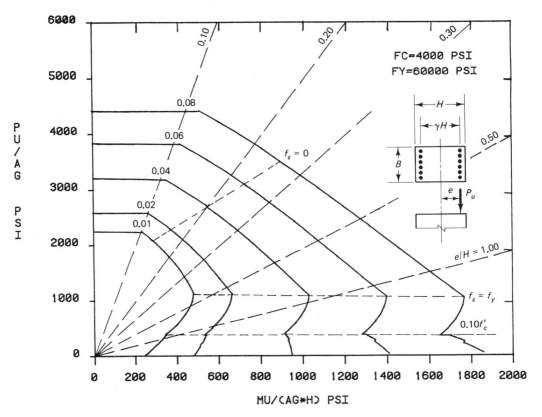

LOAD-MOMENT STRENGTH INTERACTION DIAGRAM -RECTANGULAR COLUMNS

GAMMA=0.90

Figure 13.33 Load–moment strength interaction diagram: rectangular columns.
$\gamma = 0.90$.

```
*********************************************
FC'       =4000
FY        =60000
RG        =.04
GAMMA     =.9


********COORDINATES OF INTERACTION DIAGRAM FOR RG=.04********
PU/AG(PSI)                    MU/(AG*H)(PSI)
3171.84                       0.0000
3427.88372                    241.612325

3382.3166                     262.117529

3267.71266                    316.034791

3090.41922                    395.51339
```

Figure 13.34 Set of interaction diagram coordinates printout values for $\rho_g = 0.04$
and $\gamma = 0.90$.

2906.57793	471.748139
2714.96707	545.288816
2666.65	562.978062FS=0.0
2613.84261	577.319265
2510.75764	612.737784
2404.71488	647.939712
2295.39362	683.069376
2182.425	718.292749
2065.38273	753.801675
1943.77135	789.819101
1817.01169	826.605661
1684.42235	864.467981
1545.19615	903.769293
1398.36968	944.943135
1242.78356	988.51122
1089.82143	1031.49922FS=FY
1085.28	1031.39456
1024.59	1029.1642
963.9	1025.38625
903.21	1020.0607
842.52	1013.18756
781.83	1004.76682
721.14	994.798489
660.45	983.282562
599.76	970.219039
539.07	955.607922
478.38	939.44921
417.69	921.742902
357	902.489
357.539325	926.745678
346.983915	922.893678
339.024978	926.071187
330.326453	927.503425
321.674798	929.318559

Figure 13.34 (*cont.*)

312.887529	931.014896
304.002967	932.706453
292.205621	933.105185
270.392523	930.524481
248.943617	933.453956
225.71748	935.365361
200.85596	937.472208
174.071271	939.5692
145.089125	941.688108
113.568259	943.823469
79.0936488	945.972854
41.153979	948.130438
-.887417711	950.286111

Figure 13.34 (*cont.*)

Solution

(a) *Case* $\gamma = 0.75$: The input variables are: type of column = spiral, type of problem = 2, f'_c = 4000 psi, f_y = 60,000 psi, h = 20 in., d_s = 15 in., β_1 = 0.85, A_s = 4.74 in.^{2}, number of bars = 12, c = 7.0 in., P_u = 120,000 lb, and d_{max} = 17.5 in.

The output is given in five plots in Fig. 13.35 for values of ρ_g = 0.01, 0.02, 0.04, 0.06, and 0.08 in a similar manner as that described in Ex. 13.33 for rectangular columns. The value f_s = 0 in these interaction plots denotes the point at which the stress in the tension steel closest to the extreme concrete tension fibers changes from compression to tension. The value $f_s = f_y$ represents the balanced condition. The line e/H = 0.05 represents the minimum eccentricity ratio allowed in circular columns in previous codes. The set of eccentricity ratio lines are hand-drawn to facilitate speedy use of the charts as a design tool.

Solution

(b) *Case* $\gamma = 0.90$: The input variables are: Type of column = spiral, type of problem = 2, f'_c = 4,000 psi, f_y = 60,000 psi, h = 50 in., d_s = 45 in., β_1 = 0.85, A_s = 40.0 in.^{2}, number of bars = 12, c = 20 in., P_u = 800,000 lb and d_{max} = 47.5 in.

The output is given in five plots in Fig. 13.36 for values of ρ_g = 0.01, 0.02, 0.04, 0.06, and 0.08. These plots were constructed in the same manner as in part (9) of this example using a peripheral Interactive Digital Plotter. The horizontal plateau of the diagrams at $\phi P_{n(max)} = P_{u(max)}$ has been manually drawn to intersect with the interaction curve extension upward beyond the point at f_s = 0.

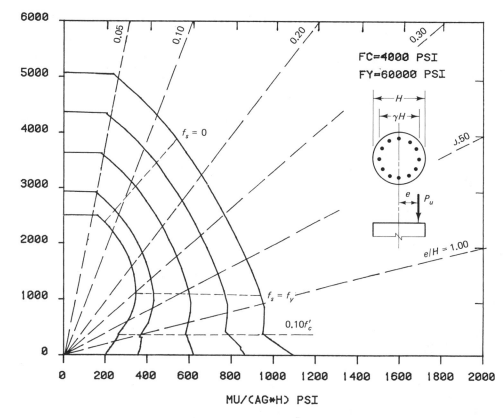

LOAD—MOMENT STRENGTH INTERACTION DIAGRAM — CIRCULAR COLUMNS

GAMMA=0.75

Figure 13.35 Load–moment strength interaction diagram: circular columns.
$\gamma = 0.75$.

13.12.5 Example 13.15 Analysis of a Biaxially Loaded Column

A corner column is subjected to a factored compression axial load $P_u = 210,000$ lb., a factored bending moment $M_{ux} = 1,680,000$ in-lb about the x-axis, and a factored bending moment $M_{uy} = 980,000$ in-lb about the y-axis. Design a rectangular tied column section to resist this factored load and factored biaxial moments, given: $f'_c = 4000$ psi and $f_y = 60,000$ psi.

Solution:

Assume a 12 in. × 20 in. cross-section with 3 #8 bars on each of the four faces of the column. Assume $\phi = 0.7$ as sufficiently reasonable for the biaxial loading.

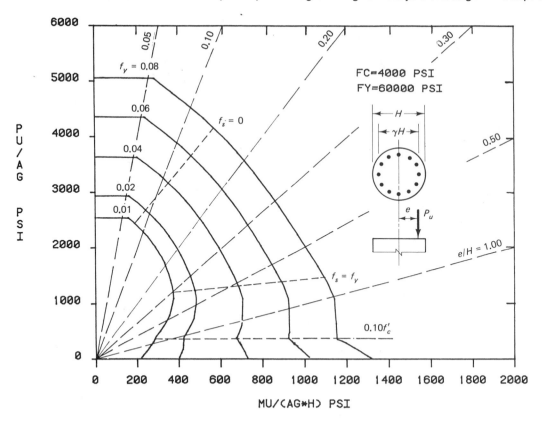

LOAD-MOMENT STRENGTH INTERACTION DIAGRAM - CIRCULAR COLUMNS

GAMMA=0.90

Figure 13.36 Load–moment strength interaction diagram: circular columns. $\gamma = 0.90$.

$$P_n = \frac{P_u}{\phi} = \frac{210,000}{0.7} = 300,000 \text{ lb.}$$

$$M_{nx} = \frac{M_{ux}}{\phi} = \frac{1,680,000}{0.7} = 2,400,000 \text{ in.-lb.}$$

$$M_{ny} = \frac{M_{uy}}{\phi} = \frac{980,000}{0.7} = 1,400,000 \text{ in.-lb.}$$

Operation type 1: Assume moment ratios β-factor $= 0.64$. The computer input data is $P_n = 300,000$ lb., $M_{nx} = 2,400,000$ in.-lb, $M_{ny} = 1,400,000$ in.-lb., $b = 12$ in., $h = 20$ in., and $\beta = 0.64$

```
*******ENGINEERING PROGRAM*******
**INPUT DATA:**
--------------
TYPE OF PROBLEM=1
PN=300000
MNX=2400000
MNY=1400000
B=12
H=20
BE=.64
**VERIFICATION OF INPUT DATA**

PN=300000
MNX=2400000
MNY=1400000
B=12
H=20
BE=.64
ARE THE DATA CORRECT? Y/NY
*******ENGINEERING PROGRAM*******




            PLEASE`WAIT`FOR`CALCULATION
   MNX/MOX =.646464647
   MNY/MOY =.633484163
   MOX=3712500
```

$\therefore$ Controlling *required* moment strength in this example is $M_{ox} = 3{,}712{,}500$ in.-lb.

Operation type 2:

(a) *Nominal Moment Strength M_{oxn}:* The computer input data is: $f'_c = 4{,}000$ psi, $f_y = 60{,}000$ psi, $b = 12$ in., $h = 20$ in., $\beta_1 = 0.85$, trial axis depth $c = 9.67$ in., $P_n = 300{,}000$ lb. $d_1 = 2.5$ in., $A_1 = 2.37$ in^2, $d_2 = 17.5$ in. and $A_2 = 2.37$ in^2.

The iterated final value of $c = 9.67$ in. for the case where the larger column cross-sectional dimension assumed perpendicular to the axis of bending was arrived at after three trials. This c value gave the actual controlling nominal moment strength M_{oxn} almost equal in value to the required controlling moment strength M_{ox} and a correct iterated contour β-factor value.

```
*******ENGINEERING PROGRAM*******
**INPUT DATA:**                          B=12
--------------                           H=20
TYPE OF PROBLEM=2                        B1=.85
FC'=4000                                 C=9.67
FY=60000                                 PU=300000
B=12                                     N=2
H=20                                      D AND A OF ROW NO. 1   ARE   2.5AND  2.37
B1=.85                                    D AND A OF ROW NO. 2   ARE  17.5AND  2.37
C=9.67                                    ARE THE DATA CORRECT? Y/NY
PN=300000                                *******ENGINEERING PROGRAM*******
NO.OF ROWS N=2
INPUT DEPTH AND AREA OF ROW NO.1
D=2.5
A=2.37
INPUT DEPTH AND AREA OF ROW NO.2                  PLEASE`WAIT`FOR`CALCULATION
D=17.5
A=2.37                                   **FINAL RESULTS**
**VERIFICATION OF INPUT DATA**           PN(LB)      =327297.6
FC'=4000                                 MN(IN-LB)   =4047893.32
FY=60000                                 E(IN)       =12.367623
```

$\therefore$ Actual nominal $M_{oxn} = 4{,}047{,}893$ in.-lb. $>$ required controlling $M_{ox} = 3{,}712{,}508$ in.-lb.; o.k.

(a) *Nominal Moment Strength* M_{oyn}: The computer input data are: $f'_c = 4{,}000$ psi, $f_y = 60{,}000$ psi, $b = 20$ in., $h = 12$ in., trial axis depth $c = 5.40$ in., $\beta_1 = 0.85$, $d_1 = 2.5$ in., $A_1 = 2.37$ in^2, $d_2 = 9.5$ in., and $A_2 = 2.37$ in^2.

The iterated final value of $c = 5.40$ for the case where the shorter column dimension was assumed perpendicular to the axis of bending was arrived at after three trials.

```
*******ENGINEERING PROGRAM*******
**INPUT DATA:**
--------------
TYPE OF PROBLEM=2
FC'=4000
FY=60000
B=20
H=12
B1=.85
C=5.4
PN=300000
NO.OF ROWS N=2
INPUT DEPTH AND AREA OF ROW NO.1
D=2.5
A=2.37
INPUT DEPTH AND AREA OF ROW NO.2
D=9.5
A=2.37
**VERIFICATION OF INPUT DATA**
FC'=4000
FY=60000
B=20
H=12
B1=.85
C=5.4
PU=300000
N=2
D AND A OF ROW NO. 1   ARE   2.5AND  2.37
D AND A OF ROW NO. 2   ARE   9.5AND  2.37
ARE THE DATA CORRECT? Y/NY
*******ENGINEERING PROGRAM*******



          PLEASE'WAIT'FOR'CALCULATION

**FINAL RESULTS**
PN(LB)        =272593.667
MN(IN-LB)     =2013462.43
E(IN)         =7.38631406
```

$\therefore$ Actual nominal $M_{oyn} = 2{,}013{,}462$ in.-lb.

$$\frac{M_{nx}}{M_{oxn}} = \frac{2{,}400{,}000}{4{,}047{,}893} = 0.593$$

$$\frac{M_{nx}}{M_{oyn}} = \frac{1{,}400{,}000}{2{,}013{,}462} = 0.695$$

The chart in Fig. 13.25 gives a contour β-factor value $\cong 0.64$. Hence, adopt the design.

Photo 70 Chicago Mercantile Exchange: a high strength concrete unique cantilever supports office tower. (Courtesy Robert B. Johnson, Alfred Benesch and Co., Chicago.)

SELECTED REFERENCES

13.1 Sippl, R. J., and Sippl, C. J., *Programmable Calculators,* Matrix Publishers, Champaign, Ill., 1978, 526 pp.

13.2 Dorf, R. C., *Introduction to Computers and Computer Science,* Boyd and Fraser, San Francisco, 1972, 628 pp.

13.3 Hewlett-Packard, *Owner's Manual for HP 41C/41CV/41CX,* Vols. 1 and 2, Hewlett-Packard, Corvallis, Oregon, August 1983, 452 pp.

13.4 Coan, J. S., *Basic BASIC—An Introduction to Computer Programming in BASIC Language,* 2nd ed., Hayden, Rochelle Park, N.J., 1978, 268 pp.

13.5 Coan, J. S., *Advanced BASIC—Applications and Problems,* Hayden, Rochelle Park, N.J., 1977, 192 pp.

13.6 Coan, J. S., *Basic FORTRAN,* Hayden, Rochelle Park, N.J., 1980, 235 pp.

13.7 Ruckdeschel, F. R., *BASIC Scientific Subroutines,* Vol. 1, McGraw-Hill, New York, 1981, 316 pp.

13.8 Apple Computer, Inc., *Apple IIe Owner's Manual,* Apple Computer, Inc. Cupertino, Calif., 1982, 140 pp.

13.9 Apple Computer, Inc., *Apple Soft BASIC Programmer's Reference Manual,* Vols. I and II, Apple Computer, Inc., Cupertino, Calif., 1982, 337 pp.

PROBLEMS FOR SOLUTION

13.1 Write a BASIC language computer program for the analysis of T-beams and L-beams with symmetrical reinforcement.

13.2 Find the design moment capacity of the T-beam in Ex. 13.7 utilizing the program in BASIC of problem 13.1.

13.3 Design the end span L-beam in Example 13.8 utilizing the program in BASIC of problem 13.1.

13.4 Write a computer program in BASIC language for the analysis of (a) deep beams, (b) corbels.

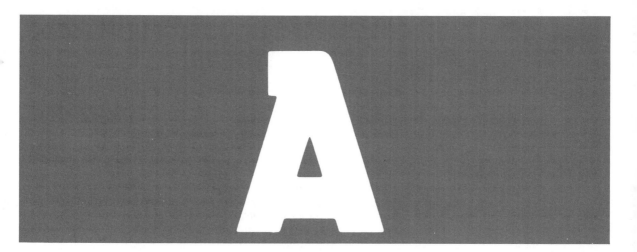

Appendix

To convert from	to	multiply by[†]
	Length	
inch	millimeter (mm)	25.4E
foot	meter (m)	0.3048E
yard	meter (m)	0.9144E
mile(statute)	kilometer(km)	1.609
	Area	
square inch	square centimeter (cm^2)	6.452
square foot	square meter (m^2)	0.09290
square yard	square meter (m^2)	0.8361
	Volume (Capacity)	
ounce	cubic centimeter (cm^3)	29.57
gallon	cubic meter (m^3)[‡]	0.003785
cubic inch	cubic centimeter (cm^3)	16.4
cubic foot	cubic meter (m^3)	0.02832
cubic yard	cubic meter (m^3)[‡]	0.765
	Force	
kilogram-force	newton (N)	9.807
kip-force	kilonewton (kN)	4.448
pound-force	newton (N)	4.448
	Pressure or Stress (Force per Area)	
kilogram-force/square meter	pascal (Pa)	9.807
kip-force/square inch (ksi)	megapascal (MPa)	6.895
newton/square meter (N/m^2)	pascal (Pa)	1.000E
pound-force/square foot	pascal (Pa)	47.88
pound-force/square inch (psi)	pascal (Pa)	6895
	Bending Moment or Torque	
inch-pound-force	newton-meter (Nm)	0.1130
foot-pound-force	newton-meter (Nm)	1.356
meter-kilogram-force	newton-meter (Nm)	9.807
	Mass	
ounce-mass (avoirdupois)	gram (g)	28.35
pound-mass (avoirdupois)	kilogram (kg)	0.4536
ton (metric)	megagram (Mg)	1.000E
ton (short 2000 lbm)	megagram (Mg)	0.9072
	Mass per Volume	
pound-mass/cubic foot	kilogram/cubic meter (kg/m^3)	16.02
pound-mass/cubic yard	kilogram/cubic meter (kg/m^3)	0.5933
pound-mass/gallon	kilogram/cubic meter (kg/m^3)	119.8
	Temperature	
deg Fahrenheit (F)	deg Celsius (C)	$t_C = (t_F - 32)/1.8$
deg Celsius (C)	deg Fehrenheit (F)	$t_F = 1.8t_C + 32$

Figure A.1 Selected conversion factors to SI units.

Bar size designa-tion	Nominal cross section area, sq. in.	Weight, lb per ft	Nominal diameter, in.
#3	0.11	0.376	0.375
#4	0.20	0.668	0.500
#5	0.31	1.043	0.625
#6	0.44	1.502	0.750
#7	0.60	2.044	0.875
#8	0.79	2.670	1.000
#9	1.00	3.400	1.128
#10	1.27	4.303	1.270
#11	1.56	5.313	1.410
#14	2.25	7.650	1.693
#18	4.00	13.600	2.257

Figure A.2 Geometrical properties of reinforcing bars.

Areas, A_s (or A'_s) in sq. in.
Columns headed **0 5** contain data for bars of one size in groups of one to ten.
Columns headed **1 2 3 4 5** contain data for bars of two sizes with from one to five of each size.

#4 bar (with #3 combinations)

n	#4: 0	#4: 5	#3: 1	#3: 2	#3: 3	#3: 4	#3: 5
1	0.20	1.20	0.31	0.42	0.53	0.64	0.75
2	0.40	1.40	0.51	0.62	0.73	0.84	0.95
3	0.60	1.60	0.71	0.82	0.93	1.04	1.15
4	0.80	1.80	0.91	1.02	1.13	1.24	1.35
5	1.00	2.00	1.11	1.22	1.33	1.44	1.55

#5 bar (with #4 and #3 combinations)

n	#5: 0	#5: 5	#4: 1	#4: 2	#4: 3	#4: 4	#4: 5	#3: 1	#3: 2	#3: 3	#3: 4	#3: 5
1	0.31	1.86	0.51	0.71	0.91	1.11	1.31	0.42	0.53	0.64	0.75	0.86
2	0.62	2.17	0.82	1.02	1.22	1.42	1.62	0.73	0.84	0.95	1.06	1.17
3	0.93	2.48	1.13	1.33	1.53	1.73	1.93	1.04	1.15	1.26	1.37	1.48
4	1.24	2.79	1.44	1.64	1.84	2.04	2.24	1.35	1.46	1.57	1.68	1.79
5	1.55	3.10	1.75	1.95	2.15	2.35	2.55	1.66	1.77	1.88	1.99	2.10

#6 bar (with #5, #4, and #3 combinations)

n	#6: 0	#6: 5	#5: 1	#5: 2	#5: 3	#5: 4	#5: 5	#4: 1	#4: 2	#4: 3	#4: 4	#4: 5	#3: 1	#3: 2	#3: 3	#3: 4	#3: 5
1	0.44	2.64	0.75	1.06	1.37	1.68	1.99	0.64	0.84	1.04	1.24	1.44	0.55	0.66	0.77	0.88	0.99
2	0.88	3.08	1.19	1.50	1.81	2.12	2.43	1.08	1.28	1.48	1.68	1.88	0.99	1.10	1.21	1.32	1.43
3	1.32	3.52	1.63	1.94	2.25	2.56	2.87	1.52	1.72	1.92	2.12	2.32	1.43	1.54	1.65	1.76	1.87
4	1.76	3.96	2.07	2.38	2.69	3.00	3.31	1.96	2.16	2.36	2.56	2.76	1.87	1.98	2.09	2.20	2.31
5	2.20	4.40	2.51	2.82	3.13	3.44	3.75	2.40	2.60	2.80	3.00	3.20	2.31	2.42	2.53	2.64	2.75

#7 bar (with #6, #5, and #4 combinations)

n	#7: 0	#7: 5	#6: 1	#6: 2	#6: 3	#6: 4	#6: 5	#5: 1	#5: 2	#5: 3	#5: 4	#5: 5	#4: 1	#4: 2	#4: 3	#4: 4	#4: 5
1	0.60	3.60	1.04	1.48	1.92	2.36	2.80	0.91	1.22	1.53	1.84	2.15	0.80	1.00	1.20	1.40	1.60
2	1.20	4.20	1.64	2.08	2.52	2.96	3.40	1.51	1.82	2.13	2.44	2.75	1.40	1.60	1.80	2.00	2.20
3	1.80	4.80	2.24	2.68	3.12	3.56	4.00	2.11	2.42	2.73	3.04	3.35	2.00	2.20	2.40	2.60	2.80
4	2.40	5.40	2.84	3.28	3.72	4.16	4.60	2.71	3.02	3.33	3.64	3.95	2.60	2.80	3.00	3.20	3.40
5	3.00	6.00	3.44	3.88	4.32	4.76	5.20	3.31	3.62	3.93	4.24	4.55	3.20	3.40	3.60	3.80	4.00

#8 bar (with #7, #6, and #5 combinations)

n	#8: 0	#8: 5	#7: 1	#7: 2	#7: 3	#7: 4	#7: 5	#6: 1	#6: 2	#6: 3	#6: 4	#6: 5	#5: 1	#5: 2	#5: 3	#5: 4	#5: 5
1	0.79	4.74	1.39	1.99	2.59	3.19	3.79	1.23	1.67	2.11	2.55	2.99	1.10	1.41	1.72	2.03	2.34
2	1.58	5.53	2.18	2.78	3.38	3.98	4.58	2.02	2.46	2.90	3.34	3.78	1.89	2.20	2.51	2.82	3.13
3	2.37	6.32	2.97	3.57	4.17	4.77	5.37	2.81	3.25	3.69	4.13	4.57	2.68	2.99	3.30	3.61	3.92
4	3.16	7.11	3.76	4.36	4.96	5.56	6.16	3.60	4.04	4.48	4.92	5.36	3.47	3.78	4.09	4.40	4.71
5	3.95	7.90	4.55	5.15	5.75	6.35	6.95	4.39	4.83	5.27	5.71	6.15	4.26	4.57	4.88	5.19	5.50

#9 bar (with #8, #7, and #6 combinations)

n	#9: 0	#9: 5	#8: 1	#8: 2	#8: 3	#8: 4	#8: 5	#7: 1	#7: 2	#7: 3	#7: 4	#7: 5	#6: 1	#6: 2	#6: 3	#6: 4	#6: 5
1	1.00	6.00	1.79	2.58	3.37	4.16	4.95	1.60	2.20	2.80	3.40	4.00	1.44	1.88	2.32	2.76	3.20
2	2.00	7.00	2.79	3.58	4.37	5.16	5.95	2.60	3.20	3.80	4.40	5.00	2.44	2.88	3.32	3.76	4.20
3	3.00	8.00	3.79	4.58	5.37	6.16	6.95	3.60	4.20	4.80	5.40	6.00	3.44	3.88	4.32	4.76	5.20
4	4.00	9.00	4.79	5.58	6.37	7.16	7.95	4.60	5.20	5.80	6.40	7.00	4.44	4.88	5.32	5.76	6.20
5	5.00	10.00	5.79	6.58	7.37	8.16	8.95	5.60	6.20	6.80	7.40	8.00	5.44	5.88	6.32	6.76	7.20

#10 bar (with #9, #8, and #7 combinations)

n	#10: 0	#10: 5	#9: 1	#9: 2	#9: 3	#9: 4	#9: 5	#8: 1	#8: 2	#8: 3	#8: 4	#8: 5	#7: 1	#7: 2	#7: 3	#7: 4	#7: 5
1	1.27	7.62	2.27	3.27	4.27	5.27	6.27	2.06	2.85	3.64	4.43	5.22	1.87	2.47	3.07	3.67	4.27
2	2.54	8.89	3.54	4.54	5.54	6.54	7.54	3.33	4.12	4.91	5.70	6.49	3.14	3.74	4.34	4.94	5.54
3	3.81	10.16	4.81	5.81	6.81	7.81	8.81	4.60	5.39	6.18	6.97	7.76	4.41	5.01	5.61	6.21	6.81
4	5.08	11.43	6.08	7.08	8.08	9.08	10.08	5.87	6.66	7.45	8.24	9.03	5.68	6.28	6.88	7.48	8.08
5	6.35	12.70	7.35	8.35	9.35	10.35	11.35	7.14	7.93	8.72	9.51	10.30	6.95	7.55	8.15	8.75	9.35

#11 bar (with #10, #9, and #8 combinations)

n	#11: 0	#11: 5	#10: 1	#10: 2	#10: 3	#10: 4	#10: 5	#9: 1	#9: 2	#9: 3	#9: 4	#9: 5	#8: 1	#8: 2	#8: 3	#8: 4	#8: 5
1	1.56	9.36	2.83	4.10	5.37	6.64	7.91	2.56	3.56	4.56	5.56	6.56	2.35	3.14	3.93	4.72	5.51
2	3.12	10.92	4.39	5.66	6.93	8.20	9.47	4.12	5.12	6.12	7.12	8.12	3.91	4.70	5.49	6.28	7.07
3	4.68	12.48	5.95	7.22	8.49	9.76	11.03	5.68	6.68	7.68	8.68	9.68	5.47	6.26	7.05	7.84	8.63
4	6.24	14.04	7.51	8.78	10.05	11.32	12.59	7.24	8.24	9.24	10.24	11.24	7.03	7.82	8.61	9.40	10.19
5	7.80	15.60	9.07	10.34	11.61	12.88	14.15	8.80	9.80	10.80	11.80	12.80	8.59	9.38	10.17	10.96	11.75

#14 bar (with #11, #10, and #9 combinations)

n	#14: 0	#14: 5	#11: 1	#11: 2	#11: 3	#11: 4	#11: 5	#10: 1	#10: 2	#10: 3	#10: 4	#10: 5	#9: 1	#9: 2	#9: 3	#9: 4	#9: 5
1	2.25	13.50	3.81	5.37	6.93	8.49	10.05	3.52	4.79	6.06	7.33	8.60	3.25	4.25	5.25	6.25	7.25
2	4.50	15.75	6.06	7.62	9.18	10.74	12.30	5.77	7.04	8.31	9.58	10.85	5.50	6.50	7.50	8.50	9.50
3	6.75	18.00	8.31	9.87	11.43	12.99	14.55	8.02	9.29	10.56	11.83	13.10	7.75	8.75	9.75	10.75	11.75
4	9.00	20.25	10.56	12.12	13.68	15.24	16.80	10.27	11.54	12.81	14.08	15.35	10.00	11.00	12.00	13.00	14.00
5	11.25	22.50	12.81	14.37	15.93	17.49	19.05	12.52	13.79	15.06	16.33	17.60	12.25	13.25	14.25	15.25	16.25

#18 bar (with #14, #11, and #10 combinations)

n	#18: 0	#18: 5	#14: 1	#14: 2	#14: 3	#14: 4	#14: 5	#11: 1	#11: 2	#11: 3	#11: 4	#11: 5	#10: 1	#10: 2	#10: 3	#10: 4	#10: 5
1	4.00	24.00	6.25	8.50	10.75	13.00	15.25	5.56	7.12	8.68	10.24	11.80	5.27	6.54	7.81	9.08	20.35
2	8.00	28.00	10.25	12.50	14.75	17.00	19.25	9.56	11.12	12.68	14.24	15.80	9.27	10.54	11.81	13.08	14.35
3	12.00	32.00	14.25	16.50	18.75	21.00	23.25	13.56	15.12	16.68	18.24	19.80	13.27	14.54	15.81	17.08	18.35
4	16.00	36.00	18.25	20.50	22.75	25.00	27.25	17.56	19.12	20.68	22.24	23.80	17.27	18.54	19.81	21.08	22.35
5	20.00	40.00	22.25	24.50	26.75	29.00	31.25	21.56	23.12	24.68	26.24	27.80	21.27	22.54	23.81	25.08	26.35

Figure A.3 Cross-sectional area of bars for various bar combinations.

Spacing, in.	Cross section area of bar, A_s (or $A_s{}'$), in.^{2} Bar size												Spacing, in.
	#3	#4	#5	#6	#7	#8	#9	#10	#11	#14	#18		
4	0.33	0.60	0.93	1.32	1.80	2.37	3.00	3.81	4.68			4
4½	0.29	0.53	0.83	1.17	1.60	2.11	2.67	3.39	4.16	6.00		4½
5	0.26	0.48	0.74	1.06	1.44	1.90	2.40	3.05	3.74	5.40	9.60	5
5½	0.24	0.44	0.68	0.96	1.31	1.72	2.18	2.77	3.40	4.91	8.73	5½
6	0.22	0.40	0.62	0.88	1.20	1.58	2.00	2.54	3.12	4.50	8.00	6
6½	0.20	0.37	0.57	0.81	1.11	1.46	1.85	2.34	2.88	4.15	7.38	6½
7	0.19	0.34	0.53	0.75	1.03	1.35	1.71	2.18	2.67	3.86	6.86	7
7½	0.18	0.32	0.50	0.70	0.96	1.26	1.60	2.03	2.50	3.60	6.40	7½
8	0.17	0.30	0.47	0.66	0.90	1.19	1.50	1.91	2.34	3.38	6.00	8
8½	0.16	0.28	0.44	0.62	0.85	1.12	1.41	1.79	2.20	3.18	5.65	8½
9	0.15	0.27	0.41	0.59	0.80	1.05	1.33	1.69	2.08	3.00	5.33	9
9½	0.14	0.25	0.39	0.56	0.76	1.00	1.26	1.60	1.97	2.84	5.05	9½
10	0.13	0.24	0.37	0.53	0.72	0.95	1.20	1.52	1.87	2.70	4.80	10
10½	0.13	0.23	0.35	0.50	0.69	0.90	1.14	1.45	1.78	2.57	4.57	10½
11	0.12	0.22	0.34	0.48	0.65	0.86	1.09	1.39	1.70	2.45	4.36	11
11½	0.11	0.21	0.32	0.46	0.63	0.82	1.04	1.33	1.63	2.35	4.17	11½
12	0.11	0.20	0.31	0.44	0.60	0.79	1.00	1.27	1.56	2.25	4.00	12
13	0.10	0.18	0.29	0.41	0.55	0.73	0.92	1.17	1.44	2.08	3.69	13
14	0.09	0.17	0.27	0.38	0.51	0.68	0.86	1.09	1.34	1.93	3.43	14
15	0.09	0.16	0.25	0.35	0.48	0.63	0.80	1.02	1.25	1.80	3.20	15
16	0.08	0.15	0.23	0.33	0.45	0.59	0.75	0.95	1.17	1.69	3.00	16
17	0.08	0.14	0.22	0.31	0.42	0.56	0.71	0.90	1.10	1.59	2.82	17
18	0.07	0.13	0.21	0.29	0.40	0.53	0.67	0.85	1.04	1.50	2.67	18

Figure A.4 Area of bars in a 1-foot-wide slab strip.

$$I_g = K_{i4}\left(\frac{1}{12}\,b_w h^3\right)$$

$$K_{i4} = 1 + (\alpha_b - 1)\beta_h^3 + \frac{3(1 - \beta_h)^2(\beta_h)(\alpha_b - 1)}{1 + \beta_h(\alpha_b - 1)}$$

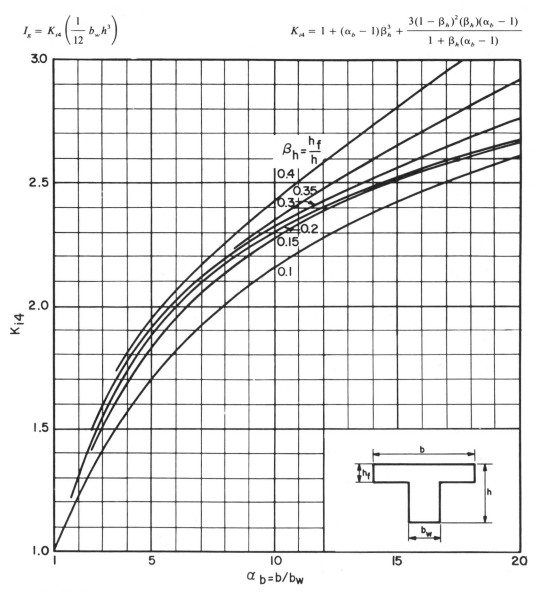

Example: For the T-beam shown, find the moment of inertia I_g:

$$\alpha_b = b/b_w = 143/15 = 9.53$$

$$\beta_h = h_f/h = 8/36 = 0.22$$

Interpolating between the curves for $\beta_h = 0.2$ and 0.3, read $K_{i4} = 2.28$

$$I_g = K_{i4}\frac{b_w h^3}{12} = 2.28\,\frac{15(36)^3}{12} = 133{,}000 \text{ in.}^4$$

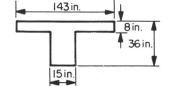

Figure A.5 Gross moment of inertia of T sections.

694

Index